PARADIGMS IN PIG SCIENCE

Proceedings of Previous Easter Schools in Agricultural Science, published by Butterworths, London

SOIL ZOOLOGY Edited by D.K. McL, Kevan (1955)
THE GROWTH OF LEAVES Edited by F.L. Milthorpe (1956)
CONTROL OF PLANT ENVIRONMENT Edited by J.P. Hudson (1957)
NUTRITION OF THE LEGUMES Edited by E.G. Hallsworth (1958)
THE MEASUREMENT OF GRASSLAND AND PRODUCTIVITY Edited by J.D. Ivins (1959)
DIGESTIVE PHYSIOLOGY AND NUTRITION OF THE RUMINANT Edited by D. Lewis (1960)
NUTRITION OF PIGS AND POULTRY Edited by J.T. Morgan and D. Lewis (1961)
ANTIBIOTICS IN AGRICULTURE Edited by M. Woodbine (1962)
THE GROWTH OF THE POTATO Edited by J.D. Ivins and F.L. Milthorpe (1963)
EXPERIMENTAL PEDOLOGY Edited by E.G. Hallsworth and D.V. Crawford (1964)
THE GROWTH OF CEREALS AND GRASSES Edited by F.L. Milthorpe and J.D. Ivins (1965)
REPRODUCTION IN THE FEMALE ANIMAL Edited by G.E. Lamming and E.C. Amoroso (1967)
GROWTH AND DEVELOPMENT OF MAMMALS Edited by G.A. Lodge and G.E. Lamming (1968)
ROOT GROWTH Edited by W.J. Whittington (1968)
PROTEINS AS HUMAN FOOD Edited by R.A. Lawrie (1970)
LACTATION Edited by I.R. Falconer (1971)
PIG PRODUCTION Edited by D.J.A. Cole (1972)
SEED ECOLOGY Edited by W. Heydecker (1973)
HEAT LOSS FROM ANIMALS AND MAN: ASSESSMENT AND CONTROL Edited by J.L. Montieth and L.E. Mount (1974)
MEAT Edited by D.J.A. Cole and R.A. Lawrie (1975)
PRINCIPLES OF CATTLE PRODUCTION Edited by Henry Swan and W.H. Broster (1976)
LIGHT AND PLANT DEVELOPMENT Edited by H. Smith (1976)
PLANT PROTEINS Edited by G. Norton (1977)
ANTIBIOTICS AND ANTIBIOSIS AGRICULTURE Edited by M. Woodbine (1977)
CONTROL OF OVULATION Edited by D.B. Crighton, N.B. Haynes, G.R. Foxcroft and G.E. Lamming (1978)
POLYSACCHARIDES IN FOOD Edited by J.M.V. Blanshard and J.R. Mitchell (1979)
SEED PRODUCTION Edited by P.D. Hebblethwaite (1980)
PROTEIN DEPOSITION IN ANIMALS Edited by P.J. Buttery and D.B. Lindsay (1981)
PHYSIOLOGICAL PROCESSES LIMITING PLANT PRODUCTIVITY Edited by C. Johnson (1981)
ENVIRONMENTAL ASPECTS OF HOUSING FOR ANIMAL PRODUCTION Edited by J.A. Clark (1981)
EFFECTS OF GASEOUS AIR POLLUTION IN AGRICULTURE AND HORTICULTURE Edited by M.H. Unsworth and D.P. Ormrod (1982)
CHEMICAL MANIPULATION OF CROP GROWTH AND DEVELOPMENT Edited by J.S. McLaren (1982)
CONTROL OF PIG REPRODUCTION Edited by D.J.A. Cole and G.R. Foxcroft (1982)
SHEEP PRODUCTION Edited by W. Haresign (1983)
UPGRADING WASTE FOR FEEDS AND FOOD Edited by D.A. Ledward, A.J. Taylor and R.A. Lawrie (1983)
FATS IN ANIMAL NUTRITION Edited by J. Wiseman (1984)
IMMUNOLOGICAL ASPECTS OF REPRODUCTION IN MAMMALS Edited by D.B. Crighton (1984)
ETHYLENE AND PLANT DEVELOPMENT Edited by J.A. Roberts and G.A. Tucker (1985)
THE PEA CROP Edited by P.D. Hebblethwaite, M.C. Heath and T.C.K. Dawkins (1985)
PLANT TISSUE CULTURE AND ITS AGRICULTURAL APPLICATIONS Edited by Lindsay A. Withers and P.G. Alderson (1986)
CONTROL AND MANIPULATION OF ANIMAL GROWTH Edited by P.J. Buttery, N.B. Haynes and D.B. Lindsay (1986)
COMPUTER APPLICATIONS IN AGRICULTURAL ENVIRONMENTS Edited by J.A. Clark, K. Gregson and R.A. Saffell (1986)
MANIPULATION OF FLOWERING Edited by J.G. Atherton (1987)
NUTRITION AND LACTATION IN THE DAIRY COW Edited by P.C. Garnsworthy (1988)
MANIPULATION OF FRUITING Edited by C.J. Wright (1989)
APPLICATIONS OF REMOTE SENSING IN AGRICULTURE Edited by M.D. Steven and J.A. Clark (1990)
GENETIC ENGINEERING OF CROP PLANTS Edited by G.W. Lycett and D. Grierson (1990)
FEEDSTUFF EVALUATION Edited by J. Wiseman and D.J.A. Cole (1990)
THE CONTROL OF FAT AND LEAN DEPOSITION Edited by K.N. Boorman, P.J. Buttery and D.B. Lindsay (1992)

Proceedings of Previous Easter Schools in Agricultural Science, published by Nottingham University Press

THE GLASSY STATE IN FOODS Edited by J.M.V. Blanshard and P.J. Lillford (1993)
RESOURCE CAPTURE BY CROPS Edited by J.L. Montieth, R.K. Scott and M.H. Unsworth (1994)
PRINCIPLES OF PIG SCIENCE Edited by D.J.A. Cole, J. Wiseman and M.A. Varley (1994)
ISSUES IN AGRICULTURAL BIOETHICS Edited by T.B. Mepham, G.A. Tucker and J. Wiseman (1995)
BIOPOLYMER MIXTURES Edited by S.E. Harding, S.E. Hill and J.R. Mitchell (1995)
MECHANISMS AND APPLICATIONS OF GENE SILENCING Edited by D. Grierson, G.W. Lycett and G.A. Tucker (1996)
PROGRESS IN PIG SCIENCE Edited by J. Wiseman, M.A. Varley and J.P. Chadwick (1998)
PERSPECTIVES IN PIG SCIENCE Edited by J. Wiseman, M.A. Varley and B. Kemp (2003)
CALF AND HEIFER REARING Edited by P.C. Garnsworthy (2004)
YIELDS OF FARMED SPECIES Edited by R. Sylvester-Bradley and J. Wiseman (2005)

Paradigms in Pig Science

J. WISEMAN
University of Nottingham

M.A. VARLEY
SCA NuTec

S. McORIST
University of Nottingham

B. KEMP
Wageningen University

NOTTINGHAM
University Press

First published by Nottingham University Press
This reissued original edition published 2023 by 5m Books Ltd www.5mbooks.com

British Library Cataloguing in Publication Data
A catalogue record for this book is available from the British Library

ISBN 9781789182965

Disclaimer

Every reasonable effort has been made to ensure that the material in this book is true, correct, complete and appropriate at the time of writing. Nevertheless the publishers and the author do not accept responsibility for any omission or error, or for any injury, damage, loss or financial consequences arising from the use of the book. Views expressed in the articles are those of the author and not of the Editor or Publisher.

Typeset by Nottingham University Press, Nottingham

EU GPSR Authorised Representative
LOGOS EUROPE, 9 rue Nicolas Poussin, 17000, LA ROCHELLE, France
E-mail: Contact@logoseurope.eu

PREFACE

The most recent meeting in the series of international Pig Science conferences was held at the University of Nottingham, Sutton Bonington Campus in July 2007 within the 'Easter School' series; the meeting represented the 62nd such event. These gatherings have now become an established part of the international pig conference calendar with symposia held every 4 to 5 years: 'Principles of Pig Science' was organised in 1993, 'Progress in Pig Science' in 1997 and 'Perspectives in Pig Science' in 2002.

The choice of 2007 was not accidental, it being the Chinese Golden Year of the Pig! Selecting 'Paradigms' as the title was also deliberate and has had many a colleague reaching for a dictionary!

A major innovation of the meeting was a greater emphasis on Health. The Sutton Bonington Campus is home to the new School of Veterinary Medicine and Science, the first new Vet School in the UK for over 60 years, and this represented a good opportunity to widen our remit. In addition, and with the increasing importance of disseminating information throughout the pig sector, we introduced Technology Transfer

We invited recognised experts as speakers from Europe, North / South America and Australia, and welcomed delegates from 16 countries (UK being the 17th). Sessions were based on:

A.   Global trends, emerging markets and future prospects
B.   Technology Transfer
C.   Health
D.   Genetics
E.   Reproduction
F.   Carcass
G.   Nutrition

We hope that the published proceedings (that are being released less than 2 months after the conference – a record!) of the meeting will appeal to all those involved in pig science and production, and will complement the previous volumes in this series.

Julian Wiseman
Mike Varley
Steve McOrist
Bas Kemp

ACKNOWLEDGEMENTS

We would like to thank all speakers for their presentations and papers, session chairs (the organisers together with Prof Paul Hughes and Dr Bill Close) and all delegates who played a key role in the meeting.

The meeting would not have gone ahead without financial sponsorship and it is a pleasure to acknowledge the support of:

    ABN
    ACMC
    Alltech
    Boehringer-ingelheim
    BPA
    BPEX
    C&H Nutrition
    Cranswick
    DANISCO
    Eurolysine/Forum
    Frank Wright
    IATC
    Intervet
    Janssen Animal Health
    JSR Genetics
    Lallemand/Biotal
    Lohmann Animal Health
    Novartis
    Novus
    PIC
    Primary Diets
    Provimi
    SCA
    Vion

Sue Golds and Emma Hooley managed the Conference Office and all administrative activities, and a large cohort of postgraduates provided invaluable support. The accommodation was of a much higher standard than ever (using our brand new student hall complex, with internet access!) and the food was great.

Dedication

This book is dedicated to the memory of Des Cole (1935-2006) who made a lasting impression on countless people worldwide, including many who attended this meeting.

CONTENTS

THE UPS AND DOWNS OF THE GLOBAL PIG MARKET

MATTHEW HARTLEY
Whole Hog, UK

Introduction

The global pig market has reacted positively to the forces of globalisation and competition in recent years. However, the global industry has seen some winners and some losers – some national pig sectors are on their way up and some are on their way down.. The current chapter summarizes the structural changes in supply and demand. Changes in pork consumption per capita and in international trade are described for different countries. Price behaviour in the global pig market has been characterized by the pig cycle over many decades and this is also examined. Whole Hog's unique price database enables analysis of this cyclical behaviour and its use to forecast price changes. The recent incidence of animal disease and the new phenomenon of "food versus fuel" are also considered in the current chapter in order that the influences on the industry's position in the global market can be fully understood.

Production and consumption – the environment for the global pig market

Pork remains the most produced meat in the world. In 2007, the FAO forecasts that 110.7 million Mt of pork will be produced globally compared to 86.2 million Mt of poultry meat, 66.6 million Mt of beef, and 13.9 million Mt of sheep and goat meat.

As Figure 1.1 shows, the balance in production between the different meat species is dynamic and the FAO predicts that, by 2030, poultry will have surpassed pork as the leading source of meat in the world. Even allowing for the rapid growth in poultry production, pork is and will continue to remain a highly important protein source for the world's population well into the current century.

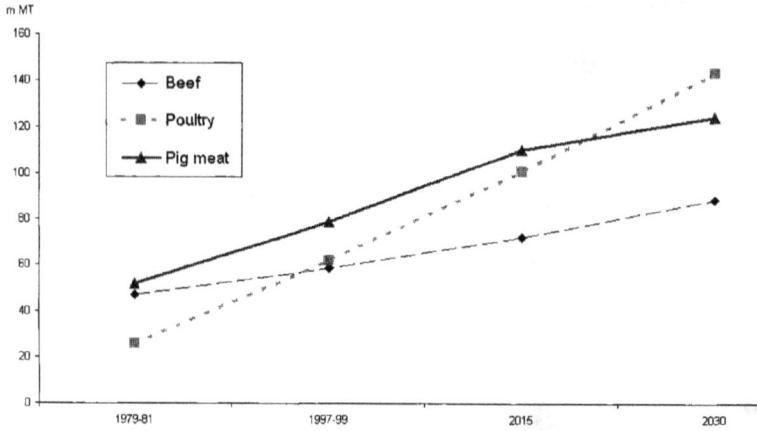

Figure 1.1 FAO world meat production forecasts, 1979-2030.

Meanwhile, as incomes rise pork consumption too will rise. Whilst major population growth is forecast in many non-pork eating countries in the Middle East and Asia, countries with a significant population and a national diet that includes pork, such as China, Brazil, and Mexico, can be expected to increase their per capita consumption as wealth levels increase.

Figure 1.2 illustrates the relationship between income levels and pork consumption amongst a sample of pork-consuming countries.

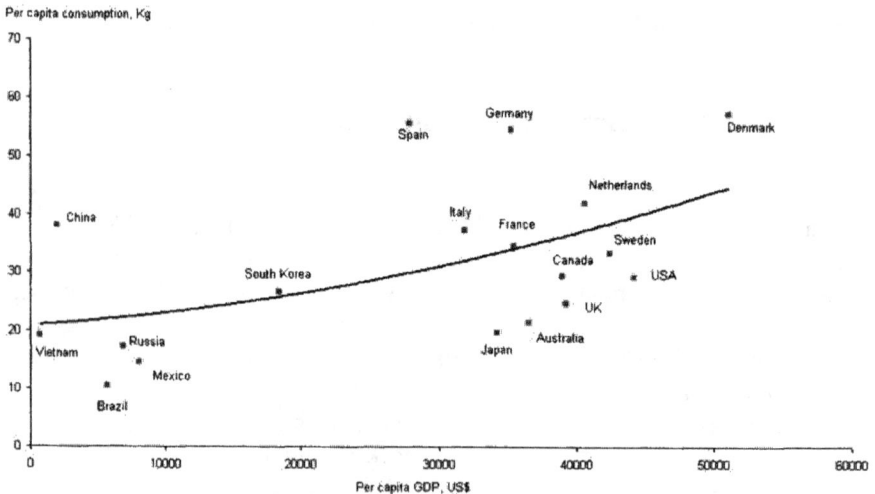

Figure 1.2 Pork consumption and income curve, 2005.

Globalisation – change is a constant

Globalisation is an established term in the economic and trade lexicon, and everyone talks about the global market for this product or that commodity. Pork is no exception and the trade in pigmeat is global, well-developed and growing. If we look at the data since the start of the new millennium it is possible to see what the effects of globalisation have been on the pig industry and get some hints on who the key global players in the pork market are going to be in the years ahead.

The world pigmeat trade has grown rapidly since the turn of the century and is forecast to continue doing so, with the EU, US, Canada and Brazil dominating pigmeat exports and Japan, Mexico, Russia, South Korea and China being key importers (Figures 1.3 and 1.4).

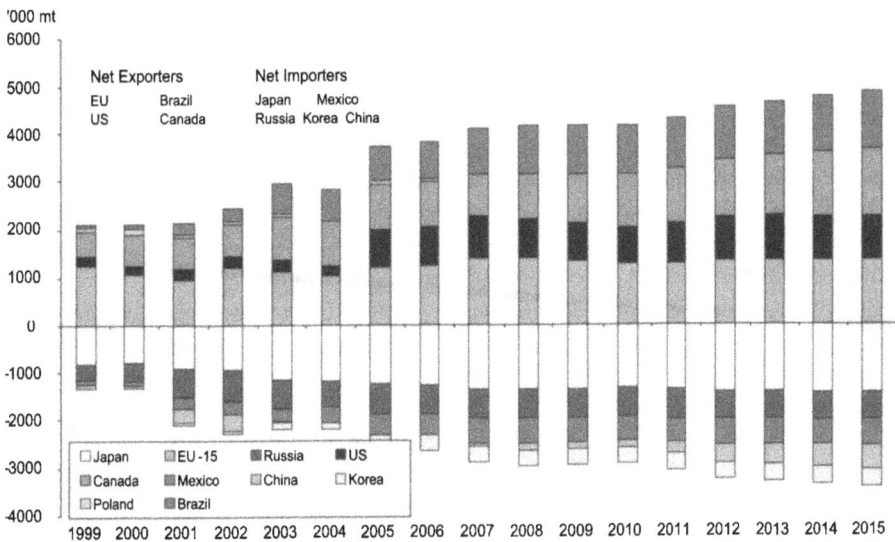

Figure 1.3 World pigmeat trade, 1999-2015 (Source: FAPRI).

It is clear that globalisation has had effects on the global pig herd – restructuring has occurred everywhere. There have been several significant changes during the 2000s – the UK's breeding herd has reduced by a fifth at the same time that the Canadian herd has increased by about the same amount. Taiwan has disappeared from the reckoning as a major exporter, Denmark has gone from strength to strength as a producer and exporter but is strongly challenged by Canada for the # 1 exporter spot, notwithstanding Canada's own restructuring programme and current difficulties with exchange rates. The USA, meantime, has contracted its herd size, although this process may be reversing, and has improved its export record.

'000 mt

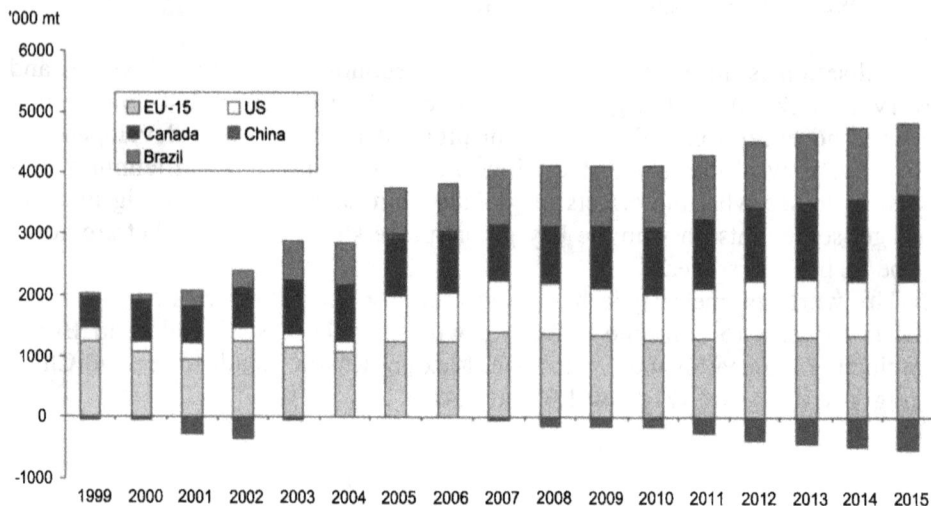

Figure 1.4 Major net pigmeat exporters, 1999-2015 (Source: FAPRI).

Table 1.1 illustrate the these changes in the size of breeding herds (sows + gilts + boars) for several countries. The data for key pig producing countries are based on changes in mid-year breeding census data in 2000 and 2006. The mid-year census point is not the same in each country.

The data illustrate that that there have been some remarkable changes in the world's pig herds. It's easy to see who has been expanding and who has been in decline.

Table 1.1 PIG BREEDING HERDS, 2000-06
(% CHANGE)

Country	% Change
Canada	+22.6%
Poland	+20.8%
Australia	+9.2%
Denmark	+7.9%
Spain	+6.9%
America	-2.8%
Germany	-3.6%
France	-14.8%
Netherlands	-20.1%
United Kingdom	-21.7%

Source: USDA / National Governments

There have been some very large percentage changes in breeding herd size in some countries – and in both directions. The UK holds the world record for cutting back on herd size – its downsizing is even larger than Taiwan's which lost 3.85 million pigs in its FMD outbreak in 1997. Taiwan has been off the major world exporters' list since then. This is another structural change – both for Taiwan and for the world pork export market.

The UK herd size has declined by an amazing 45% since December 1997 and 22% since December 2000. But it has been a roller-coaster ride with UK farmers expanding and investing in the period 1994-1997 and contracting since early 1998. The continuing contraction in the UK herd, when compared with the expansion in other herds, is a clear sign that the UK is restructuring and that this is not cyclical behaviour. That restructuring has been encouraged by low profits in the pig sector as BSE-related charges have hit pig producers; the FMD outbreak, and an overvalued exchange rate coupled with inefficient marketing have reduced UK farmers' competitiveness. The new business model for UK pig farmers is not yet clear but, whatever it is, the UK breeding herd will probably never again reach the heady heights of +800,000 sows as it did in 1997/98.

The German and Dutch pig herds have undergone their own structural changes as Germany has absorbed the inefficient pig production operations inherited from East Germany, and the Netherlands has started its adjustment to national and EU environmental regulations. In the Netherlands case there was also a significant Swine Fever outbreak in 1997 which prevented Dutch farmers from enjoying the benefits of the export market disturbances in that year. The USA herd, meanwhile, has undergone its own adjustments. Larger pig herds increasingly dominate the production scene even though the overall herd size has declined (Figure 1.5).

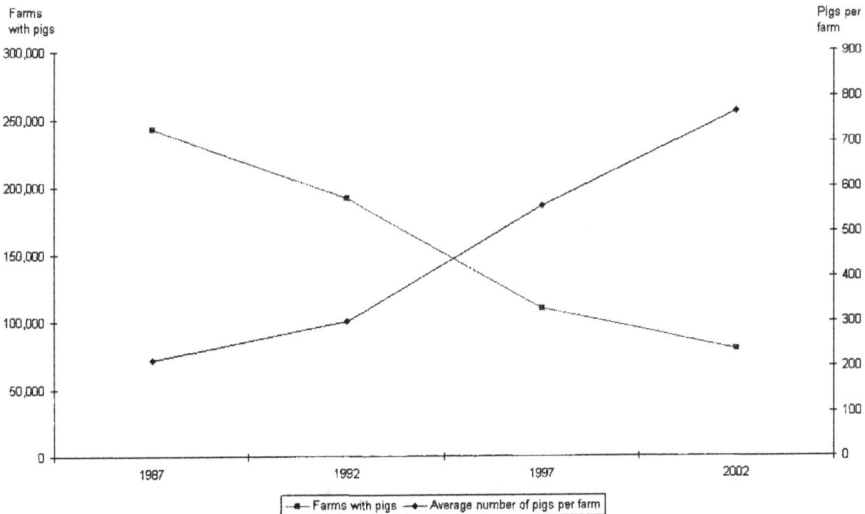

Figure 1.5 US pig farms and average herd size, 1987-2002.

In contrast to the countries that have reduced their pig herds, there are several that have maintained or increased their size. Australia, Canada, Denmark, Poland and Spain are all in this group and all play an increasingly important part in the world export trade for pork. Australian pork exports have taken off in recent years though and these export opportunities to Singapore and Japan have encouraged the Australian industry to develop specialist production for the world market.

Canada's herd is up by an impressive 23% since 2000 – although it is currently going through its own period of restructuring and the herd is broadly static at the moment. The removal of rail freight subsidies for grain in 1995 and the opening up of the western prairies for hog production and processing has driven this revitalisation of the Canadian pork industry. New investment by packers has played an important part in realising the potential of new Canadian hog barns. In Europe, Poland, Denmark and Spain have breeding herds that are up by 21%, 8% and 7% respectively since 2000. In Spain it is appropriate to describe the changes as restructuring because Spanish consumers' demand for pork has definitely shifted upwards and supply has matched it through major expansions by co-operatives and agribusiness pork operations. Poland has benefit significantly in recent years from trans-national investment, particularly from the USA, Germany and Denmark, in its previously declining industry. The post-Communist collapse of small farm pig production has been replaced by large, modern industrial production systems, supported by cheap feed and low labour costs. Finally, the position of Denmark must be acknowledged. The 8% increase in the Danish herd since 2000 is all the more remarkable when Denmark's small domestic market and land area is considered. The Danes seem to be able to increase consumption, production and exports even from a very high base level. Their economies of scale in marketing are well established and, like the Canadians, marketing plays a major part in the Danes' export success.

Restructuring of global pig production has occurred throughout the 2000s but has been spurred by different factors in different countries. The incidence of animal disease, government regulation and deregulation, and commercial enterprise has driven the restructuring activities. These all add up to a changing set of economic circumstances for pig producers in different countries. It is safe to say that change in the location of pig production is a constant factor in the world market for pork.

The pig price cycle – ups and downs

The pig price cycle and associated production cycle is well known to students of agricultural economics. There can hardly be a economics college tutor who has not handed out diagrams of the demand/supply "cobweb" to undergraduates. The pig industry has been the classic example of this cobweb as pig production shifts erratically around a price equilibrium point and produces the so-called pig price cycle over a period of years. But this is not just an academic subject. The position and direction of the pig price cycle is very important to pig producers and processors. As prices move from peak to trough in the cycle, even the most technically efficient pig producer can be swept away by a tide of losses as surplus pork is cleared from the market at rock bottom prices. Indeed, expensively-produced genetically pure and disease-free pigs are at a positive disadvantage when the market is on the floor compared with low-overhead, low-quality pigs from producers with minimal financial commitments, if the consumer does not differentiate between the two production systems.

Its easy to see that it is important to know if prices are moving from trough to peak because that is the time when the underlying cycle will help producers get the most from their investment and production plans. However most of all, producers need to know where the turning points in the cycle are. Calling the turn in the cycle is the most difficult thing for any economic forecaster to do but, of course, it is the most rewarding since it places those fortunate enough to have this information in the most profitable position.

The global pig market is very susceptible to price cycle behaviour and the cobweb theorem. Globalisation has increased this tendency. In a less open world economy the prices of producers in one country can be sheltered from the effect of changes in pig supplies in far off places. However trade liberalisation means that pig producers in one part of the globe are intimately affected by the behaviour of markets in another. Even the EU's producers, with a relatively closed market, are hit by events in, for example, Taiwan, Brazil or Singapore. As the EU's major exporters find their world markets becoming more competitive there is an inevitable downward pressure on domestic prices right across the EU – including countries where pork exports may not be important. Just as in chaos theory – when a butterfly flaps its wings in the Japanese pork market the resulting hurricane can be felt in the prairies of Western Canada, the hog barns of North Carolina or the piggeries of Jutland.

It is possible to put several different country producers prices together to get an overall view of how the different prices move together. These global market prices turn out to be highly correlated and highly cyclical. In Chart 6 it's possible to see the correlation between the three major pork exporters' prices (Canada, Denmark, and the USA). Whole Hog's office has been using these associations and some simple mathematical models to produce an overall "global price cycle" from these three key prices (and from other combinations of prices). The results of this are remarkable. Figures 1.6 and 1.7 clearly shows that the world's pig price cycle has distinct phases which relate closely to the underlying production conditions. As demand and supply got seriously out of balance in 1997 and production grew ahead of demand a turning point in the global cycle occurred in mid 1997. The subsequent downward swing of the cycle lasted for nearly two years and resulted in a pig price trough in mid

1999. The mid 1999 turning point reflected the major cutbacks in production herds across the world that occurred in the 1998/99 period. The cycle had two more upward swings in 2000/01 and 2004/05, with an intervening downturn in 2002/03, although far more muted than in 1998/99.

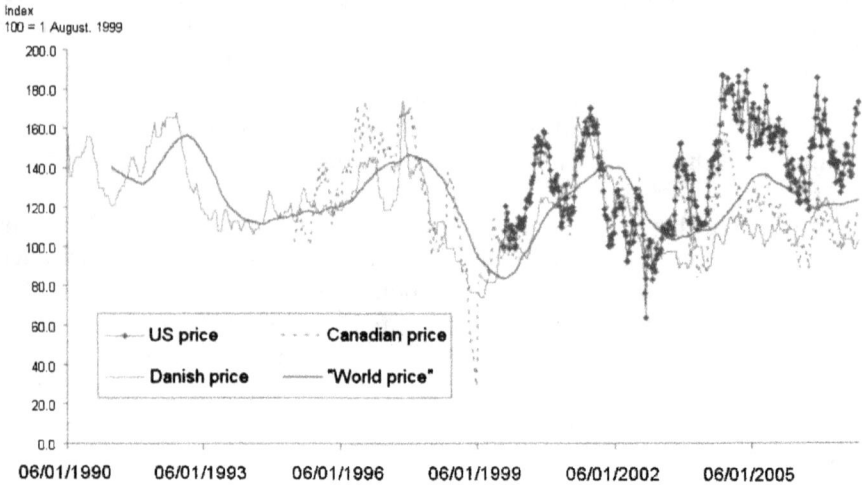

Figure 1.6 Global pig price cycle, 1990-2007.

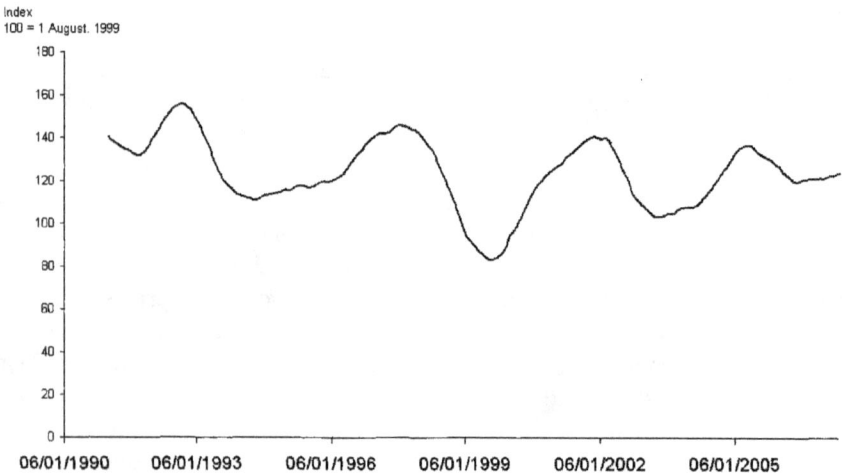

Figure 1.7 Global pig price cycle, 1990-2007.

The world's pig price cycle is now sending out a very weak bull signal. According to Whole Hog's price charts (constructed from a combination of world market pig prices) the next downward phase of the global pig price cycle may be underway. There is some evidence that the current very weak upswing has indeed peaked and that a fresh downward turn in the cycle will begin before the end of the current year. Fundamental analysis of pig production across the globe supports the technical analysis of the price charts. Whole Hog's model of global pork export supplies provided evidence for the increased supplies that preceded the price crash of 1999 - and Whole Hog's provisional forecasts of pork supply in 2007/08 put the world pork export supply significantly up on 2005/06, not least because productivity is rising amongst key producers and disease related mortality appears to be falling significantly.

What next – technical change, disease, and food vs fuel

The scientific and technical advisers to the pig industry will, if past experience is any guide, continue to provide a stream of technological advances and cost-reducing improvements to the global pig industry. The application of existing technology to major parts of the world's pig industry would increase pig farmers' productivity quite significantly and so there will be a constant demand for technical benefits in the form of genetics, disease control, feed conversion, etc. The restructuring of the world's pig industry that has occurred in the last decade will continue in the same vein in the next decade. Money and R & D efforts will be poured into genetics, disease control and other productivity gains. Geographically, the growth areas will be Brazil, central and eastern Europe, and China. All are areas where modern technology can be combined with inexpensive labour and relatively lax environmental regulations in order to produce the lowest cost product.

As the FMD outbreak in the UK showed, the pig sector can have the best bio-security but still be the most badly affected by disease spread. Despite modern advances the incidence of animal disease will increase price volatility and – in a global pig market – there may be no room for "over protective" national disease programmes. The opening and closing of trade frontiers as a result of well intentioned national disease regulations is not sustainable in the long term because these exacerbate price volatility and could produce a damaging impact on production and investment.

A very recent and alarming development for the global pig industry is the increased competition the sector faces for feed grains. Severe drought in Australia in 2006/07 and the dramatic shift, particularly in the USA, in the demand for biofuel crops has significantly increased the competition for and price of grain supplies at a global level. There is also increased demand for feed grains from China and other countries with developing livestock industries. World grain use has exceeded production in 6 of the last 7 years and carryover stocks are down to 57 days consumption. A serious weather event or crop disease outbreak in a grain-producing area would send grain prices skyrocketing and wreck all the cost of production calculations on which the modern pig industry bases its investment decisions. Even the existing situation puts an additional cost burden on the producers who do not grow their own feed stocks.

Conclusion

Whilst the fundamentals of demand and supply for the global pig industry are secure, an indestructible global price cycle, coupled with new threats from the globalisation of disease and competition for feed ensure that the roller coaster ride that is the global pig market will be with us for years to come.

2

PERSPECTIVES IN WORLD MEAT PRODUCTION, 2007 TO 2015

LUCIANO ROPPA
address

Currently there is no planet, other than Earth, with conditions to propagate life and maintain the human species. There is therefore the responsibility to keep it so for future generations. Over 78 billion people have passed through it making a contribution to human history, but there never was a moment with so many inhabitants as there is today. As each year goes by, with the population growing, the amount of land per habitant is becoming more limited. In 1980, it was 3.26 hectares of land available per person; presently it is 2.29 and in the year 2050, it will only be 1.6 hectares.

Global population is increasing. Presently, it is 6.4 billion people; in the year 2030, it will become 8.1 billion and in 2050 it will reach 9 billion. This population is growing more in developing countries than in the developed ones. In the latter, voluntary birth control has kept the population effectively constant, and in some of them it has even diminished. From 2005 to 2030, world population will grow a mean value of 26%, only 3.6% in the developed countries and 31.8% in the developing countries (Table 5.1).

Table 2.1 TRENDS IN GROWTH IN HUMAN POPULATION, 2005 TO 2030

	2005	*2030*	*%*
World	6.453	8.130	26.0
Developing Countries	5.117	6.746	31.8
Developed Countries	1.336	1.384	3.6

Source: FAO stat – Prepared by: L. Roppa, 2006.

This growing population must face new and serious challenges in the future. To underline some of these challenges, the current chapter will describe 7 of the great changes which are occurring and which will affect future generations:

1 –Available biological productivity is limited. If the planet Earth is examined as a whole, it has 51.3 billion hectares, of which 36.6 are composed of water (seas, lakes, rivers, etc) and only 14.7 of soil. Of these 14.7 billion hectares, 6.4 are unproductive and 8.3 are productive. Of these 8.3 billion hectares, 1.35 are cultivated soil, 3.35 are pastures, 3.34 are forests and 0.16 are built areas (cities, streets, roads etc.). Besides being limited, biological productivity is decreasing, for the amount of available land, per person, has diminished year on year.

2 – Water will be increasingly less available. Water is considered the natural resource which has the greatest possibility to limit sustainable growth of the planet. Our 'blue planet' has much water, but the greatest portion is salt water. In reality, 0.975 of water is formed by the Oceans and only 0.025 is fresh water. Even so, only 0.0052 is accessible fresh water (underground or in rivers and lakes). Today, 1.1 billion people have no access to drinkable water, principally in Asia and Africa. The growth in water usage per capita in the world, which doubles every 20 years, is at least twice the population growth. For the year 2025, the specialists in this area, consider that 2.7 million people will have serious problems of lack of water. In 2050, 0.45 of the world population will be living in countries which do not have the possibility of supplying the minimum daily amount of water to meet the basic needs of food and hygiene, which today are estimated at 50 litres per person.

3 – Desertification continues to grow and the areas of productive soil are diminishing. The principal regions having high desertification are located in Oceania, Africa and Asia. The Americas and Europe have the smallest proportion of desert soil or under desertification. The areas with high rates of desertification will face greater difficulty in the future in relation to the production of grain and animal production. Environmental alterations and the destruction of forests are some of the determining factors of this desertification. To reach equilibrium, to satisfy needs, every human being should not use more than the natural resources equivalent to 1.9 hectares of soil. However, the present mean value for the exploration of natural resources is already 2.3 hectares per capita.

4 - World population is migrating increasingly to the cities. In 2007, urban population will surpass the number of people living in the rural area and in the next 24 years, 0.60 of the world population will already be living in cities (Table 2.2). In the year 2030, world urban population should reach approximately 5 billion people and rural population should be reduced to 3.2 billion.

Table 2.2 PERCENT OF PEOPLE LIVING IN URBAN AREAS, 1950 TO 2030

	People living in cities (Proportion)			
	1950	1975	2005	2030
World	0.291	0.373	0.492	0.608
Developed Countries	0.525	0.672	0.749	0.817
Developing Countries	0.179	0.269	0.432	0.571

Source: FAO, 2005

5 - The proportion of the elderly has increased. Today, we have 0.10 of human population over 60 years of age. In 2050, this proportion of elderly will increase to 0.22 (Table 2.3) due to the increase in longevity. The mean age of the present human population is 28 years and in 2050 it will be 38 years.

Table 2.3 PERCENT OF PEOPLE OVER 60
YEARS OF AGE IN THE YEAR 2050

	Elderly people (proportion)
World	0.22
Developed Countries	0.33
Developing Countries	0.20

Source: United Nations, DESA, 2005 –
Prepared by: L. Roppa, 2006.

6 - The purchasing power of the population is growing. Presently, the GIP per person is 16 times greater in the developed countries than in the developing countries. In 2050, however, this difference will diminish and will be only three to six times larger. This increase is important to growth in the use of food and to reduce the number of people considered "below the poverty lines". The number of people who are hungry in the developing countries should be reduced from the 777 million today, to 440 million in 2030.

7 – The quality of the food offered to the population has improved and will continue improving. In 1960 9.87 MJ (2360 Kcal) available per person per day; today it is 11.72 MJ (2800 Kcal). In the year 2030, this will have increased to 12.76 MJ (3050 Kcal); Table 2.4.

Table 2.4 ENERGY AVAILABLE PER PERSON,
1960 TO 2030

Year	*MJ (Kcal) available per person*
1960	9.87 (2360)
2005	11.72 (2800)
2030	12.76 (3050)

Source: FAO, 2005.

In summary, in the near future, there will be a larger population, in a world with less water, living more in urban areas, with a longer mean age,, with less soil available for agriculture, with greater buying power and eating better than at present.

Therefore the challenge to accommodate the 8.1 billion people estimated for 2030 and to improve their diet, will be to increase the production of food by 50% in the next 25 years, in a planet which goes through a period of great climatic and environmental transformations.

In which regions of the world will there be a greater increase in human population?

The greatest increase in human population will occur in the developing countries. From 2005 to 2030, the population will grow by 26.1% (Table 2.5). In Asia and in Africa, it will grow more than the world mean, at 26 and 66% respectively. In Latin America it will grow 5.8%, while in Europe it will diminish 11.6%. In the year 2030, we estimate that 86% of the world population will be living in Asia, in Africa and in Latin America.

Table 2.5 PERSPECTIVES FOR THE GROWTH OF THE POPULATION, 2005 TO 2030 (BILLIONS)

	2005	*2030*	*Dif.%*
Asia	3.868	4.887	26.3
Africa	0.838	1.398	66.8
Latin America	0.565	0.711	25.8
North America	0.337	0.408	21.0
Europe	0.773	0.685	- 11.3
Oceania	0.065	0.041	- 36.9
WORLD	6.446	8.130	26.1

Source: FAO stat, 2005

In which regions of the world will there be greater increase in buying power?

Meat consumption has a strong correlation with the IGP per capita. As the buying power increases, meat consumption increases and, in consequence, meat production. In 1961, the IGP per capita in the world was US$ 2,676 and the meat consumption per person was 23 kg. In 2001, it grew to US$ 5,611.00 and consumption grew to 38 kg. In 2030, the IGP should be US$ 7,600.00 per person and consumption should grow to 45 kg (Table 2.6).

Table 2.6 GROWTH OF THE BUYING POWER AND MEAT CONSUMPTION

Year	IGP (US$ 1,995/capita)	Meat consumption, kg/person/year
1961	2,676	23.1
1971	3,714	27.8
1981	4,376	30.8
1991	4,992	34.4
2001	5,611	38.6
2030	7,600	45.3

Source: FAO – Prepared by: L. Roppa, 2006.

Meat consumption grows as salary increases, but this truth is only valid up to a certain point, known as the point of saturation. The tendency to grow is constant until reaching 80 to 100 kg of meat consumption per capita per year. From then on, it begins to stabilize. Europe, for example, already has a high consumption of animal proteins (over 200 kg per person, per year) and in spite of the growing buying power, it is difficult to increase its consumption.

In which countries is IGP per capita growing? World IGP per capita will grow by a mean value of 2.9% per year, until 2010; in developing countries it will grow 4.5%, in developed countries 2.4% and in the countries in transition, 3.8%. Therefore, the greatest growth in buying power will occur in the countries in Transition and in the developing countries. In Table 2.7 one can check the IGP of the 10 largest economies of the world presently and what is the perspective for 2030, according to Goldman Sachs and IMF data.

Table 2.7 PERSPECTIVES FOR THE GROWTH IN IGP, 2005 TO 2030

	GDP 2005 (trillion dollar)		*GDP 2050 (trillion dollar)*	
1	USA	12.8	China	49.0
2	Japan	5.0	USA	38.0
3	Germany	2.6	India	27.0
4	UK	2.3	Japan	8.0
5	China	2.3	Brazil	8.0
6	France	2.2	Mexico	7.8
7	Italy	1.8	Russia	6.2
8	Canada	1.0	Germany	5.4
9	Spain	1.0	UK	5.1
10	Korea	0.8	France	4.9
11	Brazil	0.8	Indonesia	3.9

Source: Goldman Sachs and IMF, 2005

Present meat consumption in the world

In 2005, world consumption of meat was close to 260 million tons, which corresponded to 40 kg of meat per person (Figure 2.1). In Asia the consumption per capita is 28.8 kg and in Africa, 13.6 kg. Consumption in Latin America is above the world mean value (49.4 kg), due to the high beef meat consumption in Brazil, Argentina and Uruguay. In Europe the consumption is 79.1 kg and in North America it is 141 kg.

When one multiplies the consumption per capita by the population of each continent, one sees that the largest consumption in quantity of meat occurs in Asia, with 112.9 million tons, followed by Europe (57.3) and North America (46.9).

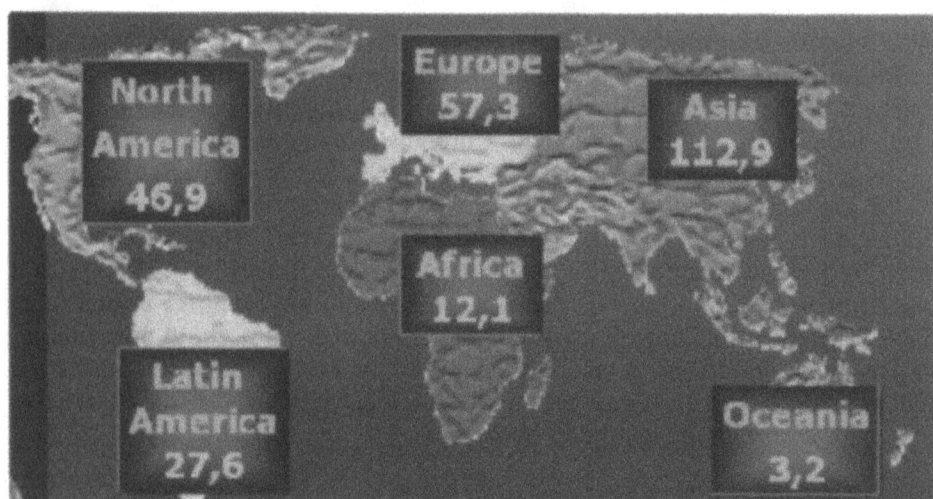

Figure 2.1 Present meat consumption, million tons (Source: L. Roppa, 2006).

Where will meat consumption grow?

The population of the developed countries, which was consuming 76 kg per capita in 1993, should increase its consumption to 83 kg in 2020, according to FAO data. In quantitative terms, consumption will grow from 99 to 115 million tons (Table 2.8).

In developing countries, consumption should increase from 21 to 30 kg per capita in the same period. To satisfy these 30 kg per person, production should increase from 88 million tons in 1993, to 188 million in 2020, due to the strong population growth of these countries.

Table 2.8 MEATS - INCREASE IN CONSUMPTION/PERSON AND TOTAL
CONSUMPTION, 1993 TO 2020

	Cons./person (kg)		*Total cons. (MMton)*	
	1993	*2020*	*1993*	*2020*
Developed countries				
Beef	25	26	32	36
Pork	28	29	36	41
Poultry	20	25	26	34
Total meat	76	83	97	115
Developing countries				
Beef	5	7	22	47
Pork	9	13	38	81
Poultry	5	8	21	49
Total meat	21	30	88	188

Source: Delgado *et al*, 1999. – Prepared by: L. Roppa, 2006.

Where will animal production be located in the future?

This will all depend on international rules. Will natural competence be judged
or will economical pressure be the main driver? The world will need to
challenge this subject. At any time, if animal production is located where the
cost of production is high, the population will pay more for food and the farmers
will need more money to subsidize production.

The main fact is that there will be an increment in production and
consumption of pigs, poultry and cattle in the future. If the logical issue is
discussed, animal protein will be concentrated in countries where the cost of
production is lower. In general, animal production will occur in countries with
favorable climate, favorable land, favorable human resources and in countries
which will produce food with quality, safety and low cost.

Forty years ago, half of all meat was produced in developing countries and
the other half in developed countries. Today, about 0.63 is produced in
developing countries. This tendency should grow, in such a manner that in
2030 over 0.70 will be produced in these countries (Figure 2.2). The same
should happen with the production of cereals: in the year 2030, 0.59 of their
production should be in developing countries and 0.41 in developed countries.

What type of meat will be more produced in the future?

The perspectives that FAO/OCDE has for 2015 are in Table 2.9. In relation to
present production, the world will have to increase production by 19% (as a
function of the increase in population and of the increase in consumption per

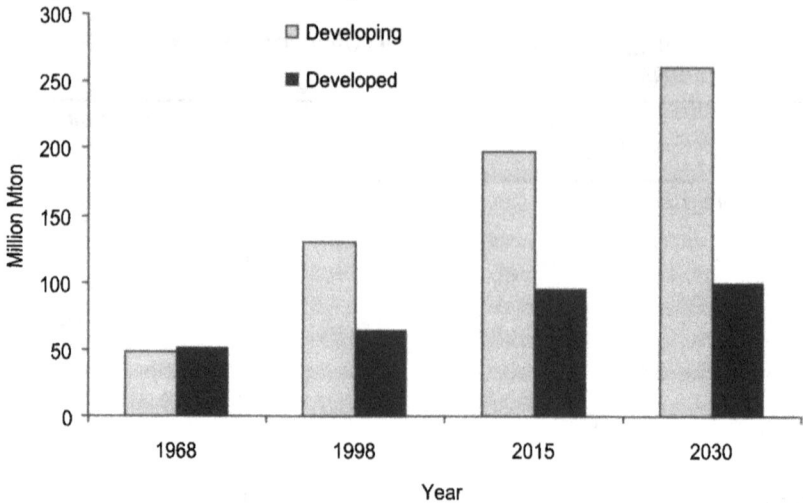

Figure 2.2 Perspectives of growth in meat production, 1968-2030 (Source: L. Roppa, 2006 – based on FAO/OCDE Statdata).

capita), going from 267 to 318 million tons. The largest growth will be in poultry meat (23.1%) and the smallest will be in pork (16.7%).

From a quantitative view, bovine meat will reach a production of 77.8 million tons, poultry meat 103 million tons and pork almost 123 million tons. In 2015, the largest production and the largest consumption per capita will continue to be in pork.

Table 2.9 MEAT PRODUCTION EXPECTED FOR 2015 (MILLION TONS)

	2006	*2015*	*%*
Poultry	83.820	103.235	23.16
Beef and Veal	65.922	77.834	18.06
Pork	105.382	122.979	16.70
Sheep	12.015	14.093	17.29
Total	267.139	318.141	19.09

Source: L. Roppa, 2006 – based on FAO/OCDE Statdata

Why could more poultry meat be consumed than pork in the future?

Poultry meat consumption has been approaching the consumption of pork for many years. Presently, it is estimated that poultry consumption should exceed that of pork around the year 2030 to 2040. The reason for this is related to religious background and to the tendency of population growth in the countries where these concepts prevail.

Today pork is not consumed by 0.33 of the world population, due to prohibition for reasons connected to religion (Table 2.10). Christians and Buddhists consume pork without any restriction. On the other hand, Moslems and Jews do not permit its consumption.

Table 2.10 RELIGION AND THE RELATIONSHIP TO PORK CONSUMPTION

Religion	People (billion)	Proportion of the world population	Pork consumption
Christian	1.9	0.327	YES
Moslem	1.1	0.197	NO
Hindu	0.78	0.133	YES/NO **
Buddhist	0.32	0.060	YES
Jewish	0.014	0.002	NO
Others *		0.281	YES

(*) Other religions, Non-religious and Atheists.
(**) The consumption of pork is not prohibited, but Hindus are vegetarian.
Source: L. Roppa, 2006 based on data from Desa, World Atlas and FAO.

Half the growth of the world population until the year 2050 will occur in only nine countries (Table 2.11). In the countries with greatest growth - India, Pakistan and Nigeria - the population is predominantly Muslim and Hindu, who do not permit consumption of pork. The population growth of China, which is a significant consumer of pork, will be limited, due to strong birth control policies, and will be compensated by the growth of Bangladeshis, who do not consume pork.

Table 2.11 COUNTRIES WITH GREATER POPULATION GROWTH IN THE FUTURE AND THEIR RESPECTIVE RELIGIOUS BACKGROUND

Countries	Population growth 2000 to 2050 (million)	Religion	Pork consumption (in 1,000 T/year)	Pork consumption (kg/person/year)
India	572	0.93 Hinduism and Muslim	499	0.46
Pakistan	162	0.97 Muslim	2	-
Nigeria	141	0.50 Muslim	215	1.19
Congo	127	0.10 Muslim	26	0.47
China	118	0.02 Muslim	48140	34.2
Bangladesh	114	0.99 Muslim and Hinduism	5	-
USA	111	0.56 Christian	8794	29.8
Uganda	103	0.16 Muslim	75	2.7
Ethiopia	102	0.50 Muslim	1.5	0.02

Source: L. Roppa, 2006 based on data from Desa, World Atlas and FAO.

The perspectives for population growth until 2050 show also that some countries will diminish their population (Russia, Ukrania, Japan, Italy, Poland, Romania, Germany), due to migration or birth control (Table 2.12). Unfortunately for pork consumption, these countries have a good tradition of consumption and have no religious barriers preventing it.

Table 2.12 COUNTRIES WHOSE POPULATION WILL DECREASE IN THE FUTURE AND THEIR RESPECTIVE RELIGIOUS BACKGROUNDS

Countries	Population growth 2000 to 2050 (million)	Religion	Pork consumption (in 1,000 T/year)	Pork consumption (kg/person/year)
Russia	- 35	Eastern Orthodox	2062	14.2
Ukraine	- 23	Eastern Orthodox	600	12.3
Japan	- 15	0.84 Buddhist	2411	19.1
Italy	- 7	0.95 Catholic	2463	42.7
Poland	- 7	0.95 Catholic	1844	47.8
Romania	- 5	0.70 Eastern Orthodox	560	25.1
Germany	- 4	0.72 Christian	4375	52.7
Belarus	- 3	0.80 Eastern Orthodox	278	27.0
Bulgaria	- 3	0.84 Eastern Orthodox	270	35.0

Source: L. Roppa, 2006 based on data from Desa, World Atlas and FAO.

In summary: world population grows in those countries where less pork is consumed, for religious reasons, and countries which have no increase in population are those which consume more pork. This is the explanation why poultry meat should overcome pork consumption between 2030 and 2040.

Which countries will produce more meat in the future?

As already mentioned, the production of beef, pork and poultry meat, will grow from 267 to 318 million tons, in the period from 2006 to 2015 (Table 2.13). The greatest quantitative growth will occur in China, an increase of 18.2 million tons. The second largest growth will happen in Brazil, 7.2 million tons. In this period, the United States will go from 3rd to 2nd largest world producer, exceeding the European Union (EU-25). China will increase its world participation from 0.272 to 0.285 of total meat produced. It is interesting to note that, of the 10 largest meat producers in 2015, six will be developing countries and together they make up 0.45 of world production.

Table 2.13 GROWTH IN MEAT PRODUCTION PER COUNTRY - 2006 TO 2015 (MILLION TONS)

	2006	*2015*	*Growth, %*	*Increase in quantity*
World	267.14	318.14	19.1	51.00
China	72.67	90.91	25.1	18.24
USA	39.71	45.06	13.4	5.35
EU 25	40.62	42.01	3.4	1.39
Brazil	21.09	28.36	34.4	7.27
India	5.79	7.52	29.8	1.73
Russia	5.43	6.98	28.7	1.55
Mexico	5.09	6.10	20.0	1.01
Argentina	4.36	5.06	16.1	0.70
Japan	2.97	2.74	(7.8)	- 0.23

Source: FAO/OCDE statdata, 2006 – Prepared by L. Roppa, 2006

Which countries will have excess production to export?

Subtractracting internal consumption of meat products in each country leaves the excess available for export or the deficit for importation. This availability of meat, positive or negative, is shown in Table 2.14, and does not consider the possible importation from other countries. As can be seen, in the year 2015 Brazil will have the largest positive value between its production and its internal consumption: 7.8 million tons (2.7 of beef, 4.3 of poultry and 0.8 of pork). In second place will be the United States, with 3.1 million tons, followed by Australia, Argentina and the European Union (EU-25). China, having the largest human population, should import bovine and poultry meat, and be an exporter of pork. India, with the second largest human population, should be an exporter of bovine meat and be practically self sufficient in the other meats, due to its low consumption. Japan, Russia, Mexico and South Korea will continue to be the largest world importers.

It is important to make clear that Table 14 does not show the future exports of each country, but rather their availability for export, considering their internal production. The United States, for example, should export more than the 0.336 million tons of pork mentioned in Table 14, for they import large quantities from Canada, which, after being industrialized, are also exported.

What can change these perspectives of meat consumption?

These perspectives in relation to increases in meat consumption in the future could be altered considering a series of factors. The main risk for future meat production is the occurrence of animal zoonotic diseases. A recent example is

Table 2.14 INTERNAL AVAILABILITY OF MEAT IN THE PRINCIPAL WORLD PRODUCERS (PRODUCTION – CONSUMPTION), 2015

	Beef	*Poultry*	*Pork*	*Total*
Brazil	+ 2.704	+ 4.369	+ 0.804	+ 7.877
USA	- 0.546	+ 3.379	+ 0.336	+ 3.169
Australia	+ 1.325	+ 0.039	- 0.110	+ 1.254
Argentina	+ 0.956	+ 0.199	+ 0.070	+ 1.225
UE 25	- 0.529	+ 0.292	+ 1.368	+ 1.131
India	+ 0.628	- 0.005	+ 0.017	+ 0.640
China	- 0.058	- 0.569	+ 0.257	- 0.370
Japan	- 1.162	- 0.658	- 1.356	- 3.176
Russia	- 0.753	- 1.049	- 0.544	- 2.346
Mexico	- 0.197	- 0.634	- 0.367	- 1.198
South Korea	- 0.393	- 0.199	- 0.301	- 0.893

Source: FAO/OCDE statdata, 2006 – Prepared by L. Roppa, 2006

Avian Flu, which reduced poultry meat consumption throughout Europe, Asia and in various other parts of the world. However it is not only Avian Flu that can compromise the rate of increase in meat consumption. BSE in cattle has already caused this and continues to cause a fall in consumption of beef meat in Japan and in other countries. Foot and Mouth Disease is another example.

Other important factors that can reduce meat consumption is the presence of undesirable substances such as dioxins, antibiotics whose use in animal diets is prohibited, etc.. Other factors which can initiate a reduction in consumption are the environmental problems related to animal production and the problems related to global or regional economy.

Could pork satisfy the growing world demands for more food?

Pork is currently the most consumed meat in the world, at present (15.9 kg per person) when compared to that of poultry (12.6 kg) and bovines (9.4 kg). It is well consumed in Europe and North America and little consumed in Africa, Latin America and in Asia (Table 2.15). It is interesting that these three last regions constitute 0.82 of the world population. Since their consumption is considerably below the world mean value, their potential as future pork markets is enormous.

The problem is that these countries are markets where most of the people have a low buying power. Most of the 2.4 billion people who exist on less than US $2 a day are located in these regions. To reach these markets, the exporting countries must produce meat at low cost. Countries having high production costs will not be competitive in these markets.

Table 2.15 ANNUAL PORK CONSUMPTION AND HUMAN POPULATION, PER CONTINENT, 2005

	Consumption, kg/person	*Population, million*
Africa	1.0	838
Latin America	9.8	565
Asia	14.5	3868
Oceania	17.4	65
North America	29.9	337
Europe	34.2	773
World	15.9	6446

L. Roppa, 2006 based on FAO data.

Europe, which produces a kilo of live swine for US$ 0.90 to 1.40, will have much difficulty to satisfy this growth of the population (Table 2.16). Brazil is a competitive country to meet these markets, for it produces pig meat with one of the lowest costs in the world.

Table 2.16 COST OF PORK PRODUCTION ACCORDING TO RASMUSSEN (2002), RABOBANK (2003) AND PIC (2005)

Countries	*Rasmussen, 2002* *(US $/kg carcass)*	*Rabobank, 2003* *(Euros /kg carcass)*	*PIC, 2005* *(US$/ kg liveweight)*
United Kingdom	1.30	-	1.39
China	-	1.35	-
Holand	-	1.30	1.10
Poland	-	1.18	0.90
Denmark	1.19	-	0.90
France	1.18	-	-
Canada	0.97	1.13	0.82
USA	1.07	1.15	0.83
Brazil	0.90	0.99	0.57

Source: L. Roppa, 2006.

Who are the large world producers of pork?

World production in 2005 was 102441 million tons, of which 0.56 was within Asia. Europe comes in second, with 0.25 and the Americas with 0.17.

Pig production has grown in the last 10 years at a rate of 2.78% a year. This growth, however, was much more concentrated in the developing countries, with a mean value of 4.45% increase a year, against an increase of only 0.83%

year in the developed countries. Therefore growth is currently strong in the developing countries.

The four largest world producers are China, with 50 million tons, the European Union (EU 25), with 21 million, the USA, close to 10 million and Brazil, with 2.7 million tons (Table 2.17). These 4 largest world producers of pork add, together, 80% of the world production.

Table 2.17 TEN LARGEST WORLD PRODUCERS OF PORK, 2005 (MILLION TONS)

1 - China	50.094
2 - EU – 25	21.440
3 – USA	9.392
4 – Brazil	2.707
5 – Canada	1.960
6 – Vietnam	1.900
7 - Russia	1.715
8 - Philipines	1.325
9 – Japan	1.260
10 - Mexico	1.175

Source: L. Roppa, 2006.

Table 2.18 compares the principal characteristics of these 4 largest world producers. China and Europe have a large number of pig per square kilometer and per habitant. This is a serious problem for Europe that, due to environmental concerns, has to limit its production and the growth of its pig production sector. On the other hand in China, due to the family basis, pig production is important to keep the population living in rural areas. The largest consumption is in the European continent, while China and the United States have similar consumptions. Brazil, comparatively, has excellent advantages: a small number of pigs per square kilometer, low cost production and a low internal consumption, which presents an excellent opportunity (and not a problem) to increase production without depending on the foreign market.

Table 2.18 CHARACTERISTICS OF THE LARGEST WORLD PRODUCERS OF PORK, 2005

	China	*UE-25*	*USA*	*Brazil*
Population, million people	1350	455	276	184
Area, million km2	9.59	3.23	9.63	8.51
Swine, million head	469.0	151.9	60.5	32.4
Swine/km2	48.9	47.0	6.3	3.8
Swine/person	0.34	0.33	0.22	0.17
Consumption, kg/person	34.2	44.2	32.0	11.3
Cost of production, US$/Kg (liveweight)	0.85 to 1.20	0.90 to 1.50	0.75 to 0.85	0.60 to 0.70

Source: L. Roppa, 2006

Could Brazil confront the three large world producers of pork?

Analyzing the comparative data with China, the European Union and the United States (Table 2.18), Brazil has fewer pigs per square kilometer, fewer pigs per person and has an excellent opportunity in its internal market, as there is a low consumption per capita.

Brazil has a historical intention, that is being increasingly expressed; to be an important world food producer. It is the largest world producer of sugarcane, oranges and coffee. It is the second largest producer of soybeans and poultry; the third largest of beef, maize and fruit; the fourth largest of pork, the fifth of milk and seventh of eggs. In other words, Brazil is among the seven largest world producers of proteins and carbohydrates, both from animals and vegetables, among over 280 countries which compose the present world (Table2.19).

Table 2.19 THE GROWING DOMINANCE OF BRAZIL IN WORLD AGRICULTURE, 2005 RANKINGS

	World rank		*Market share of global exports (%)*	*Export growth rates 2000-05 (%)*
Commodity	*Exports*	*Production*		
Sugar	1	1	42	20
Ethanol	1	1	51	79
Coffee	1	1	26	11
Orange juice	1	1	80	4
Tobacco	1	1	29	15
Beef	1	2	26	32
Poultry	1	3	35	31
Soybeans	2	2	35	22
Soymeal	2	2	25	13
Corn	4	3	35	48
Pork	4	4	14	40

Source: USDA: Foreign Agricultural Service and Global Trade Information Services data

Brazil currently is the 4th largest global producer of meat and has 0.075 of the world production. It has 0.07 of the area of the world which can be cultivated and produces 120 million tons of grains. It has 0.12 of the reserves of drinkable water and excellent natural resources to guarantee its independence in energy use. It has sufficient cultivatble land to triple its grain production and has the technology to transform vegetable proteins into animal proteins, adding value to its export products.

Due to these characteristics, it is presently the largest world exporter of meats. In 2005, Brazil exported 5.4 million tons of meats to more than 180 countries (625 thousand tons of pork, 2.1 million tons of beef and 2.7 million tons of poultry meat). Currently it is responsible for 0.347 of the poultry meat marketed in the world, for 0.261 of beef and 0.143 of pork (Figure 2.3). These numbers demonstrate the uncontested importance of Brazil to address the problem of constant growth of the world population and its demands for good quality food.

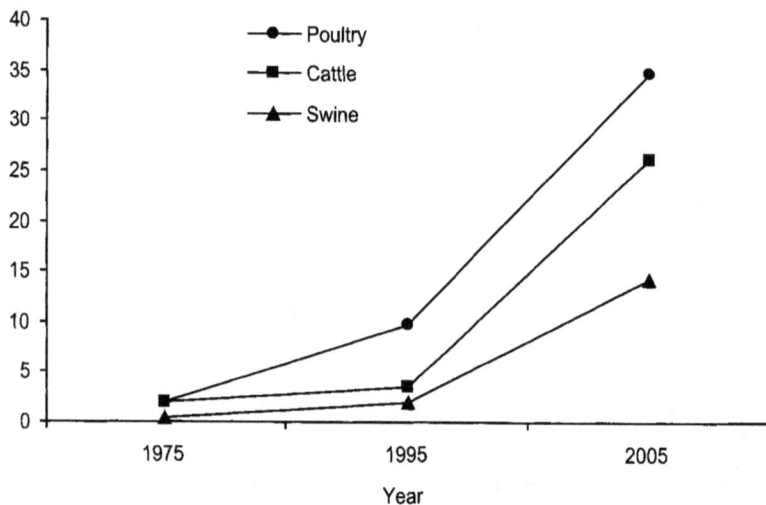

Figure 2.3 Contribution of Brazil to world exports of meats (Source: L. Roppa, 2006).

The competitive advantages of Brazil are based on: availability of land, low start-up costs, good human resources, low costs of production, good availability of grain, a strong industrial ethos, fewer diseases, favorable climate and good technology.

Present concerns include: low availability of credit, high interest rates, presence of Foot and Mouth Disease in some regions, the economical instability common to developing countries and problems in transporting products (highways and ports).

BRAZIL: CHARACTERISTICS OF THE 4TH LARGEST WORLD PRODUCER AND EXPORTER OF PORK

The fourth largest world exporter of pork, Brazil produced 2.825 million tons in 2006. Its world participation has grown significantly from 1990 and today represents 0.0268 of the total pork produced in the world (Table 2.20).

BRAZIL: EVOLUTION OF THE PIG HERD

Brazil currently has a herd of 34 million head and it is estimated that 400 thousand people depend directly on the production chain for Brazilian pig production. The value of the pig production chain is estimated at U$ 1.8 billion.

Table 2.20 EVOLUTION OF THE PARTICIPATION OF BRAZIL IN WORLD PORK
PRODUCTION, 1970 TO 2006

	Production - Brazil (million tons)	*Production - World (million tons)*	*Contribution,*
1970	0.705	35.792	0.0197
1980	1.150	52.678	0.0218
1990	1.040	69.862	0.0149
1995	1.470	78.635	0.0187
2000	2.558	89.533	0.0285
2006	2.825	105.300	0.0268

Source: L. Roppa, 2007

In 1970 the herd was 31.5 million head and production had been 705 thousand
tons. In 2006, with 32 million head, production increased to 2.825 million
tons. Therefore, in 36 years, herd growth was only 1.6%, while production
increased 300%. These numbers (Figure 2.4) clearly show the technological
evolution of the sector within this period, thanks to important contribution of
technologists and breeders in the areas of genetics, nutrition and handling.

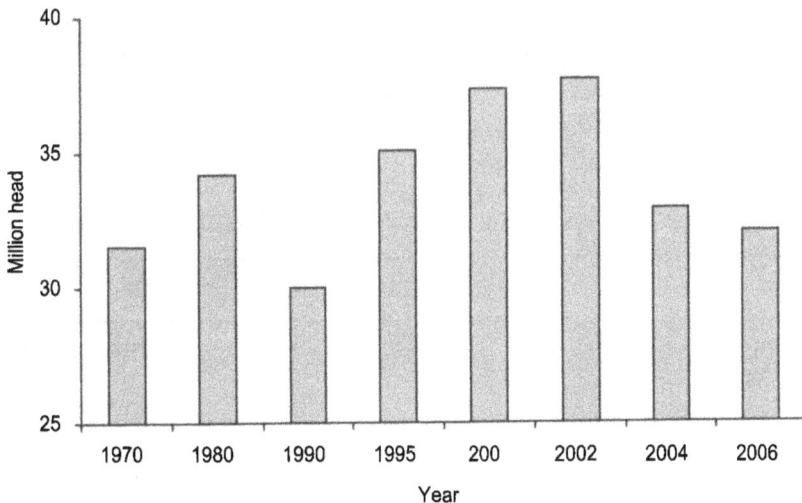

Figure 2.4 Growth of the pig herd in Brazil (Million head).

BRAZIL: BREEDING HERD

The Brazilian breeding herd currently consists of 2.428 million breeding
females. Of this total, 1.514 million are considered "improved" and are bred in

the South, Southwest and Central West regions of the country. The remaining 0.914 million breeding females are considered "unimproved" and most of them are bred in the North and Northeast regions. The principal companies that supply breeding females in Brazil are PIC, Dalland, Dan Bred, Pen Ar Lan and Genetic Pork. Official data of the evolution of the number of breeding females are shown in Table 2.21.

Table 2.21 BRAZIL: BREEDING FEMALES KEPT PER REGION, 2006

Region of the Country	Proportion of the herd	
	Improved	*Unimproved*
South	0.592	0.099
Southwest	0.200	0.065
Central West	0.110	0.140
North and Northeast	0.098	0.696
Brazil	1.00	1.00

(Source: ABIPECS, 2007).

BRAZIL: SUPPLY AND DEMAND FOR PORK - 2000 TO 2006

Table 2.22 summarises data on supply and demand of pork production in Brazil, in the period from 2000 to 2006. Production grew from 2.558 to 2.825 million tons but, with the increase of exports, internal availability fell from 2.42 to 2.35 million tons. Accordingly consumption per capita, which had reached 14.3 kg in 2000, fell to 12.5 kg.

Table 2.22 PORK: SUPPLY AND DEMAND IN BRAZIL, 2000 TO 2006 (THOUSAND TONS)

	2000	*2002*	*2004*	*2006*
Production	2558	2872	2679	2825
Importation	5	1	1	5
Internal Consumption	2561	2873	2680	2830
Exports	135	476	507	525
Availability	2426	2397	2173	2305
Consumption, kg per capita	14.3	13.8	12.2	12.5

(Source: ABCS, ABIPECS, ICEPA)

Historical evolution of pork production in Brazil is shown in Figure 2.5. The reduction in production in the years 2003 and 2004 was caused by the strong crisis in the sector, due to the excess production in 2002.

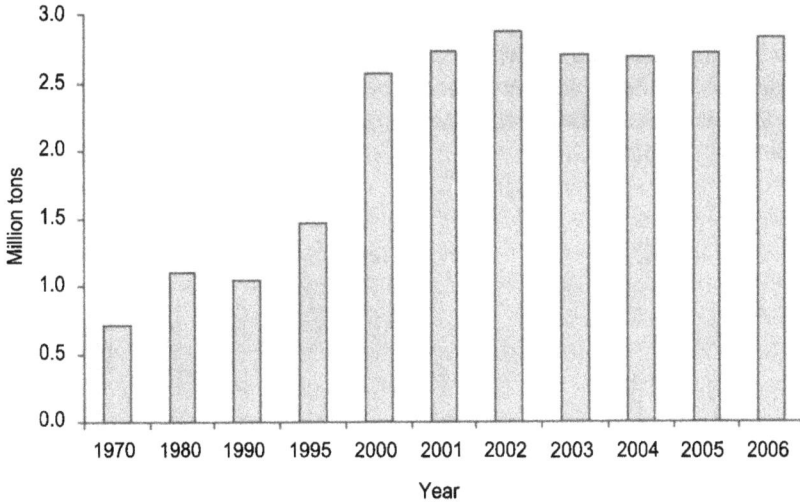

Figure 2.5 History of pork production in Brazil (million tons)

BRAZIL: PORK PRODUCTION PER GEOGRAPHICAL REGION

Table 2.23 presents pork production per geographical region, expressed in million tons produced. As can be seen, the South region has 0.58 of the country's production. This is the region where the integrated systems dominate and the strong agribusiness industrial sector is located. The Southwest region, where the independent pig breeders dominate, contributes 0.177. The Central West region, considered the new meat production and grain producing frontier in Brazil, continues to expand, participating presently with 14%.

Table 2.23 BRAZIL: EVOLUTION OF PORK PRODUCTION PER GEOGRAPHICAL REGION, 2002 TO 2006 (IN MILLION TONS)

REGIONS	*2002*	*2006*	*Change %*
SOUTH	1686.9	1655.9	- 1.8
SOUTHWEST	561.7	502.6	- 10.5
CENTRAL WEST	385.6	399.5	+ 3.6
NORTH and NORTHEAST	277.7	271.8	- 2.1
TOTAL	**2872.0**	**2829.8**	**- 1.4**

Source: ABIPECS, 2007

BRAZIL: EVOLUTION OF PORK CONSUMPTION

Pork is the most consumed meat in the world, but in Brazil it comes behind beef and poultry meat. About 0.65 of the pork consumed in Brazil is intensively produced and only 0.35 extensive, which makes it more difficult to consume in periods of economic constraint. As 0.65% of pig meat is marketed in the form of sausages and hams, which cost more that extensively-reared meat and are consumed by the wealthier salaried people, the reduction in buying power directly affects its consumption. In 2006, Brazilians consumed only 12.5 kg of pork per capita (Figure 2.6).

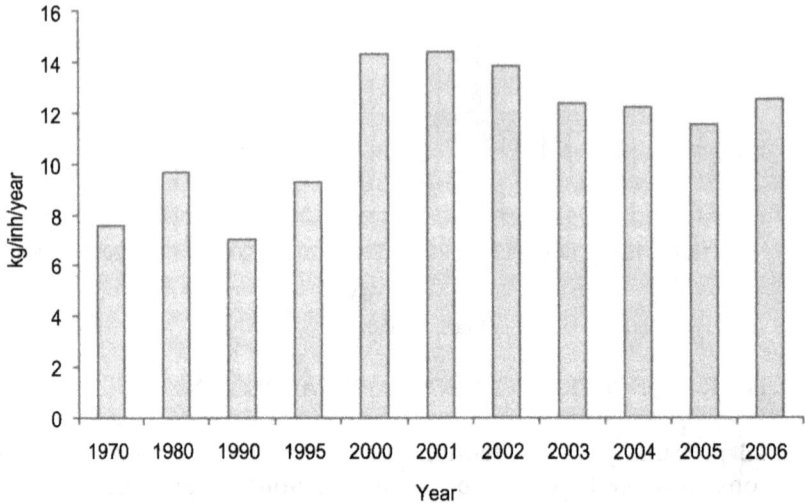

Figure 2.6 History of pork consumption in Brazil (kg/inhabitant/year) (Source: L. Roppa, 2007)

BRAZIL: EXPORTS OF PORK

Brazilian exports of pork have enjoyed considerable growth in recent years, due to excellent competitivity. As seen in Table 2.24, in 2000 Brazil was exporting 127 thousand tons. In 2006, with 525 thousand tons exported, the position of the fourth largest world exporter was confirmed, only behind Denmark, the United States and Canada.

Table 2.24 EVOLUTION OF BRAZILIAN EXPORTS OF PORK, 2000 TO 2006 (IN THOUSAND TONS)

	2000	*2001*	*2002*	*2003*	*2004*	*2005*	*2006*
Exports	127	265	475	496	507	625	525

(Source: ABIPECS, 2007)

Brazil exports 0.185 of all the pork it produces; the rest is consumed in the internal market by its 185 million inhabitants. Russia is the largest importer of

Brazilian pork, being responsible for 0.55 of total exports. This dependence on Russia was greater in the past, having reached 0.75 of the volume. The reduction in exports in 2006 was exactly due to this dependence: Russia prohibited the importation from some Brazilian states, which caused the drop in the volume exported that year.

BRAZIL: THE CHALLENGE OF INTERNAL CONSUMPTION

The greatest potential to increase pork production in Brazil is via its internal market. Today, having a population of 185 million inhabitants and with the prospect of reaching over 200 million people in 2015, any increase in consumption can represent a strong challenge for production. Table 25 describes this increase considering two situations: the first, with consumption continuing as it is currently (12.5 kg/inhabitant); in this case, in 2015, with the increase in population to 201 million, an additional 207 thousand tons pork would be needed. For this, it would be necessary to add 115000 breeding females to the present herd, considering that each female would have to produce 1.8 tons of meat per year.

In the second case, which could be called 'The Challenge', what would be needed to be produced in 2015, should Brazilian consumption per capita grow to 16 kilos, is shown. In this case, production would have to grow by another 0.9 million tons and for this it would be necessary to increase the herd by 500000 breeding females. All this increase would be based only on the increase in internal consumption, assuming that exports would continue the same as today (525000 tons).

Table 2.25 BRAZIL: PERSPECTIVE OF INCREASE IN PRODUCTION, CONSIDERING THE INCREASE IN INTERNAL CONSUMPTION

	Population	*12.5 kg* *(equal)*	*16.0 kg* *(challenge)*
2006	185	2.305	
2015	201*	2.512	3.216
Increase in Production, million tons		0.207	0.911
Increase in Breeding Females, Thousands **		115	506

(*) Growth Rate: 1.11%/ year (CIA, The World Fact Book)
(**) 1 female x 24 pigs/ year x 75 kg carcass = 1.8 t/year
Source: L. Roppa, 2007

Summary

1   World population will increase from the present 6.3 billion to 8.1 billion people in 2030. Developing countries will have 0.86 of this population.

2   World population will eat better every year: Currently it consumes 11.72 MJ (2800 Kcal) per person per day, and in 2030 it should consume 12.76 MJ (3050 Kcal) per person per day.

3   The greater part of the increase in world demand for meats in the next decade should occur in the developing countries.

4   The projection of meat production in 2015 is an increase to 318 million tons.

5   Brazil, USA, Australia, Argentina and the European Union will continue to be the great world exporters in 2015.

6   The strong buying power in Asia (Japan, Hong Kong and South Korea) will cause greater importations to this region.

7   Meat consumption per capita will increase from the present 40 kg to 45 kg in 2030.

8   Diseases which affect humans and food safety are the principal risks to the increase in meat production, for the market does not forgive the occurrence of a transmittable disease!

References

1 - "Livestock to 2020: the next food revolution"
IFPRI (International Food Policy Research Institute), USA
FAO (Food and Agriculture Organization of the United Nations)
ILRI (International Livestock Research Institute), Kenya.
Published: 1999

2 - "World agriculture towards 2015-2030"
FAO (Food and Agriculture Organization of the United Nations)
Published: 2006

3 – "Global pigmeat production"
Luciano Roppa
"Banff Pork Seminar", Canada, January 2005.

4 – "Global meat production: meeting the challenge in a changing world"
Luciano Roppa
"III Latin American Pork Congress", Brazil, October 2006.

3

HOW LARGE SYSTEMS MAKE HEALTH DECISIONS

MIKE MOHR
2805 Durham Rd, City:Wake Forest, NC 27587, USA

Making animal health decisions is a difficult task because it is full of dynamically variable aspects. Furthermore, the paradigms for making decisions as well as the supporting logic are very individual-business system specific. Although this topic is inherently complex, this chapter will attempt to demonstrate how it is done in one system as a case example.

Overview

One of the first items to remember is that the businesses are really groups of people with a common goal defined by their own sense of logic and view of the world. By definition, then large groups are subject to the same human forces as smaller businesses. Interestingly enough even publicly-traded companies are subject to this reality as are sole proprietorships.

One of the differences however is the concept of "Leveraged Business Model" and its role in the company's goal. Private companies tend to focused on use and method of use of the owners equity in the business. Decision-making is a relatively simple straight forward process: collect or observe the data, do the time-related appropriate analysis, make the decision, and then move on to the next one. Intuition plays a major role in the decision/analysis process as the decision maker is relatively close to the situation and usually has a longer-term business focus in mind. The ultimate motive is profit-based.

This is in contrast to the publically-traded firms. They must first compete for capital access, sell the story in a formal way to justify the investment, then implement in a way that conserves cash in case it is needed in other local areas. Here the focus is on cash cost and short term results to maintain investor confidence, in order to maintain fund flows into the business. Factual analyses followed by formal presentations are needed to communicate the problems and possible solutions to the production staff. Professional managers, vertical integration considerations and a "cost centre bias" present even further complications to reaching the end goal, generating overall global company profits.

Thus understanding the underlying business paradigm helps in subsequently balancing the health goal.

Balancing the health goal

Determining the health resource needs and availabilities greatly influence how health decisions are made. Most private equity groups and smaller firms do not focus on systems, thus are limited to individual efforts to make things work. For example there might be only one veterinarian for the 30,000 sow system who might have administrative and production assistants to help them accomplish tasks.

Larger system-orientated firms thus need to incorporate the leveraged model for technical assistance to the production department. In this case the veterinarian would be stretched to 130,000 sow systems which would force the reliance on assistants as well as skills in people management, computer software application use, as well as local computer hardware use knowledge. Here the production assistance must be of a higher skill base and computerized decision support systems in place.

The health goal must then be defined and balanced to the other priorities in the business. With the Intimate Model, the focus and appropriated resources are on diagnosing and curing the disease followed by preventing a reoccurrence. In the Leveraged Model, the focus is on identifying all the possible interactions involved in the health problem, defining the implicating factors, and then formulating the mitigation plan to contain the disease, not necessarily curing the disease.

For example in the intimate-modeled business, the veterinarian's role is one of being a "Doctor". This means the person must visit the sick "patients" in an effort to maximize knowledge of the situation in order to formulate an intervention plan to the patient's (many times inadvertently excluding the business and managerial interests) best interest.

Conversely in the leveraged-modeled structure, a more systems-based approach is used and is thus needed to formulate the intervention plan that addresses all the participants' needs. Hence the veterinarian's role for example is changed to one of being primarily an epidemiologist that doubles as a forensic investigator followed by having the acute senses of a businessman.

Thus one can see that the method that large production systems use to make health decisions is drastically different than businesses subscribing to the intimate model. What is shown in the current chapter is just one method to making a health decision demonstrated in a large production setting.

Solving a pneumonia problem example

A typical 2400 sow, 3 site layout system was experiencing high mortality both in nurseries as well as in finishers. This problem had been going on for 8 months to no avail. Numerous intervention plans (details in figure 3.1) were tried but under the intimate model paradigm.

Sow farm productivity was generally normal, but struggled during the outset of the project time frame. Pre weaning mortality was statistically high leading to a reduction in total number of pigs weaned. The PRRS and SIV health status was deemed quiet and under control.

Nurseries on the other hand struggled with mortality starting at 9% and steadily rose to a peak of 40% over the 8 month time frame prior to the intervention plan (Figure 2). The PRRS health status varied widely and was considered out of control despite continued success at the small nursery site.

Subsequent finishing flow also experienced rising mortalities regardless of site size and despite previously designed mitigation efforts. Mortalities started at 7% and steadily rose to a peak of 30% over the corresponding time period. As expected, the finishing PRRS health status always remained high al though the SIV status seemed to be under control with decreasing titers over the time period.

Reviewing the intervention project

The Intervention project started with a forensic investigation into the health status of the flow. Using a standardized health investigation process, serology and tissues were collected throughout the flow along with site management audits and nutritional analyses.

Histology data revealed PRRS, SIV and Mycoplasma damage throughout the various age groups. Concurrent bacterial damage also occurred within the major organs of: lungs, liver, heart, spleen and lymph nodes. Specific damage attributable to PCV was noted as well as large isolations of PCV found.

PRRS and Mycoplasma pathogens were also found, but SIV was not isolated. The list of bacterial agents was the typical suspects: *Haemopholis parasuis, Streptococcus suis, Bordatella bronchoseptica, Actinopyogenes, Ecoli* (K88), and a unique highly resistant Actinobacillus species.

Serology status confirmed the prior existence of SIV in both the H1 and H3 varieties. Mycoplasma responses were also confirmed as well as evidence that the PRRS viral load level had been too great for the immune response to overcome ie: continued exposure to various strains throughout the system prevented PRRS immune status recovery.

Since the study occurred prior to the market availability of PCV vaccine, it was decided that the intervention plan be centered around the control of bacterial

Figure 3.1 EXISTING VACCINATION PROFILE

Animal type	Vaccine type	Product name	Dosage (cc)	Timing	Effective date	Comments
Replacement boars	Influenza	Maxivac Excell (-3)	2	2 wks after placement	3/8/2006	
	Parvo/lepto/ery	SUVAXYN PLE	3	3 wks after placement	3/8/2006	
Replacement gilts	Influenza	Maxivac Excell (-3)	2	2 wks after placement	3/8/2006	
	Parvo/lepto/ery	SUVAXYN PLE	3	3 wks after placement	3/8/2006	
Boars	Lepto/erysipelas	SUVAXYN LE+B	2	Jan	9/1/2005	
	Influenza	Maxivac Excell (-3)	2	Jan	6/1/2005	
	Influenza	Maxivac Excell (-3)	2	July	6/1/2007	
	Lepto/erysipelas	SUVAXYN-LE+B	2	July	9/1/2005	
Breeding females	E. coli/clostrid	Auto Clostr-Newport	2	10 wks in gestation	9/14/2002	
	Strep/HPS	Auto STrep/HPS-Newport	2	11 wks in gestation	3/17/2003	
	Influenza	Maxivac Excell (-3)	2	12 wks in gestation	6/1/2005	
	Lepto/erysipelas	SUVAXYN LE+B	2	13 wks in gestation	9/1/2005	
	E. coli/clostrid	Auto Clostr-Newport	2	13 wks in gestation	9/14/2002	
	Strep/HPS	Auto Strep/HPS-Newport	2	14 wks in gestation	3/17/2003	
Nursery	Salmonella	Nitrosal	1	4 wks total age	3/17/2005	Run 1 bottle/100 hd orally
	E. coli	FIS E. coli	1 pkg	4 wks total age	2/29/2000	
	W/S antibiotic	Tesol-324	2.5# pkg/8 gals	5 wks total age	2/1/2005	Run with Myco. vaccination
	Mycoplasma	SUVAXYN MH	2	6 wks total age	3/1/2004	
	W/S antibiotic	Amoxicillin	6 pkg/gal	6 wks total age	2/1/2005	
	W/S antibiotic	Amoxicillin	6 pkg/gal	7 wks total age	2/1/2004	
	W/S antibiotic	Tetsol-324	2.5# pkg/8 gals	8 wks total age	1/1/2004	Run with Myco. vaccination
	Mycoplasma	SUVAXYN MH	2	8 wks total age	3/1/2004	
Finishing	W/S antibiotic	Neomycin (pint or 200g pack)	1 pkg/gal	15 wks total age	1/23/2004	
	W/S antibiotic	Tetsol-324	2.5# pkg/8 glas	16 wks total age	1/1/2004	
	W/S antibiotic	Sulmet (w/s)	1 pk/gal	20 wks total age	1/1/2004	

agents with possible mycoplasma control as well. SIV intervention plan was put on hold at the moment due to the lack of viral isolation evidence. A much larger PRRS stabilization plan was already underway which included this system and was centered on the sow herd portion of the flow.

Thus, it was decided that the intervention idea that best suited the available budget and balanced for risk would be the inclusion of Pulmotil at 363g/t during the second nursery diet phase. This treatment phase was conducted for one cycle time through the nursery, i.e. 7 weeks-worth of production.

Subsequent mortality metrics immediately returned to normal status with an interesting observation that groups not under treatment (ie: past the second nursery diet stage) on the same site improved as well. Other performance parameter metrics improved concurrently and this enhanced performance was also found in the finisher phase too.

The intervention plan improved the areas of: mortality, average daily gain, and reduced medication costs which all resulted in a reduction of total production costs per unit of gain through the nursery phase and beyond. Observations and site visit reviews all confirmed success of the intervention plan.

Calculating return on investment

The next phase of the project was to analyze and document the costs of the project in order to identify the intervention plan's decision on return on investment calculation. This process can occur several ways which is a good idea since looking at the project from various angles helps to insure that the conclusions reached are more true than simply just selecting one method.

The first step is to simply look at the raw data results to gain a first impression. Process mortality charts indicate positive statistical change had occurred in many areas but of primary concern were the mortality and total cost per pound parameters.

The next step was to apply classical statistical analysis to the project to gain insight into the data. Calculating the mean, standard deviation, minimum, maximum and ranges try to give meaning to the data points. Further calculations using regression analysis and statistical significance can also be used to identify the correlation strength of average daily gain to mortality improvement (figure 3.2).

In this case the regression equation identified 70% of the variation in average daily gain to be attributed to mortality improvement. Thus 30% of the average daily gain improvement of the project pigs was true improvement or improvement to unidentified causes. The financial impact is then calculated by taking the difference between the project pigs' average daily gain and the corresponding non-project pigs' average daily gain multiplied by the 30% true improvement correlation. This was then multiplied by the average days on feed for the project pigs groups. This total weight gain is then multiplied by the business' marginal nursery exit weight value to the system.

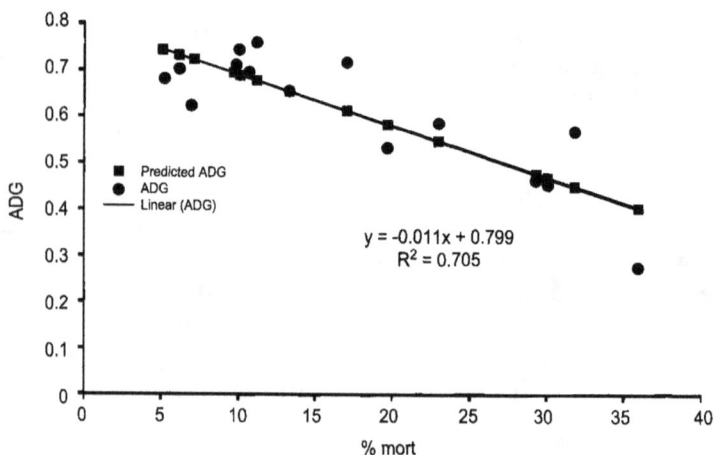

Figure 3.2 Relationship between mortality (% Mort) and average daily gain (ADG; lbs / day)

Another way to understand the data is to focus on understanding the underlying process that controls the data. In order to do this, apply the statistical process control charts to the various production parameters and then using the established rules for judging statistical change; identify the time frames of process change. Analyzing the process this way enables understanding of the cost impact to the system for both the problem as well as the solution phases. As the four rules for identifying statistical change are standardized for all manufacturing processes, objectivity during analysis is inherently maintained. A regression analysis was also applied to find the true average daily gain improvement which is needed to identify the true weight gain independent of mortality influence in the project pigs.

Now that the intervention impacts have been calculated, a simple partial budget tool is used. The standard categories are items that increase revenues, items that increase costs, items that reduce revenues and finally items that reduce costs. In this case the increased revenue items were the reduction of dead pigs (ie: saved mortality) and the additional weight gained above the normal weight associated with the saved mortality. A standard value per pig is multiplied by the number of pigs saved to find the gross financial value of reduced mortality. The weight gain is calculated by finding the difference in statistical increase in average weight per head due to the intervention plan less the normal average exit weight. This gain is then multiplied by the value the extra pounds provide to the system. Here that means first finding out what value is associated with the incremental pound in the system.

Calculating the additional costs is usually a straight forward process. Here, finding out the total cost of the nursery medication used during the project, calculating the extra cost paid the contract grower for the extra pigs saved and then adding in the remaining feed cost the saved pigs consumed.

In this project there were no items identified as causing a reduction in revenues or costs.

Then final calculations are found by first adding up the additional revenue items to the additional cost items and then subtracting the reducing revenue and cost items. This gives the net gain/loss line item for the project. Return on Investment is then simply calculated by taking this net gain/loss amount and dividing it by the total costs (additional cost items + reduced revenue items.

Summary remarks

Calculating the ROI via both classical and process statistically helps to understand true value generated by the intervention plan. I prefer to give more weight to the process calculations since this represents the fundamental reason for making the intervention plan in the first place.

A comparison of the individual system's financial performance on a cost per pound basis to the divisional average performance gives a further understanding of the intervention plan's relative value.

Thus the fundamental making of health decisions in large production systems is fairly similar to those in smaller systems, just more formal in their presentation. It is important to find the logical mathematical approach that is congruent to the way the business thinks and makes decisions. Furthermore it is imperative to thoroughly define the health challenge using a standardized forensic procedure and incorporating knowledge gained from the health monitoring system.

Since there are plenty of governmental regulatory agencies that study and validate animal health product efficacies, the real area the end user needs to be concerned about is animal health product's placement in the production process. This can be a real challenge and hence drives the need for process measurement and data information systems. For it is after the intervention plan is formulated that the real financial measurement process comes into play.

There are a number of ways to analyze the financial results from the intervention plan. Four were presented in this case study. The important point in doing intervention plan financial analysis is to help the decision makers gain confidence in their decision making ability to be used in future animal health challenges. By improving the decision making ability of the decision makers, the business will become more profitable and competitive in the global pig industry.

4

HOUSING AND ENVIRONMENT FOR GROWING PIGS

M.J.F. WILSON[1], H.G, CRABTREE[2], D. J. CHENNELLS[3], N. PENLINGTON[1]
[1]*British Pig Executive, PO Box 44, Winterhill House, Snowdon Drive, Milton Keynes, MK6 1AX;* [2]*Farmex, Pingewood Business Estate, Pingewood, Reading Berks, RG30 3UR;* [3]*Acorn House Veterinary Surgery, Linnet Way, Brickhill, Bedford MK41 7HN*

Introduction

There is a wide range of housing types for growing pigs in the UK, from outdoor-tented accommodation, through straw yards, to pens with part and fully slatted floors. Whilst overall, recent investment in growing pig facilities has been low, progressive producers have made significant financial commitments to replace or refurbish intensive systems, or, driven by consumer demand for high welfare standards, to create innovative straw-based facilities.

Hendricks *et al.* (1998) describe the process of continuous design innovation in European pig grower and finisher buildings. Design changes in EU member states appear to be influenced by climate, technology, legislation, retail and consumer demands (Table 4.1). In summary, Hendricks *et al.* (1998) suggest that systems in Northern Europe are affected by legislation regarding animal welfare and environment protection, whilst housing in Southern Europe is driven solely by technological considerations.

There is an accepted view that continental European growing pigs outperform those of the UK. Key factors proposed have been: (a) unilateral UK legislation to ban sow stalls and tethers resulting in available capital resources being diverted away from grow-out facilities in order to provide compliant gestation housing; (b) differences in pig health (particularly Post Weaning Multi-Systemic Wasting Syndrome; PMWS); (c) straw-based 'high-welfare' systems limiting performance; (d) staff and management skills failures; (e) aging buildings; (f) a recent history of low prices and serious diseases limiting the confidence to invest; (g) an aging ownership, particularly of family farms, with no next generation succession.

Production monitoring shows that good building design is primarily a matter of correctly matching the structures to the flow of pigs through them. The purpose of a pig building is to contain the pigs within a modified climate (if this were not the case, simple shelters would suffice). To modify climate effectively, the system designer needs to take account (amongst other things)

Table 4.1 HOUSING SYSTEMS FOR GROWING FINISHING PIGS IN EUROPEAN COUNTRIES ('000 HEAD). FROM HENDRICKS *et al.* (1998).

Housing system	Without/restricted straw									With straw					
	Partly slatted			Fully slatted			Solid concrete			Solid concrete			Deep litter		
	finishers	%	trend	finishers	%	trend	finishers	%	trend	finishers	%	trend	finishers	%	trend
Belgium	1092	31	↑	2291	65	↓				71	2	→	71	2	→
Denmark	2185	35	↑	3558	57	↓				312	5	→	187	3	→
Finland		?			?	↓					?	↑			
France	1099	11	→	7994	80	↑	599	6	→				300	3	↓
Germany	9385	60	↓	4693	30	↑				1251	8	↑	313	2	→
Greece	289	60	↓	193	40	↑									
Hungary	420	60	↓	280	40	↑									
Ireland	359	35	↓	615	60	↑				51	5	→			
Italy	3558	60[1]	→	1482	25	↑	890	15	↓						
Netherlands	5997	83	↑	1084	15	↓				145	2	↑			
Portugal	764	55	→	417	30	→	208	15	→						
Spain	5818	50	→	5818	50	↑									
Switzerland	135	20	↑	405	60	↓				135	20	↑			
UK	1756	37	↑	1803	38	↑				712	15	↓	475	10	→
Total	32856	47	↓↑	30633	44	↑	1697	3	↓	2677	4	↑	1346	2	↓
Total per system	Without/restricted straw						65187	94	→	With straw			4023	6	→

↑ - Increasing;　→ - Steady;　↓ - Decreasing;　↓↑ - No general trend

[1] Including system with a solid floor with a slatted alley outside

of: pig numbers, pig age, pig weight range, space requirements, feed intake, the range of ambient climate, pig temperature requirements, heat loss characteristics of the structure and acceptable levels of contaminants (such as moisture, carbon dioxide, noxious gases, dust and airborne agents of disease). In building design, regulatory constraints must necessarily be factored in. Such constraints would include; animal welfare, disease control, environmental protection, human health and safety, and employment law.

The design process follows a progression of stages, all of which need to be correct. Over the lifetime of the housing system, leaving any one of them inadequately covered will add up to many times their cost in lost performance.

- Specification
- Component selection
- System installation
- System commissioning
- End user training
- System monitoring
- System maintenance

This chapter presumes the purpose of building design to be the optimisation of pig performance within the constraints of the legislative environment and managemental practicality.

Physical performance

The UK has the highest average cost of production in Europe (Fowler, 2006) being, for the year ending December 2005, 94.5p/kg deadweight, whilst the UK cost of production was some 10% higher at 104.4p/kg deadweight. The difference could be explained by:

1. Poorer physical performance (Table 4.2) where lower finishing growth rates and fewer pigs sold per sow per year contribute to some 500 fewer kilograms of meat sold per sow per year in comparison to Belgium, Denmark, France and the Netherlands (although the significantly lighter slaughter weights in the UK is an important factor).

2. Higher costs for feed (Figure 4.1), for labour (Table 4.3) and for building, finance and miscellaneous (Figure 4.2).

Table 4.2 PHYSICAL PERFORMANCE OF PIGS IN SELECTED EU COUNTRIES (FOWLER 2006)

	Belgium	*Denmark*	*France*	*GB*	*Netherlands*
Pigs weaned per sow per year	21.39	26.09	24.16	21.50	24.52
Pigs sold per sow per year	20.93	24.29	22.45	19.42	23.36
Pigs born alive per litter	10.72	13.22	12.60	10.87	12.00
Pre weaning mortality (%)	12.4	13.1	14.4	10.9	12.3
Rearing mortality (%)	4.0	3.0	2.4	3.4	1.9
Finishing mortality (%)	4.2	4.0	4.8	6.5	2.9
Rearing DLWG (g/day)	325	427	470	509	329
Rearing FCR	1.72	1.81	1.76	1.70	1.61
Finishing DLWG (g/day)	608	855	768	639	779
Finishing FCR	2.99	2.67	2.92	2.74	2.66
Live weight at slaughter (kg)	114.6	105.0	114.6	96.9	113.8
Carcass weight (kg)	93.9	80.2	90.7	76.2	89.9
Carcass meat production per sow per year (kg)	1927	1924	1968	1441	2058

Important in this regard is that the performance of the US pig industry was once well behind European industries. Through radical restructuring, investment and adoption of Best Available Technologies (BAT), this position has been largely reversed, though they do have their own problems to address.

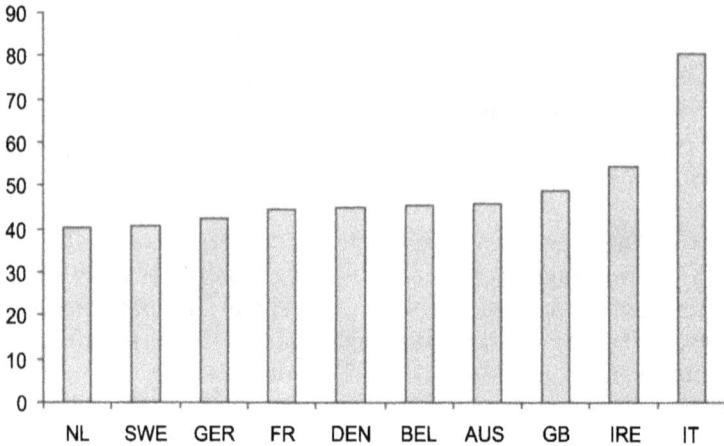

Figure 4.1 Feed costs in 2005.

Table 4.3 LABOUR COSTS IN 2005

	Aus	*Bel*	*Den*	*Fr*	*Ger*	*GB*	*Ire*	*Ita*	*NL*	*Swe*
Labour per finished pig (hours per year)	1.75	0.90	0.60	1.01	1.20	1.16	0.96	1.69	0.75	0.93
Cold carcass weight (kg)	92.0	92.1	79.2	87.7	92.3	74.2	75.1	126	88.1	85.7
Labour cost / hour (£)	7.9	10.3	12.9	11.0	10.3	8.5	7.2	9.0	12.8	14.0
Labour cost / pig (£)	13.7	9.3	7.7	11.1	12.3	9.9	6.9	15.2	9.6	13.1
Labour cost/kg carcass (£)	0.15	0.10	0.10	0.13	0.13	0.13	0.09	0.12	0.11	0.15

The UK market

UK market needs have undergone substantial change in the last decade:

- The pig gene mix has moved away from being dominated by the white breeds, and now contains higher levels of Pietrain and Hampshire lines.
- The infrastructure now comprises more large integrators interested in production efficiency *and* meat quality.
- Environment and welfare legislation has forced new (and not necessarily welcome) investment; calling on capital resource and constraining reinvestment decisions.

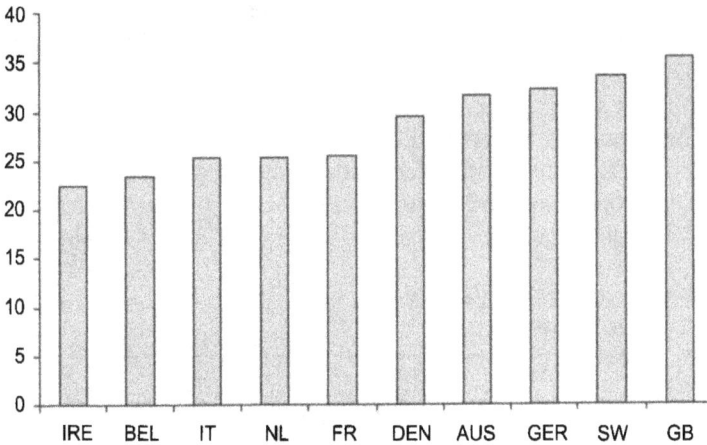

Figure 4.2 Building, finance and miscellaneous costs.

- New health challenges at both clinical and sub-clinical levels have arisen.
- Retail outlets are making increasing demands relating to production processes.
- Pigs must be grown to particular standards of food safety and animal welfare, driven through regulation and the UK Quality Assurance schemes.
- The emerging environmental agenda (Nitrate pollution, Carbon footprint, 'One Planet Living') is increasingly impacting upon production processes and animal housing.
- Feed costs are increasing.
- Availability of trained staff has diminished.
- Eating quality is replacing price as the determinant of demand.

The UK pig industry has a current and immediate economic imperative to accelerate its willingness to move with the times and take lessons from successful competitors. It must invest in research and use the results of that research by adopting new ideas. In these endeavours, the production sector has the support of industry bodies, independent and levy-supported consultants, and the allied industries (including veterinary, pharmaceutical, feed, genetic, and supply companies).

Design

In 2006, the UK British Pig Project brought together expert opinion to consider the lack of investment in grower and finisher accommodation. It has been estimated that the UK needs some 750,000 new finishing places in the next 10 years to keep pace with decommissioned buildings. The resultant report (BPEX,

2006) enables a professional assessment of the opportunities and challenges associated with construction and operation of new pig finishing buildings. It reflected a pro-active approach defining Best Practice and Best Available Technology (BAT) taking account of UK legislation and market demands. The result was four building designs and two fundamental types: a 1000 place fully slatted (35 to 110kgs) or a 560 place straw based 'wean to finish' building. The intention of the report was to highlight principles of management, environmental control, and best-practice that can be applied to any facility (Table 4.4).

Table 4.4 KEY FEATURES OF NEW STRAW AND SLATTED BUILDINGS IN THE BRITISH PIG PROJECT DESIGNS (BPEX, 2006).

Reference type	*Straw*	*Slatted*
Building design type	Stand-alone building with 560 pig places	Stand-alone building with 1000 pig places
Penning	40 pigs per pen from wean-to-finish	20-30 finished pigs per pen
Design weight range	7 – 110 kg	35 – 110 kg
Flooring	Straw bedded lying with scraped dunging areas	Fully-slatted pen area with slats conforming to EC Directive 91/630/EEC
Stocking density	Minimum of 0.545 m² of uninterrupted lying area per pig Dunging area additional	Minimum of 0.65 m² of uninterrupted lying area as per EC Directive 91/630/EEC
Ventilation	Naturally ventilated	Fan-power ventilation
Feeding	To meet UK Farm Assurance Standards	
Water provision	To meet UK Farm Assurance Sandards Water medication facility	
Lighting	To meet legislative requirements	
Hospital penning	Not included – these will be provided outside the building	

Fully slatted systems do tend to be harsher and less enriched environments, making good design more important. Problems with vice and mechanical injuries are not exclusive to these systems but, if they occur, they tend to be more severe and more difficult to control, e.g. experience suggests that lame pigs recover better and faster in strawed systems. Fully slatted accommodation allows for rooms to be "sealed" and thus used for *all in / all out* by room and so *all in / all out* by building becomes less essential. Their great advantage is their ease of cleaning and that they stay clean for longer.

Pathogens may be spread between pigs by contact with manures, nose-to-nose contact and aerial transmission between pens through gaps. There is a concern that disease may be spread more easily in straw based buildings with scrape through passages as manure is moved out past other pens. This problem is less likely to arise in slatted buildings with one slurry pit per room, though it is impossible to seal buildings totally. Any passages must be cleaned at least three times each week but preferably daily. This makes all in/all out by building so much more important. However, even with all in/all out, enteric organisms can exist in small populations and spread from pen to pen, probably much more than respiratory pathogens do. Constantly wet floors enhance organism survival, compounding disease transmission; humidity and effluent gases may also compromise air quality. Properly ventilated this should not happen.

The US has been successful with the introduction of new-start large-scale sow units of 2500 to 5000 sows. These units produce weekly batches of 1000 to 2000 pigs for off-site grow-out facilities built to a standardised modular design: 'standardised' and 'modular' being operationally important to the system. This scale of production cannot be mirrored in the UK. However, it can be emulated with the combination of batch farrowing and the bringing together of breeding units under the same management / ownership / health status. In this way weekly provisions can be assembled to fill grower/finisher units with upwards of 1000 places. Within common ownership / management, these units can be operated to standard production plans. The buildings are typically designed to operate as *all in / all out* facilities that match the flow of pigs from the supplying large-scale sow systems. This facilitates maximum bio-security and therefore high production efficiency.

The designs in the BPEX Project comply with all current welfare and Farm Assurance Certification regulations, and in particular provide, *from the outset and without subsequent adaptation:*

- Generous space allowance at all stages
- Suitable thermal environment
- Wholesome feed provision
- Facilities to provide for the appropriate number of diets to cover the chosen number of feeding phases
- Freely available water
- Straightforward cleaning (most importantly) and disinfection
- The highest bio-security standards

CONSTRUCTION

Analysis of new building design in the context of BAT suggests that two designs meet the needs of the majority of UK production.

1.   A fully straw bedded indoor grow-out facility (Figures 4.3 and 4.4) that is suitable for growing and finishing pigs, and particularly appropriate for use with outdoor reared pigs from weaning to 110kgs. It is designed as a simple, cost-effective system.
2.   A fully slatted building to house units of 1000 pigs, *all in / all out*, from 35kg to finish (Figures 4.5 and 4.6)

Figure 4.3 Interior of new straw based wean to finish accommodation. (Courtesy of BQP)

Figure 4.4 Exterior of new straw based wean to finish accommodation (courtesy of BQP)

Figure 4.5 Interior of new fully slatted accommodation. (Courtesy of . . .)

Figure 4.6 Typical UK style 1000 place fully slatted buildings. (Courtesy of . . .)

To minimise the movement of pigs and maintain a stable cohort, pigs may be taken from weaning to slaughter weight in the same pen. This removes the stress associated with moving and remixing and is particularly helpful in herds challenged by PMWS. To cope with the weight range from weaning to finishing, the pens may be partitioned to half size at the start; pigs being provided with plenty of additional straw bedding and a possibility of a warm kennel area.

Wall construction can use factory-produced panels with durable and easy-clean internal and external surfaces. The most suitable internal pen division may be achieved with plastic panel bolted to metal posts. This provides durability and easy cleaning between *all in / all* out batches.

Some pig producers may feel a need to be cautious and to maintain flexibility by the use of portal-frame structures that may be converted for other non-pig uses in years to come. This is however itself not without risk, as the purpose of a pig building is to satisfy the requirements of pigs, and compromise in this intent will lead to compromise in pig performance. In this regard, Cooper (1960) has commented:

"I have always deprecated the philosophy of flexibility in farm buildings.........If a man is to keep pigs efficiently they must be kept in the most suitable environment........There is no point in using someone's brand of balanced fattening meal to provide an expensive central heating system for some lofty building that has been designed to meet the remote possibility that it will some day be used for keeping giraffes".

MAINTENANCE CONSIDERATIONS

Design should allow easy maintenance, for instance:-

*   Hinges should be big, robust and rust proof. Hinges/bearers on flaps or control shafts should be easy to remove or split so that they can be cleaned and "freed".
*   Doors eventually always "drop", so latches should allow for this and still work easily.
*   Plastic bearers should be used so that they do not rust and seize up.
*   Pigs will always dismantle and destroy their environment, make allowance for this.
*   Multilayered plastic windows should have sealed edges so that algae cannot grow in them and reduce light. Also if these are automatically controlled flap windows they may become heavier from water and or algae entering them and compromise the control motors/activators.

STOCKING POLICY

The *all in / all out* approach is seminal to disease control and the matching of internal house environment to the requirements of specific groups of pigs of the same age entering and leaving the building at specific seasons of the year and prevailing weather patterns. The *all in / all out* pattern for production is best practised at whole unit level (as in multi-site production), and next best at whole building level. But, although not optimum, significant advantages in pig performance can be gained if *all in / all out* is practised at the level of rooms within buildings.

It is essential at the design stage to plan and decide the number of pigs and the group size that the building should stock in order that mixing and moving of pigs can be avoided and to ensure *all in / all out* by airspace. Buildings need to be fully stocked for efficient functioning of the ventilation systems needed to achieve the optimum environmental conditions for the pigs.

To facilitate *all in / all out* management at the grow-out unit, batch farrowing at the breeding unit is important to the effective management of pig flow. Control of pig flow is probably the most important determinant of success in terms of building performance (Table 4.5).

Table 4.5 WEAN TO FINISH PERFORMANCE COMPARISON *ALL IN / ALL OUT* BY SITE VERSUS CONTINUOUS FLOW

	Growth rate (g/day)	Mortality (%)
All In / All Out Average of 100 batches Straw yards	700	5.9
Continuous Flow Over 9 batches Controlled environment	720 declining to 650	7.9

Environment

The performance of modern lines of pigs raised under typical commercial conditions in the UK is generally below their genetic potential. Many environmental factors contribute to reduced appetite and poor growth rate, including; group size, stocking density, air volume per pig, air quality, air bacterial load, prevalence of disease, and the climate within the building. These stressors appear to act additively, so the removal of any one results in a positive influence on performance (Black *et al.*, 2001). Overall, the resolution of environmental challenge is dependent upon the quality of the stockmanship prevailing. Helpfully, computer technology is delivering systems to aid stockworkers in the detection of problems. The collecting of records is essential to management, but only if the data is transformed into information, and the information transformed into management decisions.

VENTILATION AND MAINTENANCE OF THE PIG ENVIRONMENT

Effective ventilation will deliver improved pig performance and enhanced working environment for staff. Compared to pigs in a dirty (carbon dioxide, dust and ammonia) environment, pigs in clean environment eat more feed, grow faster (Lee *et al.,* 2005) and have lower levels of stress.

Ventilation is the primary control mechanism for climate modification in a livestock building where the substantive heat source is provided by the animals themselves. The basics of airflow, air exchange and heat balance are known and understood as is their impact on energy metabolism in the pig (Holmes & Close 1977). Automatic and semi-automatic packaged control systems have become widely available. The undoubted benefits of the combination of knowledge and the availability of control systems to make use of that knowledge, may however lead to unfortunate consequences if choice of system is dictated by 'fashion' rather than individual requirement. Specification for ventilation system performance needs to be based upon (a) the animal's requirement for clean air (Table 4.6) ambient temperature (Table 4.7), and (b) prevailing local weather and climate conditions.

Table 4.6 THE NUMBER OF 100KG PIGS THAT CAN BE VENTILATED BY TYPICAL PROPELLER FAN SIZES TO MAINTAIN A 3°C TEMPERATURE LIFT OVER AMBIENT

Fan size – diameter in mm	Maximum capacity m³/hr	No of 100 kg pigs
450	5000	25
500	7000	35
630	10000	50
710	15000	75

Hugh Crabtree

Table 4.7 GUIDELINE MINIMUM TEMPERATURES (°C) FOR PIGS TO GIVE GOOD PERFORMANCE

Stock type	Liveweight – kg	Floor type		
		Concrete	Perforated metal	Straw
Sucking pigs	At birth	32	30	28
In creep	5	26	24	21
Weaners	5	29	28	25
	20	17	18	12
Growers	40	14	15	8
Finishers	60			
	80			
	100			
Grouped sows		17	18	12
Sows in stalls		22	21	18
Farrowing rooms		15 – 22 °C		

"Controlled Environments for Livestock" - A Farm Energy Centre Handbook

Temperature regulation should be accurate to within +/- 0.5°C and temperature differences within the building should not exceed 3°C. The ability of the ventilation system to achieve the designed performance can readily be measured using simple data-logging. Provided the planning stage has been completed properly, the ventilation system can be kept on track to provide optimum conditions within design constraints. Management responsibility covers the ensuring of (a) correct installation, (b) commissioning, and (c) operator training.

The "machine to machine" communications revolution now allows pig house ventilation systems to be permanently 'on line', yielding both (a) performance data and (b) exceptions reporting (Figure 4.7). Telemetry cannot only deliver management decision support in sustaining a return on investment, but also affects positively the welfare of housed animals.

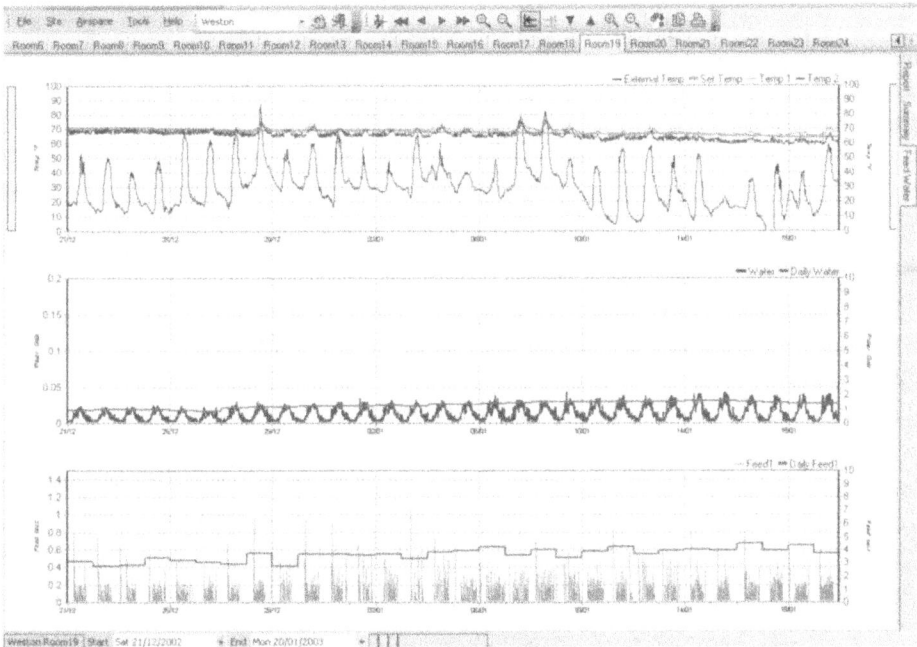

Figure 4.7 Barn report temperature, water and feed charts for a 30 day period from a finishing batch. Everything in control and working correctly. Any deviation can quickly be spotted by management and rectified.

Many mechanically ventilated intensive housing systems use an extract or negative pressure system with stage-controlled fans and mechanically controlled air inlets. Whilst the original design for this system is almost 30 years old, it has been necessary to await modern materials and controls for the potential of the system to become fully realisable.

Temperature regulation in naturally ventilated houses also benefits from automatic control of mechanically operated curtains or flaps, known as automatically controlled natural ventilation (ACNV). However, natural ventilation systems provide no control over air movement within the building, so there is a risk of unwanted draughts and wide temperature variation. Best practise with ACNV for finishing pigs is to use zone control with a maximum zone length of 10m. Hybrid systems that combine fan-based ventilation for winter control with natural ventilation for summer capacity may well find increasing markets as the impact of environmental protection and climate change regulations are felt.

Although all pig buildings are extremely dusty, straw can only make this worse. Humidity and dust together will inevitably contaminate and clog up fans, louvres, ducts and meshes etc, thus seriously and significantly reducing air flow. This is a much greater potential cause of respiratory disease than their effect on the pig where the nasal turbinates are extremely protective of the more vulnerable lower respiratory tract.

In comparison with straw-based buildings, slatted designs limit the pig's ability to choose its living conditions and therefore increase the need for the stockperson to provide the correct controlled environment. Cleaning, staff training and monitoring are essential for ventilation systems to be efficient, and are too often missed out. Especially relevant are:

- Good design and cleanliness of fan blades and outlet trunks to allow outside air to be pulled efficiently into the building. All fans, vents and grills should be easy to access, dismantle and remove. This way they can be routinely cleaned without risk to the electric motors.
- Servicing of mechanical controllers for inlets ensure their effective operation
- Tracking of prevailing in-house environmental conditions and comparison with bench marks
- Staff training in environment control operation and the following of systems protocols

High summer temperatures can cause appetite depression and resultant reductions in both slaughter weight (Wooldridge, 2007) and number of pigs slaughtered (Figure 4.8).

Predictions for climate change suggest that high temperatures are not likely to be a short term phenomenon, and producers must look to methods of ventilation and cooling that are able to cope with wider temperature variations. One such coping technology is misting to cool incoming air or achieve differential temperatures across a pen to maintain correct lying and dunging patterns, this will reduce days to reach finish weight when temperature rises (Bridges *et al.*, 2003). The extent of benefit obviously depends upon degree of temperature rise and the time over which it is maintained.

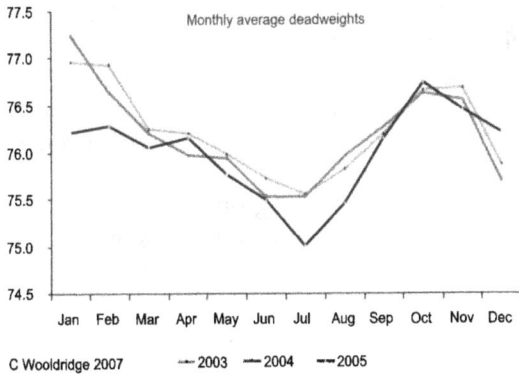

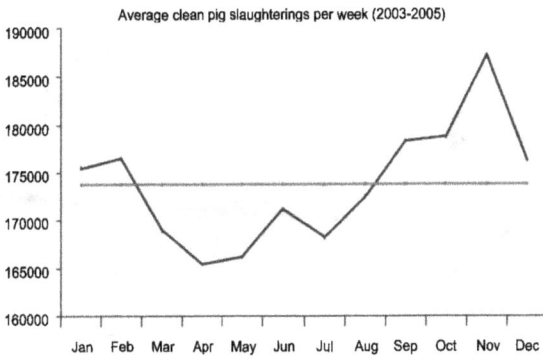

Figure 4.8 Monthly average deadweights and pig slaughterings (Wooldridge, 2007).

BEDDING AND FLOORING

Current UK regulations require manipulable materials to be available and retail demand encourages provision of straw bedding. Few experiments have been able to accurately quantify in terms of the benefit to the pig the differences between straw and slatted accommodation, but Edwards *et al.* (2005) showed that straw lead to better pig behaviour, lower risk of lameness and tail biting; but also to poorer hygiene, PMWS susceptibility and respiratory / enteric health.

FEEDING SYSTEMS

The importance of the relationship between feed and environment in pigs is well documented and understood. The greater the feed intake, the lower the ambient temperature in which the pig is comfortable. Feed energy intake is the most important factor influencing ambient temperature requirement and drives the pig's sensible heat output. This is what drives the pig's sensible heat output and its influence on the ambient environment. Other pigs in the group – dependent on group size and available space – also affect both ends of the "energy balance"; feed intake and heat loss. However, if pigs are too cold they will huddle and will not eat.

Individual feed intakes of pigs in groups is, in turn, significantly affected by feed and water delivery system design. Easy access to feed and water within an easy-to-clean environment is fundamental to the health, welfare and performance of the pig (Figure 4.9).

Figure 4.9 An easy to clean, good to work in, efficient building providing stable environment and good access to feed and water.

Conveying feed, whether in dry or liquid form, to the pen is a relatively simple process. However, ensuring adequate individual pig access at the point of delivery continues to challenge system designers. Thus it is unsurprising that *ad libitum* feeding of growing pigs is so common. Recent innovations in the design of both single and multi-space feeders for wet and dry feeding systems have included the provision of water as part of the feeder, and the introduction of level sensors to allow *ad libitum* wet feeding from a single space feeder. Unhelpfully, the numbers of pigs that can be accommodated by any particular feeder as suggested by the manufacturer is sometimes at odds with practical experience on farm. This matter is deserving of commercial trials if proper conclusions are to be drawn up concerning feed and water provision such as may allow the genetic potential of the pig to be realised.

Dry feeding is often the preferred choice of feeding system, through apparent ease of management. Recent work (MLC, 2005) comparing liquid and dry systems has shown advantages of liquid feeding. In these trials liquid feeding delivered least cost production with combined effects of improved daily gain, reduced feed intake and improved feed efficiency (Table 4.8), and 4.6p/kg deadweight lower cost of production giving effective return on investment.

If single space *ad-libitum* feeders are used then pen numbers must be small and/or space allowances greater to reduce competition. It is important to remember that pigs are "group activity" behaviourists and this is contrary to single feeding spaces. Young pigs in particular need plenty of feeder space and feeder options. Ideally when first weaned all pigs in a pen should be able to feed at the same time. Design should allow for the placement of temporary trays, bowls, and mats both for lying and floor feeding. A choice of wet and dry feed should also be considered.

Table 4.8 PIG PERFORMANCE AND CARCASS QUALITY BY FEEDING SYSTEM (MLC, 2005).

	Liquid	*Dry*	*s.e.d.*	*P*
Liveweight (kg)				
Entry (kg)	34.1	34.2	1.29	
Final (kg)	103	103	0.8	
Feed intake (kg/day)	1.75	1.85	0.021	***
Growth (g/day)	796	754	9.6	***
FCR	2.27	2.53	0.027	***
Backfat (mm)	11.5	11.4	0.30	

SUPPLY OF WATER

Manufacturers supply a choice of nipple, bite, bowl or trough drinkers. Placement is important to reduce soiling and to avoid areas where pigs may become injured by running into them for example by "pop holes" or on long walls. They should always be over the dunging channel or slats. Flow rate should be adequate and easily adjustable. Mains pressure can be too high both to cause leakage and wastage whilst drinking. Too low pressure will undoubtedly restrict intake. Restricted water intake will reduce food intake.

Ideally all pens should have at least two water points, preferably of two different types; (regardless of minimum regulations). All should be easy to clean, both as part of a stockman's daily task and for cleaning and draining between batches.

With all designs it is essential that both feed and water intake are easy and "attractive" for the pigs. These must not be the limiting factors for growth. Pigs will only grow if they eat.

MEDICATIONS

New design should allow for medications to be administered via the drinking water. There is a constant threat that in-feed medication may be prohibited or severely restricted in future. From a practical point of view, water treatment is a much more predictable and reliable method of administrating group medication, as ill pigs usually drink but may not eat.

Thus water tanks should be accessible, have lids, be large enough and of a known volume, and have individual stop cocks to allow this. It is important that they should be kept clean and have drainage points for flushing. Ideally connection points should be left in the water system for "fractionators" to be fitted. This should allow provision of medication to each individual room (ideally each pen).

Consideration should be given for space for "drop in" feeders in wet fed pigs so that a (medicated) dry feed may be provided if necessary.

LABOUR EFFICIENCY

Skilled labour in livestock production is already at a premium. Having satisfied the requirements of the pig, housing systems should be designed for maximum labour efficiency and work satisfaction. Operators need, minimally:

- Good visual contact with the animals
- An environment free of dust and aerial pollutants such as ammonia
- Penning and gates that are easy to use
- Straightforward routine pig movements
- Unimpeded service access to systems components such as controllers and augers
- Clearly documented compliance routines
- User-friendly facilities for production recording and monitoring (figure 4.10)
- High quality maintenance support.

As a general rule, labour efficiency will benefit from the mechanisation and automation of as many systems as possible without compromise to the animal. Indeed, in many cases, animal welfare and performance is better facilitated in highly automated systems as a result of operator time being saved from drudgery and elevated to animal and systems care. For example, feed delivery, temperature management and lighting pattern control can all be readily automated leaving skilled labour the time to handle issues of animal husbandry. Trained staff not only value the automation of mechanical tasks, but are also able to make better use of time released. It is unsurprising that improvements in sow and finishing herd performance have been reported to result from staff training. In the UK, a national scheme for pig production staff and professional career development has been set up: www.bpex.org.uk

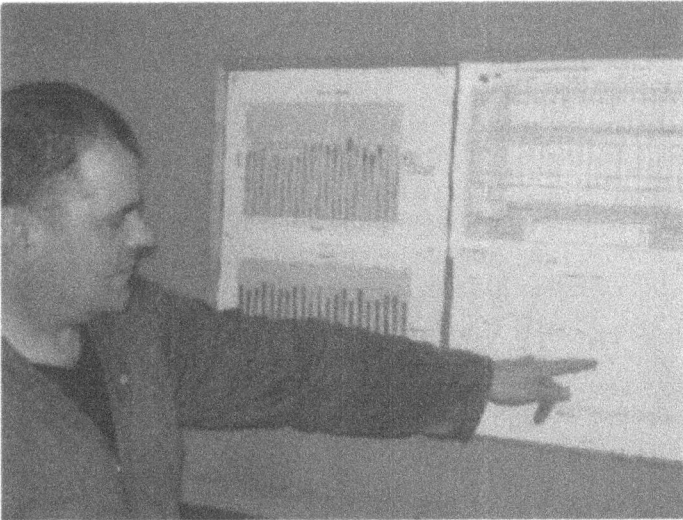

Figure 4.10

In all new designs, ease of cleaning is essential. It is a time consuming, laborious and repetitive job. Equipment for pre-soaking by overhead sprays should be installed in all new buildings, combined with detergents this will immeasurably ease the task of cleaning, and reduce washing water and result in less effluent. This will also result in less washing water being needed and thus less effluent produced. Surfaces should be impermeable as they clean faster, allow less faecal adhesion, have fewer crevices for organisms including insects and skin parasites, disinfect better, will have less "water loading" and will dry faster. This faster drying will also allow a quicker atmosphere correction and a shorter "empty time". Solid floored systems inevitably take longer to clean, produce more effluent and have the problem of drainage of the effluent water from the buildings, and may take longer to dry. All designs and pig flow projections must make allowance for "down time" for cleaning and drying.

Efficient and effective pre-soaking, use of detergents, an appropriate choice of modern lining materials and drying before restocking will lead to improved standards of cleanliness meaning that disease control is better assured and the risk of salmonella contamination in the food chain reduced.

The development of fully robotic cleaning systems is much to be welcomed. Labour costs per kg of carcass have already been identified as a cause of reduced competitiveness (Table 4.3). Mechanisation and automation will not only deliver reduced labour costs, but also the means to respond to upcoming regulatory requirements for welfare and environmental protection.

AIR QUALITY – DUST AND AMMONIA AND CARBON DIOXIDE

High levels of airborne contaminants have a negative impact on the health and productivity of animals. The level of impact depends on the concentration of pathogen microorganisms, the ability of the immune system to resist the pathogen load, and social stress. The pig has numerous defence mechanisms within the respiratory tract, nasal passages and the lungs themselves, but the animal's ability to defend itself is reduced with deteriorating air quality. Airborne gasses such as ammonia can reduce the local defence mechanisms and allow infectious pathogens to colonise and cause disease. Ammonia emissions increase with increasing ambient temperature (Rom *et al.,* 2000). On-farm epidemiological studies have demonstrated the effect that air quality has in contributing to the health and productivity of pigs (Holyoake, 2005). In many cases, improving the environment (by enhancing air quality, improving hygiene, reducing stocking density, optimising ventilation, implementing best practice manure management, and controlling temperature and humidity) is a more beneficial management strategy to improve growth performance than attempting to eliminate specific infectious pathogens.

Housed pigs are regularly exposed to ammonia and dust arising from manure, feed, bedding and the animals themselves. High levels of these aerial pollutants will adversely affect pig performance. A study by Wathes *et al.* (2004), using specialist facilities, exposed pigs to increasing levels of ammonia and dust. In general, both food intake and live-weight gain were lower for weaned pigs exposed to levels of dust above concentrations of $5.1 mg/m^3$ across ammonia concentrations up to 37 ppm (representative of typical commercial conditions). There was little evidence of an effect on pig health (Done *et al.*, 2005); but overall animal health status was high.

GROUP SIZE

UK regulations lay down minimum space requirements for pigs according to their weight. These standards are well recognised and form part of all Certification schemes. Optimisation of group size on the other hand, and the influence of animal numbers in a group upon growth is more contentious. Choice of group size is a major influence on cost of facilities, both new and adapted. Typical group sizes are between 15 and 40 pigs per pen in intensive housing, but upwards of 100 are not uncommon in straw yards. The interaction between group size and density of stocking (space per pig), appears self-evident, but is complex. A review of 20 studies comprising 22,000 animals on groups ranging from 3 to 120 pigs (Turner *et al.*, 2003) looked at the relationships of group size with: immunocompetence, clinical health status, within group variation in growth rate, carcass characteristics, and the occurrence of vice. The conclusion was that a large group size may compromise the growth of the young pigs, but the long-term consequences for other economically important traits may be slight. In practice having larger pens may reduce the costs of the facility but the management time may well be increased for inspection and selection for market, as well as making it more difficult to spot an ill pig. Recent experiments (Table 4.9) on stocking density have confirmed the understanding that decreasing group size and increasing floor and feeder space per pig will reduce morbidity and mortality and increase growth rate after weaning resulting in heavier pigs at 24 weeks (DeDecker *et al.,* 2005).

Table 4.9 INFLUENCE OF STOCKING DENSITY (DEDECKER, *et al., 2005)*

Pigs per pen	Stocking density (m²/pig)	Feeder space (cm/pig)	Morbidity and mortality (%)	Growth rate weaning to 24 weeks (g/day)	Liveweight at 24 weeks
22	0.78	4.2	8.5	688	121.8
27	0.64	3.4	10.2	660	117.1
32	0.54	2.9	12.7	635	113.1

HOSPITAL PENS

All designs and pig flow projections must make allowances for sick, disadvantaged and poor performing stock. An allowance of 0.03 of pigs needing treatment (hospital pigs), and a further 0.10 being "B" stream pigs is required. Hospital pens ideally require soft comfort flooring, which is undoubtedly best achieved by use of straw on solid floors. Pens should be small, for 4-6 pigs so that genuine *all in / all out* can be practiced for these pigs. Wipe-boards should be installed for recording treatments and responses to treatment, as well as for prognosis assessments.

"B" stream pens should be small for groups of approximately 10. "B" stream pigs should <u>never</u> need to be returned to "A" stream pens. There should be adequate pens to allow for this. Those looking after pigs and those employing people who care for pigs have regulatory responsibilities with respect to animal welfare. The Welfare of Farmed Animals (England) Regulations 2000 (S.I. 2000 No. 1870) Regulation 10 states that any person who engages a person to attend to animals should ensure that the person attending to the animals: (i) is acquainted with the provisions of all relevant statutory welfare codes relating to the animals being attended to, (ii) has access to a copy of those codes while attending to those animals, and (iii) has received instruction and guidance on those codes.

Environmental legislation

UK producers can no longer prosper by good husbandry alone, also to be understood are the complexities of environmental legislation. Within the UK the Environment Agency (www.environment-agency.gov.uk/agriculture) is the primary regulatory authority responsible for implementing the legislation and ensuring compliance by pig producers. The present tally of environmental regulations is listed in Appendix 1.

Depending on location, and more likely if buildings form part of a large-scale livestock based enterprise, the Local Planning Authority might require an Environmental Impact Assessment before granting permission for new buildings. If there are sites of environmental significance in the locality that could be damaged by ammonia or dust in the atmosphere, evidence may be required to demonstrate that any new facilities are unlikely to cause harm.   In any event, a new building development should minimise the risk of impairment to the environment, well-designed buildings should achieve efficient production and minimise emissions such as ammonia, dust and odour. All pig production operations and their management should, in the UK, adhere to the Department for the Environment, Food and Rural Affairs (Defra) Codes of Good Agricultural Practice with respect to the protection of water, (Water Code – PB0587), protection of soil, (Soil Code – PB0617) and the protection of air, (Air Code – PB0618).   These Codes are in the process of being revised and amalgamated into a combined Environmental Code.

The Integrated Pollution Prevention & Control Directive (IPPC; E.C. Directive 96/61) has introduced a more integrated approach to controlling pollution from industrial sources. For pig production IPPC regulations require any indoor pig unit with 750 or more sow places or 2000 or more finishing pig places above 30kg to obtain a permit. The legislation covers farm management including the use of animal feed, housing design, manure management (storage and spreading), emergency planning, noise and odour management and resource use (such as water and energy). The intension is to minimise emissions and damage of the environment through the use of BAT, good management and efficient use of resources.

The Water Framework Directive (WFD) was adopted by the UK in September 2000 and brings with it certain responsibilities for pig keepers and those that manage land on which manure is spread. Pig keepers, and those that spread manure on to land, must comply with current legislation and be aware that new legislation or measures introduced that are likely to be introduced with the objective of achieving good ecological status for all waters by 2015. These involve pig farms and the management of any associated land in minimising diffuse pollution of water with respect to the impact of phosphorus, nitrates, sediments and soil loss, organic wastes, pesticides, veterinary medicines and faecal microorganisms. It is significant that around 0.50 of phosphorus inputs into fresh water in the UK is attributed to agricultural sources, with leaching and soil erosion being primary causes. Management protocols need to be changed in order to reduce the likelihood of phosphorus pollution; thus phytase may be included in pig diets to increase the availability of phosphorus to the pig and thereby reduce its excretion in the slurry. Similarly there will be greater focus on protein content in animal diets.

The Nitrate Vulnerable Zone (NVZ) Action Programme implements the E.C. Nitrate Directive. The Action Programme sets upper limits for the amount of organic nitrogen in manures which may be applied to agricultural land during any twelve-month period, restricts the timing of the application of pig slurry and some other manures, and requires full allowances to be made for the available nitrogen supplied by organic manures to a crop. It is requisite therefore that pig farmers operating within one of the designated zones are able to access sufficient land for spreading organic manures and slurries in order to meet this legal requirement. Currently within NVZs there is an annual limit to the nitrogen applied on average across all of the land farmed. Presently this comprises 250kg of total nitrogen/ha on grassland and 170kg/ha on non-grass crops. These limits include manure from grazing animals and any imported organic materials. Autumn closed periods for slurry spreading apply to sandy and shallow soils. The closed periods do not apply to stackable pig manures (FYM) such as that likely to be produced from straw-based systems. Similarly dirty water with low total nitrogen content (less than $1kgN/m^3$) is also excluded under the existing rules. Annual applications of organic manures can be made on any individual field up to the limit of 250kg/ha, this excludes grazing deposition. The 250kg/ha upper limit is specified in the Water Code as the maximum considered as good agricultural practice for all land outside of NVZs.

Defra suggest that 3.08kg of N is produced as excreta for each finishing place (18 to 105kg liveweight) annually. (See 'Manure Planning in NVZs, England.';Defra PB 5504, July 2002). Therefore 1.23ha of land for spreading is required for every 100 pig places.

If the required land area is not available, surplus slurry and manures may be exported to other farms while still complying with the NVZ regulations, provided records are kept. The control of Pollution (Silage, Slurry and Agricultural Fuel Oil) Regulations 1991 requires 120 days continuous storage for slurry, this includes dirty surface water.

CLIMATE CHANGE AGREEMENT

Pig production, as with all of industry, has been targeted with reducing carbon emissions through reducing energy consumption. Pig producers are able to enter into an agreement with Defra in order to reclaim 0.80 of the climate change levy that is applied to their energy bills making for an 8% rebate. In exchange, each applicant has to agree to improve the energy efficiency of their business to agree targets. Bi-annual milestones are set. Energy efficiency is a measure of energy consumption per tonne of pig meat produced, therefore it is possible for energy usage to rise so long as the units productivity has risen proportionally further through improved efficiencies.

Conclusions

The UK has a recent history of under-investment in pig grow-out accommodation. Whilst there are some good explanations for this, including diversion of capital to cope with regulations for pregnant sow accommodation, the consequences have become burdensome for the industry. In a reversal of fortunes, UK now has poorer growth performances and efficiencies than other European nations and North America.

Reinvestment in grower / finisher housing requires attention to pig environmental requirements, building construction design, a plethora of regulatory controls, and not least the demands of retailers and consumers. The British Pig Executive, together with industry support, has launched a number of initiatives to ensure a rewarding out-turn for producers involved in capital investment into pig buildings.

References

Black, J.L., Giles, L.R., Wynn, P.C., Knowles, A.G., Kerr, C.A., Jones, M.R., Strom, A.D., Gallagher, N.L., Eamens, G.J. (2001). Factors limiting the performance of growing pigs in commercial environments. *Manipulating pig production VIII. Proceedings of the eighth biennial conference of the Australian Pig Sciences Association* pp. 9-31.

BPEX (2006). A housing blueprint for the British pig industry. British Pig Executive, Milton Keynes.

Bridges, T.C., Turner, L.W., Gates, R.S., Overhults, D.G. (2003). Assessing the benefits of misting-cooling systems for growing/finishing swine as affected by environment and pig placement date. *Applied Engineering in Agriculture* **19**: 361-366.

Controlled Environments for Livestock" - A Farm Energy Centre Handbook.

Cooper, M. McG. (1960). The pig industry from a biological standpoint. Proceedings of the Bath Conference, Pig Industry Development Authority, pp 1-7.

DeDecker, J.M., Ellis, M., Wolter, B.F., Corrigan, B.P., Curtis, S.E., Hollis, G.R. (2005). Effect of stocking rate on pig performance in a wean-to-finish production system. *Canadian Journal of Animal Science* **85**: 1-5.

Done, S.H., Chennells, D.J., Gresham, A.C.J., Williamson, S., Hunt, B., Taylor, L.L., Bland, V., Jones, P., Armstrong, D., White, R.P., Demmers, T.G.M., Teer, N., Wathes, C.M. (2005). Clinical and Pathological responses of weaned pigs to atmospheric ammonia and dust. *The Veterinary Record* July :71-80.

Edwards, S.A., Scott, K., Armstrong, D., Taylor, L.L., Gill, B.P., Chennells, D.J., Hunt, B. (2005). Finishing pig systems: health and welfare in straw-bedded or slatted housing. *The Pig Journal* **56**: 174-178.

Fowler, T. (2006) Pig Cost of Production in Selected EU Countries. Meat and Livestock Commission, Milton Keynes.

Hendricks, H.J.M., Pedersen, B.K., Vermeer, H.M., Wittmann, M. (1998). Pig housing systems in Europe: current distributions and trends. *Pig News and Information* **19**: 97N-104N.

Holmes, C. W. and Close, W. H. (1977): The influence of climatic variables on energy metabolism and associated aspects of productivity in the pig.

Holyoake, P.K. (2005). The impact of air quality on animal health. *Manipulating pig production X. Proceedings of the tenth biennial conference of the Australian Pig Sciences Association* pp. 120-131.

Lee, C., Giles, L.R., Bryden, W.L., Downing, J.L., Owens, P.C., Kirby, A.C., Wynn, P.C. (2005) Performance and endocrine responses of group housed weaner pigs exposed to the air quality of a commercial environment. *Livestock Production Science* **93**: 255-262.

MLC (2005). Finishing Systems Report No 1. Meat and Livestock Commission, Milton Keynes.

Rom, H.B., Moller, F., Dahl, P.J., Levring, M. (2000). Diet composition and modified climatic properties – means to reduce ammonia emission in fattening pig units. *Air pollution from agricultural operations. Proceedings of the Second International Conference* pp 108-115.

Turner, S.P., Allcroft, D.J., Edwards, S.A. (2003). Housing pigs in large social groups: a review of implications for performance and other economic traits. *Livestock Production Science* **82**: 39-51.

Wathes, C.M., Demmers, T.G.M., Teer, N., White, R.P., Taylor, L.L., Bland, V., Jones, P., Armstrong, D., Gresham, A.C.J., Hartung, J., Chennells, D.J., Done, S.H. (2004). Production responses of weaned pigs after chronic exposure to airborne dust and ammonia. *Animal Science* **78**: 87-97.

Wooldridge, C. (2007) It costs the industry over a million pounds a month *Pig World* March 2007 pp26-27

Appendix 1: Current UK environmental regulations

1. Control of Pollution (Silage, Slurry and Agricultural Fuel Oil) Regulations 1991 (amended) SI 1991, No 324, HMSO (ISBN 0-11-013324-2) as amended by the Control of Pollution (Silage, Slurry and Agricultural Fuel Oil (Amendment)) Regulations 1997 SI 1997, No 547, The Stationery Office (ISBN 0110640497) Environment Act 1995, HMSO (ISBN 0105425958)
2. Statutory Instrument 2000 No 1973: The Pollution Prevention and Control (IPPC) (England and Wales) Regulations 2000. SI 2000 Number 1973 IPPC The Stationary Office ISBN 0 11 099621 6
3. Environmental Protection Act 1990, Chapter 43, HMSO (ISBN 0 10 544390 5)
4. Food and Environment Protection Act 1985 HMSO (ISBN 0-10-544885-0)
5. Highways Act 1980 HMSO (ISBN 0105466808)
6. Town and Country Planning (Assessment of Environmental Effects) Regulations 1988 SI 1988, No 1199, HMSO (ISBN 0110871995) as amended by SI 1990, No 367, HMSO (ISBN 0-10-543-7743)
7. Town and Country Planning (General Permitted Development) Order 1995 SI 1995, No 418, HMSO (ISBN 011052506x)
8. Ancient Monuments and Archaeological Areas Act 1979 (as amended). HMSO (ISBN 0 10 544 679 3)
9. The Wildlife and Countryside Act 1981(England and Wales) (Amendment) Regulations 2004, Statutory Instrument 2004 No. 1487
10. Council Directive 79/409/EEC of 2 April 1979 on the conservation of wild birds
11. Council Directive 92/43/EEC of 21 May 1992 on the conservation of natural habitats and of wild flora and fauna
12. The Conservation (Natural Habitats &c.) Regulations 1994 (SI 1994 No. 2714),
13. Forestry Act 1967 and the Forestry (Felling of Trees) Regulations 1979.
14. Hedgerows Regulations 1997 Statutory Instrument 1997 No. 1160
15. Council Directive 80/68/EEC of 17 December 1979 on the protection of groundwater against pollution caused by certain dangerous substances
16. Statutory Instrument 1998 No. 2746 The Groundwater Regulations 1998
17. Council Directive 91/676/EEC of 12 December 1991 concerning the protection of waters against pollution caused by nitrates from agricultural sources
18. Statutory Instrument 2006 No. 1289 The Protection of Water Against Agricultural Nitrate Pollution (England and Wales) (Amendment) Regulations 2006
19. Disposal of carcasses. Animal By-Product Regulations 2005 (Statutory Instrument No. 2347 2005)

5

FROM SUSTAINABILITY TO SUSTAINED ABILITY: STRATEGY FOR CONTINUOUS PROFESSIONAL DEVELOPMENT WITHIN THE PIG PRODUCTION INDUSTRY IN ENGLAND

R. LONGTHORP
Address

Learning is not compulsory, but then neither is survival

W. Edwards Deming

Summary

1. The Pig Industry in England is amongst the leaders in the agricultural sector for having developed and implemented a truly "employer led" comprehensive training and personal development strategy.

2. The strategy, *"from Sustainability to Sustained Ability"*, whilst led by the pig production sector, was produced by a cross industry group involving representatives from industry, Defra, Lantra, NPTC, Landex and others.

3. Training and personal development, whilst widely acknowledged as being beneficial to individuals, businesses and industry alike, has not traditionally been seen as sufficiently "sexy" an issue for industry to enthuse over and hence engage with wholeheartedly. It can be confusing with its terminology, jargon and acronyms and therefore needs to be marketed proactively and as part of a much broader initiative than as a simple training provision alone.

4. Benefits accruing from training, whilst widely acknowledged generically, are very difficult to directly measure and attribute specifically. As a result, at individual farm level, discussion often reverts to the cost of training, which is relatively easy to define, rather than value.

5. A clear need has been identified for not just personal development and training but also a radical new and comprehensive approach that embraces all parts of the industry. The previous provision was fragmented, *ad hoc*, complex and often involved excess "paper based" activity. The new model needs to be and indeed is employer led.

6.   The 3 primary strategy objectives are:
     1. Development of a skills and qualification structure that is relevant to industry needs, accessible at all levels and flexible in structure.
     2. Provision and promotion of an attractive environment for a progressive career in the pig industry.
     3. The promotion of skills development as being central to business improvement.

7.   The 2 main personal development components of the strategy are:
     1. PIPR - Pig Industry Professional Register, a CPD[1] scheme developed by industry with help from others, owned by NPA and managed by NPTC
     2. Pig Industry Certificates of Competence developed by industry and others, accredited by QCA and assessed by industry based accredited assessors.

8.   The key to success of the strategy and achieving its planned outcomes is as much about individuals gaining **recognition** for existing skills and competencies and their efforts to develop and progress as it is about gaining qualifications.

9.   The whole strategy needs to be driven by the demands of its businesses and people but particularly so for the training and personal development elements. Any moves by others to effectively make it compulsory by incorporating it into some form of "licence to farm" should, and will, be fiercely resisted. But this is not to say that in due course, once the demand for personal development has become "endemic", that such integration will continue to be resisted. Such integration will, however, need to be industry controlled.

10. The reasons for the failure to provide an effective skills solution, previously, lie with:
     • Industry for its, at best, past ambivalence towards engaging with the process – admittedly often as a result of the complexity and confusion already referred to
     • Those organisations who carry a quasi statutory responsibility for delivering skills and training to the industry.

---

[1]The term CPD can be used to describe a large part of the generic activity associated with the strategy or it can be used to describe more specifically the Pig Industry CPD Scheme – PIPR. Where the text refers to generic CPD, the term CPD is used. Where it refers to the pig industry scheme, then PIPR is used. Furthermore, unless explicitly stated, CPD within this paper only refers to the pig industry's interpretation of the term.

- And with government who oversee and shape the activities of these organisations

11. For its part the training and skills establishment needs to acknowledge what the industry has done by redefining funding and support protocols and to encourage other sectors down the route of self determination of training and development.

12. Whilst there are encouraging noises coming from the skills establishment, specifically regarding industry led, bite sized and flexible training, the jury is still out as to whether or not the blue sky thinking that led the them down this enlightened route will actually be allowed to develop into something dynamic, exciting, user friendly and ultimately hugely beneficial to industry.

13. There is a risk that, whilst government and its agencies currently talk about the concept of skills development needing to be "employer led", the reality of this actually being delivered may well be somewhat different. There may be so many hoops to jump through and conditions imposed on this new "employer led" deal that it will turn out to be nothing more than a token gesture to keep its proponents and potential beneficiaries happy. We need to know whether any new policy coming out of the Sector Skills Agreements[2] and Leitch recommendations will in fact be simply "employer led" by name or whether it will in fact be truly "employer led" by nature.

14. The involvement of 3 government departments – Defra, DfES and Treasury – together with all the other sublevels of agencies and organisations reporting into them does not make for a simple and straightforward structure through which an ultimately effective strategy can be created and delivered.

---

[2]Sector Skills Agreements. Sector Skills Councils are licensed by the Secretary of State for Education and Skills with the remit to reduce skills gaps and improve business productivity. There are 25 Sector Skills Councils covering various Sectors. The Sector Skills Council for the land based sector is Lantra. Each Sector Skills Council was charged with developing a Sector Skills Agreement for its sector. Sector Skills Agreements are designed to *"get the right people with the right skills in the right place at the right time"*. They are created through *"a collaboration between employers, education providers, partners and government"*.

Introduction

Undertaking Continuous Professional Development (CPD) could be likened to embarking on a personal fitness and healthy lifestyle regime.

- Most people would probably agree that it is generally a "good thing" and will yield long term dividends for a relatively modest investment.

  BUT ....

- There is some inevitable up front investment of "hard yards" needed to see long term results.
- The terminology, acronyms and multitude of different systems, fads and fashions seem to many to be just the excuse they need to continue to avoid it altogether.

By eating low GI foods and following a regular programme of CV activity together with isotonic, isometric and H.I.T. routines as part of a periodization schedule, it should be possible to burn off the MJ, increase an individual's BMR & VO_2Max whilst at the same time increasing their 1RM and creating that much sought "ripped look" complete with 6 pack abs.

According to a T&F group set up by the SSC for the land-based sector, Lantra, and substantiated by several FE & HE colleges, some of them with CoVE status, by incorporating APL and mapping against the NOS via the OCF it may be possible to mitigate the GLH requirement for a QCA accredited FL2 and hence shift the funding profile of the CofC closer to the NVQ

Figure 5.1 So . . . it's all quite simple really!

- It is hardly surprising, therefore, that confusion surrounds the whole area with few within industry actually fully understanding it in its broadest context – or even being **prepared** to understand it.
- Consequently very few actually ever do anything about it; because they do not know how, have never experienced the buzz of seeing the results of effective implementation or are simply too idle and would rather let others "go for the burn".
- Finally it is seldom a topic that engenders immediate unbridled excitement and enthusiasm amongst those at who it is aimed.

So, rather like the surge in gym memberships during the first weeks of January each year, many people, who may be initially enthused with the concept of training and skills development, soon become disenchanted and give up long before results are able to accrue. There will be many amongst that band of post

Xmas bingers who appear quite happy to fork out hundreds of £ thinking that is all that is required to "shed the weight" or that simply spending the money will be adequate motivation to continue. This is similar with industry and training where there is always the temptation and risk to fork out the cash for a training scheme in a bid to get some sort of feel good factor. However, what is needed in both scenarios is an actual commitment to the whole concept, an empirical assessment of need, a clear vision of desired outcomes and a clear mechanism by which those outcomes can be achieved. Spending the money, whilst undoubtedly painful for some, is actually the easy part of the process.

Mention of the words "Continuous Professional Development" and the "Pig Industry" in the same breath often gets listeners wondering what the extramural activities of solicitors, accountants, architects and the like has to do with the pig industry. Well, of course the key is that all operate and need to operate to professional standards. Some actually **have** to demonstrate participation in CPD so as to maintain effectively their "licence" to practice. For the pig industry it is far more basic and fundamental – we need to operate to the very highest of professional standards to stay in business! For most, this is the very essence of "sustainability"!

There will undoubtedly be those (*"if in doubt, legislate"*, bureaucrats) who see an immediate opportunity to incorporate industry CPD into a potential "licence to farm". Similarly there will be those (doubting pig producers) who will use such a threat as an excuse <u>not</u> to engage with the process. Those 2 groups of people need to understand that, whilst there is an appreciation that undoubtedly industry CPD will in time move closer to such initiatives as the Better Regulation Framework[3] and Farm Assurance, in the immediate term there is an iron-willed determination to avoid any such hijacking of the scheme through <u>compulsion</u>.

Whilst the pig industry needs to demonstrate to its customers, potential recruits and legislative bodies that it acts in a professional manner and to professional standards, it is an absolute pre-requisite to the successful rollout and uptake of the strategy that it should be demand led. There should be no compulsion. It has to be led by the needs and enthusiasm of individuals, businesses and the industry. **That enthusiasm needs to be fuelled by recognition and it will simply not all stick together during rollout if the legislative stick replaces the demand driven carrot.**

Finally, notwithstanding the complexity and confusion that often surrounds the concept of CPD, industry simply has no other option but to embrace it and develop schemes to deliver it through its own efforts and vision. Industry cannot rely on others for primary input albeit it should certainly enlist the assistance and support of those other bodies such as Defra, Lantra, National Proficiency Test Council (NPTC), etc. whose very role that is. Industry must be in the

---

[3]Better Regulation Framework - a Defra initiative with the objective of reducing the regulatory burden on industry by using, amongst others, the concept of "risk based" inspection

driving seat and must remain in the driving seat. Any inertia will rapidly return the situation to a perhaps comfortable but ultimately unproductive and unsustainable status quo.

WHAT – is *from Sustainability to Sustained Ability* all about?

First of all it is not just another series of *ad hoc* training courses that industry is expected to adopt simply to prove its "training credentials". It is a comprehensive strategy incorporating not only relevant formal training and qualifications but also informal training and personal development and making an unassailable link between personal development and business development. It also aims to seek out opportunities to enhance the industry's image and improve recruitment prospects within schools, colleges and society at large.

In addition at its core is the principle, repeated elsewhere in this chapter, of **recognition**: recognition of existing competencies, recognition for undertaking professional development and recognition for enhanced competencies. Recognition may be seen as the catalyst by which much of the strategy will achieve its ultimate outcomes.

STRATEGY OBJECTIVES

Objective 1

Develop a skills and qualification structure that is relevant to industry needs and accessible at all levels and flexible in structure.

- Ensure that both training materials are referenced to the appropriate Certificates of Competence at that level at which the training is aimed and accreditation is progressively updated and relevant in line with industry development.
- Integrate technology and knowledge transfer into the system of delivery.
- Develop delivery and assessment systems that overcome barriers to access and uptake as identified in earlier section.
- Ensure that there are enough trainers available to deliver the anticipated increased demand for training and development by investing in "training the trainers".
- Both IT training and the use of IT in delivering training will be investigated and use should be made of existing IT resources such as the highly recommended BPEX "Pig Enterprise" 1 and 2 CDRoms. Consideration should be given to updating these volumes and re-launching them as indispensable support material.

- Communications will be a key element in promoting and delivering not only this strategy but also the output of the BPEX R&D and Knowledge Transfer initiatives. Serious consideration must be given to explore the role that IT can play in optimising these communications.
- Review current funding arrangements to facilitate uptake for industry approved skills development through efficient delivery systems.
- Look at mechanisms to recognise non-accredited learning.

Objective 2

Provide and promote an attractive environment for a progressive career in the pig industry.

- Exploring opportunities for education, business links and work related learning opportunities within the 14-19 curriculum
- Develop enhanced work based learning opportunities
- Produce a clear career structure with transparent development pathways
- Develop industry champions to promote the industry to potential recruits
- Develop case study material to support careers work
- Ensure careers information advice and guidance services for all ages have appropriate and relevant information about opportunities within the sector
- Develop opportunities to undertake continual professional development within the industry by establishing mechanisms for accreditation of prior learning and to recognise formally continual professional development

Objective 3

The promotion of skills development as being central to business improvement.

- To develop support for the establishment of training needs
- Promote concepts such as 'Investor in People' through encouraging businesses to develop structured training plans
- Encourage employers to invest in the skills development of themselves and their staff by demonstrating the business benefits of training through industry champions. The concept of Continuous Professional Development will explicitly include training in and development of management skills
- Encourage wider usage of National Occupational Standards as business tools for the industry
- Promote the uptake of skills development to address existing skill gaps within the industry
- Explore resource opportunities to deliver effective development within the sector

- Reward the increasing demonstration of competency through linking to other industry schemes (e.g. frequency of assurance inspection visits/ other regulatory impacts/insurance premiums levels)

Continuous professional development – what is it?

In the context of the Pig Industry Strategy for Continuous Professional Development, CPD has the aim of:

> "Increasing value through learning – to both individuals and businesses" and may be defined as:

> The ongoing enhancement of an individual's skills and knowledge in relation to their job role and profession. Development can be achieved via a range of activities including gaining qualifications and awards, attending workshops and seminars, personal study and information gathering, knowledge transfer and practitioner's guidance. Activities are recorded against a business competency framework and provide evidence of the individual's ongoing learning and skills enhancement.

CPD thus covers a wide area of activity but within the strategy there are 2 main strands:

Pig Industry Professional Register - PIPR (pronounced "piper")

This is the CPD "scheme" of the pig industry and follows established CPD principles; namely the requirement to demonstrate and have recorded, ongoing personal and professional development and to have a set of pre-determined development goals that need to be achieved over a designated time span to maintain membership of the professional register.

PIPR is "owned" by the National Pig Association but administered by NPTC[4].

The key aspects of the industry CPD scheme are to:

- record and recognise all of an individual's professional development .. ..use this to

---

[4]NPTC – formerly known as the National Proficiency Test Council but now simply NPTC and part of the City & Guilds Group

- encourage further ongoing and progressive development and training....and use the whole to
- demonstrate to fellow industry professionals, customers, suppliers, authorities and the wider public the professional nature of the industry and those who work in it

It is important to understand that it is likely that many of the activities undertaken as part of CPD will already be being undertaken by most progressive members of the industry but that such participation will currently be going unrecognised to a large extent. PIPR formalises the process and facilitates recognition.

There are 4 classes of membership:

- Skilled Stockman
- Supervisor
- Manager/Owner
- Advisor

And there are 3 main steps to the PIPR process:

1. Application for membership (Figure 5.2) of the register subject to the entry qualifications which are:

    - Formal qualifications
    - Relevant experience
    - Or a mixture of both.

2. Once registered a member is then required to attend/participate in relevant events or activities. Each event, having been registered by the organisers with NPTC, is awarded a certain number of PIPR points according to duration, type of event be it participative or not etc., etc. As well as different types of event carrying a different number of points, there is also a maximum that can be derived and qualify from a single category over a 3 year period.

3. Participation in such events/activities then needs to be recorded. This is done by event organisers providing attendance sheets on which attendees register by quoting their name and unique PIPR registration number as it appears on their membership card (see example). Or for non "organised" events by individuals filing their participation direct with NPTC. A member is required to accumulate a defined number of points over a 3 year period. Failure to accumulate the required number of points will lead to suspension as will any conduct deemed to bring the professional register into disrepute.

PIPR Membership Criteria		
	Qualification	**Experience**
Skilled Stockman		
Level 2 CoC Basic Stockmanship & Welfare (Pigs) + Level 2 CoC Pig Husbandry Skills	Yes	Not Required
NVQ Level 2 Livestock Production (Pigs) or Level 2 CoC in Livestock Husbandry	Yes	1yr
No Relevant Qualifications		4yr
Supervisor		
Level 3 CoC Pig Unit Supervision & Operation	Yes	Not Required
NVQ L3 Livestock Production **or** NCA **or** National Diploma etc.	Yes	2yr
No Relevant Qualifications		5yr
Manager/Owner		
NVQ Level 3 Livestock Production (Pigs) **or** Degree **or** HNCA in Agriculture + CoC Pig Unit Supervision & Operation Level 3	Yes	Not Required
HND in Agriculture	Yes	3yr
No Relevant Qualifications		6yr
The above list of qualifications is not exhaustive and is for example purposes only. Full details are available from NPTC or on the membership application form		
Associate/Advisor		
An agricultural qualification or verifiable evidence of suitable training for the job role undertaken and evidence that can be confirmed by two named referees that the applicant's livelihood has been substantially generated from involvement with the pig sector for a period of at least 5 years		

Figure 5.2 PIPR membership criteria

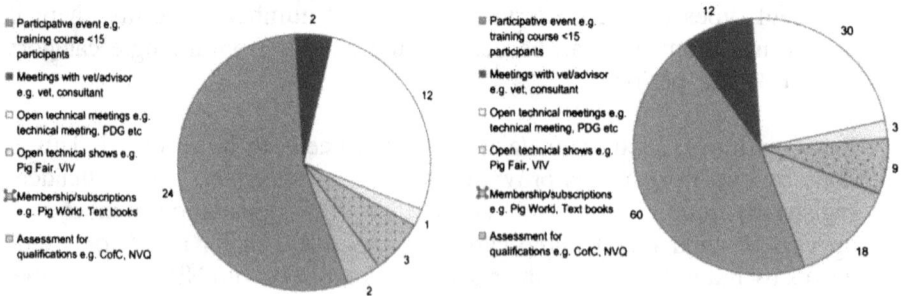

Figure 5.3 PIPR points criteria

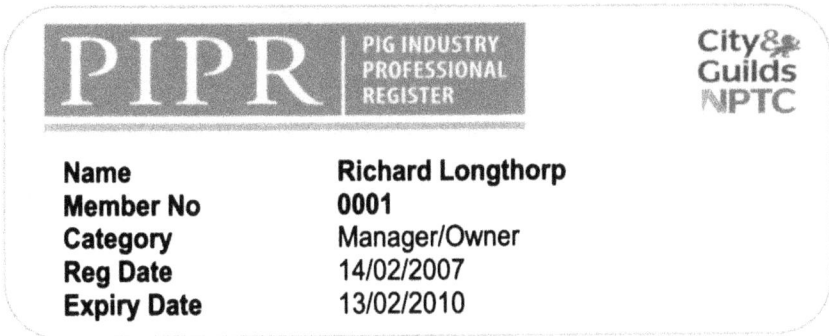

PIPR | PIG INDUSTRY PROFESSIONAL REGISTER

City&
Guilds
NPTC

Name	**Richard Longthorp**
Member No	**0001**
Category	Manager/Owner
Reg Date	14/02/2007
Expiry Date	13/02/2010

Figure 5.4 PIPR Membership Card

Pig Industry Certificates of Competence (CoC)

This is a comprehensive suite of QCA, (Qualifications & Curriculum Authority), accredited qualifications comprising 13 individual units. This type of qualification would have previously been known to many as "Proficiency Tests". The decision to follow this route of qualifications as opposed to the more heavily promoted NVQs is one that was taken quite deliberately and was based on industry's often negative experience of NVQs. **This experience showed the typical person working on a pig unit was not keen on the disproportionate amount of paperwork or "evidence" gathering that is part of the NVQ process.**

There are 3 levels of qualification at which candidates can be assessed:

1. Basic stockman
2. Skilled stockman
3. Supervisor

However, critically, candidates can pick and choose which units they are assessed on without needing to stick rigidly to the pre-defined levels. This is what is termed *"bite sized learning"*. In this case, however, a full certificate for the relevant level would not be awarded. And, as will be described in the section on challenges, this approach can have some significant negative implications for funding under the current LSC regime.

BASIC STOCKMAN

There are two compulsory units.

Candidates must pass one multiple choice examination paper for unit 1 and one practical assessment for unit 2. The pass mark for the multiple choice examination is 80%.

1.   Basic Stockmanship and Welfare Principles
2.   Basic Stockmanship and Welfare Practices

SKILLED STOCKMAN

This qualification comprises 8 units. Candidates must pass multiple choice examination papers and practical assessments in 3 or more units to achieve this qualification. The pass mark for the multiple choice papers is 80%.

Group A – Breeding Herd Operations

1.   Mating Management and Care of the Boar
2.   Care of the Dry Sow and Gilt
3.   Pig Farrowing House Operation and Care
4.   Operation of an Outdoor Pig Farrowing Unit

Group B – Weaner, Grower & Finishing Operations

5.   Care of Weaner & Grower Pigs
6.   Care of Finishing Pigs

Group C – Mechanical Operations

7.   Controlling the Environment of Pig Buildings
8.   Spreading of Farm Yard Mamure (FYM) and Slurry for Pig Units

Candidates must also have successfully achieved the Level 2 NPTC Certificate of Competence in the Safe Use of Veterinary Medicines before certification of this qualification. This CoC comprises 2 units both of which are compulsory.

SUPERVISOR

This qualification comprises 9 units; one core unit and eight optional units. Candidates must pass written and practical assessments in unit 1 plus three of the optional units to achieve this qualification. The pass mark for the written examinations is 80%.

Core unit

1.   Supervising Pig Welfare

Optional Units

2.   Boar Semen Collection and Processing
3.   Interpretation of Pig Records
4.   Management of Pig Farm Waste and Control of Pollution
5.   Pig Feed Storage, Milling and Mixing
6.   Selection of Gilts and Boars for Future Breeding
7.   Organisation of Pig Sales and Purchases
8.   Organisation of Outdoor Pig Production Sites
9.   Pig Unit Staff Supervision

Candidates must also have successfully achieved the Level 3 NPTC Certificate of Competence in Planning and Supervising the Safe Use of Veterinary Medicines before certification of this qualification. This CoC comprises 2 units both of which are compulsory.

Further details of *from Sustainability to Sustained Ability, PIPR and PICC (Certificates of Competence)* can be found on the BPEX, NPA and PIPR websites:

http://www.bpex.org/
http://www.npa-uk.net/
http://pipr.org.uk/
http://www.nptc.org.uk/

A comprehensive suite of manuals for both trainees and trainers together with supporting PowerPoint presentation material, for trainers has been developed by the industry utilising the very best of contemporary techniques and methods. Consideration is also to be given to making these manuals available in languages other than English.

WHO – is the strategy aimed at influencing?

The strategy, in its widest context, is aimed at a wide audience including:

*   Industry supply chain – individuals and businesses
*   Potential recruits and their parents
*   Government, LSC, Lantra
*   Colleges, Lecturers and Administrators

BEYOND THE FARM

One of the fundamental objectives of the strategy is to assist with recruitment:

> *"...provide and promote an attractive environment for a progressive career in the pig industry".*

Clearly the existence of a CPD strategy and its promotion outside the farm environment is a particularly useful tool with which to demonstrate the skilled and professional nature of the industry when trying to promote the industry within schools, FE colleges and society at large. It is for this reason that the concept of promoting a "professional" image through PIPR is fundamental to the integrated nature of the strategy.

TRAINING & QUALIFICATIONS

Direct participation in Certificates of Competence and PIPR are broadly aimed at people across the industry from the new entrant through the skilled stockperson to the manager or owner and beyond farmgate to advisors be they consultants, veterinarians, or supply side technical representatives.

However, whilst actual participation in Certificates of Competence or PIPR will be aimed at people employed within the industry, the concepts espoused by Certificates of Competence and PIPR generically are aimed at the much broader groups identified above. Without being cognisant of this and being proactive in promoting the whole concept of CPD within the industry to this much wider audience the full benefits and value of the strategy are unlikely to be realised.

PIG INDUSTRY CERTIFICATES OF COMPETENCE - CoC

Ultimately the qualifications structure aims to be applicable across all levels of people working within the pig production industry. Certificates of competence are currently available for technical and supervisory skills but a review will take place of how best to provide for more generic management and business skills. It may well be that, at this level, existing provision is perfectly adequate but may need additional promotion, recognition and structured integration into the overall pig industry qualifications package. Currently the *Leadership Development Group* programme available through the JSR Farming Group and Agskills is the pre eminent industry organised management training programme.

PIG INDUSTRY PROFESSIONAL REGISTER - PIPR

To maintain its integrity as a **professional** register, PIPR by definition needs to be somewhat more selective at the entry level than Certificates of Competence and so is currently designed to start at the skilled stockman level. However, investigations are currently taking place to explore how a "Pre-PIPR" category could be introduced to make the scheme appealing and open to all levels whilst at the same time maintaining scheme integrity.

Certificates of Competence and PIPR are certainly not designed to be mutually exclusive. Indeed they are designed to be complimentary. However, it is acknowledged that there may well be those who either feel more than adequately qualified and/or do not wish to undergo the assessment process but want still want to participate in the industry CPD scheme, PIPR.

WHY – do we need yet another strategy?

IN GENERAL

Before committing valuable and finite resource be it financial or human there clearly needs to be an identified need. "Need" implies that something is currently lacking and that fulfilling the need should produce benefits. That there was a general need for a strategy can be well demonstrated by some of the comments extracted from the responses to the consultation. This for a topic never having been renowned for arousing much excitement reflects the clear shift in attitude and recognition of the need for a new and comprehensive approach:

• The sooner this gets going the better."

• "The proposal should be held up as a blueprint for other parts of the agricultural industry not just the production sector."

• "I think it fair to say that in 20 years I have never seen an initiative that has quite captured the imagination of producers in the way this one has, which is quite an achievement, bearing in mind that 'training' has in the past been considered by some to be an unexciting cost, rather than an investment."

• "….the title "From Sustainability to Sustained Ability" we consider most appropriate, and indeed meaningful."

EXPERIENCE OF INDUSTRY AND OTHERS

Agskills

The experiences of Agskills have had a significant impact on the development of the strategy. Agskills was formed in 1998 with a view to improve on the effectiveness and structure of training provision. It did not seek to undertake the role of training provider or deliver training *per se*. It acted in a facilitating and coordinating role for its members. It was formed as a result of the general dissatisfaction with the then training provision within the pig sector. The training available was frequently *ad hoc*, seldom derived from a structured training needs assessment and only infrequently led to recognition. Agskills set out with bold intentions and was undoubtedly very successful in achieving its objectives, ultimately becoming the recognised leading pig industry training group in the country. This, however, was not all achieved without a huge cost of frustration along the way. It is as a result of this experience and frustrations that many of the "design components" of the new system and referred to elsewhere in this paper were incorporated.

Scoping Study 2001

This Agskills experience and the increasingly evident shortage of skilled stock people led to the NPA commissioning a scoping study by ADAS into the whole area of training provision and a competency framework. However, whilst the report, produced in November 2001, was a commendable, thorough and comprehensive piece of work and contained many excellent recommendations, the timing could not have been worse! Following hard on the heels of market collapse in the late 1990s, the industry then suffered the effects of 3 major disease outbreaks – Classical Swine Fever in 2000, Foot and Mouth Disease in 2001 and the "wasting diseases", PMWS and PDNS, concurrent with both. This led to continued financial pressures and, whilst, some may argue that this is just the very time businesses should be investing in training, the reality of marketing the concept to a "punch drunk" customer is somewhat different to say the least. The report therefore was left "on the shelf" but certainly not forgotten.

Confederation of British Industry (CBI)

Whilst giving absolutely no satisfaction, it does nevertheless give support to the conclusions drawn by the pig industry some 4 to 5 years previously that the CBI reported during 2006 their own similar conclusions on the state of skills and training provision.

- The country is being let down by a dysfunctional national skills system that is seen by many employers as irrelevant in helping them meet business training needs
- There are nearly 6,000 vocational qualifications in the UK, many of which have been designed by consultants, rather than employers, making them of doubtful economic value.
- The plethora of Government skills bodies and initiatives designed to support employers has become unwieldy and incomprehensible – less than 0.50 of employers have received useful information from many of the 49 different Government skills bodies in the UK.
- Cut the bewildering number of skills bodies and create a simplified system that offers effective support and guidance.
- Place employers in the driving seat of designing qualifications in a way that reflects the skills needed by the economy. This can be achieved by ensuring that employer-led sector skills councils are able to decide which qualifications are fit for purpose, and ensure that employers' own training can be accredited.

However, although the pig industry might have total empathy with these findings, the observation would have to be that the CBI appears to fail to acknowledge the part that industry has or, more accurately, has not, played in allowing the situation they describe to develop.

"Skills Envoy"

Finally, as evidence of need coming from experiences, government at least appears to have finally acknowledged that significant problems exist with the appointment of Sir Digby Jones, former Director-General of the CBI as the government's "Skills Envoy". Sir Digby's role will be to oversee efforts to improve skills in UK businesses.

SPECIFIC

The scarcity of skilled staff

Whilst evidence is largely anecdotal from discussions with pig farmers bemoaning the lack of suitable people to fill vacancies or indeed to staff potential business expansions, the large numbers of immigrant workers now found on UK pig farms is further evidence of the difficulty in recruiting skilled staff locally. And although a significant proportion of the immigrant workforce may well be hard working and skilled, their intentions are usually to only remain in the UK for a couple of years to accumulate sufficient pounds sterling to take back home, convert into the local currency and set themselves up with a business. This short term nature of their employment and the likely improvement in earning opportunities in their native countries mean that this option can hardly be described as a sustainable solution for UK pig production.

Structural need

The strategy underpins many other initiatives which the industry has invested heavily in but which are dependent upon a skilled workforce for an effective delivery.

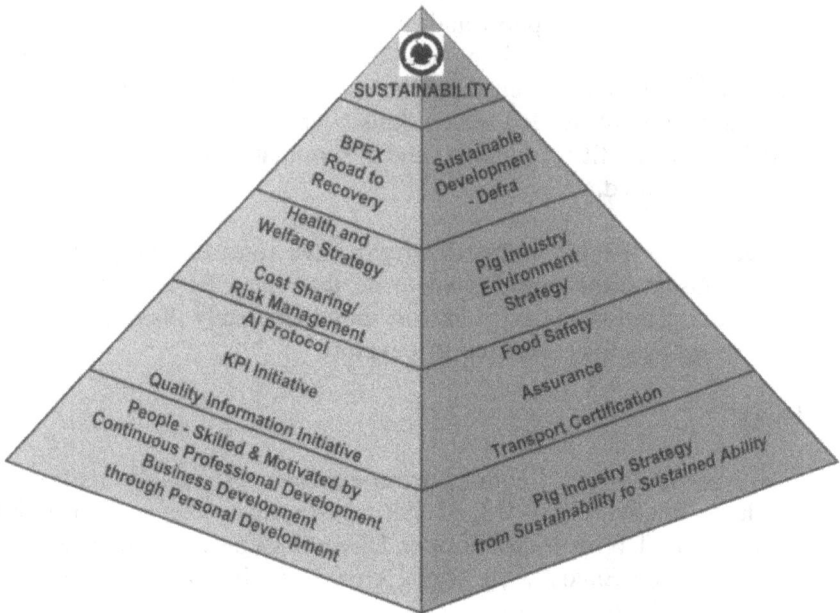

Figure 5.5 Skilled people - the foundation for delivering sustainability?

Investment in all these other (highly commendable) strategies and initiatives is unlikely to see optimal returns without adequate attention and corresponding parallel investment being given to skills development.

IMPROVED BUSINESS PERFORMANCE

Improved business performance must be the key and ultimate outcome of any training and skills strategy and yet good, hard contemporary evidence is not always that readily available. Work by English and others, though, has demonstrated the benefits. Experience by Agskills certainly substantiates their findings.

Table 5.1 INCREASED SALES PER YEAR –
1 YEAR AFTER TRAINING INITIATIVES
IMPLEMENTED

Farm	% Increase in sales
1	12.6
2	11.7
3	13.4

English, P

Figures from the Agskills triennial review required under the Vocational Training Scheme which indicate an overall increase in output of 10-15% on members' farms with the added benefit of a reduction in staff turnover of 10%.

WHEN SUCCESS ARRIVES, FROM WHERE DOES IT COME?

Another "reason" for training is that in over 30 years of business, it would be my view that it is frustratingly but frequently difficult to pinpoint the exact reasons for success when it does finally come. It is often difficult to be totally objective let alone empirical about identifying those key factors that contributed to the success. The only conclusion that I have drawn is that, on balance, if people or businesses generally do the "right things" and continue to do the "right things", then success will **at some stage** follow – but not always perhaps as directly (success doesn't always come the way you think it does - it comes from the way you think, *Robert Schuller*), or necessarily as quickly, as one would like to presume.

A word of warning here; this does not mean that by doing all the right things individuals or businesses will be immune from those distractions and challenges that affect us all. Quite the contrary. There will be times when it all may hardly seem worth it and we question - why keep doing the right things when we appear to be suffering as much as if not more than others? The key here is to remain faithful to your goal. This allows distractions to remain just that – distractions; not fundamental barriers to ultimate progress and success.

LUCK

Chance favours the prepared mind – Louis Pasteur

There is a frequent perception that others often appear to have more than their fair share of luck. The response to this is to perhaps consider one definition of "luck" as being ***"the crossroads of preparation and opportunity"*** and accept that a large part of preparation is in fact doing "the right things". In other words, whilst opportunities are all the time being presented, it is only those that are in a state of preparation of having continued to strive to do the "right things" that will be the "lucky" ones who can take advantage of those opportunities.

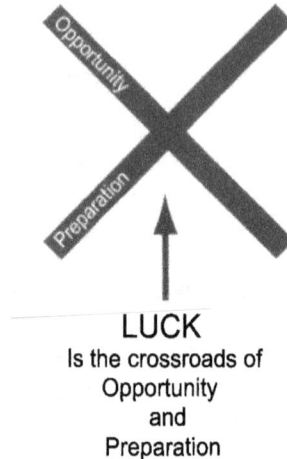

LUCK
Is the crossroads of
Opportunity
and
Preparation

Figure 5.6 Luck or 'planned opportunity'?

RECOGNITION

One of the key concepts running through much of the development and implementation of the strategy is that of **recognition.** The industry already comprises many skilled, hard working and conscientious people. However, as is often the case, their individual skills and competencies often go largely unrecognised other than being reflected in their annual pay review. And yet, as has been proposed by Herzberg, Maslow and others, whilst pay may often mistakenly be presumed to be a motivator, it is recognition and achievement that are actually the main motivators. In acknowledging that a well motivated team may and should be well remunerated, it will always be the well motivated team that outperforms the well paid team *per se.*

Recognition not only motivates but also encourages further development – a key element of any worthwhile strategy for Continuous Professional Development.

The two main "formal" elements of the strategy, Certificates of Competence and PIPR, fully reflect the importance that is placed on the concept of recognition. Existing skills and development are recognised in both Certificates of Competence and PIPR.

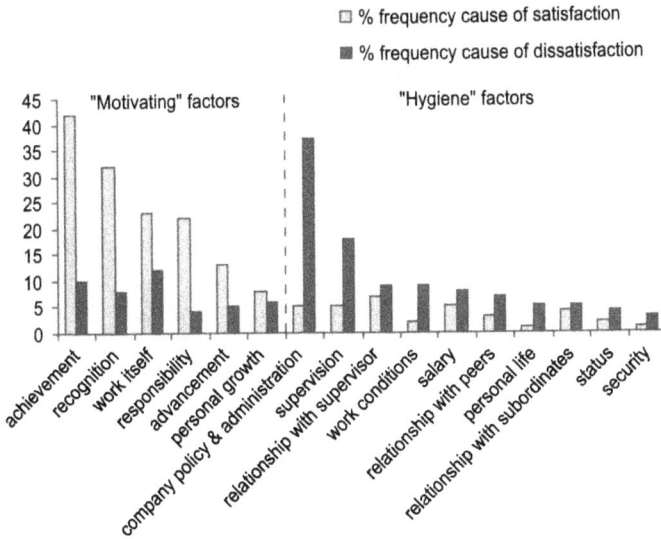

□ % frequency cause of satisfaction

■ % frequency cause of dissatisfaction

Figure 5.7 Herzberg's 'hygiene' and 'motivating factors.

As an aside, albeit a very relevant one, training of managers and owners can of course help address those hygiene factors identified above giving rise to most dissatisfaction.

Table 5.2 BENEFITS- SUMMARY

	Strategy	*PIPR*	*CoC*
Personal			
• Recognition	✓	✓	✓
• Increased Skills	✓	✓	✓
• Career Prospects	✓	✓	✓
• Demonstration of competence to customers, consumers	✓	✓	✓
• Better personal performance	✓	✓	✓
Business			
• Better business performance	✓	✓	✓
• Recruitment	✓	✓	✓
• Retention	✓	✓	✓
• Better skills for owner	✓	✓	✓
Industry			
• Promotion of Industry	✓	✓	✓
• Demonstration of competence to others	✓	✓	✓
• Support for Farm Assurance	✓	✓	✓
• Demonstration of Low Risk within concept of Better Regulation Framework	✓	✓	✓
• Assist event organisers in promoting their events	✓	✓	

ALTERNATIVES TO TRAINING

In assessing a need and how to provide for that need, it would be prudent to investigate alternative ways of that provision. However, the alternatives to training genuinely appear to be somewhat limited!

Whilst methods such as "head hunting" or indeed more traditional methods of trying to recruit a "ready made" skilled and motivated person such as placing an advertisement in *Pig World* or *Farmers Weekly* might often be quoted as alternatives to training, the reality is that there is such a small pool of available staffing resource nationally that such methods are unlikely to yield a solution let alone a sustainable one. Furthermore, the actual costs of recruitment are frequently underestimated or understated. The direct costs of recruitment such as advertising etc. are somewhat akin to the tip of an iceberg. It is what lurks beneath these visible costs that need to be taken account of:

- The organisation and execution of interviews, wasted management time when a candidate or candidates fail to turn up for interview.
- The time wasted when an applicant is offered and accepts the position but then fails to turn up on the first day of work.
- The whole process then perhaps being repeated again.
- The cost in terms of additional supervisory resource, resolution of possible conflicts etc. of induction and integration into the team.
- Problems integrating into the rest of the team and the whole recruitment process giving to start again.
- The lost production during this period.

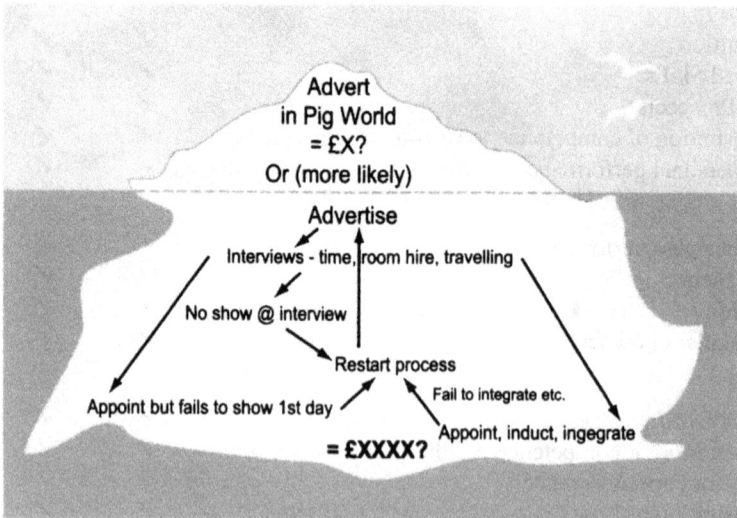

Figure 5.8 Recruitment costs - look below the water line.

HOW? – the concepts

Fundamentally, whilst all the above and more are perfect justification for the development of a strategy, it needs to be questioned why such a strategy should be necessary at all when there appears to be such a plethora of training bodies, (policy, strategic and delivery), already in existence. But it is this plethora of agencies, councils and the like, together with at best an ambivalence towards the concept from many sectors of the farming industry, that is actually part of the problem rather than part of the solution.

A GATEWAY

Where such complexity and potential for confusion, and hence rejection, exist then the strategy will seek to create a "gateway" through which producers can access anything and everything to with training and particularly funding. It will create a virtual "one stop shop". Rather like with the ubiquitous Microsoft Windows, the simple desktop, (gateway), gives access to a vast array of useful applications, (training, qualifications, funding), through millions of lines of computer code that is absolutely gobbledygook and thankfully hidden to most of us.

Gateway = Desktop   Training Infrastructure = Code

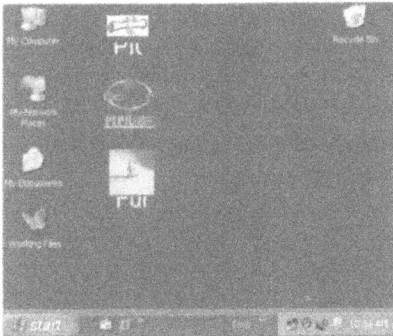

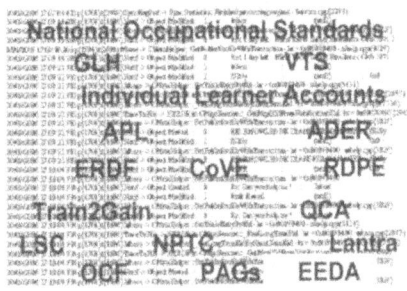

Figure 5.9 Route to training - the desktop or native code?

PARADIGM SHIFT

Whilst there is now a general acceptance by LSC, Leitch and others that training needs to be "Employer Led", the Pig Industry reached this conclusion some 3 years ago. Furthermore, thinking within the sector has already moved on and it is quite clear that for the strategy to be truly successful, sustainable and long lasting, training and personal development not only needs to be just employer led, it needs to be employee led. This may cause the cynics to raise a wry smile but there is clear evidence within the Agskills group that this is indeed happening. Members of staff are now enquiring about training opportunities and appraising all training events more critically.

For the vision to become a reality it will be necessary for a massive shift in thinking both from industry, which the pig industry has already done, and also from the skills establishment. There is a need to move from the traditional position of:

This is what we, <u>the training establishment</u>, have developed, come and get it if you want it.

to ...

This is what we, <u>the industry</u>, wants, please help us develop and deliver it and onwards to ...

This is what we, <u>the employees</u>, need and want, please develop and deliver it

And, whilst on initial reading this may appear to imply criticism of the skills establishment, in reality it is as much a criticism of industry for their failure to properly engage as it is of a skills establishment for their failure to find out what was really wanted.

For the strategy to be truly effective, long-lasting and sustainable the training ethos needs to be incorporated into mainstream everyday thinking and activity within individual pig businesses and the industry as a whole. This is a massive challenge – far greater than simply organising a few ad hoc training courses, picking up some funding and awaiting for results to follow. It needs a shift in the way we think about training and development. But the prize for rising to the challenge and beating it has to be worth the effort required.

Training and personal development needs to become part of everyday business life, it needs to become a "normal" and accepted activity, it needs to become second nature. Training and personal development needs to become a "habit".

Stephen Covey defines a habit as the intersection of knowledge, skill and desire.

- Knowledge is the *what and the why to do*
- Skill is the *how to do*
- Desire is the motivation or *want to do*

With all these three in place a habit is created.

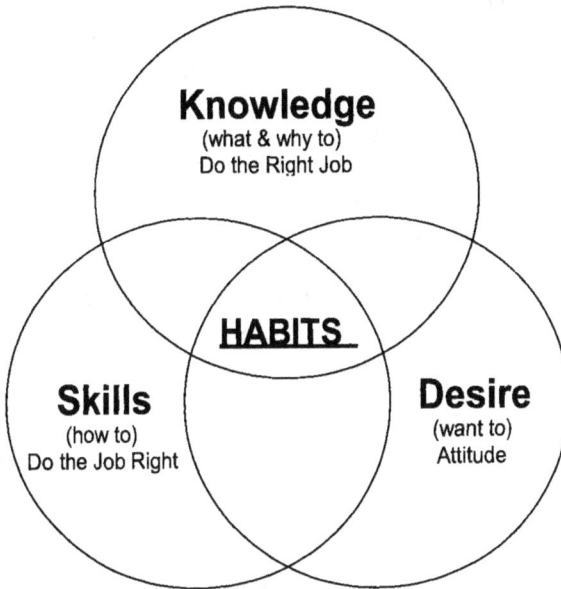

Figure 5.10 Creation of habits.

Although any individual may possess only one or two of the 3 attributes required each component, together with recognition, can be used as a motivating force and springboard to acquire or develop the others until the full set is in place to create the habit and drive forward.

Of course finding people with the "right" attitude can often be a challenge in itself but it is hoped that the soon to be introduced 14-19[6] year old diploma in environment and land based studies will help enormously in producing young people with that "right" attitude.

This should assist further in applying the concept adopted by the strategy and recommended to businesses to:

Recruit for attitude - Train for Skills

GROUPS

It is envisaged that delivery of training will be as flexible as possible but that it is likely to be best delivered via a network of training groups. These groups may be based on an agricultural college, veterinary practice, discussion group, marketing group or simply a group of similarly minded producers as was the

---

[6]The 14-19 Diploma in Environmental and Land Based Studies is a part of a DFES initiative to develop a series of 14 specialised diplomas across all major sectors of the economy and "….will give all young people the opportunity to choose a mix of learning which motivates, interests and challenges them, and which gives them the knowledge, skills and attitude they need to succeed in education, work, and life".

case with Agskills. The choice is limited only by producers' preference. The group(s) may already exist but currently serving a separate purpose such as performance benchmarking, technical discussion etc.

However, it is not envisaged that groups will be training providers *per se*, rather training facilitators looking after areas such as Training Needs Assessments and subsequent development of annual training plans based on the outcome of the TNAs; identification of suitable trainers to deliver the programme, coordinating delivery, general administration etc.

The groups may be constituted informally or formally as partnerships, companies or whatever. English Food & Farming Partnership are available to, and already have, assisted with this process.

The benefits of operating through a group structure include:

- Brings opportunities for economies of scale
- The ability to retain a group coordinator who may be responsible for TNAs, programme development an organisation etc. The coordinator, being one step removed from the normal production activities of farms, may also be able to impose more structure and discipline into the area of skills development than if left to the devices of the farm's own staff or indeed proprietor!
- Trust can be built up within a group structure. This can lead to the leveraging of additional benefits, not necessarily directly associated with the training function. Examples would include benchmarking, unit input cost comparisons, common veterinary, nutritional input etc.

WHEN – does all this happen?

The strategy really began its life at the David Black Award breakfast in November 2004 when the recipient, Gerry Brent, chose to make pig industry training and its then challenges the subject of his acceptance speech. With the then minister Lord Whitty and permanent secretary of Defra, Sir Brian Bender, listening attentively, the strategy got the support of Defra that it needed to break down some potential barriers.

After some exploratory discussions during early 2005, The Strategy Implementation Group was formed in the summer of 2005. The strategy document was produced and put out to consultation later that year. Once the strategy had received the (unanimous) support of all those responding, work started in earnest to put some of the fundamental building blocks in place. During 2006 the following progress was made:

- Developed a comprehensive suite of qualifications fully accredited by QCA

- Developed a comprehensive suite of supporting technical workbooks for both trainees and trainers
- Formed pilot groups in East Anglia, South Central and North East to pilot key elements of the programme
- Developed a team of industry assessors
- Delivered the first training sessions in working towards the industry qualifications
- Carried out the first assessments against the certificate of competence standards
- Developed the Pig Industry Professional Register, a CPD scheme for the recognition of competencies
- Appointed a full time Pig Industry Training & Development Officer
- Commenced development of promotional literature for use in schools and colleges
- Developed links with groups including FACE, Fresh Start, NFYFC, Young Pig Farmers, Farm Animal Welfare Council and Farm Assurance bodies.

WORK IN PROGRESS

With a large part of the development work now having been completed one of the main tasks remaining is in promoting the strategy and its components throughout the industry.

And, whilst work has started in making links with schools, colleges, etc. there is a significant amount of actual development work still to be undertaken in this area. This is not to say that this crucially important strand of the strategy has been overlooked or sidelined. The training, qualifications and CPD elements needed to be in place and being seen as effective first so that they can be used to promote the industry more broadly and ultimately aid in the recruiting new people into the industry.

Challenges remaining

ACCEPTANCE AND UPTAKE BY INDUSTRY

Whilst the strategy has been enthusiastically received by all who have been approached, there does remain a significant job of communicating with the rest of industry. It must always be acknowledged that here is a major risk that operating as we do in fairly closed groups we often think that we are communicating effectively with the **industry** when all we are doing at best is actually communicating within our own particular **peer group**. With this in mind an industry communications group, involving key players and specifically the BPEX Training & Development Officer and Knowledge Transfer Team, has been established. One of its aims will be to help address this very issue and ensure that this is an inclusive programme, promoted and marketed across the whole industry.

SOURCES OF EXTERNAL FUNDING

The industry fully accepts that in the medium to long term it should fund its own training and personal development programmes. However, there are certain key areas that industry believe government should provide funding:

- To ensure that the investment already made by industry is allowed to come to fruition and become predominantly self financing would benefit from some pump prime funding.
- To bring all learner's Keyskills up to a standard demanded by the workplace
- To compensate for the fact that the industry is typical of the Small and Medium Enterprise (SME) and micro business sectors and unable to make best use if economies of scale in training provision. This is even more so given the rural nature of the industry.
- Where it is broad government policy to fund to a certain level (e.g. currently Level 2 or 5 General Certificate of Secondary Education – GCSE - equivalence)

There also needs to be an overall simplification of the funding streams available and the process of applying for and managing any such bids. The pig industry is ideally placed to offer a novel solution or model to this particular problem. It has, in BPEX, a non governmental organisation with a track record, and indeed almost its *raison d'etre,* of collecting money from various sources, deciding how to spend it, spending it and then accounting to government for how it was spent. At a time when perhaps all UK agricultural levy boards are going to have to redefine themselves and their roles it would surely be a quantum leap forward if a way could be found through the mire that surrounds current skills funding to allow funds to be allocated to BPEX for further onward distribution and use for industry training. This could again serve as a useful model for other sectors.

"Bite sized" training or small individual chunks of learning, not necessarily being part of a more comprehensive qualification, is currently very much in vogue being cited in the Lantra Sector Skills Agreement as offering significant benefits and enhanced uptake. However, this is not currently reflected in funding arrangements with most core funding available through LSC being linked to NVQs or other "Full Level 2" qualification, (equivalent to 5 GCSEs). Given that the concept of bite sized learning is supported in the Sector Skills Agreement then funding should be available to support it.

Finally, there is much talk and indeed recommendations within the training and skills establishment that the whole process of skills development should become employer led. This concept is gaining so much critical support and from such a broad constituency that it must be allowed to become the key factor in determining national policy towards training and skills development. The concept should also, therefore, be fully reflected in funding policy. However, there is a serious risk that, whilst there is much talk about and indeed support for the concept of training and skills development being "employer led", the reality of this actually being delivered may well be somewhat different. There may be so many hoops to jump through and conditions imposed on this new "employer led" deal that it will turn out to be nothing more than a token gesture just to keep its proponents and potential beneficiaries happy. Furthermore unless funding follows the concept then the idea is likely to be stillborn. We need to know whether any new policy coming out of the Sector Skills Agreements and Leitch recommendations will in fact be simply "employer led" by name or whether it will in fact be "employer led" by nature.

Conclusions

Novel and challenging thinking by a few, hard work by many and cooperation between industry and its partners in the skills establishment have led to the creation of opportunities in the field of training and personal development in the pig industry not seen previously.

The launch of the Lantra Sector Skills Agreement together with recommendations in the Leitch Review and reports from the CBI, all of which point towards the importance of training and skills development being "employer led", creates a massive opportunity. An opportunity for positive change and a huge **potential** to take a quantum leap forward in the field of skills development. To ensure that this potential is realised will need a total commitment from industry, government and skills organisations alike. We must also ensure that the concept of "employer led" is allowed to become reality. There is a risk that government and some in the training and skills "establishment" will continue to use the term with gay abandon purely to tick a few politically correct boxes. But then impose a series of hoops and conditions that in reality mean that absolutely nothing will fundamentally change.

The involvement of 3 UK government departments – Defra, DfES and Treasury – together with all the other sublevels of agencies and organisations reporting into them does not make for a simple and straightforward structure through which an ultimately effective strategy can be created.

However, despite the challenges, the pig industry has already achieved a great deal in both raising awareness of the skills agenda and delivering tangible benefits. All of the hard work and enthusiasm that have got the strategy this far will not allow those remaining challenges to stand in the way of ultimate success.

References

English, P.R. (2002). Improving stockmanship through Training and Motivation. USA National Pork Board 'Pork Quality and Safety Summit' Conference, Des Moines, Iowa, USA 18-19 June 2002

English, P.R., McPherson, O., Deligeorgis, S.G., Vidal, J.M., Tarocco, C., Bertaccini, F & Sterten, H. (1998) Evaluation of the effect of training methodologies, motivational influences and staff and enterprise development initiatives on livestock performance and indices of animal welfare. Occasional Paper, British Society of Animal Science 1999

Brent, G (2006). EXIT REPORT - following funding under the Vocational Training Scheme of the English Rural Development Programme Defra Lead Region: East Midlands, Chalfont Drive, Nottingham NG8 3SN. Project Title: Skills development in the pig sector

Covey, S The 7 Habits of Highly Effective People, Free Press

Herzberg, F The Motivation to Work, Transaction Publishers

Maslow, A Maslow on Management, John Wiley & Sons

BPEX From Sustainability to Sustained Ability – Strategy for Continuous Professional Development in the Pig Production Industry in England

6

CO-OPERATIVE FARM GROUP ISSUES

I. ENTING

Wageningen University and Research Centre - Animal Sciences Group (Wageningen UR - ASG), P.O. Box 65, NL-8200 AD Lelystad, The Netherlands

Introduction

Co-operation is the practice of individuals or societies working in common with mutually-agreed goals and methods, instead of working separately in competition. Although co-operation is the antithesis of competition, the creation of a co-operating group (co-operative or co-op) is often based on the idea of forming a stronger competitive force. The International Co-operative Alliance defines a co-operative as an autonomous association of persons united voluntary to meet their common economic, social and cultural needs and aspirations through a jointly-owned and democratically controlled enterprise (IAC, 1995). The co-operative is equally owned and controlled by their members; a unique feature of the co-operative is that a member is entitled to only one vote regardless of their share in the co-op. In for-profit co-operatives, the surplus earnings are distributed amongst the members in proportion to the degree they used the service of the co-operative in that specific year. In not-for-profit co-operatives the surpluses of the enterprise are, after payment of dividends on shares, returned to the co-operative's general reserve. It can for example be used to improve the co-op's service (Gamble, 2002a).

Co-operatives have existed over almost 200 years. Robert Owen (1771 – 1858), a cotton grower and trader in Great Britain, initiated the co-operative movement and opened the first co-operative store in Scotland. His ideas were developed further by William King (1786 – 1865). He founded a periodical called 'The Cooperator' including practical advice about running a shop using the co-operative principles, of which the first edition appeared on May 1, 1828. Fifteen years later a co-operative called the Rochdale Society of Equitable Pioneers was formed. Their co-operative became the prototype for societies in Great Britain and they are most famous for the Rochdale Principles, a set of ideals for the operation of a co-operation that provide the foundation for the principles on which co-ops around the world operate to this day and which are adopted by the International Co-operative Alliance.

Most agricultural co-ops known today are so-called *traditional co-ops* in which farmers work together to strengthen their power in supply of their inputs or marketing of their outputs. They co-operate as a congenial, horizontal group and control the primary production segment of the agricultural production chain. Changes in the supply chain, e.g. more international operating markets requesting higher quantities and better time management of the product produced, challenged the traditional co-ops in their activities and structure. Some co-ops responded to the opportunities and pooled their resources to invest in successive chain segments, e.g. in processing and value-added businesses. They co-operate vertically in the chain and are so-called *new generation co-ops*.

In the current chapter the two different types of co-ops – traditional and new generation – will be discussed. Examples of them, their origin, their (dis)advantages and some success factors will be presented. It will show that the two types of co-operatives are, however, not necessarily incompatible with each other. Hybrids of the two exist and will arise in the future because co-ops, and especially the new generation co-operative, are evolving structural models adapting to market change.

Horizontal co-operation – traditional co-operatives

BACKGROUND

There are various types of traditional co-ops. In agriculture, the producer and the financial co-operative are best known and to a lesser extent the service co-operative.

Producer co-ops focus on co-operation between farmers to strengthen their purchasing, selling and marketing power for example for feed, farm supply, livestock or their products produced. The traditional co-ops focus on commodity marketing and on the selling side act as a clearinghouse for the members' products. The co-op is the legal form of grouping producer forces with the aim of being stronger as a group than as individual within the production chain. In practice farmers often operate in public as a producer organisation, producer association or producer society. These organisations can be organised in a different way, each having their own specific strategic goals (Bijman and Hendrikse, 2001):

•   Organisations of producers who within the existing market structures group their forces on the *purchasing side* to obtain better conditions of their suppliers. The co-operatives sells the input necessary to their members' economic activities. The producers operate individually on the selling side. Still many feed companies are producer co-operatives.

- On the other side, there are organisations of producers who, within the existing market structures, combine their forces on the *selling side* in order to achieve higher prices. These co-operatives market the products emerging from their members' economic activities. In the pig production chain, some slaughter/processing companies started as producer co-operatives, but subsequently they changed towards independent companies whether with shareholders or not. An example is the Vion Food Group, Europe's second largest pig slaughterhouse (see Chapter 24).
- Slightly different than the former are organisations of producers who want to operate in a more market-driven manner and bring their forces together to *tailor production* to the demands of and in close co-operation with the customer. These organisations operate in a business-to-business market and are, in contrast to the following organisation, unrecognizable in the consumer market.
- The last are organisations of producers who want to *market special products* or products of distinct quality directly to consumers. They develop their own brand, make all the investments necessary for it and operate in the consumer market. This producer co-operative operates on the edge of traditional and new generation co-ops, and shows that a precise distinction between them is not easily made.

The (reborn) popularity of producer organisations – especially the two latter forms – reflect the development of a diminished transparency of the market, a reduced number of costumers in the food sector, changing societal production demands urging for specialty products equipped with a logo or brand name representing their distinct production process or intrinsic quality, and ICT as enabling technology to ease information exchange between producers (Van der Kroon *et al*, 2001).

In the financial world co-operatives still exists. The Rabobank is a well known Dutch example of a co-operative bank. Service co-ops are rare in agriculture, but this might change in the near future with the forthcoming interest for large-scale biogas production. Examples of services that these type of co-ops provide are electricity.

EXAMPLES

An analysis of producer organisations in the Dutch horticultural sector and The Dutch pig and poultry sectors was performed by Van der Kroon *et al.* (2002) and Van Horne *et al.* (2007). Van de Kroon *et al.* (2001) interviewed 16 horticultural producer organisations. Their goals, as expressed by the members of the organisation, were predominantly better access to and service of the market through combined provision of products and marketing activities.

Although market access is essential for them, the producer co-ops did not agree to lower prices in exchange for secured market access (1.94 on a scale of 1 not agree to 5 fully agree). Therefore, in practice it could be seen that a major motive for the formation of a producer organisation was to strengthen the negotiating position on the selling side and so reduce sale costs and increase revenues through central and collective marketing of the products produced (4.25). Combined purchase of farm supply to reduce costs was not a major motive, although it facilitated the organisation. To reach the goal of a better market access and service key activities of the producer organisations were development of a brand, investing in product packaging, performing promotional activities, central selling of products, central purchasing of means of production, and having control of logistics. Not all key activities are performed by all organisations. Comparison of producer organisations which (more than) achieved their goals (group 1) to producer organisations which indicated to be less successful in this (group 2), revealed some marked insights on the organisations' goals and activities:

- Annual turn over was more than twice as large in group 1 than in group 2, while the production acres did not differ between the two groups.
- Success of the organisation was not related to the type of (horticultural) product it produces, nor was it related to geographical location.
- Group 1 organisations were more often focussed on the external market and had a businesslike attitude. Group 2 organisations were more often focussed on internal production and enhancement of mutual relationships. Investments in personnel reflected this too. Group 1 employed additional people for marketing and promotion activities; group 2 for operational production and administrative activities.
- Producer organisations in group 1 indicated more often that the ultimate success factor for their producer organisation is developing or having an innovative product. Group 2 organisations stated more often that guarantee for their organisation's success relates to the internal organisation and focuses on active co-operation between the members.

Can the advantages attributed to the traditional co-ops in horticulture be passed on to a sector which differs considerably in structure, culture and chain organisation, like the pig production sector does? If the focus is on co-operation, like the group 2 organisations, the answer is 'yes'. Many examples of alliance systems and integrator models (franchising) are present in the pig sector in which one or more pig farmers have contracts with or are owner of other pig farms (Martinez, 1999; Martinez, 2002; Pollock and Spencer, 2004). The change of small-scale farming into pig farming as a science requesting large-scale units with highly qualified personnel induced this movement. Depending on how much autonomy producers wish to retain, they fit within this new production paradigm in an integrator model (the breeder owns the animals on the nursery-finisher units) or the alliance system (the breeder has a formal agreement with the nursery-finisher units, while both remain independent businesses; Cozzarin and Westgren, 2000).

The answer to the question of passing on attributed advantages of traditional co-ops in horticulture to the pig sector, if the focus is on consumer market access like the group 1 organisations, is not a straightforward 'yes'. Particularly in sectors which require an additional processing step between primary production and consumer sales it is more difficult for a producer co-op to control the quality of the end product. Even so, product differentiation and developing new innovative products, is more complex in the pig sector. The amount of possible variations in, for example, tomatoes, is much larger and more easy to find than in pigs, and differentiation in consumer products does not take place at the primary production farm. If they already do, they are not allied to the type of the product but to the means of production (e.g. housing and handling of pigs). What makes it even more complex for the pig sector is the necessity to value (almost) all parts of the pig carcass (Van der Kroon, 2001). To be successful as a group 1 producer organisation in the pig sector, it seems logical to gain more control over the successive steps in production chain and gain vertical co-operation within the chain.

Vertical co-operation – new generation co-operatives

BACKGROUND

In new generation co-ops, farmers control or coordinate more than one segment within the production chain. The continuum of organisational forms to adjust successive chain segments moves from 'open markets' or 'spot markets' in which all segments are independent, to 'total vertical integration' in which all segments are managed and owned by one company. Many intermediate forms of agreements can be found in between the two ends, for example marketing contracts, strategic alliance, formal co-operation, production contracts, and quasi-vertical integration.

In an open market the position of the producer stays open and sales of the products takes place just before production has been terminated. Producers do not commit themselves to a specified market until the product is ready for sale. Price is the only coordination mechanism between chain segments and it provides the stimulus to adapt quantity and quality. High prices lead to higher production. This can lead to overproduction and consequently low prices, resulting in an adjustment of production to a lower level. Actually it is an inefficient mechanism, while in practice high production is lagging in time with high prices. In open markets the farmer is responsible for all farm decisions for example what type of breed, feed, and housing and the way this is financed.

Total vertical integration is the coordination form that has total control over production in terms of quantity, quality and timing of the product. One company owns all segments in the chain and management guidelines of the company determine the product stream within the chain. The producer (or farm manager) is compensated for his contribution in management skills and labour time.

The other organisational forms residing between the two extremes (marketing contracts, strategic alliances, formal co-operation, production contracts and quasi-vertical integration) all apply to a lesser or higher degree contracts. Before production has been terminated (or even started), producers oblige themselves to sell or hand over products to a buyer or owner in the next chain segment. The several chain segments are partly dependent and connected to each other. The degree of interaction between segments, the control and ownership over the chain segments differs between the different forms. In a quasi-vertical integration or a franchise model a pig production corporation owns, for example, the breeding, gestation and farrowing unit. The corporation has contracts with the two downstream stages of production (nursery and finishing) and retains ownership over the pigs until they are marketed. This form of contract production does not appeal to producers willing to stay independent and willing to be a self-employed entrepreneur. Forms like marketing contracts, in which agreements are made on product amount and product price and stating no interference of the contractor in farm management, might be more attractive to them. In this case, however, the contract only assures market access for the producer. New generation co-ops are often formal co-operations, but might be based also on (equity-based) strategic alliances or production contracts.

The trend towards increasing vertical coordination reflects, according to Cozzarin and Barry (1998), "the growing influence of consumers in controlling the agri-food agenda, the increasing marketing power of large food companies, and technological changes in information system development that necessitate coordination.". In this way the drivers for new generation co-operates do not differ from the ones for traditional co-ops. Additional theoretical explanations for conversions of vertical chain coordination within time can be found in economic literature. Reduction of transaction costs within the supply chain is often quoted as the most important one (e.g. Hayenga *et al.*, 1996; Martinez, 1999; Martinez, 2002; MacDonald *et al.*, 2004). Transaction costs include costs related to assure market access by producers and to guarantee supplies by processors and packers, searching for the best price, negotiations to obtain or pay this price, the enforcement of contracts, the transfer of information from one chain segment to another, and losses due to badly tuned transactions as a consequence of incomplete information within the chain. Hayenga *et al.* (1996) concluded, however, that transaction cost minimization is helpful but quite incomplete in explaining the coordination changes taking place in the US pig sector. Reduction of price and production risks amongst the chain, financing of production inputs to reduce capital investments by producers, advancement in technologies which can only economically be applied in large-scale production, and combining of knowledge and skills are other incentives for networking within the chain (Hayenga *et al.*, 1996; Cozzarin and Barry, 1998; Pollock and Spencer, 2004).

NEW GENERATION VERSUS TRADITIONAL CO-OPS

New generation co-ops, in the form of formal co-operations, share many of the attributes of traditional co-ops, however, the following make new generation co-ops different from the traditional ones (Gamble, 2002b):

- Membership is not open but limited to those who purchase delivery rights. They can be seen as a two-way contract between the producer (member) and the co-op and are tied to the level of investment. They oblige the member to deliver amount x of product y containing quality z to the co-op, and oblige the co-op to accept the delivery. This assures market access by the member and guarantees supply by the co-op. If the member can not meet the quantity and quality requirements from their own production, they must purchase the product elsewhere or the co-op will charge the member for the difference.
- Membership is restricted to a maximum and closed for new members when the targeted delivery rights are sold. New members are from that moment only allowed if existing members sell (part of) their delivery rights to them.
- New generation co-ops demand higher investments from members.

Advantages of new generation co-ops over traditional co-ops are therefore efficiency due to integration of production and processing reducing transaction costs between chain segments, and increased member commitment due to equity investments by members. Disadvantages of new generation co-ops over traditional co-ops are the requirements for high levels of expertise to run the co-op, higher levels of initial investments and higher levels of economical risk (Gamble, 2002b).

EXAMPLES

Because of the recent history of new generation co-operatives, not much research has been undertaken in analysis of these co-operatives in agriculture in general or in pig production specifically. Waner (2000) present results of an nation wide analysis of new generation co-operatives in the United States, including 117 surveys with half of the approached co-ops responding (n=60). Just under a quarter of the responses concerned livestock production, which was not further specified into species. On a scale of 1 (not important) to 5 (very important) capturing more of the added value of crops (4.9) and low commodity prices (4.6) were the dominant factors in formation of this type of co-op. Obstacles in the formation and the operation of the co-op concerned marketing of the product (3.8 on a scale of 1 insignificant to 5 significant), financial matters (borrowing funds from local financial institutions, score 3.5) and to attract enough members to participate from the beginning (3.4). Once the co-operative has proven its existence, it was no problem to replace members who (partly) quit and sold their rights. On a scale of 1 (not important) to 5 (important) success of the co-operative was mainly attributed to successful marketing of the product (4.8), accurate perception of the need for the product (4.6), and financial commitment of the members (4.6).

Conclusions

New generation co-operatives have emerged as a response to the transformation of agriculture that has been occurring over the last ten to fifteen years. Like traditional co-operatives, new generation co-operatives are playing an important role in keeping the agricultural sector and rural areas healthy, prosperous and efficient (Fulton, 2001). This chapter shows that sector structure (high degree of vertical coordination or not), the type of products produced (producing consumer ready products as in horticulture or producing products for further processing as in pig production) and the goal of the co-op (strengthen negotiating position in relation to purchase and sale prices, better gearing of the market through specialty products, or gaining higher returns on investment through expansion in processing) will determine whether a traditional co-op will be sufficient and a well-fitted organisation or whether a new generation co-op will be more suitable.

References

Bijman, W.J.J., and G.W.J. Hendrikse. (2001) *Opkomst van telersvereningingen in de Nederlandse voedingstuinbouw.* Maandblad voor Accountancy en Bedrijfseconomie, 6: 256-266.
Cozzarin, B.P., and P.J. Barry. (1998) *Organizational structure in agricultural production alliances.* International Food and Agribusiness Management Review, 1(2): 149-165.
Cozzarin, B.P., and R.E. Westgren. (2000) Rent sharing in multi-site hog production. *American Journal of Agricultural Economics*, 82: 25-37.
Fulton, M. (2001) New *Generation Co-operative Development in Canada.* Centre for the Study of Co-operatives, University of Saskatchewan, Canada. 31 pp.

Gamble, R.W. (2002a) *How to form a co-operative*. Ministry of Agriculture, Food and Rural Affairs, Agriculture and Rural Division, Factsheet Order No. 02-019.

Gamble, R.W. (2002b) *New Generation Co-operatives*. Ministry of Agriculture, Food and Rural Affairs, Agriculture and Rural Division, Factsheet Order No. 02-017.

Hayenga, M.L., Rhodes, V.J., Grimes, G.A., and J.D. Lawrence. (1996) *Vertical coordination in hog production*. US Department of Agriculture, Packers and Stockyards Programs, Grain Inspection, Packers and Stockyards Administration Research Report (GIPSA-RR) 96-5.

MacDonald, J, Perry, J, Ahearn, M, Banker, D, Chambers, W, Dimitri, C, Key, N, Nelson, K, and L. Southard. (2004) *Contracts, Markets and Prices: Organizing the production and use of agricultural commodities*. US Department of Agriculture, Economic Research Service, Agricultural Economic Report No. 837.

Martinez, S.W. (1999) *Vertical coordination in the pork and broiler industries : implications for pork and chicken production*. US Department of Agriculture, Economic Research Service, Food and Rural economics Division, Agricultural Economic Report No. 777.

Martinez, S.W. (2002) *Vertical coordination of marketing systems: lessons from the poultry, egg and pork industries*. US Department of Agriculture, Economic Research Service, Food and Rural economics Division, Agricultural Economic Report No. 807.

Pollock, G., and A. Spencer. (2004) *A pig production alliance system – suitable for smaller herds*. Department of Primary Industries and Fisheries, Queensland, Australia, www.dpi.gld.gov.au/pigs.

Van Horne, P., Kortstee, H., Oosterkamp, E., and T. Veldkamp. (2007) *Producentenorganisaties in de varkens- en pluimveehouderij*. LEI, Den Haag, in press.

Van der Kroon, S.M.A., Tacken, G.M.L., Van Uffelen, R.L.M., Van Paassen, R.A.F., Poot, E.H., and A.J. de Buck. (2002) *Producentenverenigingen in beeld*. LEI, Den Haag, Rapport 5.02.15.

Waner, J. (2000) *New* Generation Cooperatives and The Future of Agriculture: An Introduction. In: *New Generation Cooperatives: Case Study*. Illinois Institute for Rural Affairs, Macomb, IL, USA.

7

LIQUID FEEDING: A TECHNOLOGY THAT CAN DELIVER BENEFITS TO PRODUCER PROFITABILITY, PIG HEALTH & WELFARE, ENVIRONMENT, FOOD SAFETY AND MEAT QUALITY

B. P. GILL
Meat and Livestock Commission, Milton Keynes, UK

Introduction

As a feed processing, mixing and delivery technology, liquid feeding holds considerable potential for delivering benefits to many areas of competitive and strategic importance to the pig sector. Unlike dry diets, a liquid medium provides greater scope for processing feed ingredients using existing and emerging technologies to improve pig health and performance, reduce nutrient excretion, and enhance the safety and quality of pork and pork products. The application and development of these technologies are currently limited but could progress in parallel with consumer expectations and policy pressure for higher standards of food safety and quality, animal welfare and the environment. At present liquid feeding is mainly used by around 0.20 to 0.25 of the GB finishing sector, including a number of specialist operators who derive a competitive advantage from bulk inclusion of high moisture food industry co-products to pig diets. Liquid feeding is not widely used in diets for weaned piglets and sows due to cost and operational reasons and is likely to represent less than 0.05 of pig numbers. In this chapter, the aim is to consider the wider benefits that liquid feeding can bring to the pig industry, such as improved health & welfare, enhanced food safety and quality and reduced environmental impact as well as improved physical and financial performance. This is based on published literature, including results from a major programme of multidisciplinary research on liquid feeding finishing pigs co-ordinated by the UK Meat and Livestock Commission (MLC).

Pig performance

FINISHING PIGS

Liquid feeding probably has its origins in the finishing of pigs on co-products, such as whey and skim milk, from local dairy factories. This symbiosis enabled factories to dispose fresh whey without excessive transportation and at minimal cost to pig producers who in turn gained economic benefit from receiving large volumes of a high value feed ingredient at low cost.

With the drive to reduce cost of production, and the dairy processing sector recovering a greater value from its co-products, the list of high moisture co-products has extended to include those from the brewing, distillery, vegetable processing and confectionary industries. In contrast to producers who specialise in the use of co-products, there are producers who use liquid feeding merely as a delivery system for dry diets either milled and mixed on farm or purchased from the compound feed sector. For these producers the cost benefits of liquid over dry feeding are likely to be lower than for those who are able to use food industry co-products at a scale that maximises their economic benefits.

One of the first objectives in the MLC's research programme was to establish if there were any benefits, performance and financial, from adding water to a dry home milled-and-mixed meal diet and feeding it through a liquid feeding system over feeding the same diet *ad libitum* in pellet form. The trial (MLC, 2004), involving 1024 pigs, showed that *ad libitum* liquid feeding improved daily gain and feed conversion ratio (FCR) by 5.6 and 10.3% respectively, without adversely affecting carcase fatness, and giving an improved net return of 4.6p/kg dead weight (Table 7.1), allowing capital investment in a liquid feeding system to be recovered within 2.6 years (Table 7.2; MLC, 2006). These performance benefits are consistent with a review of the literature by Jensen and Mikkelsen (1998), who reported an average improvement of 4.4% and 6.9% in daily gain and FCR from liquid feeding growing and finishing pigs.

Table 7.1 PERFORMANCE AND COST COMPARISONS OF DRY AND LIQUID FEEDING GROWING AND FINISHING PIGS (MLC, 2004)

Feeding	*Dry*	*Liquid*	*s.e.d.*	*P[1]*
Entry weight (kg)	34.2	34.1	1.29	
Final weight (kg)	102.9	103.0	0.75	
Feed intake (kg/day)	1.85	1.75	0.021	***
Daily gain (g)	754	796	9.6	***
Feed conversion ratio	2.53	2.27	0.027	***
Carcase backfat depth (P_2, mm)	11.4	11.5	0.30	
Cost of production (p/kg dead weight)	99.2	94.6	-	-

*** $P < 0.001$

As mentioned, it is generally accepted that food industry liquid co-products can be more cost competitive sources of energy and amino acids for pigs than conventional ingredients such as cereals and soya bean meal, although they are price-competitive linked and both are subject to the economics of supply

Table 7.2 COST SAVINGS AND ESTIMATED PAYBACK PERIOD FOR LIQUID FEEDING (MLC, 2006)

Feeding	Dry	Liquid
Physical performance:		
Daily gain (g)[1]	754	796
Allowed days to gain 75 kg[2]	102	96
Days per batch (including wash down)	109	103
Pigs per year[3]	6697	7087
Capital investment:		
Feeding system	£5,300[4]	£64,420[5]
Cost of production[6] (p/kg dead weight)	99.2	94.6
Liquid feeding compared with dry:		
CoP saving per kg dead weight	4.6p	
CoP saving per pig	£3.54	
CoP saving per year	£25,088	
Payback period for liquid feeding system	2.6 years	

[1] Trial 1 Finishing Pigs Systems Research (MLC, 2004).
[2] From 30 kg to slaughter at 105 kg liveweight.
[3] Within a 2000 pig place unit.
[4] Dry feeding system with bin, centreless auger and feeders.
[5] Liquid feeding system with hammer mill, plus elevator and installation (£8,500), central processing unit of bins and augers for 3 cereals, 2 proteins and oil, with processing tank and controls installed in a new building (£42,300), tanks, pipeline and feeders (£13,620).
[6] Cost of production (CoP) calculations include weaners, feed, labour, power, water, straw, mortality, waste management and capital investment, and assumes the following based on current trade quotes: (1) Building costs of £193/m² for straw base and £227/m² for fully slatted housing; these are the same for dry and liquid feeding systems. Building capital costs are depreciated over 25 years at 6% interest, with repair/maintenance costs at 2%. (2) Capital cost of feeding equipment is depreciated over 20 years at 6% interest, with repair/maintenance costs at 4%.

and demand. Provided co-products are judiciously used, risks are spread by using 3 or more different types in the diet, and formulations take into account variability in composition, they can offer significant cost savings. In the MLC's research, a comparison of the results across trials showed that in diets where liquid co-products were used, a net saving of around 9p/kg deadweight could be achieved in cost of production. The benefits of liquid over dry feeding combined with co-product inclusion places liquid feeding at a competitive advantage of nearly 14p/kg dead weight over feeding dry pelleted diets.

WEANED PIGLETS

Liquid feeding holds the greatest potential for the newly-weaned piglet than any other class of pig, as the delivery of nutrients and water in a single package avoids the immediate need to establish separate drinking and feeding behaviours, and reduces the risk of food and water deprivation, a key factor associated with growth check and loss of enteric function and integrity after weaning (Kelly *et al.*, 1991; Pluske *et al.*, 1996). This is supported by studies conducted under controlled research comparing dry and liquid feeding with results showing significant and substantial improvements in intake and gain from liquid feeding (Table 7.3) and the protective effects of liquid feeding on intestinal integrity and absorptive capacity as measured by villus height and surface area (Deprez *et al.*, 1987; Hurst *et al.*, 2001). In a review of 10 studies (Jensen and Mikkelsen, 1998), the post-weaning growth advantage of liquid over dry feeding averaged 12.3%.

Table 7.3 FEED INTAKE AND GROWTH IN WEANED PIGLETS ON DRY OR LIQUID FEEDING

Study	Pigs per treatment	Weaning age (d)	Duration (weeks)	Intake (g/d) Dry	Liquid	Gain (g/d) Dry	Liquid
(Partridge *et al.*, 1992)	100	23	3	310	351	281	312
(Russell *et al.*, 1996)	24	23	4	443	807	343	428
(Russell *et al.*, 1996)	48	23	4	545	654	397	454
(Kim *et al.*, 2001)	60	11	2	292	369	257	397
(Kim *et al.*, 2001)	60	11	2	309	337	268	358
(Hurst *et al.*, 2001)	18	9.6 kg at weaning	4	434	675	425	585
(Han *et al.*, 2006)	24	21	3	445	510	334	367

Despite encouraging research results, liquid feeding is not widely used for weaned piglets on commercial farms. Although capital investment is a potential barrier to uptake, the risks attached to liquid feeding weaned piglets are considered to outweigh the benefits, possibly based on the experiences of individual producers who have reported problems with feed hygiene, undesirable or over fermentation, feed refusal, feed wastage, loss of health and performance, technical complexity and reliability and excessive labour input.

Manufacturers of liquid feeding systems are addressing these problems and there are examples where design, innovation and the introduction of new technology has led to systems delivering the benefits of liquid feeding weaned piglets found under controlled research. The success of these systems is based on the following key features:

- Feed freshness and hygiene is maintained by mixing little and often.
- Piglets can be fed *ad libitum* without deterioration in feed freshness.
- High feed turnover limits the development of natural fermentation.
- The system can mix and deliver liquid feed of a high dry matter content (280 to 300 g/kg) to meet requirements at a time when appetite and volumetric capacity may limit daily gain.
- Feed is delivered noise-free to troughs, avoiding stress and disturbance from the operation of pumps and valves.
- Trough design reduces feed wastage and improves pen hygiene.

The installation of a system with these innovations has resulted in considerable performance benefits on one commercial farm (personal communication: D J & J Witherick & Sons), with gain 150g/d above the national weaner herd average of 449g/d (MLC, 2005d).

SOWS

There is an absence of published scientific literature on the effects of liquid feeding on sow performance. Most of the information, such as in trade journals and conference presentations, is based on field observations, comments and anecdotal evidence.

One potential application of liquid feeding is to improve the rebreeding potential of weaned sows by reducing their weight loss in lactation through increased feed intake. The importance of nutritional management during lactation is often overlooked as shortfalls can have a profound effect on the ability of the sow to return to oestrus after weaning. Excessive maternal weight loss during lactation, resulting from inadequate feed intake and nutrient supply, is associated with an extended weaning to first oestrus or service interval, and a reduction in the size of the subsequent litter, particularly in first litter sows (Eissen *et al.*, 2003; King and Dunkin, 1986; King and Williams, 1984; Kirkwood *et al.*, 1987; Koketsu *et al.*, 1996). A small-scale study (Demeèková *et al.*, 2003) showed that liquid compared with dry feeding resulted in a significant and progressive increase in feed intake over 3 weeks after farrowing in first litter gilts (Figure 7.1). Although subsequent reproductive performance was not reported, this study demonstrates the principle of using liquid feeding to protect maternal body mass in lactation by increasing intake with potential benefits to rebreeding and litter productivity. The benefits of liquid feeding could be greater under higher temperatures, such as in summer and hotter climates, where the addition of cool or chilled water to feed could reduce heat stress and increase intake, provided that the system is designed and managed to limit mal-fermentation.

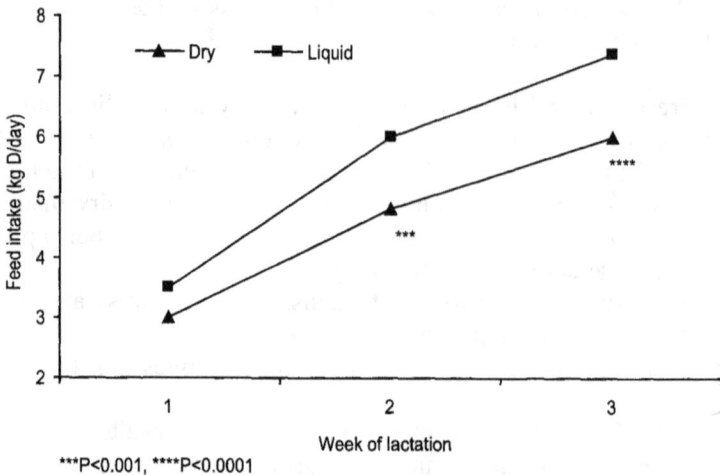

Figure 7.1 Feed intake in first litter lactating sows on dry and liquid feed (Demeèková *et al.*, 2003)

The development and commercial use of trough sensor technology could bring future advantages to the use of liquid feeding lactating sows as the delivery of feed into each trough is triggered by the appetite demands of the animal. This could avoid manual overloading of troughs, a major cause of appetite loss in early lactation due to the presence of rejected and staled feed.

Health and welfare

There are an increasing number of studies adding to our understanding of the positive impact of liquid feeding on the health and welfare of pigs.

In growing and finishing pigs, liquid compared with dry pellet feeding shifted the distribution of gastric ulcer scores at slaughter in favour of lower scores (Table 7.4) on an increasing severity scale of 0 to 5, resulting in a significant reduction in mean score (Scott *et al.*, 2007). Ileal, caecal and colon digesta samples from liquid fed pigs had reduced coliform counts resulting in significantly higher lactic acid to coliform ratios (Figure 7.2) as an indicator of improved gut health status (Hillman *et al.*, 2004; MLC, 2004).

In the newly-weaned piglet, as a result of higher intake, liquid feeding has been found to protect gut integrity and absorptive capacity as shown by measurements of villus height, crypt depth and surface area (Figures 7.3 and Table 7.5; Deprez *et al.*, 1987; Hurst *et al.*, 2001; Yang *et al.*, 2001). Deprez *et al.* (1987) found that differences in small intestinal morphology between dry and liquid fed weaned piglets were independent of faecal shedding of haemolytic *E. coli*, though field observations suggest significantly fewer cases of post-weaning *E. coli* enterotoxaemia on farms using liquid feeding in comparison to farms feeding dry diets.

Table 7.4 FREQUENCY OF GASTRIC LESIONS (PROPORTION OF PIGS) AND MEAN GASTRIC LESION SCORES AT SLAUGHTER IN FINISHED PIGS FED DRY AND LIQUID FEED AND HOUSED ON FULLY SLATTED AND STRAW-BASED FLOORING SYSTEMS (SCOTT *et al.*, 2007)

| Housing | Fully slatted | | Straw-based | |
| Feeding | Liquid | Dry | Liquid | Dry |
Lesion score				
0	0.188	0.022	0.417	0.044
1	0.312	0.044	0.375	0.244
2	0.250	0.267	0.104	0.156
3	0.146	0.244	0.104	0.222
4	0.104	0.200	0	0.111
5	0	0.222	0	0.222
Mean	0.017[b]	0.032[a]	0.009[c]	0.028[a]

[a,b,c,d] different superscripts in the same row are significantly different (P<0.001)

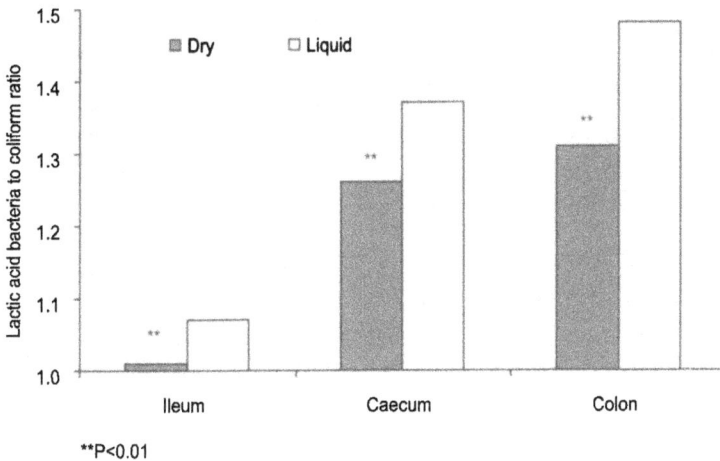

Figure 7.2 Lactic acid bacteria to coliform ratios at slaughter in digesta samples from finished pig (MLC, 2004)

There has been limited research on the effects of liquid feeding on sow and litter health, with one study (Demeèková *et al.*, 2002) showing that a partially sterilised liquid diet fermented with *Lactobacillus plantarum*, significantly decreased faecal coliform shedding in lactating sows. Piglets nursed by these sows and also those fed the partially sterilised non-fermented liquid diet had increased faecal loading of lactic acid bacteria and reduced faecal coliform counts compared with piglets nursed by dry fed sows (Figure 7.4). Colostrum from sows fed the partially sterilised and *L. plantarum* fermented liquid diets

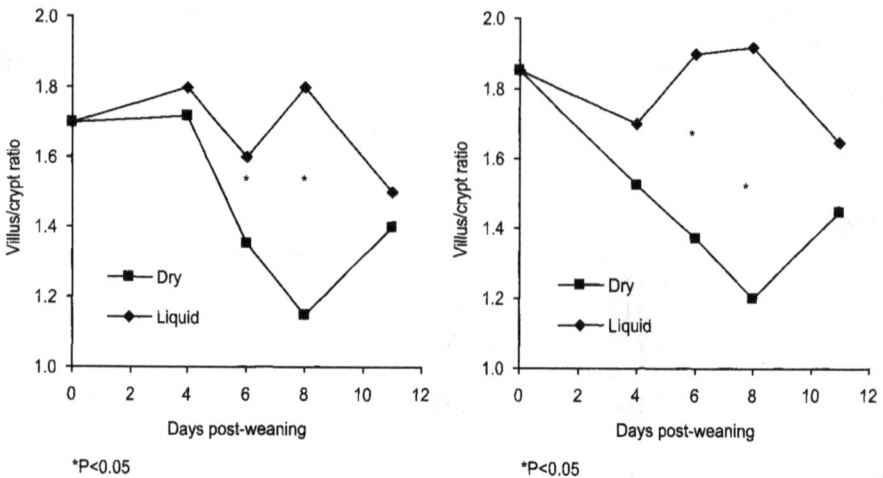

Figure 7.3 Villus height/crypt depth ratio in the ileum (a) and distal jejunum (b) of weaned piglets on dry and liquid feeding (Deprez *et al.*, 1987)

Table 7.5 EFFECTS OF DRY AND LIQUID FEEDING WEANED PIGLETS ON VILLUS AND MICROVILLUS HEIGHT AND MICROVILLUS SURFACE AREA 28 DAYS AFTER WEANING (HURST *et al.*, 2001)

Distance (proportion) along small intestine	*Villus height (μm)*		*Microvillus height (μm)*		*Microvillus surface area (μm²)*		*P[1]*
	Dry	*Liquid*	*Dry*	*Liquid*	*Dry*	*Liquid*	
0.02	340	414	0.66	0.98	0.220	0.334	***
0.25	294	368	0.89	1.11	0.331	0.411	***
0.50	340	362	1.26	1.23	0.470	0.481	
0.75	278	350	1.10	1.37	0.371	0.502	***

[1]Effect of diet: *** $P<0.001$

was found to exhibit greater mitogenic activity on isolated intestinal epithelial cells. Although the mechanisms regulating this process were unclear, the results from the study suggest that intestinal growth and function may be enhanced in piglets from sows fed liquid diets.

Liquid feeding may offer some welfare advantages over dry feeding, for example within the MLC research programme, Scott *et al.* (2007) observed increased levels of restful behaviour and reduced oral activity directed at pig and pen parts in growing and finishing pigs fed liquid diets (Table 7.6). This was attributed to increased gut fill and improved sense of satiety from intake of large volumes of liquid feed, though the presence of 0.3% ethanol as an end product of natural fermentation may have had an additional calming influence.

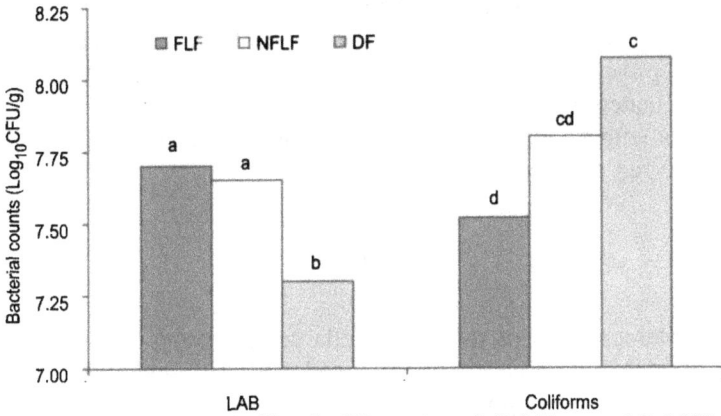

Means with different superscript are significantly different (a vs. b P<0.01; c vs. d P<0.001)

Figure 7.4 Lactic acid bacteria (LAB) and coliforms in the faeces of sows fed fermented liquid feed (FLF), non-fermented liquid feed (NFL) and dry feed (DF), (Demeèková *et al.*, 2002).

Table 7.6 MEAN PROPORTOIN OF OBSERVATIONS WHERE DRY AND LIQUID FED PIGS HOUSED ON FULLY SLATTED AND STRAW-BASED FLOORING WERE RECORDED PERFORMING VARIOUS BEHAVIOURS (SCOTT *et al.*, 2007)

Housing (H)	*Fully slatted*		*Straw-based*		*P*	
Feeding (F)	*Liquid*	*Dry*	*Liquid*	*Dry*	*H*	*F*
Lesion score						
Standing	0.194	0.257	0.276	0.312	***	*
'Sleeping'	0.650	0.564	0.523	0.486	***	**
Eating	0.039	0.044	0.046	0.050		
Drinking	0.004	0.012	0.007	0.012		***
Investigating	0.152	0.219	0.276	0.309	***	*
Oral behaviour towards:						
Straw	-	-	0.143	0.140		
Pen parts	0.069	0.106	0.057	0.076	*	***
Another pig	0.070	0.100	0.079	0.088		**

* P<0.05, ** P <0.01, *** P<0.001. There were no significant feeding by housing interactions.

The beneficial effects of liquid feeding in promoting restful behaviour have also been reported in field observations on dry sows by Baynes *et al.* (1999) who found the majority (0.70) to be settled following one hour after feeding. In contrast, in Sweden where the use of liquid feeding group housed dry sows in long troughs has become very common, high levels of aggression have been reported during the delivery of restricted amounts of feed (Olsson, 1997). The system was considered undesirable for sow welfare due to a number of inadequate design and operational features such as lack of control over feed delivery, under feeding, uneven distribution of feed in troughs due to poor flow, lack of physical barriers to separate sows during feeding, dunging in troughs, and inadequate demarcation of dunging and lying areas. In this case weakness rests with a lack of awareness over welfare needs rather than whether the sows were fed a liquid or dry diet.

Food safety

Since the implementation of the Salmonella control programme in Denmark in 1995, the Danish pig industry has focused on a risk management approach from feed-to-food, including work by the National Committee for Pig Production in identifying on-farm factors which could increase the risk of Salmonella carriage in finished pigs. Research on feed concluded that whilst ingredients were themselves not a source of Salmonella, feed formulation, feed processing and feed form influenced the percentage of pigs tested positive for Salmonella at slaughter (Table 7.7; Danskeslagterier, 1999). Subsequent work has led to the development of a Salmonella control diet containing coarsely ground barely, 100g sugar beet pulp/kg and a mixture of 10g lactic acid and 10g formic acid/kg (Danskeslagterier, 2003).

Table 7.7 EFFECT OF FEED FORM AND PROCESSING ON SALMONELLA IN FINISHER PIGS (DANSKESLAGTERIER, 1999)

	Fine pellets	*Coarse pellets*	*Cold-pressed pellets*	*Expanded feed*	*Meal*	*Partially heat-treated feed*
Salmonella positive (proportion)	0.129	0.056	0.086	0.046	0.028	0.046
Production value						
Gross Margin per pen place year (DKK)	379	315	318	241	253	270
Index	100	83	84	63	67	71

These studies indicate that the dietary mechanism controlling Salmonella centres on the delivery of acids to the gut through enhanced microbial fermentation or by direct supplementation of the diet with organic acids. This is parallel to the mechanism of using liquid feeding to control Salmonella and enteropathogens by dietary acidification through controlled fermentation using selected strains of lactic acid bacteria or by natural fermentation before feeding (e.g. Boesen *et al.*, 2004; Canibe and Jensen, 2003; Hojberg *et al.*, 2003; Jensen and Mikkelsen,

1998) . This has been demonstrated by the addition of lactic and acetic acids to sterilised liquid diets in molar concentrations equivalent to those generated under controlled fermentation using *L. plantarum* , with both diets achieving similar levels of *S. typhimurium* inhibition (Figure 7.5; van Winsen *et al.*, 2000).

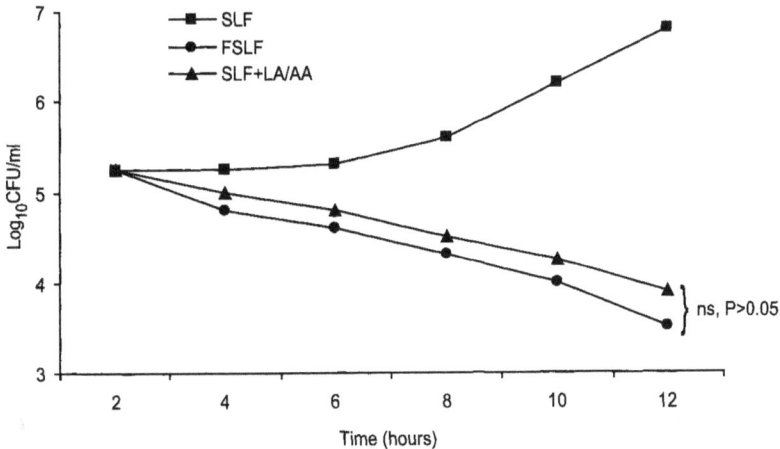

Figure 7.5 Survival of *S. typhimurium* in heat sterilised liquid feed (SLF), fermented SLF (FSLF) and SLF supplemented with 200 mmol/litre of lactic acid and 30 mmol/litre of acetic acid (SLF + LA/AA) (van Winsen *et al.*, 2000)

Under farm conditions most, if not all, liquid diets will have some background natural fermentation to a variable degree and analysis of 76 samples over the lifespan of the MLC's research programme were found to contain 118 (s.e.m. 6.9) and 38 (s.e.m. 3.8) mmol/litre of lactic and acetic acid respectively with a pH mean of 4.83 (s.e.m. 0.040) measured in 145 samples. These compare with169 and 250 mmol/litre for lactic acid and 26 and 25 mmol/litre for acetic acid and 4.36 and 3.8 for pH reported in experimentally fermented liquid diets by Canibe and Jensen (2003) and van Winsen *et al.* (2000) respectively. van Winsen *et al.* (2000) concluded that Salmonella could be controlled by the acidification of liquid feed to a pH of less than 4.5 by the addition of either lactic or acetic acid at minimum levels of 150 and 80 mmol/litre respectively, and this was as effective as feed fermentation. A target of 100 to 150 mmol/ litre of lactic acid has been recommended by Brooks *et al.* (2002) to prevent the proliferation of Salmonella in fermented feed.

In the MLC's research (MLC, 2005b), fermentation of the cereal fraction of liquid feed using *Pediococcus acidlactici* did not result in a major difference in the number of finished pigs tested positive for Salmonella at slaughter (Table 7.8), though the level of carriage was generally low in this study to establish a statistical effect. Feeds with and without fermented cereals had high lactic acid bacteria (LAB) counts and were similar in pH and in lactic and acetic acid content, suggesting that natural fermentation can generate favourable conditions for Salmonella control in the absence of fermenting with a selected strain of bacteria.

Table 7.8 SALMONELLA AT SLAUGHTER IN PIGS FED LIQUID DIETS WITH AND WITHOUT FERMENTED CEREALS AND LACTIC ACID BACTERIA (LAB) COUNTS, LACTIC AND ACETIC ACID CONTENT OF THE DIET (MLC, 2005b)

	Liquid diet (- fermented cereals)	Liquid diet (+ fermented cereals)
Salmonella positive (proportion)		
ELISA	0.06	0.03
Caecal carriage	0.03	0.01
LAB in liquid feed (Log_{10} CFU/g)	9.17	9.31
Lactic acid in liquid feed (mmol/litre)	84	98
Acetic acid in liquid feed (mmol/litre)	36	27

The potential benefits of liquid feeding to food safety were initially recognised in Denmark from the results of an epidemiological study (Dahl, 1997) that showed a lower prevalence of Salmonella infections in herds using liquid feeding compared with herds feeding dry diets. Salmonella seroprevalence in bloods samples collected randomly from finished pigs slaughtered in Dutch commercial abattoirs was lower in pigs derived from herds using liquid feed (van der Woolf *et al.*, 2001). In a survey of Canadian farms using either liquid (21) or dry (61) feeding for growing and finishing pigs (Farzan *et al.*, 2006), at least one pig tested sero-positive for Salmonella antibodies on 0.98 of the dry feeding farms and 0.84 of the liquid feeding farms. Salmonella was isolated from 25 out of 420 faecal samples (0.06) from dry fed pigs compared with 3 out of 400 (0.008) samples from liquid fed pigs. At farm level, 0.38 of the dry feeding farms tested positive for Salmonella compared with 0.15 of the liquid feeding farms.

Salmonella reduction in finishing pigs by liquid feeding has been confirmed by the results of the MLC's study (MLC, 2004) comparing dry pellet and liquid feeding in straw based and fully-slatted housing (Figure 7.6). Evidence of natural fermentation in the liquid feed included the presence of a high lactic acid bacteria count (9.2 Log_{10} CFU), a low pH (4.9) and lactic (147 mmol/litre) and acetic (38 mmol/litre) acid production.

In liquid diets, protein content has been identified as an additional risk factor for Salmonella carriage in finished pigs (MLC, 2005c). A reduction in crude content from 213g to 166 g/kg dry matter of liquid diets formulated to meet nutrient standards (BSAS, 2003) using synthetic amino acids resulted in a significant reduction in the proportion of pigs tested positive for caecal carriage at slaughter (0.056 vs. 0.017).

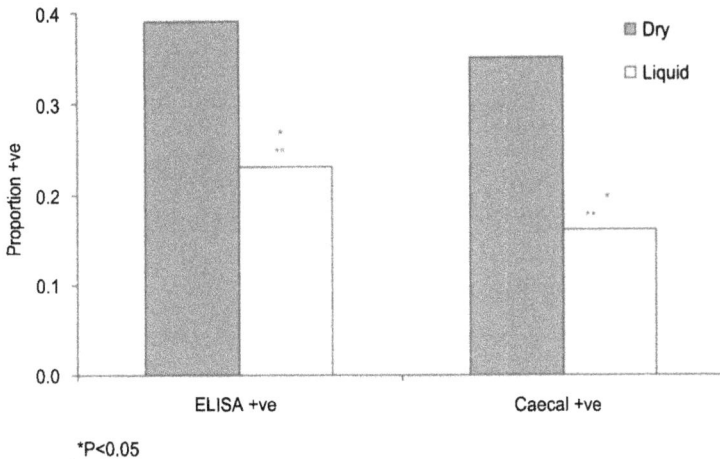

Figure 7.6 Proportion of slaughtered pigs tested positive for Salmonella fed dry and liquid feed during growing and finishing (MLC, 2004)

It is worth mentioning that many of the dietary modifications developed to control Salmonella under dry feeding have been found to incur a cost penalty (see for example Table 7.7) due to increased feed costs and a loss in pig growth performance. In the case of liquid feeding there are no conflicts between Salmonella control, growth performance and cost saving, which are all improved.

Meat quality

The initial question concerning meat quality is to establish whether liquid feeding has any undesirable effects given that fermentation may present a risk in the production of compounds such as skatole, a principle component of boar taint. In the MLC's study there was no evidence of a loss in the eating quality of meat from pigs fed liquid compared with dry pellet feed, including skatole and indole content (Table 7.9; MLC, 2004). Similarly Andersson *et al.* (1997) found no differences in fat skatole content of liquid and dry pigs, though the addition of whey to liquid diets significantly reduced skatole levels. The beneficial effects of including whey in reducing skatole levels is probably linked to the presence of lactose as a dietary substrate for gut fermentation which has also been observed by the addition of other fermentable ingredients, such as sugar beet pulp and chicory roots to dry diets (Gill *et al.*, 1993; Hansen *et al.*, 2006).

Table 7.9 MEAT QUALITY IN PIGS FED DRY AND LIQUID FEED (MLC, 2004)

	Dry	*Liquid*
Fresh Meat Quality		
Drip loss (%)	6.31	5.79
Taints (ppm)		
Indole	0.028	0.034
Skatole	0.049	0.056
Cooking loss (%)	22.89	22.93
Eating quality		
Lean		
Juiciness	14.51	15.01
Tenderness	15.77	15.76
Pork flavour	13.14	13.34
Abnormal flavour	3.19	2.73
Boar flavour	1.76	2.10
Overall acceptability	13.57	13.63
Fat		
Pork flavour	11.39	11.98
Abnormal flavour	2.00	1.93
Boar flavour	1.40	1.46
Pork odour	10.11	10.22
Abnormal odour	3.07	2.81
Androstenone odour	1.68	1.64
Skatole odour	1.38	1.42

The effects of feeding fermented liquid diets on meat quality have been examined by Hansen *et al.* (2000) and MLC (2005b), though the approaches differed with MLC using *Pediococcus acidilactici* to pre-ferment the cereal fraction of the liquid diet at 35°C for 24h and the Danish method based on accelerating natural fermentation by holding the complete diet at 30°C for 8h then mixing with an equal quantity of fresh feed before feeding. The MLC treatment comparison was liquid feed with and without fermented cereals and the Danish study compared fermented liquid with non-pelleted dry feed. In both studies fermented liquid feed did not elevate fat skatole and indole levels, though Hansen *et al.* (2000) reported a deterioration in sensory scoring of cooked loins for pig, rancid and off-flavours, whereas MLC (2005b) observed no undesirable effects on sensory flavour and odour attributes of cooked loin and fat samples. A previous study in Denmark, Jensen *et al.*(1997) found a significant reduction in fat skatole content of pigs fed fermented liquid feed. These inconsistencies indicate that fermentation is not a uniform or standard process and differences in method, microbial diversity, fermentation pattern, end products and end product content may have complex and diverse effects on odour and flavour attributes of pig meat.

An interesting finding in the MLC study was that tenderness and juiciness of cooked loin from pigs fed the diet containing fermented cereals was significantly improved. The underlying mechanism for this is unknown, though improved loin tenderness and juiciness in response to the addition of sugar beet pulp to dry diets reported by Gill *et al.* (1993) points to a role involving end products of microbial fermentation.

The effect of reducing protein content (213g to 166g/kg dry matter) in liquid diets on meat quality has been investigated by MLC (2005c) and the results showed no differences in fat skatole content and sensory scoring of meat for juiciness, tenderness, flavour and odour.

Environment

By far the greatest, though often overlooked, contribution to the environment made by liquid feeding is the recycling of high moisture food industry co-products within the food chain. Materials such as whey, yogurt, bread, distillers grains and solubles, steamed potato peelings and confectionary products are readily accepted by growing / finishing pigs and sows, with an estimated total of 1.3 m tonnes recycled annually through the UK pig industry. It is worth noting that the practice of swill feeding catering waste and materials that contain, have been or are at risk of contact with meat and meat by-products has been banned in the UK since 2001 following the outbreak of foot-and-mouth disease.

As liquid feeding improves the conversion of feed to gain, a simple mass balance calculation based on the results presented in Table 7.1 would indicate that liquid feeding offers a saving of between 0.5 and 0.6 kg/pig in N excretion over dry feeding. For a growing and finishing unit with 2000 pig places, this represents a total reduction of around 4 tonnes/year in N emissions.

One approach to reducing N emission by growing and finishing pigs, that has not been widely tested under liquid feeding, is the strategic use of crystalline amino acids (AA) in the formulation of low protein diets. The success of this strategy has been demonstrated in dry diets (Verstegen and Jongbloed, 2003), but in liquid diets a loss in growth performance may be associated with microbial degradation of supplementary crystalline AA during controlled or natural fermentation as reported from field studies (Pedersen *et al.*, 2002a; Pedersen *et al.*, 2002b). In the MLC's research the use of AA to reduce crude protein content from 213 to 166 g/kg dry matter in liquid diets formulated to meet recommended nutrient standards (BSAS, 2003), reduced effluent ammonium and total N and ammonia emissions from growing and finishing pigs (Table 7.10; MLC, 2005c). A simple mass balance calculation suggests a reduction approaching 1 kg/pig of total N emission to the environment using this strategy, which factors up to nearly 8 tonnes per year for a 2000 place growing and finishing unit. However, the environmental benefits may have to be balanced against a loss in daily gain and FCR found in the MLC study and this could not be explained by AA loss as feed sample analyses showed that they were present inline with expected formulation targets (Table 7.11), despite a high level of microbial activity generating lactic and acetic acid and ethanol as end products of natural fermentation (Table 7.12).

Table 7.10 EFFECTS OF DIETARY CRUDE PROTEIN REDUCTION ON EFFLUENT N CONTENT AND AMMONIA EMISSIONS FROM PIGS FED LIQUID FEED (MLC, 2005c)

	Control diet	Low protein diet	P
Effluent composition			
Ammoniacal nitrogen (mg NH_4-N/kg)	4150	3530	*
Total nitrogen (mg N/kg)	5690	5210	*
Ammonia emission (g NH3-N per lu hour)	1.11	0.73	***

$^*P<0.05$, ** P <0.01, *** P<0.001

Table 7.11 ANALYSED AND EXPECTED AMINO ACID CONTENT OF CONTROL AND LOW PROTEIN LIQUID DIETS (MLC, 2005c)

	Control diet		Low protein diet	
Total amino acids (g/kg)	Actual	Expected	Actual	Expected
Lysine	10.11	9.90	9.30	9.60
Methionine	2.88	3.20	2.72	3.00
Threonine	7.60	7.30	6.57	6.50
Tryptophan	2.71	2.60	2.19	2.10
Free amino acids (g/kg)				
Lysine	0.24	0	3.51	3.56
Methionine	0.17	0	0.61	0.50
Threonine	0.20	0	1.51	1.44

Table 7.12 MICROBIAL COUNTS, pH AND ETHANOL, LACTIC AND ACETIC ACID CONTENT OF CONTROL AND REDUCED CRUDE PROTEIN LIQUID DIETS (MLC, 2005c)

	Control diet	Low protein diet
Lactic acid bacteria (Log_{10} CFU/g)	8.69	8.04
Yeast (Log_{10} CFU/g)	5.65	5.86
Enterobacteraciae (Log_{10} CFU/g)	3.68	3.18
pH	5.24	5.00
Ethanol (mmol/litre)	53	55
Lactic Acid (mmol/litre)	144	130
Acetic Acid (mmol/litre)	44	37

Another environmental benefit, which could be derived from liquid feeding, is the release of phosphorous from phytate hydrolysis by the activation of endogenous phytases in cereals during steeping. Increasing the availability of phosphorous in ingredients which lack phytases such as soyabean and rapeseed meal by co-steeping with milled wheat and wheatfeed provides a scope for reducing dietary supplementation with inorganic phosphates and reductions in phosphorous emissions. *In vitro* studies at the University of Plymouth have shown that it is possible to release over 0.90 of the phosphorous in wheat and in a mixture containing 600g milled wheat, 300g wheatfeed and 100g soyabean meal / kg by steeping at 50°C for 7 hours (Figure 7.7; MLC, 2005a).

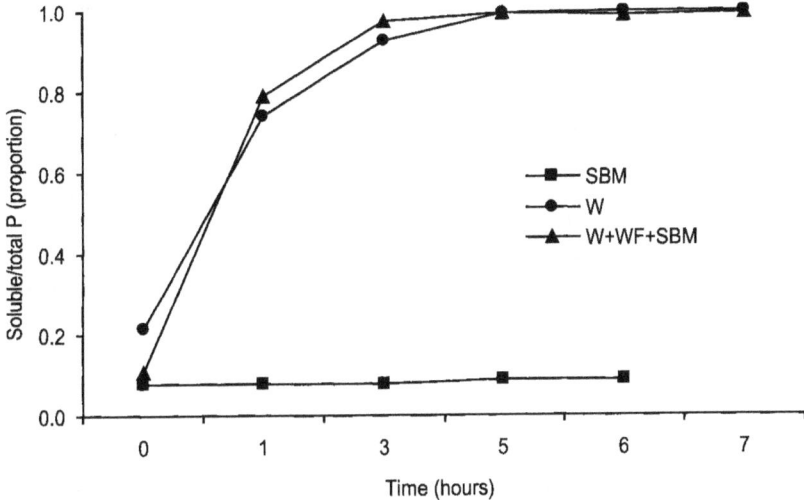

Figure 7.7 Release of soluble P as a proportion of total P from soyabean meal (SBM) and wheat (W) steeped alone or co-steeped with wheatfeed (WF) (MLC, 2005a)

Work on particle size showed that hammer milling the cereal fraction of a liquid diet using screen sizes of 5, 4, 3 and 2 mm and double milling through a 2 mm screen did not improve dry matter, energy and protein digestibility in growing pigs (Table 7.13; Thompson *et al.*, 2004). These results suggest that with liquid feeding, milling energy can be saved by the use of a larger screen size without adversely affecting pig performance, unlike dry diets where an increase in particle size results in a loss of digestibility and feed conversion (Wondra *et al.*, 1995).

An area yet to be fully explored is the use of processing, enzymes and ingredients with unique physico-chemical properties to improve the storage, handling, mixing and pumping characteristics of ingredients, co-products and complete diets in liquid feeding. The rapid sedimentation of dry meal particles in mixing tanks, pipes and troughs is a particular problem in liquid feeding

Table 7.13 COEFFICIENT OF TOTAL TRACT APPARENT DIGESTIBILITY IN GROWING PIGS OF A LIQUID DIET CONTAINING CEREALS MILLED THROUGH DIFFERENT HAMMER MILL SCREEN SIZES (THOMPSON *et al.*, 2004)

| | Hammer mill screen size (mm) | | | | | |
	5	4	3	2	(2 x 2)	P
Dry matter	0.825[a]	0.815[abc]	0.804[c]	0.811[bc]	0.821[ab]	**
Gross energy	0.812[a]	0.802[b]	0.790[d]	0.775[cd]	0.808[ab]	*
Crude protein	0.853[a]	0.826[b]	0.818[b]	0.838[ab]	0.856[a]	*
Oil (A)	0.753	0.720	0.743	0.782	0.785	

[a,b,c,d] different superscripts in the same row are significantly different

systems as this causes blockages, feed rejection and waste and requires frequent stirring and recirculation to maintain homogeneity. Energy savings and the rheology of feeds could be improved for example by the use of co-products such as distillers solubles with good particle suspension and slip properties and enzymes to reduce viscosity in gelatinised food industry co-products such as steamed potato peelings.

Conclusion

At present liquid feeding primarily offers an opportunity for pig producers to reduce costs resulting from improved performance in growing and finishing pigs and the use of competitively purchased food industry co-products. The potential benefits are however much wider and extend to pig health and welfare, food safety, meat quality and the environment. These benefits will eventually materialise with the shift from dry to liquid feeding and developments in liquid feeding technology driven by cost competitiveness and societal demands for continually higher standards of animal health and welfare, food safety and quality and the environment.

Acknowledgements

The Finishing Systems Research Programme was funded by the UK Defra, MLC/BPEX.

The UK research participants were:

Acorn House Veterinary Surgery
Liquid Feeders Research Group
Meat and Livestock Commission

Scottish Agricultural College
Silsoe Research Institute
University of Newcastle upon Tyne
University of Nottingham
University of Plymouth
Veterinary Laboratories Agency

References

Andersson, K., Schaub, A., Andersson, K., Lundstrom, K., Thomke, S. and Hansson, I. (1997). The effects of feeding system, lysine level and gilt contact on performance, skatole levels and economy of entire male pigs. *Livestock Production Science* **51**: 131-140.

Baynes, P. J., J, H. E., Guise, H. J. and Penny, R. H. C. (1999). The effect of liquid feeding in a deep straw system on sow behaviour and welfare. In *Proceedings of the British Society of Animal Science Annual Meeting*, Scarborough, Paper 184.

Boesen, H. T., Jensen, T. K., Schmidt, A. S., Jensen, B. B., Jensen, S. M. and Moller, K. (2004). The influence of diet on *Lawsonia intracellularis* colonization in pigs upon experimental challenge. *Veterinary Microbiology* **103**: 35-45.

Brooks, P. H., Beal, J. D., Niven, S. J. and Campbell, A. (2002). *Fermented liquid feed for pigs: Potential for improving productivity and reducing environmental impact.* A report produced for Defra and the MLC, Project Number LS0812, University of Plymouth.

BSAS. (2003). *Nutrient Requirement Standards for Pigs.* p 28. BSAS, Midlothian, UK.

Canibe, N. and Jensen, B. B. (2003). Fermented and non-fermented liquid feed to growing pigs: Effect on aspects of gastrointestinal ecology and growth performance. *Journal of Animal Science* **81**: 2019-2031.

Dahl, J. (1997). Feed-related risk factors for subclinical salmonella infections. *Veterinary Information*: 17-20.

Danskeslagterier. (1999). *The National Committee for Pig production, Annual report 1999.* The Danish Bacon & Meat Council, Copenhagen.

Danskeslagterier. (2003). *The National Committee for Pig Production Annual Report 2003.* The Danish Bacon & Meat Council, Copenhagen.

Demeèková, V., Kelly, D., Coutts, A. G. P., Brooks, P. H. and Campbell, A. (2002). The effect of fermented liquid feeding on the faecal microbiology and colostrum quality of farrowing sows. *International Journl of Food Microbiology* **79**: 85-97.

Demeèková, V., Tsourgiannis, C. A. T. and Brooks, P. H. (2003). Effect on average daily feed intake during lactation and piglet growth during the first 2 weeks of life of feeding fermented liquid feed, non-fermented liquid feed or dry feed. In *Proceedings of the British Society of Animal Science Annual Meeting*, York,   p70.

Deprez, P., Deroose, P., Hende, V. d., Muylle, E. and Oyaert, W. (1987). Liquid versus dry feeding in weaned piglets: The influence on small intestinal morphology. *Journal of Veterinary Medicine* **34**: 254-259.

Eissen, J. J., Apeldoorn, E. J., Kanis, E., Verstegen, M. W. A. and Greef, K. H. d. (2003). The importance of a high feed intake during lactation of primiparous sows nursing large litters. *Journal of Animal Science* **81**: 594-603.

Farzan, A., Friendship, R. M., Dewey, C. E., Warriner, K., Poppe, C. and Klotins, K. (2006). Prevalence of *samonella spp.* on Canadian pig farms using liquid or dry-feeding. *Preventive Veterinary Medicine* **73**: 241-254.

Gill, B. P., Hardy, B., Perrott, J. G., Wood, J. D. and Hamilton, M. (1993). The effect of dietary fibre on the meat eating and fat quality of finishing pigs fed *ad libitum. Animal Production* **56(3)**: 421.

Han, Y. K., Thacker, P. A. and Yang, J. S. (2006). Effects of the duration of liquid feeding on performance and nutrient digestibility in weaned pigs. *Asian-Australian Journal of Animal Science* **19**: 396-401.

Hansen, L. L., Mejer, H., Thamsborg, S. M., Byrne, D. V., Roepstorff, A., Karlsson, A. and Hansen-Moller, J. (2006). Influence of chicory roots (cichorium intybus l) on boar taint in entire male and female pigs. *Animal Science* **82**: 359-368.

Hansen, L. L., Mikkelsen, L. L., Agerhem, H., Laue, A., Jensen, M. T. and Jensen, B. B. (2000). Effect of fermented liquid food and zinc bacitracin on microbial metabolism in the gut and sensoric profile of *M. Longissimus dorsi* from entire male and female pigs. *Animal Science* **71**: 65-80.

Hillman, K., Hunt, B., Davies, R. and Gill, B. P. (2004). The microbial status of the pig and its environment under different housing and feeding systems: 1. Liquid versus dry feeding in fully slatted and straw-bedded housing. In *Proceedings of the British Society of Animal Science Annual Meeting*, York, March 2004,   p 44.

Hojberg, O., Canibe, N., Knudsen, B. and Jensen, B. B. (2003). Potential rates of fermentation in digesta from the gastrointestinal tract of pigs: Effect of feeding fermented liquid feed. *Applied and Environmental Microbiology* **69**: 408-418.

Hurst, D., Lean, I. J. and Hall, A. D. (2001). The effects of liquid feed on the small intestine mucosa and performance of piglets at 28 days postweaning. In *Proceedings of the British Society of Animal Science Annual Meeting*, Scarborough, March 2001,   paper 162.

Jensen, B. B. and Mikkelsen, L. L. (1998). Feeding liquid diets to pigs. In: *Recent Advances in Animal Nutrition*   (ed. P. C. Garnsworthy and J. Wiseman), pp 107-126. Nottingham University Press, Loughborough, UK.

Jensen, M. T., Jensen, B. B., Agergaard, N., Hansen, L. L., Mikkelsen, L. L.

and Laue, A. (1997). Effect of liquid feed on microbial production of skatole in the hind gut, skatole absorption to portal blood and skatole deposition in backfat. In *Proceedings of the EAAP Working Group: Production and utilisation of entire male pig, boar taint in entire male pigs*, Stockholm, Sweden, 1-3 October 1997, p 84-87.

Kelly, D., Smyth, J. A. and McCracken, K. J. M. (1991). Digestive development of the early-weaned pig. *British Journal of Nutrition* **65**: 169-180.

Kim, J. H., Heo, K. N., Odle, J., Han, I. K. and Harrell, R. J. (2001). Liquid diets accelerate the growth of early-weaned pigs and the effects are maintained to market weight. *Journal of Animal Science* **79**: 427 - 434.

King, R. H. and Dunkin, A. C. (1986). The effect of nutrition on the reproductive performance of first-litter sows. *Animal Production* **42**: 119-125.

King, R. H. and Williams, I. H. (1984). The effect of nutrition on the reproductive performance of first-litter sows. 1. Feeding level during lactation, and between weaning and mating. *Animal Production* **38**: 241-247.

Kirkwood, R. N., Mitaru, B. N., Gooneratne, A. D., Blair, R. and Thacker, P. A. (1987). The influence of dietary energy intake during successive lactations on sow prolificacy. *Canadian Journal of Animal Science* **68**: 283-290.

Koketsu, Y., Dial, G. D., Pettigrew, J. E. and King, V. L. (1996). Feed intake pattern during lactation and subsequent reproductive performance of sows. *Journal of Animal Science* **74**: 2875-2884.

MLC. (2004). *Finishing pigs: Systems research. Production Trial 1 Dry versus liquid feeding in two contrasting finishing systems (fully slatted versus straw based housing).* p 59. Meat and Livestock Commission, Milton Keynes.

MLC. (2005a). *Finishing pigs - system research. Final report report to Defra.* Project LS 3601.

MLC. (2005b). *Finishing pigs: Systems research. Production Trial 3 Controlled fermentation of cereals in liquid diets fed to pigs in two contrasting finishing systems (fully slatted versus straw based housing).* p 35. Meat and Livestock Commission, Milton Keynes.

MLC. (2005c). *Finishing pigs: Systems research. Production Trial 4 Reducing the protein content of liquid diets fed to pigs in two contrasting finishing systems (fully slatted versus straw based housing).* p 45. Meat and Livestock Commission, Milton Keynes.

MLC. (2005d). *Pig Yearbook.* Meat and Livestock Commission, Milton Keynes.

MLC. (2006). *Maximising returns from finishing pigs.* p 32. Meat and Livestock Commission, Milton Keynes.

Olsson, A.-Ch. (1997). Liquid feeding for non-lactating group housed sows: Function and feed competition. In *Proceedings of the 5th International Symposium Livestock Environment*, Minnesota, USA, p 626-631.

Partridge, G. G., Fisher, J., Gregory, H. and Prior, S. G. (1992). Automated wet feeding of weaner pigs versus conventional dry diet feeding - effects on growth rate and feed conversion. *Animal Production* **54:** 484 (Abstr.).

Pedersen, A. O., Maribo, H., Aaslyng, M. A., Jensen, B. B. and Hansen, I. D. (2002a). *Fermented grain in liquid feed for heavy pigs.* Report Number: 547, The National Committee for Pig Production, Denmark.

Pedersen, A. O., Maribo, H., Kranker, S., Canibe, N., Hansen, I. D. and Aaslyng, M. A. (2002b). *Fermented liquid feed for finishers - pelleted feed.* Report Number: 567, The National Committee for Pig Production, Denmark.

Pluske, J. R., Williams, I. H. and Aherne, F. X. (1996). Maintenance of villous height and crypt depth in piglets by providing continuous nutrition after weaning. *Animal Science* **62:** 131-144.

Russell, P. J., Geary, T. M., Brooks, P. H. and Campbell, A. (1996). Performance, water use and effluent output of weaner pigs fed ad libitum with either dry pellets or liquid feed and the role of microbial activity in the liquid feed. *Journal of the Science of Food and Agriculture* **72:** 8-16.

Scott, K., Chennells, D. J., Armstrong, D., Taylor, L., Gill, B. P. and Edwards, S. A. (2007). The welfare of finishing pigs under different housing and feeding systems: Liquid versus dry feeding in fully-slatted and straw-based housing. *Animal Welfare* **16:** 53-62.

Thompson, J. E., Wiseman, J. and Gill, B. P. (2004). Physico-chemical aspects of liquid feed: The effect on component digestibility in growing/finishing pigs of 1) dietary dry matter concentration and 2) dietary fineness of grind. In *Proceedings of the British Society of Animal Science Annual Meeting,* York, March 2004, p 41.

van der Woolf, P. J., Wolbers, W. B., Elbers, A. R. W., van der Heijden, H. M. J. F., Koppen, J. M. C., Hunneman, W. A., van Schie, F. W. and Tielen, M. J. M. (2001). Herd husbandry factors associated with the serological salmonella prevalence in finishing pig herds in the Netherlands. *Veterinary Microbiology* **78:** 205-219.

van Winsen, R. L., Lipman, L. J. A., Biesterveld, S., Urlings, B. A. P., Snijders, J. M. A. and van Knapen, F. (2000). Mechanism of salmonella reduction in fermented pig feed. *Journal of the Science of Food and Agriculture* **81:** 342-346.

Verstegen, M. W. A. and Jongbloed, A. W. (2003). Crystalline amino acids and nitrogen emission. In: *Amino acids in animal nutrition* (ed. J. P. F. D'Mello), pp 449 - 458. CABI Publishing, Wallingford, UK.

Wondra, K. J., Hancock, J. D., Behnke, K. C., Hines, R. H. and Stark, C. R. (1995). Effects of particle size and pelleting on growth performance, nutrient digestibility, and stomach morphology in finishing pigs. *Journal of Animal Science* **73:** 757-763.

Yang, J. S., Lee, J. H., Ko, T. G., Kim, T. B., J, C. B., Kim, Y. Y. and Han, I. K. (2001). Effects of wet feeding of processed diets on performance, morphological changes in the small intestine and nutrient digestibility in weaned pigs. *Asian-Australian Journal of Animal Science* **14:** 1308-1315.

8

A PRACTICAL APPROACH TO HEALTH CONTROLS AND RISK MANAGEMENT IN THE PIG BREEDING PYRAMID

J.D. MACKINNON
Pig Health and Production Consultancy Cheneys Cottage East Green Kelsale Saxmundham Suffolk IP17 2PH

Introduction

Over the past 30-40 years, pig production has intensified and has become more structured in its approach to pig breeding with the establishment of breeding pyramids headed by genetic nucleus herds. Intensification and structured breeding have in turn meant that large-scale movements of animals take place, often over long distances. In the context of animal health, a breeding pyramid is a single population of animals, but it has theoretically no geographical constraint and therefore there is always the risk that infectious disease can enter and become established within the pyramid at different locations. Disease eradication policies in various countries have had varying degrees of success in controlling the economically important notifiable diseases. Relaxation of border controls in the European Union, allowing increased traffic of live pigs, semen and carcasses in and out of member states, has meant that infectious agents may be transported over thousands of miles and appear unexpectedly in areas where hitherto they are unknown (Madec and Rose, 2003). Unexpected disease may take time to diagnose, which in turn can give rise to a potentially long high risk period during which occult dissemination could take place. In the outbreak of classical swine fever in the UK in 2000, the high risk period was 68 days (Mackinnon, 2001). Production diseases, which are non-notifiable but nevertheless of economic importance, have been widely disseminated in pig breeding pyramids via live pigs and semen, exemplified in recent years by the porcine reproductive and respiratory syndrome virus (PRRSv) and porcine circovirus Type 2 (PCV2).

Management of the risk of infection arising from the introduction of live pigs or semen requires a basic understanding of the epidemiological and clinical characteristics of the important production diseases. This knowledge must be used to determine testing protocols and biosecurity arrangements, for neither of which are there currently any industry standards. The classification of herds

129

as negative for specific pathogens cannot be achieved easily because of the economic and practical constraints of extensive testing. Commercial pressures are such that perceived freedom from specific diseases is highly desirable, but to confirm a herd totally free of a disease, it would be necessary to subject every animal, and possibly selected biological niches, to tests that have perfect sensitivity and specificity. Such a notion is highly unlikely on a commercial scale, both currently and for the foreseeable future.

The veterinarian is usually the guardian of herd health and must be prepared to assess the risks arising from the introduction of live pigs or semen, to estimate the potential impact of disease on productivity and to offer advice on mitigation of these risks. However, whilst the veterinarian has control over the herds for which responsibility is held, there is usually no control over the source herds higher in the breeding pyramid. In advising on matters of health control, they therefore are almost totally reliant on information gained second hand, often verbally and often weeks or months out of date. Because there are no industry standards of evidence for health status, this system of "vet-to-vet" contact prior to animal movement is built entirely on professional trust and confidence. The veterinarian advising clients on the purchase of genetic material requires as much information as possible, which must be as contemporary as possible. These assessments should be based on the most advanced and effective methodologies available at the time. In the field of pollution control, Council Directive 96/61/EC (Anon, 1996) uses the concept of "best available technique" (BAT) to prevent or control emissions of pollutants. Perhaps a similar approach could be taken when attempting to define the health status of a herd supplying pigs or semen. It is important to appreciate that regardless of what systems are in place, a level of risk will always exist and in the development of standardised protocols for clinical assessment, testing procedures and reporting, this must be recognised by all parties concerned - the vendors, the purchasers and the advisers. In today's concerned society, zoonotic diseases are also of major interest.

Alexander (1998) lists 43 possible porcine zoonoses, 16 of which are considered to be a real concern. The point is made that people travel more frequently and more widely, and exotic zoonoses should therefore not be ignored. Tucker (2006) divides zoonotic diseases into those that are food-borne, occupational or emerging and considers that the food-borne zoonoses such as salmonellosis to be the most significant. Statutory measures are already in place in some countries to control salmonella species through abattoir monitoring (Davies *et al.*, 2004). Other zoonotic pathogens attract increasing attention such as Yersinia enterocolitica (Kwaga and Iversen, 1993), Hepatitis E (Chandler *et al.*, 1999; Meng *et al.*, 2000; Drobinec *et al.*, 2001) and influenza (Campitelli *et al.*, 1997), especially when two-way flow of infection between humans and pigs has been demonstrated (de Jong *et al.*, 1988; Yu *et al.*, 2007). Recently, the zoonotic potential of *Streptococcus suis* was demonstrated in China (Lun *et al.*, 2007) and it was given widespread media coverage. It is entirely possible that the monitoring of zoonotic agents other than salmonella will become the remit of veterinarians working in the pig industry. A leap of faith is always involved when accepting that a herd is totally free of any disease.

Diseases considered to be of economic important in pig production

A number of porcine diseases are currently considered to be notifiable by l'Office International des Epizooties (OIE) as follows (Anon, 2007a):

Multi-species	*Specifically porcine*
Anthrax (Bacillus anthracis)*	African Swine Fever*
Aujeszky's Disease (pseudorabies)*	Classical Swine Fever*
Brucella suis	Nipah virus encephalitis
Foot-and-Mouth Disease*	Porcine cysticercosis
Japanese B-encephalitis	Porcine reproductive and
Leptospirosis	respiratory syndrome virus
Rabies*	Swine vesicular disease*
Rinderpest	Transmissible gastro-enteritis
Trichinellosis	Vesicular stomatitis*

Diseases notifiable in Britain in the above lists are marked with a '*' (Anon, 2007b). In addition, Teschen disease is notifiable in Britain and transmissible gastro-enteritis is notifiable in Northern Ireland.

It is assumed that national biosecurity protocols are in place to minimise the risk of introduction of notifiable diseases, though vulnerability was painfully apparent in the UK in 2000 and 2001 with outbreaks of classical swine fever and foot-and-mouth disease occurring respectively.

On a day-to-day basis, many common production diseases of economic importance occur and it is these that often become endemic in commercial herds as a result of inadequate biosecurity and monitoring protocols within breeding and commercial production pyramids.

Though by no means exhaustive, the following list contains infectious diseases with the greatest potential impact on pig production because of their propensity to establish themselves as chronic disease:

Enzootic pneumonia (Mycoplasma hyopneumoniae)
Glässer's Disease (*Haemophilus parasuis*)
Ileitis (Lawsonia intracellularis)
Leptospirosis (especially serogroup Australis)
Mange (Sarcoptes scabei var suis)

Pleuropneumonia (*Actinobacillus* pleuropneumoniae)

Porcine circovirus Type 2 (PCV2) disease: Post-weaning multi-systemic wasting disease (PMWS); Porcine dermatitis and nephropathy syndrome (PDNS); interstitial pneumonia; hepatitis; infertility

Porcine reproductive and respiratory syndrome (PRRS)

Progressive atrophic rhinitis (Toxigenic *Pasteurella* multocida Type D)

Salmonellosis (S. enterica: Serotypes S. typhimurium, S. enteritidis and S. derby - public health importance)

Streptococcal disease (arthritis, pneumonia and meningitis)

Swine dysentery (Brachyspira hyodysenteriae; Other Brachyspira spp.)

In commercial terms, pleuropneumonia and enzootic pneumonia are considered to be the two most important respiratory diseases of pigs (Loeffen *et al.*, 1999). Recent experiences with PCV2 infections may have overshadowed this opinion for several years but, in the UK, these have abated whilst the aforementioned remain as major concerns, especially when potentiated by PRRS virus (PRRSv).

The breeding pyramid - an example of vulnerability

Figure 8.1 depicts the typical pathways of movement of genetic material in the pig breeding pyramid. Live breeding animals, both male (♂) and female (♀), are moved from genetic nucleus herds to multiplication herds and from multiplication herds to commercial herds. Live males tend to be moved directly from nucleus herds to commercial herds, either as terminal sires or as parent damline sires if replacement females are bred in situ. Live males are also moved to boar studs for semen production for artificial insemination in all three categories of breeding herds. These tiers of the pyramid may well be widely distributed, often across national borders.

Commercial herds tend to operate in one of three possible modes: Farrow-to-finish on one site, farrow-to-finish on separate sites for breeding and rearing (known as multi-site production) or as part of an integrated production scheme where pigs for growing and finishing are either purchased or are produced on breeding farms under the production scheme's control. Integrated production schemes, in common with breeding pyramids, may operate over relatively wide geographical areas and sometimes across national borders.

Assuming that the genetic nucleus herds are closed to the intake of live animals once established, there is within the pyramid one potential pathway for introduction of disease - semen from boar studs. The boar studs in return generally have two potential disease pathways - live boars from both the nucleus and multiplication herds. Multiplication herds also have two potential disease pathways - live animals from the genetic nucleus and semen from the boar stud. If, as in some cases, separate nucleus herds exist for the production of males and females, the number of potential disease pathways increases to three.

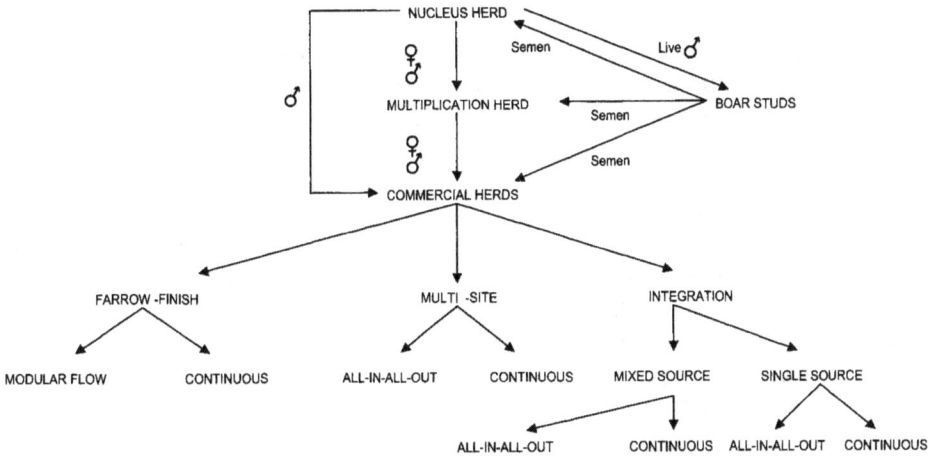

Figure 8.1 Typical flow pathways in pig production

The multiplication herds are one tier below the apex of the pyramid so the direct pathways of entry for the nucleus herds become indirect pathways for the multiplication herds, bringing the total to four.

Primary or direct pathways for commercial herds include the movement of males from the genetic nucleus herd, males and females from multiplication herds and semen from boar studs, making a total of three possibilities, four if male and female multiplication is separated. Some commercial herds take replacements from more than one multiplier and thus the number of disease pathways is further increased pro rata. All pathways into the multiplication herds act as indirect pathways for the commercial herds, giving rise to at lease six possible routes by which disease might filter downwards. It becomes clear that effective biosecurity and early warning systems in herds and studs at the top of the pyramid are essential, bearing in mind that disease may also pass backwards up the pyramid via fomites. True modular flow production in farrow-to-finish herds will separate pigs into sub-populations by age or by time periods and avoid contact between modules. Buildings can be emptied, cleaned and disinfected thereby minimising the impact of disease. However, if production flow is continuous, diseases can be perpetuated.

Multi-site production is designed to separate groups of pigs more completely. Each site can be operated on an "all-in-all-out" basis or it can be used continuously. In the latter case, buildings on site can be operated as for modular flow in farrow-to-finish units.

In integrated schemes, pigs are either moved at weaning or at some point between 18 and 45kg liveweight, depending on national or regional practice. They will either be reared on farms with pigs from other sources or they may be reared alone. In either case, production can be all-in-all-out or modular with cleaning and disinfection, or it can be continuous.

Disease from sources extraneous to the breeding pyramid

Infection is most likely to enter a herd *de novo* via live pigs or semen but, depending on location, any point within the breeding pyramid is potentially at risk from the introduction of disease from extraneous sources via animate, mechanical or biological vectors, inanimate vectors (fomites) and aerosols. For example and with specific regard to pig production, one of the greatest threats may come from increasing numbers of wild boar at a time when outdoor production is also increasing. The author has observed piglets with brown and yellow striping typical of wild boar born in an outdoor herd of white phenotype. Albina *et al.* (2000) carried out serological tests on 12,025 serum samples from wild boar in France between 1991 and 1998 and found 423, 249, 80 and 33 samples to be sero-positive for Aujeszky's disease, Border disease, classical swine fever and PRRS respectively. A similar survey carried out in Switzerland between 2001 and 2003 found no positive samples for classical swine fever, but Brucella suis was cultured from one sero-positive animal (Leuenberger *et al.*, 2007). There is clearly a need for systematic screening of wild animals utilising new genetic methods for identifying potentially dangerous viruses (Vilcek and Nettleton, 2006).

Avian wildlife represents potential carriers of Brachyspira hyodysenteriae (Glock *et al.*, 1978; Jensen *et al.*, 1996; Jansson *et al.*, 2004). A novel enteropathogenic spirochaete, Brachyspira suanatina, differing genetically from *B. hyodysenteriae* has recently been isolated from a mallard duck, which produced changes consistent with swine dysentery in a pig challenge study (Råsbäck *et al.*, 2007). Arthropod transmission of PRRSv is possible (Otake *et al.*, 2002a; Otake *et al.*, 2004; Schurrer *et al.*, 2004), flies being able to transport the virus over a distance equal to or more than 1.7 kilometres and mosquitoes up to 10 kilometres.

With the exception of foot-and-mouth disease virus, it has been difficult to demonstrate air-borne transmission of important pig pathogens. *Actinobacillus* pleuropneumoniae is species specific and is a fragile organism that does not survive well away from its host (Sheehan *et al.*, 2003). Survival in air has not been documented but in aerosols with 28% humidity, 0.22 will survive for 5 minutes and 0.08 for 45 minutes. When humidity is increased to 79%, the number of organisms surviving for 5 minutes increases to 0.60 (Taylor, 1999). In infected farms, *A. pleuropneumoniae* can be carried by aerosol over short distances of between 1 and 2.5 metres only (Torremorell *et al.*, 1997; Jobert *et al.*, 2000; Savoye *et al.*, 2000) but attempts to simulate transmission between closely located farms have indicated that this route of entry is most unlikely under natural conditions (Kristensen *et al.*, 2002).

In contrast, it has been speculated that airborne spread of Mycoplasma hyopneumoniae can occur over distances of up to 3.2 kilometres (Goodwin, 1985) and proximity of farms to each other or to roads used for pig transport is considered to be a risk factor for infection (Jorsal and Thomsen, 1988). Under experimental and field conditions, *M. hyopneumoniae* has been detected in air samples (Stark *et al.*, 1998) and using an experimental model originally designed to demonstrate aerosol transmission of viruses (Dee *et al.*, 2000), Cardona *et al.* (2005) have demonstrated aerosol transmission of *M. hyopneumoniae* over 150 metres. However, they did not test the infectivity of the recovered organism. Czaja *et al.* (2002) found it possible to reproduce porcine enzootic pneumonia with a single dose of an aerosol of cultured *M. hyopneumoniae* created by passing industrial-grade compressed air at a rate of 25 litres per minute through a nebuliser containing 10ml of culture into a chamber containing pigs. Fano *et al.* (2005a) placed sentinel pigs in a trailer and exposed them to exhaust air for seven days from a building containing pigs that had been artificially infected with *M. hyopneumoniae* and PRRSv. *M. hyopneumoniae* was identified in air samples from the building and from the sentinel pigs exposed to it.

Airborne transmission of infective PRRSv has been demonstrated over short distances ranging from 1 metre (Kristensen *et al.*, 2004) to 150 metres via air pipes (Otake *et al.*, 2002b; Dee *et al.*, 2005a), but there was a 50% reduction in the log concentration of PRRSv RNA for every 33 metres. Recent work (Hermann *et al.*, 2006) indicates that temperature has a greater effect than humidity on the survival of airborne PRRSv, with increasing stability at lower temperatures. At 5°C, the half-life of PRRSv is 193 minutes. Dee *et al.* (2005a) have suggested that the possibility of airborne introduction of PRRSv *de novo* is remote and should not be used as an excuse for inadequate investigation of other biosecurity risks.

Cargill and Davies (1999) consider Sarcoptes scabei var suis to be ubiquitous in pig populations unless specifically eradicated. Adult mites can survive off pigs for up to 96 hours at temperatures below 25°C and for 12 days at 7-18°C providing they are not exposed to direct sunlight or dessication, so it is possible that viable adult mites or their eggs could be transmitted adhering to fomites.

Contaminated fomites may be the route of entry for enteric pathogens such as *B. hyodysenteriae*. Brachyspira species can survive for long periods of time in pig feces (China and Taylor, 1978) ranging from 48 days at 0-10°C down to 7 days at 25°C. Dilution with water tends to enhance survival. Boye *et al.* (2001) demonstrated that *B. hyodysenteriae* will survive in soil for 10 days, in soil mixed with 10% porcine feces for 78 days and in pure porcine faeces for 112 days. Mechanical transmission of infection via contaminated footwear is a distinct possibility (Windsor and Simmons, 1981) and it is highly likely that vehicles contaminated with infected faeces are a major risk factor in spreading infection.

Although Pirtle and Beran (1996) considered indirect transmission of PRRSv via fomites an unlikely event, Otake *et al.* (2002c) demonstrated transmission of infection via boots, coveralls and hands. Infective PRRSv can be recovered from scale models of pig trailers unless they are subject to thermal drying (Dee *et al.*, 2004; Dee *et al.*, 2005b) suggesting that contaminated vehicles may also be a source of infection in the field if not thoroughly dried out after washing and disinfecting.

A. pleuropneumoniae can survive for a few days if protected by mucus or other organic material and will survive in water for up to 30 days at 4°C (Taylor, 1999), thus contaminated fomites, particularly vehicles, could be a source of infection. Mycoplasma hyopneumoniae is a delicate organism and highly susceptible to dessication (Taylor, 2006a) but under suitable conditions and in suitable media, it can be preserved for 2-4 weeks (Nagatomo *et al.*, 2001). There are no known reservoir hosts for *M. hyopneumoniae*; it does not appear to persist in the human respiratory passage (Goodwin, 1985) and it has not been possible to demonstrate mechanical transmission by personnel in close contact with pigs (Batista *et al.*, 2002). Whilst airborne transmission of *M. hyopneumoniae* into a susceptible population is likely, it appears very unlikely that mechanical transmission can occur.

Maintenance of disease from sources within the breeding pyramid

The primary routes for transmission of porcine pathogens within the breeding pyramid are by direct pig-to-pig contact or via semen. Secretions, discharges, feces and urine, or fomites freshly contaminated by them, serve as secondary routes of transmission. Once established, the different pathogens employ different means of persistence within their host population, often making surveillance and monitoring difficult. Latency is a typical feature of Aujeszky's disease and swine dysentery (Rziha *et al.*, 1982; Griffin and Hutchings, 1980). The herpesvirus causing Aujeszky's disease persists in the trigeminal ganglia and the tonsils (Cheung, 1995) and recrudesces during stressful events. In the meantime it may be asymptomatic. *B. hyodysenteriae* may be dormant in the crypts of Lieberkuhn under certain dietary conditions possibly asymptomatically, (Pluske *et al.*, 1996; Siba *et al.*, 1996; Durmic *et al.*, 1998), emerging to precipitate clinical dysentery when dietary changes are made that alter fermentable substrates in the hind gut. The author has experience of asymptomatic *B. hyodysenteriae* infection of a herd of 1200 sows kept outdoors in East Anglia, a situation which led to the loss of a contract for the sale of store pigs for finishing. Following a complaint from the purchaser, pooled dry swabs taken from growing pigs at 8, 10 and 12 weeks of age gave positive polymerase chain reaction tests (PCR) for the 10 and 12-week old age groups, though clinical dysentery had never been observed in the herd before, nor has it been subsequently.

In addition to the presence of carrier pigs with latent infection, *B. hyodysenteriae* can persist in the mouse population on the farm. Mice have been shown experimentally to shed *B. hyodysenteriae* for 180 days (Joens, 1980) and it has also been identified in rats, dogs and flies (Taylor, 2006b). Mice are prolific breeders. They reach sexual maturity at 5-6 weeks of age and breed throughout the year producing up to 8 litters of between 4 and 7 pups. Large colonies of mice can exist on farms providing a potentially constant flow of *B. hyodysenteriae* to pigs via fecal contamination of the food supply.

Latent infection may also occur vertically with *Streptococcus suis* Type 2 following oral infection from maternal vaginal secretions during parturition (Amass *et al.*, 1995; Amass *et al.*, 1996). The organism colonises the palatine tonsil (Williams *et al.*, 1973) and certain conditions will precipitate a bacteraemia causing clinical disease at any time up to around 6 months of age.

Tonsillar colonisation with *A. pleuropneumoniae* can occur from around 11 days of age onwards and transmission from dam to offspring increases as colostral immunity wanes (Vigre *et al.*, 2002). Asymptomatic infection can occur depending on strain virulence, the organism residing in the upper respiratory tract. Carrier animals can transmit infection to other susceptible animals so detection of asymptomatic carriers is an essential means of preventing transmission into a susceptible herd (Chiers *et al.*, 2002).

The most likely indication of an infectious pig is an *A. pleuropneumoniae*-positive tonsil and the level of infectivity is dependent on the number of colony forming units (cfu) in the nasal cavities (Velthuis *et al.*, 2002). Since *A. pleuropneumoniae* is primarily a parasite of the respiratory tract and since there is no published evidence for transmission by non-porcine species or humans, discovery of infection must indicate a result of a breakdown in biosecurity somewhere within the pyramid.

M. hyopneumoniae is maintained within herds by vertical and horizontal transmission, the latter being the most common in chronically infected populations. Infection by direct contact has been confirmed using "seeder" and "sentinel" pigs, evidence for infection determined 28 days after mixing by PCR and after 35 days by serology (Fano *et al.*, 2000b). Infection was shown to persist for 185 days and direct contact transmission may take several weeks to complete. Transmission studies show that the rate of infection can be high (Meyns *et al.*, 2004). One piglet infected before weaning will infect on average one pen-mate during the nursery phase. Non-infected pigs in direct contact with infected pigs are seven times more likely to sero-convert than those in indirect contact (Morris *et al.*, 1995). Vaccination does not necessarily prevent infection. Pieters *et al.* (2006) detected *M. hyopneumoniae* by a PCR test 200 days post-infection in both vaccinated and non-vaccinated pigs that were subjected to a 14-day period of exposure.

The PRRS virus is an enveloped RNA virus and is highly unstable in the presence of detergents (Plagemann, 1996), at pH below 6 or above 7.5 (Benfield *et al.*, 1992) and at high ambient temperatures (Hermann *et al.*, 2006). It is therefore unlikely to survive in piggery environments for long periods. Infectivity has been shown to persist for 1-6 days at 20-21°C, 3-24 hours at 37°C and 6-20 minutes at 56°C (Benfield *et al.*, 1999). Rather than survive in the environment, the virus is maintained within persistently infected animals, which excrete virus for several months (Wills *et al.*, 1997a). This explains long-term herd infection and transmission via clinically healthy carriers. Infected animals shed virus in saliva (Wills *et al.*, 1997b), nasal secretions (Christianson *et al.*, 1992), urine (Wills *et al.* 1997b), semen (Sorensen *et al.*, 1994; Christopher-Hennings *et al.*, 1995; Christopher-Hennings *et al.*, 1996), feces (Yoon *et al.*, 1993; Pirtle and Beran 1996) and mammary secretions (Wagstrom *et al.*, 2001). The latter route of excretion occurs if sows are infected in late gestation but vertical infection as a result is thought to be uncommon.

Iatrogenic transmission via hypodermic needles is undoubtedly a significant means by which PRRSv can be perpetuated in herds (Otake *et al.*, 2002d). The parenteral transmission of PRRSv potentially means that infection can occur during husbandry procedures such as ear-notching, teeth-clipping and tail-docking in addition to routine inoculations (Zimmerman, 2006).

Semen as a vector of disease in artificial insemination

Boar semen is a vehicle for many viruses, particularly during the viraemic stages of disease (Guérin and Pozzi, 2005) including classical swine fever virus (de Smit *et al.*, 1999; Floegel *et al.*, 2000), PRRSv (Christopher-Hennings *et al.*, 1995a), Aujeszky's disease virus, foot-and-mouth disease virus and swine vesicular disease virus (van Rijn *et al.*, 2004) and artificial insemination is therefore a major potential route of infection within breeding pyramids.

In their excellent review of PRRSv infection in the boar, Prieto and Castro (2005) describe the complex epidemiology of the virus and the significance of the shedding of the virus in semen. There is huge potential for the widespread dissemination of infection throughout breeding pyramids via artificial insemination, natural insemination and the natural courtship behaviour of boars. It is possible to detect PRRSv in semen from experimentally infected boars up to 92 days post-infection (Christopher-Hennings *et al.*, 1995b). Both artificial and natural insemination with PRRSv-infected semen can cause sero-conversion of the recipient females, but the finding is not consistent (Swenson *et al.*, 1994a; Swenson *et al.*, 1994b; Gradil *et al.*, 1996; Prieto *et al.*, 1997a).

Imprtant bacteria carried in the genital tracts of boars include Brucella suis (MacMillan, 1999) and Leptospira bratislava (Australis groups) (Ellis *et al.*, 1986), both of which have the potential to be spread venereally.

Maintenance of infection within breeding pig pyramids by transplacental infection

Transplacental infection occurs typically with viruses and may give rise to congenitally infected piglets, notably with PRRSv, Aujeszky's disease and classical swine fever. Infection of fetuses follows exposure of gilts and sows during gestation, particularly when exposure occurs in the later stages (Mengeling *et al.*, 1994; Prieto *et al.*, 1996; Prieto *et al.*, 1997a; Prieto *et al.*, 1997b). The birth of viraemic piglets will then give rise to widespread infection of susceptible pigs after weaning and mixing in nursery accommodation.

Biodiversity and virulence - diagnostic and prognostic dilemmas

Huge biodiversity and differences in virulence exist within many species of porcine pathogens, creating difficulty in diagnosis and prognosis.

Of the 22 species of *Actinobacillus* known, 19 are associated with animals, but only *A. pleuropneumoniae* is considered to be a primary pathogen (Christensen and Bisgaard, 2004). Based on classification by capsular antigens, 15 serotypes of *A. pleuropneumoniae* are currently recognised (Taylor, 2006c). Virulence is determined by the possession of one or more of three cytotoxins, which cause varying severity of lung lesions and are designated Apx I-III (Bertram, 1990; Nicolet, 1990; Euzéby, 1998; Taylor, 2006c: Table 8.1). All species possess a fourth toxin designated Apx IV (Frey *et al.*, 1994; Jansen *et al.*, 1995).

Table 8.1 DIVERSITY AND VIRULENCE OF ACTINOBACILLUS PLEUROPNEUMONIAE SEROTYPES

Toxin	*Molecular size (kDa)*	*Serotypes*	*Activity*
Apx I	105-110	Biotype 1; serotypes 1, 5a, 5b, 9, 10, 11	Strongly haemolytic and cytotoxic Kills macrophages
Apx II	103-105	All biotype 1 except serotype 10 All biotype 2	Weakly haemolytic and moderately cytotoxic Kills macrophages
Apx III	120	Biotype 1, serotypes 2, 3, 4, 6, 8	Non-haemolytic and strongly cytotoxic
Apx IV	202	All biotype 1 and 2	Not produced *in vitro*

Regional and national differences in serotype profiles exist but, as more live animals cross national borders, more diverse serotypes will be found in more pig-producing areas. Non-pathogenic, or at least apparently non-pathogenic isolates of *A. pleuropneumoniae* have been identified in pigs that are antigenically and biochemically similar to *A. pleuropneumoniae* serotypes 1 and 9 (Gottschalk *et al.*, 2003) which can confuse diagnosis and prognosis. To confuse matters further, *A. pleuropneumoniae* Biotype 2 shares characteristics in common with *Actinobacillus* suis and *Pasteurella*

(*Mannheimia*) *haemolytica* (Euzéby, 1998), which produce similar gross pathological changes in lung tissue. The author has experience of respiratory disease emerging in a high health herd of replacement gilts in which the only agent discovered was *A. pleuropneumoniae* serotype 14, although this is considered to be non-pathogenic.

There is currently no single test that can distinguish all serotypes of *A. pleuropneumoniae*. A strong humoral response occurs to all three main toxins, irrespective of serotype and, since they are antigenically related, there is no correlation between serotype and secretion of exotoxin in the enzyme-linked immunosorbent assay (ELISA) (Nielsen *et al.*, 2000). Modifications of the ELISA using specific lipopolysaccharides have been developed for serotype 2 (Klausen *et al.*, 2001), serotype 12 (Andresen *et al.*, 2002) and a mix-ELISA for serotypes 2, 6 and 12 (Grøndahl-Hansen *et al.*, 2003). The authors consider these tests to be adequate for herd screening. Blocking-ELISAs have been developed for serotype 6 (Klausen *et al.*, 2001) and serotype 5 (Klausen *et al.*, 2002) but these cross-react with serotypes 3 and 8 and with 7 and 12 respectively in natural infections. Enøe *et al.* (2001) combined the ELISA with a complement fixation test (CFT) to provide a means of specifically detecting serotype 2 serologically.

Screening for *A. pleuropneumoniae* in herd surveillance or the diagnosis of infection should always include the objective of clarifying the toxin profile, which is currently best achieved by a combination of culture and PCR using paired tonsillar swabs from individual animals (J.R. Thomson. Personal communication). It is possible to carry out direct PCR tests on tonsillar biopsies and tracheobronchial lavage without a culture step (Savoye *et al.*, 2000) but this technique is not entirely practical for routine herd screening.

B. hyodysenteriae is currently divided into 11 serogroups based on lipopolysaccharide antigens, which can be further divided into electrophoretic types (Lymbery *et al.*, 1990). Although it is recognised that a cytolytic haemolysin may be a virulence determinant in the pathogenesis of swine dysentery (Lysons *et al.*, 1991), it is not currently possible to predict virulence of a field isolate other than by artificial infection (J.R. Thomson. Personal communication). Restriction fragment length polymorphism PCR (16S-RFCP-PCR) for *B. hyodysenteriae* has high sensitivity and can detect 102-103 cells/ gramme of feces (Calderaro *et al.*, 2006). PCR methodology can also be used on tissue samples embedded in paraffin (Weissenböck *et al.*, 2005). It was demonstrated that B. murdochii, normally considered to be innocuous, can multiply extensively in single infections and cause enterocolitis.

The choice between PCR and traditional culture and biochemistry (CBT) for the detection of *B. hyodysenteriae* should be governed by the presence or absence of clinical dysentery. CBT methodology has been shown to have greater sensitivity than PCR in clinical samples, (Råsbäck *et al.*, 2006) but it may not be sufficiently sensitive to pick up atypical or asymptomatic infection. Experience in the field has shown that asymptomatic infection with *B.*

hyodysenteriae can occur (Moller *et al.*, 1998) but the diagnostician cannot be sure whether the absence of clinical disease is due to dietary conditions (Pluske *et al.*, 1996; Siba *et al.*, 1996; Durmic *et al.*, 1998), to the absence of synergistic co-infection (Harris and Lysons, 1992), a novel type (Råsbäck *et al.*, 2007), suppression by antibacterials or to lack of virulence. Therefore, based on current knowledge, it would be unwise to consider tolerating the presence of *B. hyodysenteriae* in a breeding pyramid, regardless of apparent virulence.

Even greater diversity, virulence and, therefore, potential diagnostic difficulty exist amongst *Streptococcus* and *Haemophilus* species. At least 35 serotypes of *Strep. suis* Type 2 (Perch *et al.*, 1983; Higgins *et al.* 1995) and 15 serotypes of *Haemophilus parasuis* (Kielstein and Rapp-Gabrielson, 1992) have been identified, several of which may occur together at any one time. However, it has been shown to be possible to discriminate between virulent and non-virulent strains of *Strep. suis* Type 2 by ELISA (Vecht *et al.*, 1993) and to detect virulent strains of *Strep. suis* Type 2 and highly-virulent strains of *Strep. suis* Type 1 in tonsillar samples by PCR (Wisselink *et al.*, 1999).

The detection of *M. hyopneumoniae* infection is relatively straightforward. Nested PCRs have been described for screening nasal swabs (Mattson *et al.*, 1995; Calsamiglia *et al.* 1999) and tracheobronchial washings (Calsamiglia *et al.*, 2000; Verdin *et al.*, 2000), the detection of *M. hyopneumoniae* being considered more effective after DNA amplification, and they were found to be significantly more sensitive than blocking-ELISAs for sera and immunofluorescence for lung tissue ($p<0.05$). The nested PCR technique has application in confirming the absence of infection, but the results of Mattson *et al.* (1995) suggest that *M. hyopneumoniae* can only be detected in nasal cavities for a limited period during infection. Nested PCR results correlate well with the presence of characteristic histopathological lesions and with lesions that are less characteristic of *M. hyopneumoniae* infection (Calsamiglia *et al.*, 2000).

Serology using ELISAs would be a cheaper and more practical method of screening herds for *M. hyopneumoniae* infection. An indirect-ELISA has been compared with a blocking-ELISA, the latter being more sensitive at 28, 43 and 62 days post-vaccination (Ameri-Mahabadi *et al.*, 1005). Serum ELISAs for the detection of field infection caused by *M. hyopneumoniae* would be expected to have relatively low sensitivity but high specificity because cell mediated immunity is of greater significance than humoral immunity. Routine serology alone is therefore not the most reliable means of confirming that a herd is free of *M. hyopneumoniae*.

PRRS viruses are broadly thought of as being divided into European and North American genotypes, and initially it appeared that the latter showed greater diversity than the former. It is likely that the European genotype entered the pig population more than 10 years before its epidemic emergence (Forsberg *et al.*, 2001). Over time in Spain there has been continuing adaptation of the virus to pig cells with decreasing similarity to the originally isolated Lelystad virus (Mateu *et al.*, 2006) and it is now thought that, contrary to previous assumptions, genetic diversity is at least as high in the European genotype as it is in the North American genotype and recombination has occurred (Forsberg *et al.*, 2002). Genetic variability has also been reported from Austria, where it is believed that the North American genotype entered via live pig imports (Indik *et al.*, 2005); in Italy, where it is believed that the most frequent source of infection in PRRS-positive farms is the introduction of animals carrying a new variant (Pesente *et al.*, 2006) and in eastern Europe where the diversity of the European genotype is thought to exceed that of the North American genotype (Stadejek *et al.*, 2006). Genetic distances between different strains of European PRRS viruses have certainly increased over time, but this is not believed to be due to geographical location, timing of sample collection or the use of vaccination (Pesch *et al.*, 2005). The need for effective and rigorously applied biosecurity in conjunction with strain surveillance increases as the diversity of the viruses increases.

The development of reverse transcription nested PCR (RT-nPCR) for the detection of virus in semen has overcome problems of interference from cytotoxins that are naturally present (Prieto and Castro, 2005) allowing semen to be effectively screened. In choosing the type of RT-nPCR, consideration must be given to the RNA extraction methods because boar semen is viscous, non-homogeneous and has a high cell count and volume. Two of six commercial RT-nPCR systems tested were found to be equally best, one of which was superior in detecting natural as opposed to "spiked" infection in semen (Christopher-Hennings *et al.*, 2006). PCR testing of semen has been able to detect boar studs positive for PRRS that were thought to be negative following serum ELISA (Revilla-Fernández *et al.*, 2005). Serological tests can however be used for large scale screening. A double blocking ELISA can distinguish between antibodies produced by the European and American genotypes of PRRSv from 7 days post-infection (Sørensen *et al.*, 1998) and in detecting natural infection, ELISA is more sensitive than immunofluorescence (Yahara *et al.*, 2002). The immunoperoxidase monolayer assay (IPMA) is more sensitive than commercial ELISAs (Drew, 1995) and is often used to confirm ELISA results in herd screening tests. However, since there is no difference in antibody response between carriers and non-carriers (Horter *et al.*, 2002), RT-nPCR testing of oropharyngeal swabs is considered to be the best option currently available for detecting carrier pigs.

Clearly there are many diagnostic tests available for the important porcine pathogens, each with their own advantages and disadvantages, and there is therefore a need for agreement on a validated and standardised best available technique (BAT) approach to the classification of the health status of herds in the breeding pyramid.

Making the decision to move breeding stock - the management of risk

Veterinarians working in pig production will be faced with the necessity to make decisions about the health of pig herds as sources of breeding or finishing stock on an almost daily basis, but it is often very unlikely that he or she will have been involved in the actual health assessment unless employed within a fully integrated system with clinical control throughout the pyramid. Therefore a number of searching questions must be asked about how disease risks are assessed and controlled and there must be an appreciation of the application and interpretation of diagnostic tests.

Risk assessment and risk management are part of a continuum as depicted in Figure 8.2, which overlaps both source and recipient farms (Anon, 1993. Adapted from Hurd, 2003). The process commences with identification of hazards, an estimate of the likelihood that they will have an effect and to what degree. Risk assessment procedures have application in national disease control but could equally apply to the pig breeding pyramid. Risk management is aimed at preventing hazard becoming reality and mitigating the effect using a scenario tree of all mutually exclusive outcomes that can occur (Miller *et al.*, 1993; MacDiarmid and Pharo, 2003).

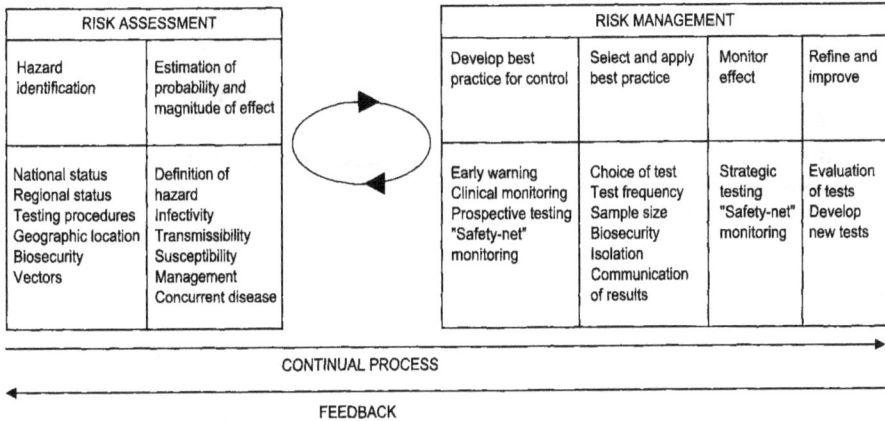

RISK ASSESSMENT			RISK MANAGEMENT			
Hazard identification	Estimation of probability and magnitude of effect		Develop best practice for control	Select and apply best practice	Monitor effect	Refine and improve
National status Regional status Testing procedures Geographic location Biosecurity Vectors	Definition of hazard Infectivity Transmissibility Susceptibility Management Concurrent disease		Early warning Clinical monitoring Prospective testing "Safety-net" monitoring	Choice of test Test frequency Sample size Biosecurity Isolation Communication of results	Strategic testing "Safety-net" monitoring	Evaluation of tests Develop new tests

CONTINUAL PROCESS

FEEDBACK

Figure 8.2 Risk assessment and risk management - a continuum

Hazard identification

In simplistic terms, any list of important pig diseases forms the basis of hazard identification, which must then be modified according to national and regional status and to the presence of any means by which disease might be carried into a herd. Can wild boar, for example, enter the farm and have contact with the domestic stock? Is a disease already endemic in the herd?

Estimation of probability that a herd will become infected

This is a far more complex exercise because it must take into account many of the features of the pathogens discussed earlier. These would particularly include the survivability, transmissibility and the virulence of the pathogen, the infective dose and the susceptibility of the stock. It must also take into account the biosecurity arrangements and the location of the herd and the methods used for genetic replacement, i.e. introduction of live animals or use of artificial insemination as depicted in Figure 8.1. Probabilities can be calculated for the introduction of infection from simple equations, but mathematical models can never take into account all naturally occurring biological variations. Using foot-and-mouth disease virus as an example of aerosol transmission of disease, Manuel-León and Casal (2001) point out that the number of pathogens present in a droplet is unknown and state that the probability that an outbreak occurs depends on the fact that at least one animal receives an infective dose. Should this happen, the probability that the herd becomes infected follows a binomial distribution, which depends entirely on the number of animals in the herd according to the equations:

$$P_{(herd)} = 1 - (1-P_{(i)})^n \qquad \text{Equation 1}$$

where $P_{(herd)}$ is the probability of infection for the herd, $P_{(i)}$ is the probability that one animal receives at least one droplet containing an infective dose and n is the number of susceptible animals in the herd. Equation 1 has been used in alternative forms to calculate the risk of introduction of disease via fomites (MacDiarmid, 1993):

$$T = 1 - (1-p)^n \qquad \text{Equation 2}$$

where T is the probability of introduction, p is the probability that the fomites contain or carry infectious pathogens and n is the number of occasions on which susceptible animals are exposed. Equation 2 can also be adapted to calculate the probability of a herd becoming infected *de novo* or of receiving infectious pigs from herds where the prevalence of disease is known by substituting prevalence for p. The latter might apply in a situation where the commercial need to introduce a genetic line overrides the risk and impact of introducing a particular disease. Incidence rates vary for different diseases depending on their means of transmission. Between 1994 and 1998, 34 of 297 (0.114) farms surveyed in Denmark became infected with *A. pleuropneumoniae*, with an average annual incidence of 0.34 (Zhuang *et al.*, 2007). The annual incidence of *M. hyopneumoniae* infection has been estimated at between 2 and 0.14 (Harris and Alexander, 1999).

Potentiation and amplications - the magnitude of effect

Risk can be amplified by herd management factors and potentiated by the presence of other diseases as illustrated by Figure 8.3. There are many examples of this in the literature, examples being: Zhuang *et al.* (2007) showed that herds comprising more than 500 breeding sows had approximately three times higher estimated risk of infection than smaller herds; Rose *et al.* (2003) found that together with management factors that enhance the spread of disease, the chance of post-weaning multi-systemic wasting syndrome of pigs (PMWS) increased when pigs tested positive for porcine parvovirus and PRRS virus; Feng *et al.* (2001) demonstrated that infection *in utero* by PRRSv increased the susceptibility of piglets to challenge with *Streptococcus suis* type 2 and Goldberg *et al.* (2000) have shown that with PRRSv infection, increasing herd size is a significant risk factor for death of sows and for respiratory disease in nursery pigs.

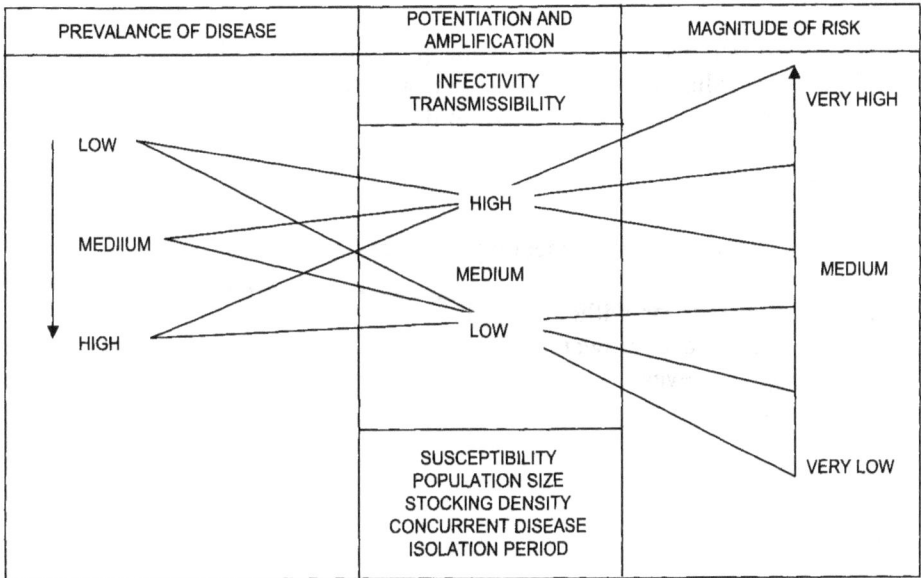

Figure 8.3 Magnitude of risk - potentiation and amplification

Assessing risk - best practice options

EARLY WARNING

De novo infection in herds which, by definition, would have greater susceptibility than herds in which the infection already exists, is likely to cause a range of clinical signs that are a departure from the norm. It is thus essential that a robust clinical monitoring and reporting system is in place and the frequency of observations is great enough to detect change as an early warning system. In nucleus and multiplication herds, the frequency of clinical inspection by an experienced veterinarian should ideally be commensurate with disease incubation periods but this is likely to be impractical and costly in practice. The author considers that the most practical approach would be to co-ordinate the sale of breeding stock with a herd inspection, perhaps monthly. Boar studs must under EU legislation be inspected within 72 hours of semen collection and thus frequency of observations is not generally an issue (Anon, 1999).

A change in clinical status should never be ignored. A standardised approach should be taken in the reporting of prevalence of clinical signs in nucleus and multiplication herds, together with monitoring for significant changes in incidence of selected parameters such as mortality, cough index, etc. by the cumulative sum (cusum) technique (Sard, 1979). Whilst it is relatively easy to identify the emergence of a previously unobserved clinical sign in a herd free of important diseases, it becomes more difficult in a situation where incidence and prevalence of an existing disease are increasing, denoting the possibility of a change in pathogen genotype or the presence of an additional pathogen.

Although not exhaustive, the following is a list of clinical signs of importance in the monitoring of the breeding pyramid:

Respiratory

Coughing in any age group, especially finishers and adults
Laboured breathing
Nasal discharge, with or without blood
Sneezing in piglets and growers
Discharge from the eyes
Facial distortion

Gastrointestinal

Diarrhoea in growers, finishers or adults
Blood in faeces
Persistent diarrhoea in piglets or weaners

Central nervous sytem

Convulsions in piglets and growers
Tremors in piglets
Staggering or lack of co-ordination in any age group
Paraplegia in any age group

Skin

Cyanosis of extremities, especially ears and udder line
Vesicles around snout, mouth or feet
Purple or brick-red patches

Itchiness
Head-shaking

Limbs

Acute lameness
Swollen joints

Reproduction

Abortion
An increase in stillbirths or mummified fetuses
An increase in irregular returns to service
Phantom pregnancies

General

Weight loss or wasting in growers and finishers
Anaemia in weaners and growers
Weakness and low viability in piglets
Sudden increases in mortality rates in any age group

Confirmation of clinical observations and surveillance

LABORATORY TESTS

Clinical observations should always be considered alongside laboratory tests in reaching a diagnosis, but it must be borne in mind that infection may be asymptomatic in latent or carrier states. In establishing a testing protocol it is therefore necessary to define the purpose of testing: diagnosis or surveillance. A change in clinical observations may require diagnostic sampling as a matter of urgency and a range of samples can be taken from affected live or dead pigs. These procedures are relatively straightforward, but agreement on protocols for surveillance is rather more difficult to achieve since it requires careful consideration of the type of test or tests to be employed, the sample size, sample selection and the frequency of testing. Economic and commercial interests must also be borne in mind. The objective is to determine whether or not a hazardous pathogen exists in the population and, if so, to what extent (prevalence). It is also helpful to know the incidence of disease in herds known to be infected since this is an indicator of infectiousness. Serological screening is relatively inexpensive when compared with techniques such as PCR testing or culture and biochemistry but a major disadvantage is that it does not differentiate between animals that have the disease and those that do not. There is therefore always the probability of false positives and false negatives when trying to establish freedom of disease in an animal population (Baldock, 1998). Serological testing should therefore be combined with attempts to detect or isolate the pathogen.

In choosing a test, certain attributes should be considered (Altman and Bland, 1994a; Altman and Bland, 1994b; Gresham 2002; Burr and Snodgrass, 2004: Table 8.2). Values in Table 8.2 are based on individual tests and are therefore not necessarily appropriate for determining herd status, which is more difficult because it depends not only on the sensitivity and specificity of the first, but also on the number of animals tested, the true within-herd prevalence and a second threshold decision of cut-off value in terms of tolerance of the occurrence of positive tests (Christensen and Gardner, 2000). Reasons for positive and negative tests include the following (Baysinger, 1999):

Table 8.2 SENSITIVITY, SPECIFICITY AND TEST VALUES

Test results	True individual status	
	Diseased	*Non-diseased*
Number positive	a (true positive)	b (false positive)
Number negative	c (false negative)	d (true negative)
Total number of results	a + c	b + d
Value of test	**Definition**	**Calculation**
Sensitivity	Proportion of diseased animals testing positive	$a/(a+c)$
Specificity	Proportion of non-diseased animals testing negative	$d/(b+d)$
Positive predictive value	The probability that a positive result is correct	$a/(a+b)$
Negative predictive value	The probability that a negative result is correct	$d/(c+d)$
Accuracy of the test	The proportion of all results that are correct	$\dfrac{a+d}{a+c+b+d}$
Likelihood ratio	The probability that a positive result is obtained from a diseased animal rather than from a non-diseased animal	$\dfrac{Sensitivity}{1 - Specificity}$

Positive test:
   Actual infection
   Group cross-reactions
   Passively acquired immunity
   Non-specific test (false positive)

Negative test:
  Absence of exposure to antigen
  Improper handling of specimens
  Improper timing (recent infection prior to sero-conversion)
  Improper selection of test
  Laboratory error
  Toxic substances (interfering with test)
  Insensitive test (false negative)
  Inadequate sample size

When screening herds, it is worth considering the following points (Baldock, 1998):

1. As the number of animals tested in a particular herd increases, herd-level sensitivity increases.
2. As the number of animals used to classify the herd as positive is increased, there is a corresponding increase in specificity.
3. As herd-level sensitivity increases, herd level specificity decreases and vice versa.
4. Individual and herd-level screening test characteristics are not equivalent.
5. If the specificity of a test at individual animal level is perfect (100% or 1), all non-infected animals will be correctly classified as diseased, i.e. herd specificity is 100%.
6. However, the converse is not true when only a sample of animals from a herd is tested. Herd-level sensitivity will be less than 100%.

For detection of disease within a population, the sample size (n) depends on an estimation of the likely minimum prevalence, the size of the population and the desired level of confidence of finding at least one positive pig. Either standard tables or the following formula can be used:

$$n = H - (\frac{D+1}{2}) \; x \; (1 - (1 - (1 - C))\frac{1}{D}))) \qquad \text{Equation 3}$$

where is H is the population size, D is the estimated disease prevalence (or expected number of positive samples) and C is the confidence level required (Cannon and Roe, 1982; Done and Wilesmith, 1988; Pointon *et al.*, 1990). The author's view is that cost constrains the taking of sufficient samples in most cases and therefore there is either a case for storing sera for testing at a later date if appropriate or pooling samples to increase sample size cost-effectively. Pooled samples are clearly only of value for detection.

If it is desired to detect one positive animal in a herd of 2000 pigs with a confidence level of 95 ± 5% and where a disease has a prevalence of 10%, 138 samples would be required (Pointon *et al.*, 1990). However, this assumes that the test has 100% sensitivity and 100% specificity, which is unlikely. Cameron

Table 8.3 SOME EXAMPLES OF SENSITIVITY AND SPECIFICITY OF DIAGNOSTIC TESTS

Pathogen	Test	Specimen	Sensitivity	Specificity	Authors
Aujeszky's disease virus	ELISA (anti-gE)	Meat juice	93.2	98.3	Le Potier *et al.* (1998)
Actinobacillus pleuropneumoniae					
Serotype 2	Blocking - ELISA	Serum	100.0	92.8	
	Complement fixation	Serum	90.6	98.6	Enøe *et al.* (2001)
	ELISA + CFT	Serum	90.2	99.9	
Serotype 6	Blocking - ELISA	Serum	100.0	97.0	Klausen *et al.* (2001)
Serotype 12	Blocking - ELISA	Serum	77.0	100.0	Andresen *et al.* (2002)
Serotypes 2, 6, 12	MIX - ELISA	Serum	98.0	95.0	Grøndahl-Hansen *et al.* (2003)
PRRSv	ELISA Herd Check™	Serum	100.0	95.5	Manufacturer (IDEXX)
PRRSv	ELISA 2XR™	Serum	97.4	99.6	Manufacturer (IDEXX)
Mycoplasma hyopneumoniae	Nested - PCR	Serum	43.4		
	Nested - PCR	Lung	83.1		Verdin *et al.* (2000)
	Blocking - ELISA	Serum	32.6		
	IFAT	Lung	44.6		
	Indirect ELISA	Serum (vaccinated pigs, 62 days)	92.6	99.2	Ameri-Mahabadi *et al.* (2005)
	Blocking ELISA	Serum (vaccinated pigs, 62 days)	97.0	99.2	
Sarcoptes	ELISA	Serum	87.9	99.5	Bornstein and Wallgreen (1997)
	Microscopy	Skin scrapes	100.0	32.8	
	ELISA	Meat juice	71.0	77.0	Vercruysse *et al.* (2006)
	ELISA	Serum	73.0	81.0	
	Microscopy	Ear scrapes	86.0	100.0	

and Baldock (1998) have developed a formula for calculation of herd size that takes into account the sensitivity and specificity of the test to be used. The software for the calculation (FreeCalc. Version 2) can be downloaded from www.ausvet.org (accessed 9th April 2007). Some examples of how sensitivity and specificity vary with test and type of pathogen are shown in Table 8.3.

In addition to the taking of an appropriate number of samples, the sampling must be correctly structured and not biased by convenience. Random sampling compiles the sample by using random numbers to select individual animals or pens; systematic sampling chooses every n^{th} pen or animal commencing at a point selected by random number; stratified sampling divides the population into exclusively targeted groups which are then randomly sampled. In herds deemed to be free of disease, the latter procedure might for example be used to detect maternal antibody in young piglets, since its presence can be considered a reliable indicator of maternal exposure.

The frequency of testing should take into account incubation periods, time from infection to sero-conversion and so on, but again there are cost constraints so a compromise must be reached. In the UK, the industry norm is to test some animals in nucleus and multiplication herds every three months or so, but sample size tends to be ill-defined and constrained by cost. Furthermore, the frequency of sampling may not be conducive to the detection of new infection at an early enough stage, i.e. prior to the delivery of some batches of replacement breeding stock or semen. This places significant onus on the purchaser, or his veterinarian, with them alone responsible for assessing the quality of the purchase before buying. This may not always be fully appreciated by the purchaser.

Monitoring - the use of a safety net

Slaughter-house surveys provide valuable information about the prevalence of pathological signs of disease and are therefore useful for monitoring the effectiveness of control measures. However, for the purposes of detection of specific disease, lesions seen on the slaughter-line have low sensitivity and specificity unless they are genuinely pathognomonic. In some cases, lesions are absent because they have resolved by the time the observations are made, or they are missed because of the line speed. Furthermore, lesions seen at slaughter would indicate that the disease has been present in the herd for some time.

Routine testing in the slaughter-house, such as serological tests on muscle exudates ("meat-juice"), is also flawed by inappropriate timing. However, the procedure can be used for specific monitoring programmes to determine national progress in disease control, such as for Aujeszky's Disease (Le Potier *et al.*, 1998) and PRRS (Mortensen *et al.*, 2001) or it can be used for public health in the monitoring of zoonotic agents such as salmonella (Davies *et al.*, 2003). Serological tests of meat-juice correlate only with exposure and, in the case of salmonella, there is no significant correlation between positive ELISA results and intestinal carriage or carcass contamination (Davies *et al.*, 2003; Davies *et al.*, 2004; Zheng *et al.*, 2007; Nowak *et al.*, 2007). Beloiel *et al.* (2003) concluded that it is during the first 8-61 days of life that shedding and infection take place as a random effect of sow factors such as parity, the degree of shedding during parturition and the level of maternal antibody in colostrum. Sero-conversion occurs in the last third of the finishing phase as a result of random effect of litter of origin and contamination of the pen in which the pigs are reared and thus sero-positive clusters are found at slaughter.

It is worth noting that high levels of pathogen contamination potentially exist in slaughterhouses and therefore results of highly sensitive tests, particularly PCR tests, must be viewed with caution because of the possibility of cross-contamination of samples.

Control - isolate in-coming breeding stock or be dammed

Laboratory tests do not have perfect sensitivity and specificity and, in their application and interpretation, assumptions have to be made about minimal disease prevalence, otherwise the sample size selected would have to include all animals in the target population. Clinical signs provide an early warning of disease, but they are not always apparent in latent infections or infections of low virulence. With current knowledge, the ultimate in risk management as far as the transfer of live animals within the breeding pyramid is concerned has to be effective isolation.

A recent survey of 429 pig producers carried out in the UK (Anon, 2007) identified some serious shortcomings in this aspect of risk management. Of these, 0.52 had purchased live animals within a period of 12 months, of which only 0.79 isolated them on arrival. Isolation facilities were within 100 metres of the main herd for 0.55, between 100 and 500 metres for 0.27, between 500 and 1000 metres for 0.08 and greater than 1000 metres for 0.10. Only 0.39 of producers interviewed isolated animals for a period of five weeks or more and only 0.39 consulted a veterinarian on policies for the introduction of replacement breeding stock. Just 0.09 considered co-operating with neighbours over matters of biosecurity.

The basic requirements for isolation facilities and procedures should aim for the following:

Not contiguous with the recipient herd.

A distance from the recipient herd of 500 metres or more.

Protection from wildlife, rodents and birds.

Operated all-in-all-out.

Adequate control of arthropods.

Arrival of animals from source herd to coincide with appropriate testing.

Transfer of animals to recipient herd after further testing as appropriate - perhaps 100% sampling and pooled testing.

Separate personnel or visited at the end of the working day.

Boots and overalls provided.

Hand-washing facilities as a minimum hygiene requirement.

Avoidance of perpetuation of disease by creating carriers through the so called "bugging-up" procedures whereby in-coming stock are intentionally infected.

Maximum possible period of isolation bearing in mind shedding periods for pathogens - at least 6 weeks.

Strategic use of vaccination - unlikely to eliminate carriage of infection.

Strategic use of antibiotics - also unlikely to eliminate carriage of infection.

Where isolation facilities are lacking, it is often advised that as few animal transfers as possible are made since common sense would seem to suggest that the risk of introduction of infection reduces with declining frequency of addition of groups of replacement breeding stock into recipient herds. The probability that a recipient herd will become infected will depend on the prevalence of infection in the in-coming animals, the number of pathogens carried, the infective dose, the overall infective load, the method of transmission and the degree of exposure of recipient animals. It is thus difficult to calculate the probability of infection, but if equation 2 is applied for an assumed prevalence of infection, it would appear mathematically that the frequency of introductions makes little difference to the probability of infection being transmitted to the recipient herd (L. E. Green and G. Medley. Personal Communication).

Conclusions

For the control of health of pig herds and risk management in transferring breeding stock or semen, there needs to be a greater understanding of virulence determination, latency, and disease potentiation and amplification. At the same time there needs to be development of multi-valent, inexpensive and rapid tests, possibly for use on-farm. Multiplex PCRs for example have been shown to be effective at simultaneously detecting several pathogens (e.g. Lee *et al.*, 2007) and therefore this technology has enormous potential in health control.

In the shorter term there is an urgent need to agree standardised industry protocols for monitoring clinical signs, tests and sampling procedures, surveillance and reporting of results down to commercial levels of production in the breeding pyramid.

Acknowledgement

Assistance from Julie Lennox in preparing the manuscript is greatly appreciated.

References

Albina E., Mesplède A., Chenut G., LePotier M.F, Bourbao G., LeGal S. and Leforban Y. (2000). A serological survey on classical swine fever (CSF), Aujeszky's disease (AD) and porcine reproductive and respiratory syndrome (PRRS) virus infections in French wild boars from 1991 to 1998. *Veterinary Microbiology*, **77**, 43-57.

Alexander T. (1998). Zoonoses. In: Proceedings of the 15th IPVS Congress, Birmingham, England. 5th-9th July 1998. pp.167-174.

Altman D.G., and Bland J.M. (1994a). Statistics Notes: Diagnostic tests 1: Sensitivity and specificity. *British Medical Journal*, **308,** 1552.

Altman D.G. and Bland J.M. (1994b). Statistics Notes: Diagnostic tests 2: Predictive value. *British Medical Journal*, **309**, 102.

Amass S.F., Clark L.K. and Wu C.C. (1995). Source and timing of *S. suis* infection in neonatal pigs: Implications for early weaning procedures. *Journal of Swine Health and Production*, **3,** 189-193.

Amass, S.F., Clark L.K., Knox K., Wu C.C. and Hill M.A. (1996). *Streptococcus suis* colonization of piglets during parturition. *Journal of Swine Health and Production*, **4**, 269-272.

Ameri-Mahabadi M., Zhou E.M. and Hsu W.H. (2005). Comparison of two swine *Mycoplasma hyopneumoniae* enzyme-linked immunosorbent assays for detection of antibodies from vaccinated pigs and field serum samples. *Journal of Veterinary Diagnostic Investigation*, **17**, 61-64

Andresen L.O., Klausen J., Barfod K. and Sørensen V. (2002). Detection of antibodies to *Actinobacillus pleuropneumoniae* Serotype 12 in pig serum using a blocking enzyme-linked immunosorbent assay. *Veterinary Microbriology*, **89**, 61-67.

Anon (1993). Adapted from Hurd (2003). In: Risk Analysis methods applied to swine disease controls. Swine Disease Conference for Swine Practitioners, 2003. Iowa State University, USA. pp.90-94.

Anon (1996). Council Directive 96/61/EC (24 September 1996). Concerning integrated pollution prevention and control. Official Journal L157, 10/10/1996, 0026-0040.

Anon (1999). Council Directive 90/429/EC (26 June 1990). Laying down the animal health requirements applicable to intra-community trade in the imports of semen of domestic animals of the porcine species. Official Journal L224, 18/8/1990, 0062-0072.

Anon (2007a). Diseases notifiable to the OIE. http://www.oie.int/eng/maladies/en_classification 2007.htm?eld7 (Accessed 18 March 2007).

Anon (2007b). Notifiable diseases. http://www.defra.gov.uk/animalh/diseases/notifiable/index.htm (Accessed 18 March 2007).

Anon (2007c). An independent Evidence Baseline for Farm Health Planning in England. Report for Defra by ADAS UK Ltd. March 2007. 146 pages.

Baldock F.C. (1998). What constitutes freedom from disease in livestock? *Australian Veterinary Journal*, **76**, 544-545.

Batista L., Pijoan C., Ruiz A. and Utrera V. (2002). Evaluation of *Mycoplasma hyopneumoniae* transmission from an infected to a näive herd. In: Proceedings of an International Symposium on Swine Disease Eradication, 2002. University of Minnesota Swine Disease Prevention Centre, Minnesota, USA. 043.

Baysinger A.K. (1999). Use of descriptive statistics in the interpretation of population serology. In: Proceedings of the American Association of Swine Practitioners. 30th Annual Meeting, St. Louis, Missouri, USA. 27th February-2nd March 1999. pp. 345-355.

Beloeil P.A., Chauvin C., Proux K., Rose N., Queguiner S., Eveno E., Houdayer C., Rose V., Fravelo P. and Madec F. (2003). Longitudinal serological responses to *Salmonella enterica* of growing pigs in a subclinically infected herd. *Preventive Veterinary Medicine*, **60**, 207-226.

Benfield D.A., Collins J.E., Dee S.A., Halbur P.G., Joo H.S., Lager K.M., Mengeling W.L., Murtaugh M.P., Rossow K.D., Stevenson G.W. and Zimmerman J.J. (1999). Porcine Reproductive and Respiratory Syndrome. In: Diseases of Swine 8th Edition. Ed. B.E. Straw, S. D'Allaire, W.L. Mengeling and D.J. Taylor. Blackwell Science, Oxford, UK. ISBN 0-632-05256-2, pp. 201-232.

Benfield D.A., Nelson, E., Collins J.E., Harris L., Goyal S., Robinson D., Christianson W.T., Morrison R.B., Gorcyca D. and Chladek D. (1992). Characteristics of swine infertility and respiratory syndrome (SIRS) virus (isolate ATCC VR-2332). *Journal of Veterinary Diagnostic Investigation*, **4**, 127-133.

Bertram T.A. (1990). *Actinobacillus pleuropneumoniae*: Molecular aspects of virulence and pulmonary injury. *Canadian Journal of Veterinary Research*, **54**, S53-S56.

Bornstein S. and Wallgren P. (1997). Serodiagnosis of sarcoptic mange in pigs. *Veterinary Record*, **141**, 8-12.

Boye M., Baloda S.B., Leser T.D. and Møller K. (2001). Survival of *Brachyspira hyodysenteriae* and *B. pilosicoli* in terrestrial microcosms. *Veterinary Microbiology*, **81**, 33-40.

Burr P. and Snodgrass D. (2004). Demystifying diagnostic testing: Serology. *In Practice*, **26**, 498-502.

Calsamiglia M., Pijoan C. and Trigo A. (1999). Application of a nested polymerase chain reaction assay to detect *Mycoplasma hyopneumoniae* from nasal swabs. *Journal of Veterinary Diagnostic Investigation*, **11**,

246-251.

Calsamiglia M., Collins J.E. and Pijoan C. (2000). Correlation between the presence of enzootic pneumonia lesions and the detection of *Mycoplasma hyopneumoniae* in bronchial swabs by PCR. *Veterinary Microbiology,* **76,** 299-303.

Calderaro A., Bommezzadri S., Gorrini C., Piccolo G., Peruzzi S., Dettori G. and Chezzi C. (2006). Comparative evaluation of molecular assays for the identification of intestinal spirochaetes from diseased pigs. *Veterinary Microbiology,* **118,** 91-100.

Cameron A.R. and Baldock F.C. (1998). A new probability formula for surveys to substantiate freedom from disease. *Preventive Veterinary Medicine,* **34,** 1-17.

Campitelli L., Donatelli I., Foni E., Castrucci M.R., Fabiani C., Kawaoka Y., Krauss S. and Webster R.G. (1997). Continued evolution of H1N1 and H3N2 influenza viruses in Italy. *Virology,* **233,** 310-318.

Cannon R.M. and Roe R.T. (1982). Livestock Disease Surveys: A field manual for veterinarians. Australian Government Publishing Service, Canberra, Australia. 35 pages.

Cardona A.C., Pijoan C. and Dee S.A. (2005). Assessing *Mycoplasma hyopneumoniae* aerosol movement at several distances. *Veterinary Record,* **156,** 91-92.

Cargill C. and Davies P.R. (1999). External Parasites. In: Diseases of Swine 8th Edition. Ed. B.E. Straw, S. D'Allaire, W.L. Mengeling and D.J. Taylor. Blackwell Science, Oxford, UK. ISBN 0-632-05256-2, pp. 669-683.

Chandler J.D., Riddell M.A., Li F., Love R.J. and Anderson D.A. (1999). Serological evidence for swine hepatitis E virus infection in Australian pig herds. *Veterinary Microbiology,* **68,** 95-105.

Cheung A.K. (1995). Investigation of pseudorabies virus DNA and RNA in trigeminal ganglia and the tonsil tissue of latently infected swine. *American Journal of Veterinary Research,* **56,** 45-50.

Chia S.P. and Taylor D.J. (1978). Factors affecting the survival of *Treponema hyodysenteriae* in dysenteric pig feces. *Veterinary Record,* **103,** 68-70.

Chiers D., Donne E., van Overbeke I., Ducatelle R. and Haesebrouck F. (2002). Evaluation of serology, bacteriological isolation and PCR for the detection of pigs carrying *Actinobacillus pleuropneumoniae* in the upper respiratory tract after experimental infection. *Veterinary Microbiology,* **88,** 385-392.

Christensen H., and Bisgaard M. (2004). Revised definition of *Actinobacillus sensu stricto* isolated from animals. *Veterinary Microbiology,* **99,** 13-30.

Christensen J. and Gardner I.A. (2000). Herd-level interpretation of test results for epidemiological studies of animal diseases. *Preventive Veterinary Medicine,* **45,** 83-106.

Christianson W.T., Collins J.E., Benfield D.A., Harris L., Gorcyca D.E., Chladek D.W., Morrison R.B. and Joo H.S. (1992). Experimental reproduction of swine infertility and respiratory syndrome in pregnant sows. *American*

Journal of Veterinary Research, **53**, 485-488.

Christopher-Hennings J., Nelson E.A., Nelson J.K., Hines E.J., Swendon S.L., Hill H.T., Zimmerman J.J. Katz J.B.,Yaeger M.J., Chase C.C. and Benfield D.A. (1995a). Detection of porcine reproductive and respiratory syndrome virus in boar semen by PCR. *Journal of Clinical Microbiology*, **33**, 1730-1734.

Christopher-Hennings J., Nelson E.A., Hines R.J., Nelson J.K., Swenson S.L., Zimmerman J.J., Chase C.L., Yaeger M.J. and Benfield D.A. (1995b). Persistence of porcine reproductive and respiratory syndrome virus in serum and semen of adult boars. *Journal of Veterinary Diagnostic Investigation*, **7**, 456-464.

Christopher-Hennings J., Nelson E.A. and Benfield D.A. (1996). Detecting PRRSv in boar semen. *Journal of Swine Health and Production*, **4**, 37-39.

Christopher-Hennings J., Dammen M., Nelson E., Rowland R. and Oberst R. (2006). Comparison of RNA extraction methods for the detection of porcine reproductive and respiratory syndrome virus from boar semen. *Journal of Virological Methods*, **136**, 248-253.

Czaja T., Kanci A., Lloyd L.C., Markham P.F. Whithear K.G. and Browning G.F. (2002). Induction of enzootic pneumonia in pigs by the administration of an aerosol of *in vitro* cultured *Mycoplasma hyopneumoniae*. *Veterinary Record*, **150**, 9-11.

Davies R.H., Heath P.J., Coxon S.M and Sayers A.R. (2003). Evaluation of the use of pooled serum, pooled muscle tissue fluid (meat-juice) and pooled feces for monitoring pig herds for salmonella. *Journal of Applied Microbiology*, **95**, 1016-1025.

Davies R.H., Dalziel R., Gibbens J.C., Wilesmith J.W., Ryan J.M., Evans S.J., Byrne C., Parba G.A., Pascoe S.J. and Teale C.J. (2004). National survey for salmonella in pigs, cattle and sheep at slaughter in Great Britain (1999-2000). *Journal of Applied Microbiology*, **96**, 750-760.

Dee S.A, Deen J. Otake S. and Pijoan C. (2000). An experimental model to evaluate the rôle of transport vehicles as a source of transmission of porcine reproductive and respiratory syndrome virus to susceptible pigs. *Canadian Journal of Veterinary Research*, **68**, 128-133.

Dee S.A., Jacobson L., Rossow K.D. and Pijoan C. (2005a). A laboratory model to evaluate the rôle of aerosols in the transport of porcine reproductive and respiratory syndrome virus. *Veterinary Records*, **156**, 501-504.

Dee S., Torremorell M. Thompson R., Dean J. and Pijoan C. (2005b). An evaluation of thermo-assisted drying and decontamination for the elimination of porcine reproductive and respiratory syndrome virus from contaminated livestock transport vehicles. *Canadian Journal of Veterinary Research*, **69**, 58-63.

de Jong J.C., Paccaud M.F., de Rond-Verloop F.M., Huffels N.H., Verwei C.,

Weijers T.F., Bangma P.J., van Kregten E., Kerckhaert J.A.M., Wicki F. and Wunderli W. (1988). Isolation of swine-like influenza A (H1N1) viruses from man in Switzerland and The Netherlands. *Annales de l'Institut Pasteur Virologie*, **139**, 429-437.

de Smit A.J., Bouma A., Terpstra C. and van Oirschot J.T. (1999). Transmission of classical swine fever virus by artificial insemination. *Veterinary Microbiology*, **67**, 239-249.

Done J.T. and Wilesmith J.W. (1988). Immunology applied: Monitoring and sampling. *Pig Veterinary Society Proceedings*, **20**, 61-70.

Drew T.W. (1995). Comparative serology of porcine reproductive and respiratory syndrome in eight European laboratories using immunoperoxidase monolayer assay and enzyme-linked immunosorbent assay. *Revue Scientifique et Technique de l'Office International des Epizooties*, **14**, 761-775.

Drobinec J., Favorov M.O., Shapiro C.N., Belle B.P., Mast E.E., Dadu A., Culver D., Iarovoi P., Robertson B.H. and Margolis H.S. (2001). Hepatitis E virus antibody prevalence among persons who work with swine. *Journal of Infectious Diseases*, **184**, 1594-1597.

Durmic Z., Pethick D.W., Pluske J.R. and Hampson D.J. (1998). Changes in bacterial populations in the colon of pigs fed different sources of dietary fibre and the development of swine dysentery after experimental infection. *Journal of Applied Microbiology*, **85**, 574-582.

Ellis W.A., McParland P.J., Bryson D.G. and Cassells J.A. (1986). Boars as carriers of leptospires in the Australis serogroup on farms with an abortion problem. *Veterinary Record*, **118**, 563.

Enøe C., Andersen S., Sørensen V. and Willeberg P. (2001). Estimation of sensitivity, specificity and predictive values of two serological tests for the detection of antibodies against *Actinobacillus pleuropneumoniae* serotype 2 in the absence of a reference test ("gold standard"). *Preventive Veterinary Medicine*, **51**, 227-243.

Euzéby J.P. (1998). Dictionnaire de Bactériologie Vétérinaire. Société de Bacteriologie Systématique et Vétérinaire (SBSV). www.bacterio.cict.fr./bacdico/aa/pleuropneumoniae.html (Accessed 15 March 2007).

Fano E., Pijoan C. and Dee S. (2005a). Evaluation of the aerosol transmission of a mixed infection of *Mycoplasma hyodysenteriae* and porcine reproductive and respiratory syndrome virus. *Veterinary Record*, **157**, 105-108.

Fano E., Pijoan C. and Dee S. (2005b). Dynamics and persistence of *Mycoplasma hyopneumoniae* infection in pigs. *Canadian Journal of Veterinary Research*, **69**, 223-228.

Feng W., Laster S.M., Tompkins M., Brown T., Xu J.S., Altier C., Gomez W., Benfield D. and McCaw M.B. (2001). *In utero* infection by porcine reproductive and respiratory syndrome virus is sufficient to increase susceptibility of piglets to challenge by *Streptococcus suis* Type II.

Journal of Virology, **75**, 4889-4895.

Floegel G., Wehrend A., Depner K.R., Fritzmeier J., Waberski D. and Moennig V. (2000). Detection of classical swine fever virus in semen of infected boars. *Veterinary Microbiology,* **77**, 109-116.

Forsberg R., Oleksiewicz M.B., Petersen A.M., Hein J., Bøtner A. and Storgaard T. (2001). A molecular clock dates the common ancestor of European-type porcine reproductive and respiratory syndrome virus at more than 10 years before the emergence of disease. *Virology,* **289**, 174-179.

Forsberg R., Storgaard T., Nielsen H.S., Oleksiewicz M.B, Cordioli P., Sala G., Hein J. and Bøtner A. (2002). The genetic diversity of European-type PRRSv is similar to that of the North American type but is geographically skewed within Europe. *Virology,* **299**, 38-47.

Frey J., Kuhn R. and Nicolet J. (1994). Association of the CAMP phenomenon in *Actinobacillus pleuropneumoniae* with the RTX toxins Apx I, Apx II and Apx III. *FEMS Microbiology Letters,* **124**, 245-251.

Glock R.D., Kinyon J.M. and Harris D.L. (1978). Transmission of *Treponema hyodysenteriae* by canine and avian vectors. In: Proceedings of the 5th IPVS Congress, Zagreb, 13-15th June, 1978. KB63

Goldberg T.L., Weigel R.M., Hahn E.C. and Scherba G. (2000). Associations between genetics, farm characteristics and clinical disease in field outbreaks of porcine reproductive and respiratory syndrome virus. *Preventive Veterinary Medicine,* **43**, 293-302.

Goodwin R.F.W. (1985). Apparent re-infection of enzootic pneumonia-free pig herds: Search for possible causes. *Veterinary Record,* **116**, 690-694.

Gottschalk M., Broes A., Mittal K.R., Kobisch M., Kuhnert P., Lebrun A. and Frey J. (2003). Non-pathogenic *Actinobacillus* isolates antigenically and biochemically similar to *Actinobacillus pleuropneumoniae*: A novel species? *Veterinary Microbiology,* **92**, 87-101.

Gradil C., Dubuc C. and Eaglesome M.D. (1996). Porcine reproductive and respiratory syndrome virus: Seminal transmission. *Veterinary Record,* **138**, 521-522.

Gresham A.C.J. (2002). The interpretation of laboratory test results for pigs. *The Pig Journal,* **49**, 113-133.

Griffin R.M. and Hutchings D.A. (1980). Swine dysentery: Observations on the frequency of latent infection. *Veterinary Record,* **107**, 559.

Grøndahl-Hansen J., Barfod K., Klausen J., Andresen L.O., Heegaard P.M.H. and Sørensen V. (2003). Development and evaluation of a mixed long-chain lipopolysaccharide-based ELISA for serological surveillance of infection with *Actinobacillus pleuropneumoniae* serotypes 2, 6 and 12 in pig herds. *Veterinary Microbiology,* **96**, 41-51.

Guérin B. and Pozzi N. (2005). Viruses in boar semen: Detection and clinical as well as epidemiological consequences regarding disease transmission by artificial insemination. *Theriogenology,* **63**, 556-572.

Harris D.L. and Alexander T.J.L. (1999). Methods of Disease Control. In: Diseases of Swine 8th Edition. Ed. B.E. Straw, S. D'Allaire, W.L. Mengeling and D.J. Taylor. Blackwell Science, Oxford, UK. ISBN 0-632-05256-2. pp. 1077-1110.

Harris D.L. and Lysons R.J. (1992). Swine Dysentery. In: Diseases of Swine 7th Edition. Ed. A.D. Leman, B.E. Straw, W.L. Mengeling, S. D'Allaire, D.J. Taylor. Iowa State University Press, Ames, USA. ISBN 0-8138-0442-6. pp. 599-616.

Hermann J.R., Munoz-Zanzi C.A., Roof M.B., Burkhart K. and Zimmerman J.J. (2005). Probability of porcine reproductive and respiratory syndrome (PRRS) virus infection as a function of exposure and route. *Veterinary Microbiology*, **110**, 7-16.

Hermann J., Yoon K-J., Hoff S., Munoz-Zanzi C. and Zimmerman J. (2006). An update on PRRSv aerobiology research. In: Proceedings of the Allen D. Leman Swine Conference, 2006. University of Minnesota, USA. pp.102-104.

Higgins R., Gottschalk M., Boudreau M., Lebrun A. and Henricksen J. (1995). Description of six new *Streptococcus suis* capsular types. *Journal of Veterinary Diagnostic Investigation*, **7**, 405-406.

Horter D.C., Pogranichny R.M., Chang C.-C., Evans R.B., Yoon K.-J. and Zimmerman J.J. (2002). Characterisation of the carrier state in porcine reproductive and respiratory syndrome virus infection. *Veterinary Microbiology*, **86**, 213-228.

Inkik S., Schmoll F., Sipos W. and Klein D. (2005). Genetic variability of PRRS virus in Austria: Consequences for molecular diagnostics and viral quantificiation. *Veterinary Microbiology*, **107**, 171-178.

Jansen R., Briare J., Kamp E.M., Gielkens A.L.J. and Smits M.A. (1995). The CAMP effect of *Actinobacillus pleuropneumoniae* is caused by Apx toxins. *FEMS Microbiology letters*, **126**, 139-143.

Jansson D.S., Johansson K.E., Olofsson T., Råsbäck T., Vagsholm I., Pettersson B., Gunarsson A. and Fellström C. (2004). *Brachyspira hyodysenteriae* and other strongly beta-haemolytic and indole-positive spirochaetes isolated from mallards (*Anas platyrhynchos*). *Journal of Medical Microbiology*, **53**, 293-300.

Jensen N.S., Stanton T.B. and Swayne D.E. (1996). Identification of the swine pathogen *Serpulina hyodysenteriae* in rheas (*Rhea americana*). *Veterinary Microbiology*, **52**, 259-269.

Jobert J.L., Savoye C., Cariolet R., Kobisch M. and Madec F. (2000). Experimental aerosol transmission of *Actinobacillus pleuropneumoniae* to pigs. *Canadian Journal of Veterinary Research*, **64**, 21-26.

Joens L.A. (1980). Experimental transmission of *Treponema hyodysenteriae* from mice to pigs. *American Journal of Veterinary Research*, **41**, 1225-1226.

Jorsal S.E. and Thomsen B.L. (1988). A Cox regression analysis of risk factors related to *Mycoplasma suipneumoniae* re-infection in Danish SPF herds. *Acta Veterinaria Scandinavica*, **84**, (Supplement), 436-438.

Kielstein P. and Rapp-Gabrielson V..J. (1992). Designation of 15 serovars of *H. parasuis* on the basis of immunodiffusion using heat stable antigen extracts. *Journal of Clinical Microbiology*, **30**, 862-865.

Klausen J., Andresen L.O., Barfod K. and Sórensen V. (2001). Blocking enzyme-linked immunosorbent assay for detection of antibodies against *Actinobacillus pleuropneumoniae* serotype 6 in pig serum. *Veterinary Microbiology*, **79**, 11-18.

Klausen J., Andresen L.O., Barfod K. and Sórensen V. (2002). Evaluation of an enzyme-linked immunosorbent assay for serological surveillance of infection with *Actinobacillus pleuropneumoniae* serotype 5 in pig herds. *Veterinary Microbiology*, **88**, 223-232.

Kristensen C.S., Angen Ø., Andreasen M., Takai H., Nielsen J.P. and Jorsal H.E. (2004). Demonstration of airborne transmission of *Actinobacillus pleuropneumoniae* serotype 2 between simulated pig units located at close range. *Veterinary Microbiology*, **98**, 243-249.

Kristensen C.S., Bøtner A., Takai H., Nielsen J.P. and Jorsal S.E. (2004). Experimental airborne transmission of PRRS virus. *Veterinary Microbiology*, **99**, 197-202.

Kwaga J. and Iversen J.O. (1993). Plasmids and outer membrane proteins of *Yersinia enterocolitica* and related species of swine origin. *Veterinary Microbiology*, **36**, 205-214.

Lee C-S., Moon H-J., Yang J-S., Park S-J., Song D-S., Kang B-K. and Park B-K. (2007). Multiplex PCR for simultaneous detection of pseudorabies virus, porcine cytomegalovirus and porcine circovirus in pigs. *Journal of Virological Methods*, **139**, 39-43.

LePotier M.F., Fournier A., Houdayer C., Hutet E., Auvigne V., Hery D., Sanaa M. and Toma B. (1998). Use of muscle exudates for the detection of anti-gE antibodies to Aujeszky's disease virus. *Veterinary Record*, **143**, 385-387.

Leuenberger R., Boujon P., Thür B., Miserez R., Garin-Bastuji B., Rüfenacht J. and Stärk K.D.C. (2007). Prevalence of classical swine fever, Aujeszky's disease and brucellosis in a population of wild boar in Switzerland. *Veterinary Record*, **160**, 362-368.

Loeffen W.L.A., Kamp E.M., Stockhofe-Zurwieden N., van Nieuwstadt A.P.K.M., Bongers J.H., Hunneman W.A., Elbers A.R.W., Baars J., Neth J. and van Zijderveld F.G. (1999). Survey of infectious agents involved in acute respiratory disease in finishing pigs. *Veterinary Records*, **145**, 123-129.

Lun Z-R., Wang Q-P., Chan X-G., Li A-X. and Zhu X-Q. (2007). *Streptococcus suis*: An emerging zoonotic pathogen. *The Lancet, Infectious Diseases*, **7**, 201-209

Lymbery A.J., Hampson D.J., Hopkins R.M., Combs B. and Mhoma J.R.L.

(1990). Multilocus enzyme electrophoresis for identification and typing of *T. hyodysenteriae* and related spirochaetes. *Veterinary Microbiology*, **22**, 89-99.

Lysons R.J., Kent M.A., Bland A., Sellwood R., Robinson W.F. and First A. (1991). A cytolytic haemolysin from *T. hyodysenteriae*: A probable virulence determinant in swine dysentery. *Journal of Medical Microbiology*, **34**, 97-102.

MacDiarmid S.C. (1993). Risk analysis and the importation of animals and animal products. *Revue Scientifique et Technique de l'Office International des Epizooties*, **12**, 1093-1107.

MacDiarmid S.C. and Pharo H.J. (2003). Risk analysis: Assessment, management and communication. *Revue Scientifique et Technique de l'Office International des Espizooties*, **22**, 397-408.

Mackinnon J.D. (2001). Some clinical and epidemiological aspects of the outbreak of classical swine fever in East Anglia in 2000. *State Veterinary Journal*, **11**, 2-7.

MacMillan A.P. (1999). Brucellosis. In: Diseases of Swine 8th Edition. Ed. B.E. Straw, S. D'Allaire, W.L. Mengeling and D.J. Taylor. Blackwell Science, Oxford, UK. ISBN 0-632-05256-2. pp. 385-393.

Madec F. and Rose N. (2003). How husbandry practices may contribute to the course of infectious diseases in pigs. In: Proceedings of the 4th International Symposium on Emerging and Re-emerging Pig Diseases, Rome, 29th June- 2nd July, 2003. pp. 9-18.

Mateu E., Diaz I., Darwich L., Casal J., Martin M. and Pujols J. (2006). Evolution of ORF5 of Spanish porcine reproductive and respiratory syndrome virus strains from 1991-2005. *Virus Research*, **115**, 198-206.

Mattson J.G., Bergstrom K., Wallgreen P. and Johansson K.E. (1995). Detection of *Mycoplasma hyopneumoniae* in nose swabs from pigs by *in vitro* amplification of the 165 r RNA gene. *Journal of Clinical Microbiology*, **33**, 893-897.

Meng X-J. (2000). Zoonotic and xenozoonotic risks of the hepatitis E virus. *Infectious Disease Review*, **2**, 35-41.

Meyns T., Maes D., Dewulf J., Vicca J., Haesebrouck F. and de Kruif A. (2004). Quantification of the spread of *Mycoplasma hyopneumoniae* in nursery pigs using transmission experiments. *Preventive Veterinary Medicine*, **66**, 265-275.

Mengeling W.L., Lager K.M. and Vorwald A.C. (1994). Temporal characterisation of transplacental infection of porcine fetuses with porcine reproductive and respiratory syndrome virus. *American Journal of Veterinary Research*, **55**, 1391-1398.

Miller L., McElvaine M.D., McDowell R.M. and Ahl A.S. (1993). Developing a quantitative risk assessment process. *Revue Scientifique et Technique de l'Office International des Epizooties*, **12**, 1153-1164.

Møller K., Jensen T.K., Jorsal S.E., Leser T.D. and Carstensen B. (1998).

Detection of *Lawsonia intracellularis*, weakly â-haemolytic spirochaetes (WBHS), *Salmonella enterica* and haemolytic *E. coli* from swine herds with and without diarrhoea amongst growing pigs. *Veterinary Microbiology*, **62**, 59-72.

Morris C.R., Gardner I.A., Hietala S.K., Carpenter T.E., Anderson R.J. and Parker K.M. (1995). Seroepidemiologic study of natural transmission of *Mycoplasma hyodysenteriae* in a swine herd. *Preventive Veterinary Medicine*, **21**, 323-337.

Mortensen S., Strandbygaard B., Bøtner A., Feld N. and Willeberg P. (2001). Monitoring porcine reproductive and respiratory syndrome virus infection status in swine herds based on analysis of antibodies in meat-juice samples. *Veterinary Research*, **32**, 441-453.

Nagatomo H., Takegahara Y., Somoda T., Yamaguchi A., Uemura R., Hagiwara S. and Sueyoshi M. (2201). Comparative studies of the persistence of animal mycoplasmas under different environmental conditions. *Veterinary Microbiology*, **82**, 223-232.

Nicolet J. (1990). Overview of the virulence attributes of the HAP group of bacteria. *Canadian Journal of Veterinary Research*, **54**, S12-S15.

Nielsen R., van den Bosch J.F., Plambeck T., Sørensen V. and Nielsen J.P. (2000). Evaluation of an indirect enzyme-linked immunosorbent assay (ELISA) for detection of antibodies to the Apx toxins of *Actinobacillus pleuropneumoniae*. *Veterinary Microbiology*, **71**, 81-87.

Nowak B., van Müffling T., Chaunchom S. and Harting J. (2007). Salmonella contamination in pigs at slaughter and on the farm: A field study using an antibody ELISA test and a PCR technique. *International Journal of Food Microbiology*. In Press.

Otake S., Dee S.A., Rossow K.D., Moon R.D. and Pijoan C. (2002a). Mechanical transmission of porcine reproductive and respiratory syndrome virus by mosquitoes, *Aedes vexans*. *Canadian Journal of Veterinary Research*, **66**, 191-195.

Otake S., Dee S.A., Jacobson L., Torremorell M. and Pijoan C. (2002b). Evaluation of aerosol transmission of porcine reproductive and respiratory syndrome virus under controlled field conditions. *Veterinary Record*, **150**, 804-808.

Otake S., Dee S.A., Rossow K.D., Deen J., Joo H.S. Molitor T.W. and Pijoan C. (2002c). Transmission of porcine reproductive and respiratory syndrome virus by fomites (boots and coveralls). *Journal of Swine Health and Production*, **10**, 59-65.

Otake S., Dee S.A., Rossow K.D., Joo H.S., Deen J. and Molitor T.W. (2002d). Transmission of porcine reproductive and respiratory disease virus by needles. *Veterinary Record*, **150**, 114-115.

Otake, S., Dee. S.A., Moon R.D., Rossow K.D., Trincado C. and Pijoan C. (2004). Studies on the carriage and transmission of porcine reproductive and respiratory syndrome virus by individual houseflies (*Musca domestica*). *Veterinary Record*, **154**, 80-85.

Perch B., Pedersen K.B. and Henrichsen J. (1983). Serology of capsulated streptococci pathogenic for pigs: Six new serotypes of *Streptococcus suis*. *Journal of Clinical Microbiology*, **17**, 993-996.

Pesch S., Meyer C. and Ohlinger V.F. (2005). New insights into the genetic diversity of European porcine reproductive and respiratory syndrome virus (PRRSv). *Veterinary Microbiology*, **107**, 31-48.

Pesente P., Rebonato V., Sandri G., Giovanardi D., Ruffoni L.S. and Torriani S. (2006). Phylogenetic analysis of ORF5 and ORF7 sequences of porcine reproductive and respiratory syndrome virus (PRRSv) from PRRS-positive Italian farms: A showcase for PRRSv epidemiology and its consequences on farm management. *Veterinary Microbiology*, **114**, 214-224.

Pieters M., Fano E., Dee S. and Pijoan C. (2006). Transmission of *Mycoplasma hyopneumoniae*: How long can it last? Proceedings of the Allen D. Leman Swine Conference, 2006. University of Minnesota College of Veterinary Medicine, Minnesota, USA. 79-81.

Pirtle E.C. and Beran G.W. (1996). Stability of porcine reproductive and respiratory syndrome virus in the presence of fomites commonly found on farms. *Journal of American Veterinary Medical Association*, **208**, 390-392.

Plagemann P.G.W. (1996). Lactate dehydrogenase-elevating virus and related viruses. In: Fields Virology. 3rd Edition. Ed. B.N. Fields, D.M. Knipe, P.M. Howley, R.M. Charnock, T.P. Monath, J.L. Melnick, B. Roizman and S.E. Straus. Lippincott, Williams and Wilkins, Philadelphia, USA. ISBN 10: 0781702534.

Pluske J.R., Pethick D.W., Durmic Z., Mullan B.P. and Hampson D.J. (1996). The incidence of swine dysentery in pigs can be reduced by feeding diets that limit the amount of fermentable substrate entering the large intestine. *Journal of Nutrition*, **126**, 2920-2933.

Pointon A.M., Morrison R.B., Hill G., Dargatz D. and Dial G. (1990). Monitoring pathology in slaughtered stock: Guidelines for selecting sample size and interpreting results. National Animal Health Monitoring System (NAHMS). United States Department of Agriculture, Fort Collins, Colorado, USA. 21 pages.

Prieto C. and Castro J.M. (2005). Porcine reproductive and respiratory syndrome virus infection in boars: A review. *Theriogenology*, **63**, 1-16.

Prieto C., Suárez P., Simarro I., Solana A., Castro J.M., Sánchez R. and Martín-Rillo S. (1996). Exposure of gilts in early gestation to porcine reproductive and respiratory syndrome virus. *Veterinary Record*, **138**, 536-539.

Prieto C., Suárez P., Simarro I., García C., Martín-Rillo S. and Castro J.M. (1997a). Insemination of susceptible and pre-immunised gilts with boar semen containing porcine reproductive and respiratory syndrome virus.

Theriogenology, **47**, 647-654.

Prieto C., Suárez P., Simarro I., García C., Fernández A. and Castro J.M. (1997b). Transplacental infection following exposure of gilts to porcine reproductive and respiratory syndrome virus at the onset of gestation. *Veterinary Microbiology*, **57**, 301-311.

Råsbäck T., Fellström C., Gunnarsson A. and Aspán A. (2006). Comparison of culture and biochemical tests with PCR for detection of *Brachyspira hyodysenteriae* and *B. pilosicoli*. *Journal of Microbiological Methods*, **66**, 347-353.

Råsbäck T., Jansson D.S., Johansson K-E. and Fellström C. (2007). A novel enteropathogenic, strongly haemolytic spirochaete isolated from pig and mallard, provisionally designated *Brachyspira suanatina* sp. nov. *Environmental Microbiology*, **9**, 983-991.

Revilla-Fernández S., Wallner B., Truschner K., Benczak A., Brem G., Schmoll F., Mueller M. and Steinborn R. (2005). The use of endogenous and exogenous reference RNAs for qualitative and quantitative detection of PRRSv in porcine semen. *Journal of Virological Methods*, **126**, 21-30.

Rose N., Larow G., Le Diguerher G., Eveno E., Jolly J.P., Blanchard P., Oger A., Le Dimna M., Jestin A. and Madec F. (2003). Risk factors for porcine post-weaning multisystemic wasting syndrome (PMWS) in 149 French farrow-to-finish herds. *Preventive Veterinary Medicine*, **61**, 209-225.

Rziha H.J., Doller P.C. and Wittmann G. (1982). Detection of Aujeszky's disease virus and viral DNA in tissues of latently infected pigs. In: Current Topics in Veterinary Medicine and Animal Science. Ed. G. Wittmann and S.A. Hall, **17**, 205-211.

Sard D.M. (1979). Dealing with data: The practical use of numerical information: 14, Monitoring changes. *Veterinary Record*, **105,** 323-328.

Savoye C., Jobert J.L., Berthelot-Hérault F., Keribin A.M., Cariolet R., Moran H., Madec F. and Kobisch M. (2000). A PCR assay used to study aerosol transmission of *Actinobacillus pleuropneumoniae* from samples of live pigs under experimental conditions. *Veterinary Microbiology*, **73**, 337-347.

Schurrer J.A., Dee S.A., Moon R.D., Rossow K.D., Mahlum C., Mondaca E., Otake S., Fano E., Collins J.E. and Pijoan C. (2004). Spatial dispersal of porcine reproductive and respiratory syndrome virus-contaminated flies after contact with experimentally infected pigs. *American Journal of Veterinary Research*, **65**, 1284-1292.

Sheehan B.J., Bossé J.T., Beddek A.J., Rycroft A.N., Kroll J.S. and Langford P.R. (2003). Identification of *Actinobacillus pleuropneumoniae* genes. *Infection and Immunity*, **71**, 3960-3970.

Siba P.M., Pethick D.W. and Hampson D.J. (1996). Pigs experimentally infected with *S. hyodysenteriae* can be protected from developing swine dysentery by feeding them a highly digestible diet. *Epidemiology and Infection*, **116**, 207-216.

Sørensen K.J., Strandbygaard B., Bøtner A., Madsen E.S., Nielsen J. and Hare P. (1998). Blocking ELISAs for the distinction between antibodies against European and American strains of porcine reproductive and respiratory syndrome virus. *Veterinary Microbiology*, **60**, 169-177.

Stadijek T., Oleksiewicz M.B., Potapchuk D. and Podgorska K. (2006). Porcine reproductive and respiratory syndrome virus strains of exceptional diversity in Eastern Europe support the definition of new genetic subtypes. *Journal of General Virology*, **87**, 1835-1841.

Stärk K.D., Nicolet J. and Frey J. (1998). Detection of *Mycoplasma hyopneumoniae* by air sampling with a nested PCR assay. *Applied and Environmental Microbiology*, **64**, 543-548.

Swenson S.L., Hill H.T., Zimmerman J.J., Evans L.E., Landgraf J.G., Wills R.N., Sanderson T.P., McGinley M.J., Brevik A.K. and Ciszewski D.K. (1994a). Excretion of porcine reproductive and respiratory syndrome virus in semen after experimentally induced infection in boars. *Journal of the American Veterinary Medical Association*, **204**, 1943-1948.

Swenson S.L., Hill H.T., Zimmerman J.J., Evans L.E., Willis R.W. and Yoon K.J. (1994b). Artificial insemination of gilts with porcine reproductive and respiratory syndrome (PPRS) virus-contaminated semen. *Journal of Swine Health and Production*, **2**, 19-23.

Taylor D.J. (1999). Actinobacillus pleuropneumonia. In: Diseases of Swine 8th Edition. Ed. B.E. Straw, S. D'Allaire, W.L. Mengeling and D.J. Taylor. Blackwell Science, Oxford, UK. ISBN 0-632-05256-2. pp. 343-354.

Taylor D.J. (2006a). Enzootic pneumonia. In: Pig Diseases. 8th Edition, 2006. D.J. Taylor. Book Production Consultants Ltd., Cambridge, UK. ISBN 0-9506932-7-8. pp.178-187.

Taylor D.J. (2006b). Pleuropneumonia (*Actinobacillus pleuropneumoniae* infection). In: Pig Diseases. 8th Edition, 2006. D.J. Taylor. Book Production Consultants Ltd., Cambridge, UK. ISBN 0-9506932-7-8. pp. 207-214.

Taylor D.J. (2006c). Swine dysentery. In: Pig Diseases. 8th Edition, 2006. D.J. Taylor. Book Production Consultants Ltd., Cambridge, UK. ISBN 0-9506932-7-8. pp. 156-164.

Torremorell M., Pijoan C., Janni K., Walker R. and Joo H.S. (1997). Airborne transmission of *Actinobacillus pleuropneumoniae* and PRRS in nursery pigs. *American Journal of Veterinary Research*, **58**, 828-832.

Tucker A. W. (2006). An update and overview of zoonotic infections of pigs. *The Pig Journal*, **57**, 178-191

van Rijn P.A., Wellenberg G.J., Hakze-van der Honing R., Jacobs L., Moonen P.L.J.M. and Feitsma H. (2004). Detection of economically important viruses in boar semen by quantitative Real Time PCR™ technology. *Journal of Virological Methods*, **120**, 151-160.

Vecht U., Wisselink H.J., Anakotta J. and Smith H.E. (1993). Discrimination between virulent and non-virulent *Streptococcus suis* Type 2 strains by enzyme-linked immunosorbent assay. *Veterinary Microbiology*, **34**, 71-82.

Velthuis A.G.J., de Jong M.C.M., Stockhofe N., Vermeulen T.M.M. and Kamp E.M. (2002). Transmissionof *Actinobacillus pleuropneumoniae* in pigs is characterised by variation in infectivity. *Epidemiology and Infection*, **129**, 203-214.

Vercruysse J., Geurden T. and Peelaers I. (2006). Development and Bayesian evaluation of an ELISA to detect specific antibodies to *Sarcoptes scabei var suis* in the meat juice of pigs. *Veterinary Record*, **158**, 506-508.

Verdin E., Saillard C., Labbé A., Bore J.M. and Kobisch M. (2000). A nested PCR assay for the detection of *Mycoplasma hyopneumoniae* in tracheobronchiolar washings from pigs. *Veterinary Microbiology*, **76**, 31-40.

Vigre H., Angen Ø., Barfod K., Lavritsen D.T. and Sørensen V. (2002). Transmission of *Actinobacillus pleuropneumoniae* in pigs under field-like conditions: Emphasis on tonsillar colonisation and passively acquired colostral antibodies. *Veterinary Microbiology*, **89**, 151-159.

Vilèek Š. and Nettleton P.F. (2006). Review: Pestiviruses in wild animals. *Veterinary Microbiology*, **116**, 1-12.

Wagstrom E.A., Chang C.C., Yoon K.J. and Zimmerman J.J. (2001). Shedding of porcine reproductive and respiratory syndrome virus in mammary gland secretions of sows. *American Journal of Veterinary Research*, **62**, 1876-1880.

Weissenböck H., Maderner A., Herzog A-M., Lussy H. and Nowotny N. (2005). Amplification and sequencing of *Brachyspira* spp. specific portions of *nox* using paraffin-embedded tissue samples from clinical colitis in Austrian pigs shows frequent solitary presence of *Brachyspira murdochii*. *Veterinary Microbiology*, **111**, 67-75.

Williams D.M., Lawson G.H.K. and Rowland A.C. (1973). Streptococcal infections in piglets. The palatine tonsil as a portal of entry for *S. suis*. *Research in Veterinary Science*, **15**, 352-362.

Willis R.W., Zimmerman J.J., Yoon K-J., Swenson S.L. McGinley M.J., Hill H.T., Platt K.B., Christopher-Hennings J. and Nelson E.A. (1997a). Porcine reproductive and respiratory syndrome virus: A persistent infection. *Veterinary Microbiology*, **55**, 231-240.

Wills, R.W. Zimmerman J.J., Yoon K-J., Swenson S.L., Hoffman L.J., McGinley M.J., Hill H.T. and Platt K.B. (1997b). Porcine reproductive and respiratory syndrome virus: Routes of excretion. *Veterinary Microbiology*, **57**, 69-81.

Windsor R.S. and Simmons J.R. (1981). Investigation into the spread of swine dysentery in 25 herds in East Anglia and assessment of its economic significance in five herds. *Veterinary Records*, **109**, 482-484.

Wisselink H.J., Reek F.H., Vecht U., Stockhofe-Zurwieden U., Smits M.A. and Smith H.E. (1999). Detection of virulent strains of *Streptococcus suis*

Type 2 and highly virulent strains of *Streptococcus suis* Type 1 in tonsillar specimens of pigs by PCR. *Veterinary Microbiology*, **67**, 143-157.

Yoon K.J., Joo H.S., Christianson W.T., Morrison R.B. and Dial G.D. (1993). Persistent and contact infection in nursery pigs experimentally infected with porcine reproductive and respiratory syndrome (PRRS) virus. *Journal of Swine Health and Production*, **1**, 5-8.

Yu H., Zhang G-H., Hua R-H., Zhang Q., Liu T-Q., Liao M. and Tong G-Z. (2007). Isolation and genetic analysis of human origin H1N1 and H3N2 influenza viruses from pigs in China. Biochemical and Biophysical Research Communications. In Press.

Zimmerman J. (2006). A review: PRRS virus transmission. In: Proceedings of the Swine Conference for Swine Practitioners, 2006. Iowa State University, Iowa, USA. pp 19-29.

Zheng D.M., Bonde M. and Sørensen J.J. (2007). Associations between the proportion of salmonella-seropositive slaughter pigs and the presence of herd level risk factors for introduction and transmission of salmonella in 34 Danish organic outdoor (non-organic) and indoor pig finishing farms. *Livestock Production Science*, **106,** 189-199.

Zhuang Q., Barfod K., Wachmann H., Mortensen S., Lauritsen D.T. Ydesen B. and Willeberg P. (2007). Risk factors for *Actinobacillus pleuropneumoniae* serotype 2 infection in Danish genetic specific pathogen-free herds. *Veterinary Record*, **160**, 258-262.

9

PIG RESPIRATORY BACTERIAL PATHOGENS: BACTERIAL OVERVIEW AND VACCINE TRENDS

SIMONE OLIVEIRA

University of Minnesota, College of Veterinary Medicine, Veterinary Diagnostic Laboratory, Veterinary Population Medicine Department, 1333 Gortner Ave, Saint Paul MN 55108, USA

Introduction

Haemophilus parasuis, Streptococcus suis, Actinobacillus suis, and *Actinobacillus pleuropneumoniae* are colonizers of the upper respiratory tract of pig that can potentially cause severe disease and high mortality in susceptible herds. These pathogens are of special concern in herds co-infected with viral pathogens such as porcine reproductive and respiratory syndrome virus and porcine circovirus type 2. Both viruses debilitate the innate immune system, which may facilitate secondary bacterial infections (Thacker et al, 2001; Segales, Domingo, and Chianini, 2004).

In most North American pig herds, *H. parasuis* and *S. suis* are the main agents involved in nursery mortality, whereas *A. suis* and *A. pleuropneumoniae* affect mainly finishing and adult pigs. Naïve populations are an exception, since disease caused by these pathogens may occur at any age (Oliveira, Pijoan and Morrison, 2002; Oliveira and Pijoan, 2004; Oliveira 2007).

Control of these pathogens has always been a challenge. Attempts to eliminate *H. parasuis* and *S. suis* from weaning pigs have been unsuccessful. *Actinobacillus pleuropneumoniae* eradication has been accomplished by using depopulation-repopulation strategies. Segregate early weaning and medicated early weaning, combined with test and removal of positive sows, has also been successfully used to produce *A. pleuropneumoniae*-free weaning pigs. Although *A. suis* eradication has not yet been attempted, many U.S. herds are considering using strategies similar to those used for *A. pleuropneumoniae* to eliminate *A. suis*.

In endemically-infected herds, prevention and control of disease caused by *H. parasuis, S. suis, A. suis,* and *A. pleuropneumoniae* has traditionally been performed using antibiotic treatments and/ or vaccination. Results may vary considerably between herds, and custom-made strategies are the ones that usually succeed. New diagnostic techniques for detection and characterization of these pathogens have considerably improved our understanding of dynamics of infection, and this information has been critical for the development of control strategies (Oliveira and Pijoan, 2004)

This chapter aims to review the current knowledge on *H. parasuis, S. suis, A. suis* and *A. pleuropneumoniae* diagnosis, epidemiology, and control. Trends in vaccine development will also be addressed.

Haemophilus parasuis

Haemophilus parasuis may be isolated from the nasal cavity, tonsil, and trachea of healthy carrier pigs. Under favorable conditions (naïve pigs, stress, viral confections) this pathogen can cause systemic infection and death within 24 to 72 hours. Severe outbreaks with 60 % mortality have been reported in naïve populations. In conventional herds, mortality caused by *H. parasuis* occurs mainly in the nursery, between 7 and 8 weeks of age (Oliveira and Pijoan 2004).

Haemophilus parasuis is highly variable regarding its genotype and phenotype (serovar). There are at least 15 different serovars described in the literature, and many field isolates are non-typable using the techniques available (Kielstein and Rapp-Gabrielson, 1992; Del Rio, Gutierrez and Rodriguez, 2003; Tadjine, Mittal and Bourdon, 2004). The University of Minnesota Veterinary Diagnostic (MN VDL) genomic database (https://mvdl.auxs.umn.edu/vetlabs/genomics.html) has identified at least 42 different genotypes of *H. parasuis* among 116 field isolates (information retrieved on March 2007). *Haemophilus parasuis* has the highest Simpson's index of diversity (0.93) among all the pig bacterial pathogens that are genotyped at the MN VDL, meaning that a random isolate has a 93 % chance of being included in a unique genotype group (Simpson, 1949). This information suggests that *H. parasuis* has a dynamic genome. This may be the reason why the development of a universal vaccine has not yet been accomplished. Even though there is high genotypic and phenotypic variability among *H. parasuis* field isolates, homologous protection (serovar or strain specific) is usually satisfactory. As expected, heterologous protection is either partial or non-existent (Oliveira and Pijoan, 2004). Exposure of 5-day old piglets to a low dose (10^4 CFU/ml) of live virulent *H. parasuis* in the presence of maternal immunity has been shown to reduce nursery mortality by 50% compared with inactivated vaccines (commercial or autogenous) (Oliveira, Pijoan, and Morrison, 2004). These results suggest that *H. parasuis* may express important antigens *in vivo* that are apparently not produced during *in vitro* culture.

Significant advances in *H. parasuis* genomics have been recently reported. One of the most important scientific contributions is the development and validation of a transformation system that can be used to generate *H. parasuis* mutants targeting specific genes (Bigas, Garrido and Rozas, 2005). Differential display and microarray analysis have been successfully used to identify genes that are up or down regulated as a result of *in vitro* conditions mimicking the *in vivo* environment. Some of these genes have DNA sequences similar to those involved in virulence in other bacterial species from the pasteurellaceae family (Hill, Metcalf and MacInnes, 2003; Melnikow, Dornan and Sargent, 2005). The description of potential virulence factors and the availability of a transformation system will certainly advance our knowledge on vaccine development for *H. parasuis*. However, parallel to the evaluation of vaccine candidates, the study of the stability of *H. parasuis* genome and the conditions that may affect it are equally important.

Actinobacillus suis

Actinobacillus suis is also a colonizer of the upper respiratory tract. Although this organism has been reported to colonize the vaginal mucosa, nasal cavity, and tonsils of healthy (carrier) pig, our laboratory has only confirmed its presence in the tonsils of sows and piglets by isolation and PCR testing. Similarly to *H. parasuis*, *A. suis* can also cause systemic infection. Although some of the *A. suis* virulence factors are known (Apx I $_{var. suis}$ and Apx II $_{var. suis}$), the factors that may trigger systemic infection still remain to be defined. *Actinobacillus suis* may affect pigs from varied ages, including neonatal, nursery, finishing, and adult pigs. Lesions associated with *A. suis* infection are highly variable, and may include pneumonia, pleuritis, pericarditis, enteritis, metritis, pethechial hemorrhages, and presence of bacterial emboli in several organs, including the skin. The differential diagnosis of *A. suis* infection includes *H. parasuis*, *Actinobacillus pleuropneumoniae*, *Erysipelothrix rhusiopathiae,* and other bacterial pathogens that can cause systemic infection. Traditional diagnosis is performed by isolating *A. suis* from clinical samples. PCR tests are also available for the differential diagnosis of these pathogens (Oliveira, 2006, 2007).

Actinobacillus suis has recently emerged as a new treat to the pig industry in the United States. Finishing pigs are the main population affected by *A. suis,* especially between 12 and 20 weeks of age. Antibiotics and autogenous vaccines are being used as the first lines of defense by most pig producers (Oliveira 2006, 2007). However, there are very limited data on efficacy of autogenous vaccines (Lapointe, D'Allaire and Lacouture, 2001). *Actinobacillus suis* differs from other commensal organisms colonizing the upper respiratory tract of pig regarding strain variability. The Simpson's index of diversity for *A. suis* genotypes is 0.64, meaning that a random isolate has a 64% chance of being included in a unique genotype group using BOX-PCR, for example (Simpson, 1949; Versalovic, Koeuth and Lupski, 1991;Oliveira et al, 2007). Genotyping of *A. suis* isolates recovered from clinical cases in the North American pig herds revealed a limited genetic variability, with only seven strains being identified among 88 isolates recovered from 26 different herds. *Actinobacillus suis* is relatively clonal compared with *H. parasuis*, for example, which has a diversity index of 0.93 (Oliveira et al, 2007). Phenotypic diversity of *A. suis* is also limited. Only 2 serovars have been described so far, namely O1 and O2 (Rullo, Papp-Szabo and Michael, 2006). Pathogenicity studies suggest that isolates from serogroup O2 tend to be more virulent than O1 isolates (Slavic, DeLay and Hayes, 2000). Serotyping of *A. suis* isolates used for autogenous vaccine production also confirms that a higher percentage of O2 isolates were associated with clinical disease compared with O1 isolates (Slavic, Toffner, and Monteiro, 2000). *Actinobacillus suis* from serovar O1 share antigens with environmental organisms, such as yeast, for example. Consequently, most pigs have some level of immunity to this serovar (Monteiro, Slavic, and Michael, 2000).

Considering the limited genotypic and phenotypic variability among *A. suis* isolates involved in disease, the development of a vaccine that will potentially protect against most isolates in the field seams feasible. However, important data regarding the association between genotype, serovar, toxin, and protein profiles, for example, still remain to be defined. This information is relevant for vaccine development, as are factors that are involved in the pathogenesis of *A. suis* infection. Another question that needs to be answered is the potential cross-reaction and cross-protection between Apx toxins I and II from *A. pleuropneumoniae* and *A. suis*. A commercial vaccine against *A. pleuropneumoniae* containing Apx I, Apx II, and Apx III is available in Europe, but not in the United States.

Actinobacillus pleuropneumoniae

Actinobacillus pleuropneumoniae also colonizes the upper respiratory tract. It can be detected in the tonsils of healthy carrier pigs by isolation or PCR (Gagne, Lacouture, and Broes, 1998; Fittipaldi, Broes and Harel, 2003). Similarly to *H. parasuis*, 15 serovars of *A. pleuropneumoniae* have been reported (Blackall, Klaasen and van den Bosch, 2002). Isolates from different serovars may carry different sets of Apx toxin genes (Sthitmatee, Sirinarumitr and Makonkewkeyoon, 2003). *Actinobacillus pleuropneumoniae* can cause severe necrotizing and hemorrhagic pleuropneumonia, and these lesions are mainly associated with the production of the Apx toxins (Bosse, Janson and Sheehan, 2002). Apx I is strongly hemolytic and cytotoxic, Apx II is moderately cytotoxic, and Apx III is nonhemolytic, but strongly cytotoxic. Usually, serovars that express Apx I are more virulent than those that express Apx II and/or III, and serovars that express 2 toxins are more virulent than those that express only one Apx toxin (Frey, 1995). There are exceptions to this rule, especially depending on the host background immunity. A recently described toxin, Apx IV, has been reported to be expressed *in vivo*, but not *in vitro*. This toxin is produced by all *A. pleuropneumoniae* serovars, and is thought to be species-specific (Schaller, Kuhn and Kuhnert, 1999). Detection of antibodies against this toxin may be used as an indicator of *A. pleuropneumoniae* infection (Dreyfus, Schaller and Nivollet, 2004). The role of Apx IV in virulence still remains to be defined.

Protective immunity to *A. pleuropneumoniae* is usually serovar-specific. Although strain variability within serovar groups has been demonstrated, the importance of strain-specific immunity versus serovar-specific immunity is less evident than for *H. parasuis*, for example. The Apx toxins produced by *A. pleuropneumoniae* are one of the main virulence factors associated with the development of pleuropneumonia, and immunity to these toxins has been shown to generate protective immunity. More importantly, vaccines containing the Apx toxins can potentially generate cross-protective immunity against different *A. pleuropneumoniae* serovars (Frey, 1995; Haesebrouck, Pasmans and Chiers, 2004).

Even though immunity to the Apx toxins has been demonstrated to be protective, vaccine failure may still occur. *Actinobacillus pleuropneumoniae* interacts closely with epithelial cells in the lower respiratory tract and toxins produced during infection are delivered directly to the cell surface. This close association between *A. pleuropneumoniae* and respiratory epithelial cells may impair the binding of specific antibodies to the Apx toxins, resulting in development of necrosis and hemorrhage (Haesebrouck *et al,* 2004).

Several modified live *A. pleuropneumoniae* vaccines have been developed and evaluated, with partial protective immunity across serovars being reported (Inzana, Todd and Veit, 1993; Fuller, Thacker and Duran, 2000; Tonpitak, Baltes and Hennig-Pauka, 2002; Tumamao, Bowles and van den Bosch, 2004; Bei, He and Yan, 2005; Maas, Jacobsen and Meens, 2006; Xu, Chen and Shi, 2006). Although results using modified live vaccines are promising, these vaccines are not yet commercially available.

Streptococcus suis

Streptococcus suis may be isolated from the nasal cavity, tonsils, and vaginal mucosa of healthy carrier animals. *Streptococcus suis* has the highest phenotypic diversity among the pathogens addressed in this review, with 35 different serovars being reported. Serovars 1, 2, 1/2, 7 and 9 are most frequently isolated from lesions in pig. Among these, serovar 2 is one of the most prevalent and virulent, and most of the information available in the literature is based on experiments using this serovar. Potential virulence factors that have been reported in the literature include the capsular polysaccharide, the muramidase-released cell-wall protein, and an extracellular protein factor. *Streptococcus suis* also secretes a hemolysin, named suilysin, which is able to damage epithelial cells. The role of these factors in virulence still needs further clarification. *Streptococcus suis* mutants lacking the muramidase-released cell wall protein and the extracellular protein factor, for example, have been shown to be as virulent as the wild type strains containing these factors (Haesebrouck *et al.*, 2004).

Streptococcus suis is an early colonizer of the upper respiratory tract of pigs, and eradication of this pathogen is unlikely to be successful using strategies that eliminate *A. pleuropneumoniae*, for example. Disease caused by *S. suis* is characterized by meningitis, arthritis, endocarditis, pericarditis, and septicemia with polyserositis. *Haemophilus parasuis* may cause lesions similar to those caused by *S. suis*. Furthermore, the timing of systemic infection and peak of mortality caused by these pathogens is also similar, around 4-6 weeks after weaning. The differential diagnosis of the infection by *H. parasuis* and *S. suis* is important for disease control. Detection of *H. parasuis* in clinical samples by PCR has been successful in defining the role of this pathogen in mortality (Oliveira, Galina, Pijoan, 2001; Oliveira *et al.*, 2002).

At least one commercial vaccine containing a serovar 2 *S. suis* is available in the North American market. This particular product is currently not licensed in Europe. Field veterinarians have reported varied results with commercial and autogneous vaccines against *S. suis*. Considering that there is strain variability within serovar groups and that several different serovars and strains may be isolated from affected pig herds, genotyping can be used to further characterize isolates recovered from clinical cases and to define the prevalent strains involved in mortality. This information can be useful for selection of strains to be included in autogenous vaccines. As for any autogenous vaccine, monitoring of new cases and update of the strains included in the vaccine are important to reduce the chances of vaccine failure. Other factors that may affect vaccine efficacy are timing of vaccination, potential interference of maternal immunity with pig vaccination, and concurrent viral infections. PRRS virus, for example, has been shown to facilitate *S. suis* systemic infection in experimentally infected pigs (Galina, Pijoan and Sitjar, 1994; Thanawongnuwech, Brown and Halbur, 2000).

Modified live *S. suis* vaccines have been evaluated, with varying results being reported in different studies. Protective immunity was mainly evaluated against homologous (wild-type) challenge, and heterologous protection has not been evaluated. Clinical signs following vaccination with modified live prototype vaccines were reported, and most studies concluded that further evaluation of modified live vaccine candidates is needed. (Busque, Higgins and Caya, 1997; Wisselink and Stockhofe-Zurwieden, 2002; Fittipaldi, Harel and D'Amours, 2007)

Final considerations

Haemophilus parasuis, *S. suis*, *A. suis*, and *A. pleuropneumoniae* are economically important to the pig industry. These pathogens are frequently involved in sudden death and high mortality and their differential diagnosis is critical for disease control. New molecular-based tools are available for detection and characterization of these pathogens (www.vdl.umn.edu). Strategies used by most pig herds to control these pathogens include antibiotic treatments and vaccination. Commercial vaccines are available for *H. parasuis, S. suis* (U.S.) and *A. pleuropneumoniae*. Autogenous vaccines have also been used to prevent mortality caused by these pathogens. Vaccines containing known virulence factors have a better chance of providing cross-protective immunity. Inactivated vaccines containing specific serovars may provide good homologous protection. Modified live vaccines are being evaluated for *H. parasuis, S. suis,* and *A. pleuropneumoniae*, with varying results. These vaccines are not yet commercially available, but they hold promise considering that infection by these pathogens usually generates protective immunity.

References

Bei W, He Q, Yan L, Fang L, Tan Y, Xiao S, Zhou R, Jin M, Guo A, Lv J, Huang H, Chen H. (2005) Construction and characterization of a live, attenuated apxIICA inactivation mutant of *Actinobacillus pleuropneumoniae* lacking a drug resistance marker. *FEMS Microbiol Lett.* **243**(1):21-7.

Bigas A, Garrido ME, de Rozas AM, Badiola I, Barbe J, Llagostera M. (2005) Development of a genetic manipulation system for *Haemophilus parasuis. Vet Microbiol.* **105**(3-4):223-228.

Blackall PJ, Klaasen HL, van den Bosch H, Kuhnert P, Frey J. (2002) Proposal of a new serovar of *Actinobacillus pleuropneumoniae*: serovar 15. *Vet Microbiol.* **84**(1-2):47-52.

Bosse JT, Janson H, Sheehan BJ, Beddek AJ, Rycroft AN, Kroll JS, Langford PR. (2002) *Actinobacillus pleuropneumoniae*: pathobiology and pathogenesis of infection. *Microbes Infect.* **4**(2):225-235. Review.

Busque P, Higgins R, Caya F, Quessy S. (1997) Immunization of pigs against *Streptococcus suis* serotype 2 infection using a live avirulent strain. *Can J Vet Res.* **61**(4):275-279.

Del Rio ML, Gutierrez CB, Rodriguez Ferri EF. (2003) Value of indirect hemagglutination and coagglutination tests for serotyping *Haemophilus parasuis. J Clin Microbiol.* **41**(2):880-882.

Dreyfus A, Schaller A, Nivollet S, Segers RP, Kobisch M, Mieli L, Soerensen V, Hussy D, Miserez R, Zimmermann W, Inderbitzin F, Frey J. (2004) Use of recombinant ApxIV in serodiagnosis of *Actinobacillus pleuropneumoniae* infections, development and prevalidation of the ApxIV ELISA. *Vet Microbiol.* **99**(3-4):227-238.

Fittipaldi N, Broes A, Harel J, Kobisch M, Gottschalk M. (2003) Evaluation and field validation of PCR tests for detection of *Actinobacillus pleuropneumoniae* in subclinically infected pigs. *J Clin Microbiol.* **41**(11):5085-5093.

Fittipaldi N, Harel J, D'Amours B, Lacouture S, Kobisch M, Gottschalk M.(2007) Potential use of an unencapsulated and aromatic amino acid-auxotrophic *Streptococcus suis* mutant as a live attenuated vaccine in pig. *Vaccine.* [Epub ahead of print]

Frey J. (1995) Virulence in *Actinobacillus pleuropneumoniae* and RTX toxins. *Trends Microbiol.* **3**(7):257-261. Review.

Fuller TE, Thacker BJ, Duran CO, Mulks MH. (2000) A genetically-defined riboflavin auxotroph of *Actinobacillus pleuropneumoniae* as a live attenuated vaccine. *Vaccine.* **18**(25):2867-2877.

Gagne A, Lacouture S, Broes A, D'Allaire S, Gottschalk M. (1998) Development of an immunomagnetic method for selective isolation of *Actinobacillus pleuropneumoniae* serotype 1 from tonsils. *J Clin Microbiol.* **36**(1):251-254.

Galina L, Pijoan C, Sitjar M, Christianson WT, Rossow K, Collins JE. (1994) Interaction between *Streptococcus suis* serotype 2 and porcine reproductive and respiratory syndrome virus in specific pathogen-free piglets. *Vet Rec.* **134**(3):60-64.

Haesebrouck F, Pasmans F, Chiers K, Maes D, Ducatelle R, Decostere A. (2004) Efficacy of vaccines against bacterial diseases in pig: what can we expect? *Vet Microbiol.* **100**(3-4):255-268. Review.

Hill CE, Metcalf DS, MacInnes JI. (2003) A search for virulence genes of *Haemophilus parasuis* using differential display RT-PCR. *Vet Microbiol.* **96**(2):189-202.

Inzana TJ, Todd J, Veit HP. (1993) Safety, stability, and efficacy of noncapsulated mutants of Actinobacillus pleuropneumoniae for use in live vaccines. *Infect Immun.* **61**(5):1682-1686.

Kielstein P, Rapp-Gabrielson VJ. (1992) Designation of 15 serovars of *Haemophilus parasuis* on the basis of immunodiffusion using heat-stable antigen extracts. *J Clin Microbiol.* **30**(4):862-865.

Maas A, Jacobsen ID, Meens J, Gerlach GF. (2006) Use of an *Actinobacillus pleuropneumoniae* multiple mutant as a vaccine that allows differentiation of vaccinated and infected animals. *Infect Immun.* **74**(7):4124-4132.

Melnikow E, Dornan S, Sargent C, Duszenko M, Evans G, Gunkel N, Selzer PM, Ullrich HJ. (2005) Microarray analysis of *Haemophilus parasuis* gene expression under in vitro growth conditions mimicking the in vivo environment. *Vet Microbiol.* **110**(3-4):255-263.

Oliveira S, Galina L, Pijoan C. (2001) Development of a PCR test to diagnose Haemophilus parasuis infections. *J Vet Diagn Invest.* **13**(6):495-501.

Oliveira S., Pijoan C. Morrison R. (2002) Role of *Haemophilus parasuis* in nursery mortality. Proceedings of the Allen D. Leman Pig Conference, Minneapolis, USA, p.111.

Oliveira S, Pijoan C. (2004) Haemophilus parasuis: new trends on diagnosis, epidemiology and control. *Vet Microbiol.* **99**(1):1-12. Review.

Oliveira S., Pijoan C., Morrison R. (2004) Comparison of *Haemophilus parasuis* control in the nursery using vaccination and controlled exposure. *J Pig Health and Prod.* **12**(3):123-128.

Oliveira S. (2006) *Actinobacillus suis*: an update on genotyping and antibiotic susceptibility. Proceedings of the Allen Leman Pig Conference, Minneapolis, MN. p. 93.

Oliveira S. (2007) *Actinobacillus suis* diagnostics, epidemiology and control – on the path from good to great. Proceedings of the 38th American Association of Pig Veterinarians Annual Meeting, Orlando FL, p. 371-376.

Rullo A, Papp-Szabo E, Michael FS, Macinnes J, Monteiro MA.(2006) The structural basis for the serospecificity of *Actinobacillus suis* serogroup O:2. Biochem Cell Biol. **84**(2):184-90.

Segales J, Domingo M, Chianini F, Majo N, Dominguez J, Darwich L, Mateu E. (2004) Immunosuppression in postweaning multisystemic wasting syndrome affected pigs. Vet Microbiol. **98**(2):151-158. Review.

Schaller A, Kuhn R, Kuhnert P, Nicolet J, Anderson TJ, MacInnes JI, Segers RP, Frey J. (1999) Characterization of apxIVA, a new RTX determinant of *Actinobacillus pleuropneumoniae*. *Microbiology.* **145**:2105-2116.

Simpson E.H. (1949) Measurement of diversity. *Nature.* **163**:688

Slavic D, DeLay J, Hayes MA, MacInnes JI. Related (2000) Comparative pathogenicity of different *Actinobacillus suis* O/K serotypes. Can J Vet Res. **64**(2):81-87.

Slavic D, Toffner TL, Monteiro MA, Perry MB, MacInnes JI. (2000) Prevalence of O1/K1- and O2/K3-reactive *Actinobacillus suis* in healthy and diseased pig. J Clin Microbiol. **38**(10):3759-3762.

Sthitmatee N, Sirinarumitr T, Makonkewkeyoon L, Sakpuaram T, Tesaprateep T. (2003) Identification of the *Actinobacillus pleuropneumoniae* serotype using PCR based-apx genes. *Mol Cell Probes.* **17**(6):301-305.

Tadjine M, Mittal KR, Bourdon S, Gottschalk M. (2004) Development of a new serological test for serotyping *Haemophilus parasuis* isolates and determination of their prevalence in North America. J Clin Microbiol. **42**(2):839-840.

Thacker EL. (2001) Immunology of the porcine respiratory disease complex. *Vet Clin North Am Food Anim Pract.* **17**(3):551-565. Review.

Thanawongnuwech R, Brown GB, Halbur PG, Roth JA, Royer RL, Thacker BJ. (2000) Pathogenesis of porcine reproductive and respiratory syndrome virus-induced increase in susceptibility to *Streptococcus suis* infection. *Vet Pathol.* **37**(2):143-52.

Tonpitak W, Baltes N, Hennig-Pauka I, Gerlach GF. Construction of an *Actinobacillus pleuropneumoniae* serotype 2 prototype live negative-marker vaccine. *Infect Immun.* 70(12):**7120**-7125.

Tumamao JQ, Bowles RE, van den Bosch H, Klaasen HL, Fenwick BW, Storie GJ, Blackall PJ. Comparison of the efficacy of a subunit and a live streptomycin-dependent porcine pleuropneumonia vaccine. *Aust Vet J.* 82(6):**370**-374.

Versalovic J, Koeuth T, Lupski JR. (1991) Distribution of repetitive DNA sequences in eubacteria and application to fingerprinting of bacterial genomes. *Nucleic Acids Res.* **19**(24):6823-31.

Wisselink HJ, Stockhofe-Zurwieden N, Hilgers LA, Smith HE. (2002) Assessment of protective efficacy of live and killed vaccines based on a non-encapsulated mutant of *Streptococcus suis* serotype 2. *Vet Microbiol.* 84(1-2):**155**-168.

Xu F, Chen X, Shi A, Yang B, Wang J, Li Y, Guo X, Blackall PJ, Yang H. Characterization and immunogenicity of an apxIA mutant of *Actinobacillus pleuropneumoniae. Vet Microbiol.* **118**(3-4):230-239.

10

TRENDS IN PIG FARMING IN CHINA AND SOUTH-EAST ASIA

STEVEN MCORIST[1] AND XIANJIN YANG[2]

[1]*School of Veterinary Medicine and Science, University of Nottingham, Sutton Bonington, Leics LE12 5RD, England;* [2]*China Animal Industry Association, Beijing, China*

Introduction

Swine production in China and south-east Asia continues to move ahead in terms of consolidation, expertise and production levels. The size and scale of these pig industries takes your breath away. For example, China has 1.3 billion consumers who eat their way through 620 million slaughter pigs, produced from 50 million sows and Vietnam has 30 million slaughter pigs, produced from 4 million sows. Significant swine industries occur throughout south-east Asia, with intensification most advanced in Thailand and The Philippines and least advanced in Cambodia, Laos and Myanman.

Pig raising in this region is usually split into 3 distinct sectors: Traditional back-yard producers operating with 1 to 10 sows, small family producers with an average herd size of 40 sows and developing integrators with commercial or intensive farms (defined as more than 50 sows per farm), which are basically the western pig farming model. The latter are quickly increasing in number and overall integration throughout the region, particularly around major expanding cities such as Beijing (pop. 14m), Shanghai (pop. 17m) and Guangzhou (pop. 12m), where urbanisation (the marked shift in population from rural areas to the cities) is most active. In some of these peri-urban areas, the commercial farm sector is considered 60 to 80 % of pig populations.

These intensification processes towards the western model include housing of specific age groups into separate larger units composed of breeding, nursery or fattening pigs. These farms may also have dedicated breeder pig suppliers; complex commercial feed suppliers and may use off-site, contractor finisher farms. Pig housing, vaccinations, antibiotic usage, all-in, all-out and disinfection procedures are often improved. The Wens Animal Husbandry Co., based in Guangdong is regarded as China's largest integrated pig company, with 110,000 sows. The Charoen Pokphand group operates throughout south-east Asia, based in Bangkok, and has 100,000 sows, and many other major operations are developing.

Community policies

The key broader benefits of swine farm intensification include more effective land usage for food production – China has only 7% of the world's cultivable land, but has 22% of the world's population. Development of an efficient farm sector also addresses two other main community policies - food security for the entire population and increasing incomes in rural areas to attempt to offset or arrest urbanisation. Agriculture remains an important part of Asian economies; producing 15% of GDP and accounting for 40% of employment in China. There is therefore still considerable support for backyard farms, where small numbers of pigs are fed crop residues, table scraps and forage in simple housing. This type of farm probably still represents 70 to 80% of the total number of pigs raised in China and several other south-east Asian countries. These pigs add value to subsistence farming, as over 80% of these household farms have less than 0.6 hectare – not enough to support a family by crop production alone. In this system, crop production and animal raising are complementary and the low-cost pigs are viewed as a "savings bank". Chinese policies will support new technologies, including GM crops, but business development proposals, which utilise less labour, will not be of interest.

Despite this, urbanisation continues at a rapid pace in China and throughout Asia. On average, families that move to the city will show a considerable rise in income, perhaps joining the manufacturing workforce. Given the scale of the population in China, this translates to a considerable growth in urban, middle class families in places like Shanghai or Beijing. Figure 10.1 illustrates the creation of an extra 40 million middle class Chinese households in the past decade, a trend which is predicted to continue. These newly-enriched neo-urban families will typically purchase more meat protein. An average Chinese city population growth of 4% per year will directly lead to a similar increase in pork sales. The Chinese cultural preference for pork over other forms of meat has further translated this into an increased proportion of pork purchased as a percentage of the increased overall meat purchased, see Figure 10.2. Based on income growth alone, pork consumption will grow by a further 3% per year in urban areas (total 7%) and 1.5 to 2% in rural areas in China. Provinces of China such as Hong Kong SAR that already have western levels of wealth show pork consumption levels of nearly 60 kg per person per year, indicating that even more growth may occur.

While the full communist command structure was removed from pig farming in China in 1975, state quotas and price controls remained until 1985, then allowing for a free-market farm and product distribution system to develop. However, direct government intervention into pig farming and/or markets remains an important issue. Figure 10.3 illustrates the ups and downs of pig production in China. From 1985 to 1996, steady and rapid expansion of production was matched by increased marketing and demand. In mid-1997, low feed prices and high pork prices led to very high profits. The extremely

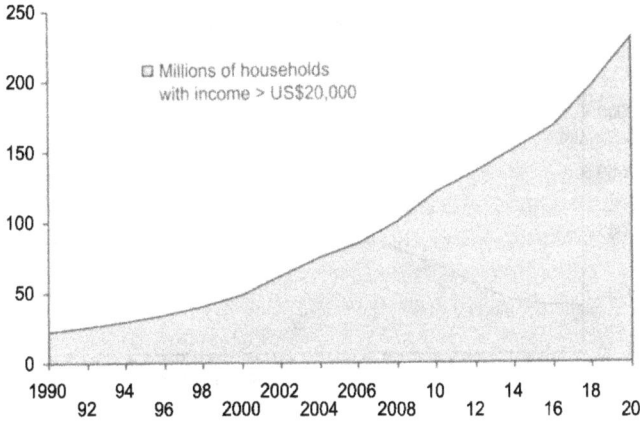

Figure 10.1 The rapid rise of the Chinese middle classes.

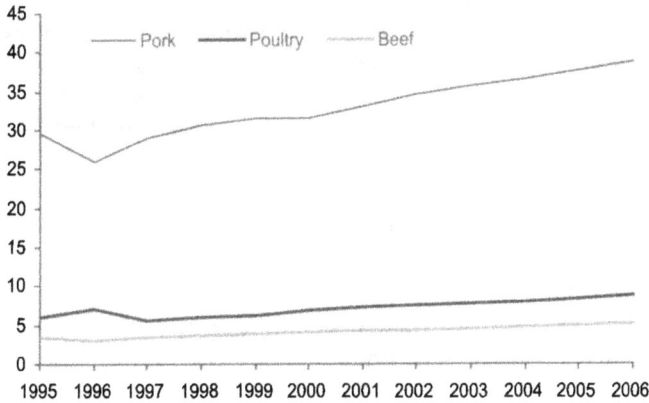

Figure 10.2 China meat consumption – Asians prefer pork.

large number of small farms present in China and the concurrent expansion of larger farms meant that markets were rapidly flooded in an attempt by farmers to capture profit. Over-supply ensued, prices plummeted and many farms went out of business in 1998 and 1999. In response in 2003, the Chinese government attempted to reorganise pig farming to attempt to reduce risk and volatility of the market. Larger farms and market stability were encouraged, again forcing many small backyard farms out of business. One aim was to smooth the risky economic nature of operating in an industry with so many small operators.

The continued expansion of the industry and the inherent supply-demand risks in the pig industry has led to further state interventions being considered and even encouraged by some policy groups. Data on the pig industry outputs/ prices etc is limited in China, making risk assessment and price control difficult,

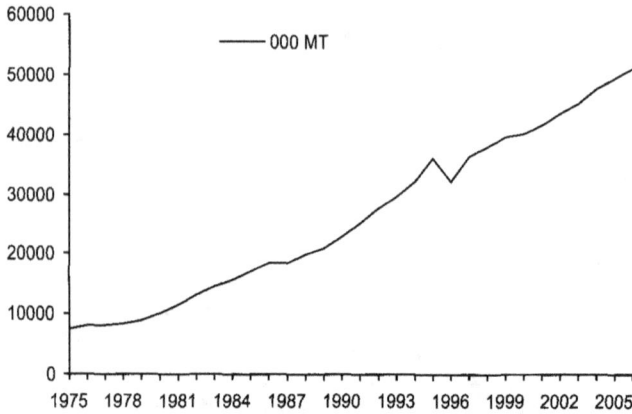

Figure 10.3 China pork production is growing steadily.

even with an interventionist policy. Farmer organisations are poorly formed, with no equivalent of the NPPC or similar bodies. Related policy ideas have included imposition of quotas on the numbers of sows each farm is allowed; collection of restaurant waste or other industrial by-products from cities for feeding to pig farms. One area that remains greatly under-explored in the expansion of the industry in China is swine breeding – regional breeding centres have limited European or local x European stock. There are therefore few farm pigs with improved feed conversion characters, typical of western pig breeding advances in the past decade. Artificial insemination from off-site, dedicated boar studs is largely unavailable.

Despite some of them joining WTO, China and other south-east Asian countries typically still have high trade barriers to pork imports. The price paid for any eventual imports is also low. The most likely import - frozen pork, remains of low perceived nutritive value compared to fresh pork throughout Asia. Besides pork import tariff barriers of 12% in China, another barrier consists of restricting imports to those originating in licensed overseas abattoirs. Obtaining licenses is a long and prohibitive process and very few abattoirs in Europe or elsewhere have taken the opportunity. This has perhaps led to the existence of an acknowledged pork smuggling issue, with pork crossing land borders from Vietnam and elsewhere into southern China. In contrast, China has been keen to develop its already significant pork export business. Cooked pork products leave China for Ukraine, Russia and other destinations, totalling 330 million tonnes in 2005 (5th ranked globally).

Pork markets in China

Fresh pork cuts account for around 85% of pork sold in China. Of the 15%

remaining processed pork, most is sold as packaged small cuts. The percentage of the overall market now taken up by "western" or pink pig pork and the amount retained by the 40 or so popular native pig meat breeds, such as Jinhua, Tai Hu, Su Tai or Xiang breeds tends to vary. Since the early 1990's, increased food sales in cities have been dominated by the growth of western-style supermarkets, which have typical meat cabinets, pleasingly predominated by pork. Overall supermarket share of the retail market in China is approximately 20% - compared to 60 to 80% in the west. Foreign retailers such as Carrefour, Walmart, Metro and Tesco are operating in some of these urban markets. This retail market has middle-class consumers that demand quality, traceability and no residues. There is even a thriving "organic" pork market in urban China.

However, the scale of the local fatty pork market should not be underestimated. The widespread backyard farm sector tends to produce fatty local-breed pigs, sold to local markets, especially in more rural areas. But Jinhua hams, Xiang pig roasts and fatty pork cuts retain a notable presence in small stores, restaurants and supermarkets throughout China. Even in Shanghai, market share of traditional local meats went from 24% up to 60% during 1993 to 1998, as prices for western-style pork rose.

The main retail chains for pork in China remain the traditional 'wet' markets, wholesalers, food processors and retailers. Small farms may kill and process their pigs on-farm or in small local butchers, for sale at local roadside or regional farm markets. A larger wet market chain may consists of pigs being trucked and slaughtered at a medium-size abattoir during the night, ½ or ¼ carcases being taken to processors or distributors and sold to retailers within the ensuing day. Food retailers or household consumers would be expected to cook and serve the "hot" meat within a short time frame. Refrigeration often does not play a role in these processes. The trade remains highly fragmented, with numerous small wholesalers and distributors, often providing their own transport for the pigs and pork.

There has been considerable concern about the quality and legality of butchers and abattoirs throughout China. Despite the Hog Slaughtering Act passed in 1998, problems remain, with the recent and serious *Streptococcus suis* outbreak in illegal butchers in Jilin an example. One reason for the persistence of small butchers is the higher cost of slaughtering in medium or larger abattoirs.

Farm profits and costs

This size and scale of the Asian industry is not enough to guarantee a profit; many pig producers are emerging from a period of sustained losses. Current farm-gate prices in a major city like Guangzhou (Canton) are up to US$ 1.50 per kg. Retail prices in a Guangzhou or other city supermarket are around US$ 2.50 per kg – China has low slaughter and distribution costs. Current average cost of production (COP) in commercial units is around US$ 0.95 per kg of liveweight pig, with 75% or more of that being feed costs. Local farm worker wages on commercial farms are around US$ 0.60 per hour. These prices are at the high end for the diverse provinces and cities around China. Even the intensive sector industry economics are less developed in northern and central China. The Hong Kong SAR market provides a price premium of a US$ 2.00 price compared to the local price of US$ 1.50. Extra inputs to reach the Hong Kong market include necessary auditing of medications and vaccines as well as extra transport costs, even from the south-east Guangdong region.

While pig farms are being developed all over China, there were some extra increases in the Northern provinces, due to enhanced proximity to local Chinese corn supplies. However these Northern provinces (Jilin, Shandong etc) suffer from lack of technical expertise and lack of easy access for their pork products to the major markets, such as Beijing, Shanghai and Guangzhou. There is also currently a rapid rise in corn usage for biofuel production in northern China, diverting animal feed into ethanol production, with consequent availability and price issues, identical to those occurring in North America. This suggested "northern shift" has therefore stalled and most recent pig farm consolidation continues to occur in the southern and eastern coastal belt.

Manufactured commercial feeds are now widely used and the actual diet content is usually a standard corn-soy base diet. Feed normally arrives on farm in bags, which allows accurate data on feed consumption and costs. Even backyard farms typically use some manufactured feed as supplement. Most commercial pigs are fed and managed under normal western-style finisher pig conditions, except that they may be limit-fed. This limit feeding of pigs is commonly used to allow farmers to keep feed costs lower, improve leanness and to juggle pig outputs to perhaps wait for better prices. On some farms, high feed prices and market conditions meant that pigs targeted for only a 100 kg liveweight may not reach it until 24 or more weeks-old, even with western origin "pink" pigs.

Sources of pigs

Most commercial farms with western pigs derive their key breeding stock from provincial swine breeding centres, either Government or privately run. These centres have usually had on-going imports of western breeding stock in air-freighted batches of 40 to 100 sows. The largest one is the state-backed China National Animal Breeding Stock Corporation based in Beijing. But all provinces have their own swine breeding centre: typical examples include the Yisheng centre with 3,000 sows located in Yentai, Shandong, which imported Pietrain and Large Whites from North America and the Guoshou development Co. with 2,000 sows in Xiamen, Fujian province, which imported pigs from nearby Taiwan. Overseas breeding companies have a small presence in China; PIC has made considerable efforts to set up 2 separate nucleus herds, and a new multiplier herd in Hubei, with well-run facilities and a range of PIC line pigs. Topigs and Hypor also have nucleus herd operations in China. The Thai Charoen Pokphand swine farming group has set up a large number of breeder herd start-up operations in China and Vietnam over the past 20 years, typically as joint ventures with local groups. Pigs for these farms have usually originated from the Thai-based grandparent CP line 40 or 51, which are based on Large White/Landrace crosses.

Again, the size of the local fatty breed market should not be underestimated. Some larger breeder operations exist, which carry the most popular breeds. For example, the Zhejiang centre has 400 Jinhua sows; the Suzhou City centre offers both meat breed sows, such as SuTai and mothering sow breeds such as the Meischan and Fengjing. In these and many more centres, local breed sows are cross-bred to form new lines such as a Tai Hu/Duroc cross production pig line.

Live on-farm boars remain popular due to the difficulties with establishing a boar stud program with a proper distribution network of semen supplies. Unlike Thailand, semen imports are rare in China. Most commercial farms make a considerable effort to purchase quality boars from the provincial breeding centres, with occasional live boar auctions occurring, for example at the Guangzhou swine breeding centre.

Disease patterns

Alterations in disease patterns due to intensification in Asia have been reviewed elsewhere by Ranald Cameron (FAO, 2000). Among the many complex changes in disease transmission with increasing intensification, the more limited contact between sows and their children may lead to reduced early transmission of some agents, with consequent susceptibility to these agents if introduced to them at a later age. For some well-adapted porcine agents, such as *Haemophilus parasuis*, *Mycoplasma hyopneumoniae* and *Lawsonia intracellularis*, this later contact can lead to enhanced clinical signs.

On backyard farms in south-east Asia, old favourites such as foot and mouth disease (FMD) and classical swine fever (CSF) remain very common, with little biosecurity and few vaccinations occurring, due to the low-cost business model. These pigs and other infected livestock (such as feral goats) represent a clear danger for clean commercial and breeder herds. The Government-controlled CAHIC vaccine group has a range of modern factories around China. These can supply a range of vaccines including FMD and CSF. There are also several smaller local vaccine companies in China producing PRRS and other vaccines. Supply, potency, handling and storage of locally produced vaccines all remain important issues. The government has made considerable strides to improve registration procedures for western origin vaccines. Many western-origin swine vaccines from Merial or Boehringer are now centrally registered and freely available for use. PRRS vaccines have proved especially popular. However, the price of vaccines (both imported and local) means that they often remain limited in use to more progressive farms for swine farm improvements. Registration of foreign-origin FMD vaccines remains blocked in China. In other countries, such as Thailand, imported vaccines are widely available.

Conclusions

Swine production in China and south-east Asia continues to move ahead in terms of consolidation, expertise and production levels. Intensification of swine farming addresses the important community issues of effective land usage for food production, food security for the entire population and increasing incomes in rural areas. Increased food sales in cities have been dominated by growth of supermarkets, which have typical fresh meat cabinets, pleasingly predominated by pork. However, overall supermarket share of the retail market in China is only 20% and the main retail chains for pork remain the traditional 'wet' markets. While the western or "pink" pig farming model is increasing in size and strength, the scale of the local fatty pork market should therefore not be underestimated. Many small farms retain fatty breed pigs for local and wider markets of fresh pork and specialised products such as Jinhua hams.

11

IDIOSYNCRASIES OF PIGLET LIPID METABOLISM AND THEIR RELATIONSHIP TO POSTNATAL MORTALITY

XI LIN, BENJAMIN CORL AND JACK ODLE

Department of Animal Science, North Carolina State University, Raleigh, NC, USA

Introduction

Postnatal mortality represents a significant problem for the global pig industry (NAHMS, 1997, 2000; Herpin, Damon and Le Dividich, 2002), claiming 0.15 - 0.20 of piglets born alive. In some less-developed countries, the rate may exceed 0.30 (Lay, Matteri, Carroll, Fangman and Safranski, 2002; Losinger, 2005). While the number of piglets born alive and the number weaned per litter have increased since 1998, the mortality rate has remained unchanged (Figure 11.1). On average 0.75 of the deaths occur within three days of birth, emphasizing the need for early intervention. The underlying etiology is complex, but data indicate that inadequate nurture is among the leading causes, accounting for 0.20 of deaths (Figure 11.2). The impact of inadequate nutrition is even greater than direct estimates suggest because piglets suffering from starvation are more susceptible to crushing by the dam which is reported as the predominant cause of death (Figure 11.2). While the importance of early nutrition has been recognized (Varley, 1995), practical improvements in mortality rate are yet to be realized possibly due to technical difficulties in implementation and in potential costs/benefit constraints.

Starvation, although a consequence of inadequate nutrient supply, also can result from inadequate nutrient utilization by the neonates. Because newborn pigs have very low energy reserves (< 0.02 body fat; Mannaert and McCrea, 1963) with no immediate and appropriate nutrition at birth (Varley, 1995), milk lipids become the principal substrate for oxidative metabolism after birth, comprising 0.60 of dietary energy (Girard, Ferré and Duée, 1992). Furthermore, the energy requirement of the piglet is maximum in the immediate postnatal period due to their high relative growth rate, high relative surface area and the need for thermoregulation (Le Dividich and Sève, 2000). Therefore, piglet survival greatly depends on their rapid metabolic adaptation to utilize milk fat as their primary postnatal fuel. However, metabolic studies suggest that neonatal piglets have a limited capacity to catabolize fatty acids. Limitations may stem

187

from low gene expression as well as from unique characteristics of key enzymes in the fatty acid oxidative pathway, suggesting that regulation of fatty acid oxidation may differ in pigs compared with other mammals.

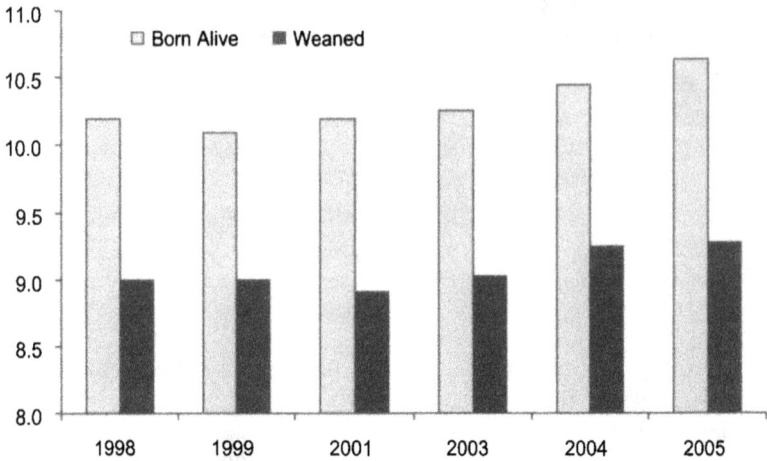

Figure 11.1 Estimated piglet mortality rates in the USA. While the number of piglets born alive and the number weaned per litter show slight upward trends, mortality rates have not changed appreciably over the past eight years. Data compiled from the *Pigchamp* database.

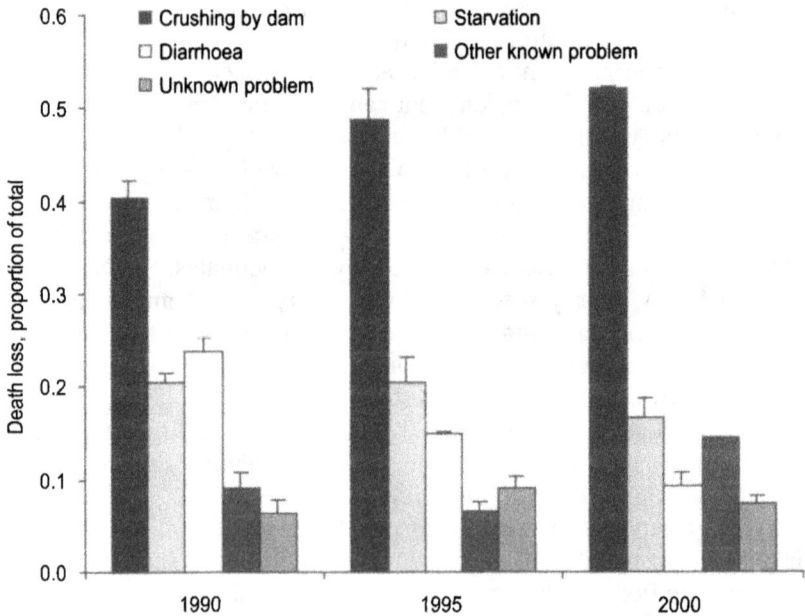

Figure 11.2 Suggested causes of postnatal piglet mortality in the USA. This book chapter examines a biochemical component of "starvation", estimated to account for about 20% of deaths. Data compiled from reports by the National Animal Health Monitoring System (NHAMS).

Recent developments in cellular and molecular biology have motivated detailed investigation of the attenuated metabolic development in neonatal piglets and have illuminated interesting molecular idiosyncrasies that merit further examination. Nutritional and pharmacologic modulation of key lipid-metabolizing enzymes that likely regulate development of postnatal fatty acid oxidation may eventually lead to improved efficiency of energy utilization and reduced postnatal mortality. Accordingly, the curent chapter will highlight the nutritional and metabolic idiosyncrasies of lipid metabolism that underlie piglet postnatal morbidity and mortality.

Timely energy support is vital for survival and growth

Abrupt and dramatic changes in the bioenergetics and nutrition of the piglet at birth necessitate immediate adaptation if the animal is to thrive. The piglet's metabolic rate (energy demand) increases by 2 fold at birth, congruent with an increase in the relative body surface area, the need to evaporate birth fluids, and the need to retain homeothermy in an environment that is often more than 10 °C below its lower critical temperature (Stanier, Mount and Bligh, 1984). Survival therefore hinges on adequate nurture, including both 1) adequate fuel intake and 2) adequate fuel metabolism. Regarding the first, while nature endows the piglet with modest glycogen stores to temporarily buffer the postnatal energy demand, the animal simply cannot sustain the elevated metabolic rate without exogenous fuel intake (i.e., milk consumption). Regarding the second, newborn piglets have a limited protein oxidative capacity in the first week after birth (Marion and Le Dividich, 1999), even under conditions of starvation or cold stress (Benevenga, Steinman-Goldsworthy, Crenshaw and Odle, 1989; Herpin, Le Dividich and van Os, 1992). Therefore, rapid biochemical adaptations (described later) are needed to utilize milk-fat for energy which adds further "metabolic stress" to the piglet. Both of these important stressors are related to the post mortality directly and merit further discussion.

INADEQUATE INTAKE

Immediately after birth, the piglet must successfully engage instinctive suckling behavior in a competitive environment with multiple littermates. Behavioral research (Rohde Parfet and Gonyou, 1988) has shown that > 30 minutes may lapse between birth and first-suckling. Lower birth weight piglets with lower glycogen reserves take longer to establish a successful suckling rhythm, thereby further predisposing them to morbidity and mortality. Inadequate consumption of milk predictably lowers piglet heat production and body temperature drops accordingly (Le Dividich & Noblet, 1984). Inadequate intake of colostrum-derived IgG compromises passive immunity as well. In a weakened state, the piglet is less able to compete for milk, so intake further declines leading to the negative spiral that ultimately results in death (Mount, 1968; Tuchscherer, Puppe, Tuchscherer and Tiemann, 2000). To combat this problem of inadequate energy intake, we investigated two direct intervention approaches to supply supplemental energy to the piglets. In the first approach (see Veum & Odle, 2001 for review), previous efforts (Braude, Mitchell, Newport, and Poter, 1970; Lecce, 1975) were extended to supply supplemental milk replacer to needy piglets. While this approach shows some promise, especially when supplemental milk is supplied to heat-stressed dams (Azain, Tomkins, Sowinski, Arentson and Jewell, 1996), adoption by the industry has been hampered by high cost of milk ingredients and the mechanical challenges of operating the feeding systems in a production environment. In the second approach, the use of medium-chain triglycerides (MCT) as a supplemental exogenous fuel (Odle, Lin, van Kempen, Drackely and Adams, 1994; Odle, 1997) that can be directly gavaged to compromised piglets was explored extensively. This research led to identifying and understand an underlying impairment in the piglet's ability to oxidize fatty acids and produce ketone bodies (discussed later).

INADEQUATE METABOLISM

In addition to inadequate fuel intake, the problem of negative energy balance may be exacerbated by inadequate fuel metabolism by the neonate. Indeed, the piglet's ability to switch rapidly /adapt from carbohydrate to fat oxidation is paramount for survival. The permeability of the placenta to fatty acids is limited and, therefore, the primary energy substrate for the developing fetal pig is glucose (Battaglia and Meschia, 1988). However, during the postnatal period, milk lipids become the principal substrate for oxidative metabolism, comprising 0.60 of dietary energy (Girard *et al.*, 1992). To buffer this transition from carbohydrate- to lipid-based metabolism, the piglet is born with a reserve of hepatic glycogen that precipitously declines by 48 h (Pégorier, Duée, Assan, Peret and Girard, 1981). This fuel utilization profile is further confirmed by reports of high respiratory quotients immediately postpartum (Noblet and Le Dividich, 1981). In addition, the piglet lacks an appreciable fat depot (<0.02 of body weight; Mannaerts and McCrea, 1963). Therefore, piglet survival also hinges on their rapid metabolic adaptation to utilize milk fat as their primary postnatal fuel. Although postnatal increases in fatty acid oxidation were reported (Wolfe, Maxwell and Nelson, 1978), they must be interpreted in light of the general increase in metabolic rate which occurs after birth (Odle, Benevenga and Crenshaw, 1991a). For example, the increase is consistent with an increase of mitochondrial proliferation and respiration which is supported by the increased oxygen consumption rate. Odle, Benevenga and Crenshaw (1991b) also observed that the increase of hepatic fatty acid oxidation occurred in both small and normal birth-weight pigs during the first 48 h of life.

Idiosyncrasies of hepatic fatty acid oxidative metabolism in neonatal pigs

Lipid metabolism in rodents has been well documented, and is often used as a reference standard when characterizing the pathways involved; however, important species differences do exist. Indeed, research has revealed a number of idiosyncrasies in the neonatal pig that make extrapolation from rodent data inappropriate. These differences and their ramifications for fuel homeostasis in the piglet will be highlighted.

Like other mammalian neonates the pig must rapidly adapt to a change in fuel source that is initiated at birth. These adaptations include an up-regulation of gluconeogenesis, fatty acid oxidation and ketogenesis all of which serve to ensure adequate energy supply. The neonatal pig is extremely susceptible to hypoglycaemia within 24 h after birth. During this time, liver glycogen in both the fed and starved animal decreases precipitously and, unless suckling ensues, animals' glycogen will become exhausted within 48 h (Pégorier *et al.*, 1981). Therefore, the use of milk fat as energy substrate is a prerequisite of utmost importance for successful transition to extra-uterine life.

However, at least two lines of evidence confirm that the neonatal piglet is limited in its capacity to catabolize dietary fatty acids. First, the capacity of the one-day-old piglet to oxidize fatty acids is only 0.32 of the rate of the 24-d-old piglet (Bieber, Markwell, Blair and Helmarath, 1973), and the rates of ß-oxidation in liver preparations are markedly lower compared to that in liver from other species such as rabbits (Duée, Pégorier, Manoubi, Herbin, Kohl and Girard, 1985; Pégorier, Garcia-Garcia, Prip-Buus, Duée, Kohl, and Girard, 1989) and rats (Duée, Pégorier, Quant, Herbin, Kohl and Girard, 1997). This low hepatic capacity to oxidize fatty acids has been contrasted to a high capacity for esterification. Indeed, Pégorier, Duée, Girard and Peret (1983) reported that 0.90 of oleate taken up by piglet hepatocytes was re-esterified with limited flux through ß-oxidation, regardless of age or nutritional status. Second, suckling piglets do not display a hyperketonemia despite elevated plasma non-esterified fatty acid (Bengtsson, Gents, Harkarainen, Hellström and Persson, 1969; Pégorier *et al.*, 1981, Adams, Lin, Yu, Odle and Drackley, 1997a). This starkly contrasts other mammalian species (e.g., rats, rabbits, etc) which show pronounced hyperketoemia during suckling (Foster and Bailey, 1976; Figure 11.3). For example, plasma ketones may exceed 2 mM in the rat (Robles-Valdes, McGarry and Foster, 1976), but the concentration in newborn pigs is less than 0.25 mM (Pégorier *et al.*, 1981). The lower ketone concentration is attributed to reduced synthesis and not increased utilization as Tetrick, Adams, Odle and Benevenga (1995) showed that, at physiological concentrations, ß-hydroxybutyrate contributes less than 0.05 to the piglet's energy requirement. Because ketone bodies provide important glucose-sparing carbon, aiding otherwise glucose-dependent tissues (e.g., neural tissues), their absence may be detrimental to the survival of the piglet which is keenly susceptible to hypoglycaemia (Swiatek, Kipnis, Mason, Chao and Cornblath, 1968). Furthermore, insofar as fatty acid oxidation also is required to support active gluconeogenesis, impaired fat oxidation also could contribute indirectly to hypoglycemia.

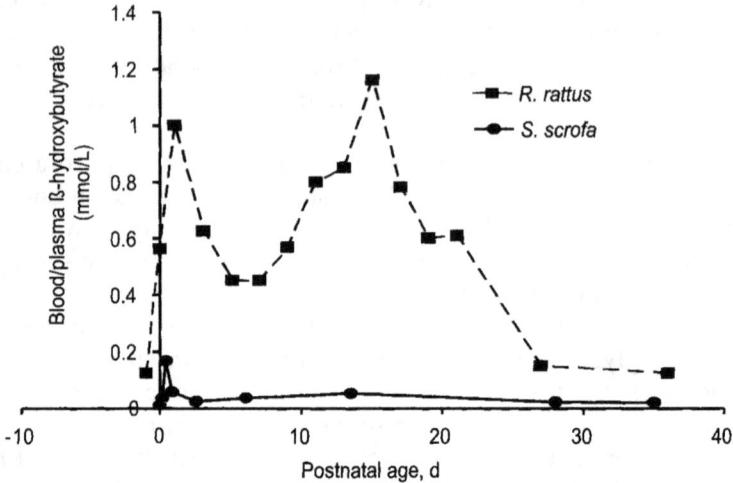

Figure 113 Suckling hyperketonemia in *R. rattus* (rats) exceeds that in *S. scrofa* (pigs) by greater than 10-fold. Adapted from Foster & Bailey, 1976 and Bengtsson *et al.*, 1969, respectively.

The partitioning of dietary fatty acids towards esterification and away from oxidation (Pégorier *et al.*, 1983) suggests that carnitine palmitoyltransferase I (CPT I) activity is connected to the low oxidative capacity of the newborn pig liver (Figure 11.4). This enzyme is well known for its central role in the regulation of oxidative lipid metabolism (McGarry, 2001), controlling the shuttle of long-chain fatty acids into the mitochondrion where they are subjected to the pathway of ß-oxidation. During physiological states in which lipogenesis is occurring, acetyl-CoA carboxylase is activated and the attendant high level of malonyl-CoA inhibits CPT I thereby ablating the simultaneous (and futile) fatty acid oxidation by preventing entry into the mitochondria. Therefore, regulation of CPT I is thought to function in directing fatty acyl-CoA between esterification and oxidation fates. Indeed, we were able to redirect carbon flux of palmitate through the oxidative pathway and concomitantly reduce the rate of esterification by addition of L-carnitine, a co-factor for CPT I (Odle, Lin, van Kempen, Drackely and Adams, 1995; Figure 11.5). Evidence indicates that the activity of CPT I develops rapidly in the neonatal pig, and this increase coincides with increased tissue carnitine concentrations (Beiber *et al.*, 1973).

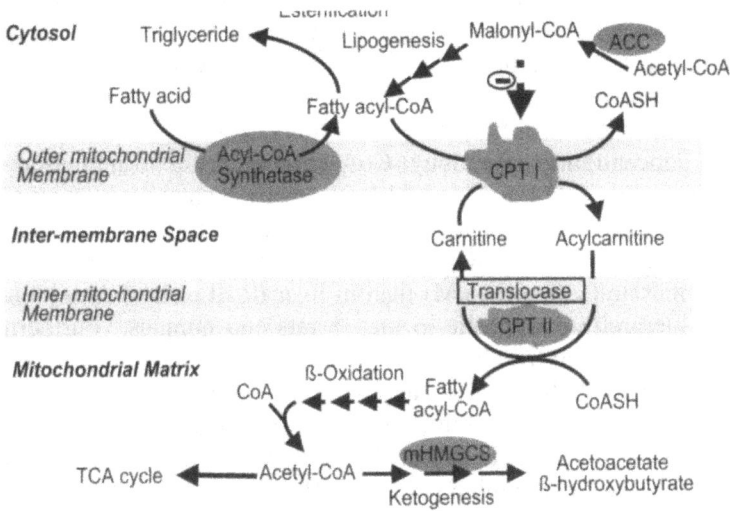

Figure 11.4 Central dogma of hepatic long-chain fatty acid metabolism. Following activation to their Co-A esters by *acyl-CoA synthetase*, fatty acids may be esterified in the cytosol, or may be shuttled into the mitochondria via the *carnitine palmitoyltransferase* (CPT) system. The shuttle is subject to acute allosteric regulation in that CPT1 is inhibited by malonyl-CoA which is the first committed metabolite formed in the cytosolic pathway of lipogenesis by *acetyl-CoA carboxylase* (ACC). Within the mitochondria, fatty acids are subjected to the pathways of ß-oxidation and ketogenesis. As discussed in the text, piglets show the following idiosyncrasies: 1) negligible lipogenesis, with low concentrations of malyonyl-CoA, 2) comparatively substantial esterification capacity, 3) a chimeric form of CPT1 that is highly sensitive to malonyl-CoA inhibition but has a high affinity for carnitine, 4) limited ketogenesis owing to low activity of *3-hydroxy-3-methylglutaryl-CoA synthase* (mHMGCS), and 5) comparatively high rates of peroxisomal ß-oxidation (not illustrated). Adapted from Heo, 2000.

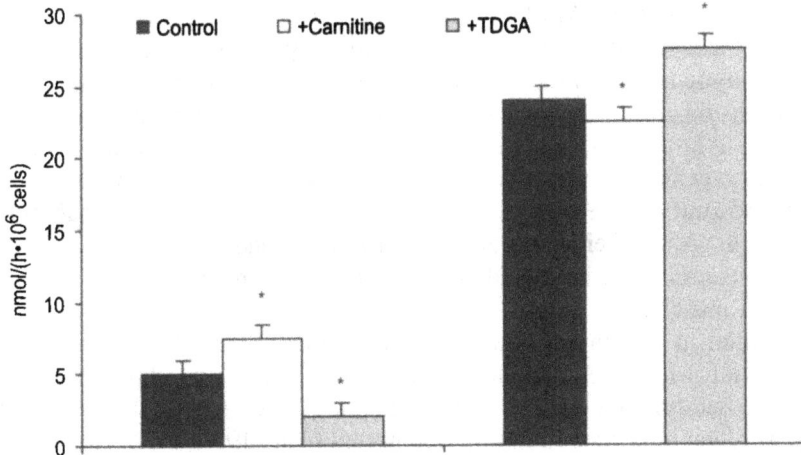

Figure 11.5 Radiolabeled palmitate metabolism by piglet hepatocytes. As a co-substrate for CPT1, supplemental carnitine increased oxidation and reciprocally reduced esterification; whereas, TDGA (a specific inhibitor of CPT1) produced the opposite effects. These data illustrate the ability of CPT1 to channel fatty acids between catabolic and anabolic fates. Adapted from Odle *et al.*, 1995.

In pigs, like other mammalian neonates, CPT I is reversibly inhibited by malonyl-CoA. Malnonyl-CoA is produced by acetyl-CoA carboxylase (ACC), specifically the α isoform in lipogenic tissues and the β isoform, with a mitochondrial leader sequence, in non-lipogenic tissues (Kim, 1997). In non-lipogenic tissues like muscle, malonyl-CoA presence may serve only to regulate CPT I activity. In muscle, the concentration of malonyl-CoA required for 0.50 inhibition of CPT I (IC_{50}) is significantly less than that of liver in many mammals. The pig, however, is considerably different in regards to CPT I regulation by malonyl-CoA in both liver and muscle. In fact, the IC_{50} for inhibition of CPT I by malonyl-CoA in pigs is greater in muscle (5.48–6.34 μM) than in liver (0.10 μM) (Schmidt and Herpin, 1998). This hierarchy is opposite to that in rats and humans. Furthermore, this high degree of sensitivity occurs in both fed and fasted piglets (Duée *et al.*, 1994; Lin and Odle, 1995; Schmidt and Herpin, 1998). The finding that pig liver was 20 times more sensitive to inhibition by malonyl-CoA ultimately lead to the characterization of pig CPT I. Nicot, Hegardt, Woldegiorgis, Haro and Marrero (2001) were the first to determine that pig CPT I is structured as a natural chimera of rat liver and muscle isoforms. The hybrid pig protein contains elements resembling both liver- (L) and muscle-(M)-CPT I isotypes, with the L-CPT I binding site for acyl-CoA and the M-CPT I binding sites for carnitine and malonyl-CoA. Accordingly, pig L-CPT I exhibits saturation kinetics similar to the rat for carnitine (126 μM) and palmitoyl-CoA (35 μM); however, the IC_{50} for malonyl-CoA (140 nM), is similar to human M-CPT I (Nicot *et al.*, 2001). Although the liver is not a lipogenic organ for the pig (Mersmann, Goodman, Houk and Anderson, 1973), which in part could account for a higher sensitivity to malonyl-CoA, it is unknown what the direct consequences of this increased sensitivity are toward fuel homeostasis of the neonatal pig. Related observations by Dyck, Cheng, Stanley, Barr, Chandler, Brown, Wallace, Arrhenius, Harmon, and Yang, Nadzan and Lopaschuk (2004) showed that inhibiting malonyl-CoA decarboxylase (MCD) increased myocardial malonyl-CoA concentrations and decreased fatty acid oxidation in pig hearts *in vivo*, suggesting that MCD might have a role in the regulation of fatty acid oxidation by changing malonyl-CoA concentrations. Furthermore, gene knockout studies also have shown that blood ketone bodies were significantly increased by overnight fasting of rats lacking acetyl-CoA caboxylase ß, (Abu-Elheiga, Matzuk, Abo-Hashema and Wakil, 2001), suggesting that ACCß controls fatty acid oxidation in liver and its regulation may be different in liver and muscle.

In addition to the unique structure and regulation of CPT I, the activity of mitochondrial 3-hydroxy-3-methylglutaryl-Coenzyme A synthase (mHMGCS), the putative rate limiting enzyme of ketogenesis, is extremely low in piglets, being 70 % lower than in neonatal rabbits (Adams and Odle, 1993) or adult rats (Duée *et al.*, 1994). Ketogenesis represents significant carbon flux during the neonatal period of most mammalian neonates. In rats, ketogenesis increases rapidly during postnatal development (Figure 11.3) (Girard *et al.*, 1992) or fasting (McGarry and Foster, 1980). The increase is ascribed in part to increased mRNA abundance, protein abundance and activity of mHMGCS (Thumelin, Forestier, Girard and Pégorier, 1993; Quant, 1994). The critical role of mRNA abundance also has been demonstrated in sheep during development (Lane, Baldwin and Jesse, 2002) and in rats by using a ketogenic diet (Cullingford, Eagles and Sato, 2002). In contrast, there is no substantial mRNA detected in piglets during the first two weeks after birth, although the abundance does increase later in life (Adams, Alho, Asins, Hegardt and Marrero, 1997b). This observation suggests that the unusually low ketogenic capacity and lack of hyperketonemia in postnatal piglets is associated with low expression of the mHMGCS gene during development. In addition, the gene expression can be highly induced by fasting during this stage, while suckled rat pups have no such changes in mRNA level of mHMGCS when fasted (Adams *et al.*, 1997b). Consistent with the increase in mRNA abundance, enzyme activity also is increased by 27-fold, but it still remains 50 fold lower than observed in fasted adult and suckling rats. These data imply that a post-transcriptional mechanism could be involved in the expression of mHMGCS. Indeed, Barrero, Alho, Ortiz, Hegardt, Haro and Marrero (2001) demonstrated that the attenuated activity of mHMGCS in piglets is not associated with low mRNA expression per se, but rather a low rate of translation.

Peroxisomal ß-oxidation in pigs

Peroxisomal ß-oxidation is an alternate pathway of metabolism of fatty acids that has been well characterized in rodents and other species. In general, oxidation occurring in the peroxisome is considered ancillary to mitochondrial oxidation. The peroxisome is, however, essential for the oxidation of medium- and very-long-chain fatty acids that are not oxidized very well by the mitochondrion (Osmundsen, Bremer and Pedersen, 1991; Van den Bosch, Schrakamp, Hardeman, Zomer, Wanders and Schutgens, 1993; Wanders, Vreken, Ferdinandusse, Jansen, Waterham, van Roermund and Van Grunsven, 2001). Because the peroxisome lacks an electron transport chain, electrons are transferred directly to molecular oxygen (Mannaerts, Debeer, Thomas and De Schepper, 1979). Thus, the first step in peroxisomal ß-oxidation, catalyzed by fatty acyl-CoA oxidase (FAO) (ACO), is not coupled to ATP production, but rather, heat is generated and released. Therefore, peroxisomal ß-oxidation may play an important role in thermogenesis (Goglia, Liverini, Lanni, Lossa and Barletta, 1989 ab). In addition, ACO activity is greatest with longer carbon chain lengths (Vanhove, van Veldhoven, Fransen, Denis, Wanders, Eyssen, and Mannaerts, 1993). Thus, fatty acids are only chain shortened (not fully combusted to acetyl-CoA) and peroxisomal ß-oxidation produces approximately 30% less energy compared to mitochondrial ß-oxidation (Reddy and Mannaerts, 1994).

Because of the characteristics of peroxisomal ß-oxidation, its ontogeny and putative role in postnatal thermoregulation has received research attention in last 10 years. Interestingly, we have showed that the pig has a high proportion of peroxisomal fatty oxidation that represents a greater percentage (40-50%) of total ß-oxidation, independent of the administration of peroxisome proliferators (Yu, Drackley, Odle and Lin, 1997a; Yu, Drackley and Odle, 1997b). The induction of peroxisomal oxidation occurs immediately postpartum, is greater in the suckled versus fasted piglet (Yu *et al.*, 1997b; Yu, Drackley and Odle, 1997c), and is reliant on the initiation of suckling (Yu *et al.*, 1997c). The control of peroxisomal ß-oxidation appears to be correlated with the developmental pattern of fatty acyl-CoA oxidase activity. The rapid postnatal development of peroxisomal ß-oxidation that we have measured in piglets and that others have determined in rats (Tsukada, Mochizuki and Konishi, 1968; Veerkamp and van Moerkerk, 1986) may be a physiologically adaptive mechanism for neonatal survival and growth. In addition, peroxisomal oxidation may have a great contribution to the acetate pool (70%) which predominates the acid soluble products pool (ASP) during in vitro ß-oxidation (Lin *et al.*, 1995), presumably because of high acetyl-CoA hydrolase activity present in peroxisomes. However, we have noticed that inhibition of CPT I significantly reduces acetate production (Lin *et al.*, 1995). Also, induction of peroxisomal ß-oxidation by stimulating peroxisome proliferator activated receptor α (PPARα) resulted in a great increase of CPT I and ACO activities and ß-oxidation in both peroxisome and mitochondria (Yu et., 1997a; Lyvers Peffer, Lin, and Odle, 2005; Figure 11.6). From this we infer that the regulation of fatty acid oxidative metabolism is different in pigs compared to other mammals. The physiological significance of peroxisomal ß-oxidation of fatty acids and the functional coordination between peroxisomal and mitochondrial ß-oxidation in neonates merits further investigation.

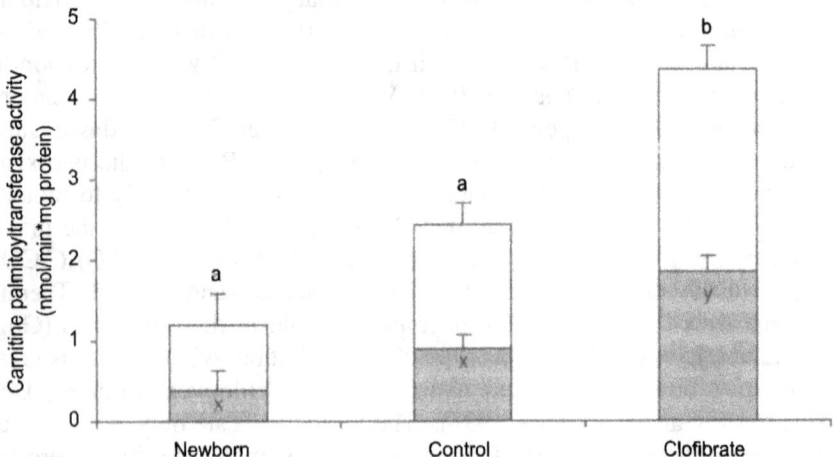

Figure 11.6 Malonyl-CoA sensitive and malonyl-CoA insensitive hepatic carnitine palmitoyltransferase activity (CPT) in piglet liver. Activity was measured at the end of a 10-12 d feeding period of ± clofibrate (a PPARα agonist). Total activity is represented by summation of malonyl-CoA sensitive (CPT I) and insensitive activities. Different letters (a,b) denote significant difference between treatments for malonyl-CoA sensitive CPT activity ($p < 0.05$). Different letters (x,y) denote significant difference between dietary treatment for malonyl-CoA insensitive CPT activity ($p < 0.05$). Modified from Lyvers Peffer *et al.*, 2005.

The role of PPAR in regulation of fatty acid oxidation in neonatal pigs

The key regulatory enzymes in hepatic peroxisomal and mitochondrial ß-oxidation (ACO, CPT I, mHMGCS and MCD) all are under PPARα regulation. A member of the nuclear receptor superfamily, this transcription factor was first discovered as an inducer of peroxisome proliferation (as reviewed by Schoonjans, Staels and Auwerx, 1996). This involves the up-regulation of peroxisomal genes (mainly those involved with ß-oxidation), an increase in peroxisomal size, and the production of more peroxisomes (Purdue and Lazarow, 2001). In mammals, peroxisomal proliferation is induced by both physiological stimuli (fatty acids) and a variety of xenobiotics through their direct binding with PPARα (Kliewer, Xu, Lambert, and Willson, 2001; Totland, Madsen, Klementsen, Vaagenes, Kryvi, Frøyland, Hexeberg and Berge, 2000). In mice and rats, PPARα is highly expressed in cells that have high fatty acid catabolic rates including the liver, kidney, heart, and skeletal muscle (Braissant, Foufelle, Scotto, Dauca and Wahli, 1996). However, only the liver and to a lesser degree, the kidney, undergo peroxisome proliferation (Schoonjans *et al.*, 1996). Proliferation also is species dependent; while the rat is quite responsive to peroxisome proliferation, humans and pigs remain unresponsive despite the peroxisome being an essential organelle (Vamecq and Draye, 1989; Cheon, Nara, Band, Beever, Wallig and Nakamura, 2005).

Located upstream of both L-CPT I and mHMGCS genes is the response element (PPRE; Ortiz, Mallolas, Nicot, Bofarull, Rodriguez, Hegardt, Haro and Marrero, 1999), implying that expression of both genes may be regulated through the use of peroxisome proliferators. We (Yu, Odle and Drackley, 2001) showed a >2 fold increase in total CPT activity when pigs were fed clofibrate for 2 wk, and a 3 fold increase in peroxisomal ß-oxidation of the liver. The increase of hepatic CPT activity and peroxisomal ß-oxidation appears to be associated with an increase of mRNA expression by clofibrate (Luci, Giemsa, Kluge and Eder, 2007)

Most recently, the PPAR distribution in pig tissues was investigated at 1 day, 5 weeks and 25 weeks of age. The deduced amino acid sequence of PPARα in pigs revealed an evolutionary distance to rodent species, suggesting that the difference could account for the species-dependent response to peroxisome proliferators, such as clofibrate. Pig PPARα mRNA is predominately expressed in kidney and liver (Sunvold, Grindflek and Lien, 2001) and expression of PPARα mRNA and its target gene ACO also were observed in adipose tissue from two genetic populations of pigs (Ding, Schinckel, Weber and Mersmann, 2000). Clofibrate induces genes involved in mitochondrial fatty acid oxidation and ketogenesis in 21-day old fasted pigs (Cheon *et al.*, 2005) and 56-day old pigs (Luci *et al.*, 2007) without liver hyperplasia or hepatomegaly. However, liver glycogen was reduced by clofibrate. These data suggested that PPARα plays a central role in metabolic adaptation in pig liver (Cheon *et al.*, 2005).

Conclusion

Inadequate nurture remains one of the leading causes of piglet mortality. Increasing energy supply and utilization will play an essential role in preventing the serious losses stemming from inadequate nurture. At birth, the piglet is poorly adapted to use milk lipids, and the attenuated production of ketone bodies exacerbates the problem for extrahepatic tissues. Because ketone bodies function as glucose-sparing carbon for otherwise glucose-dependent tissues, lack of ketone bodies may be detrimental to the survival of piglets. In addition, because fatty acid oxidation is required to support active gluconeogenesis, impaired fatty acid oxidation also could contribute indirectly to hypoglycemia. The enzymologic basis for attenuated ketogenesis and low hepatic ß-oxidation in piglets remains an active area of research. Modulation of CPT I by malonyl-CoA inhibition is generally accepted as the predominate control site for ketogenesis and fatty acid oxidation in liver. This system may be important in neonatal swine because pig CPT-I is particularly sensitive to malonyl-CoA inhibition which appears to be associated with the unique structure of the pig enzyme. An intramitochondrial regulation site of ketogenesis also has been suggested which is an idea strongly supported by reports of negligible activity of the ketogenic enzyme mitochondrial HMG-CoA synthase in pig liver. The low activity of this enzyme results from suppressed expression in neonatal pigs and may be related to post transcriptional modification and /or specific differences in enzyme kinetics. It remains unclear if the low mitochondrial HMG-CoA synthase activity in pigs controls the hepatic ß-oxidation flux in this species. Because PPARα is largely expressed in liver and induces genes involved in both peroxisomal and mitochondrial fatty acid oxidation, PPARα is now considered to be an essential transcription factor in regulating hepatic fatty acid oxidation. Transcriptional regulation of enzymes by PPARα activation is required in the ß-oxidative pathway of the mitochondria and peroxisomes including CPT I and CPT II (Chatelain, Kohl, Esser, McGarry, Girard and Pégorier, 1996; Mascaró, Acosta, Ortiz, Marrero, Hegardt and Haro, 1998), mHMGCS (Rodriguez, Gil-Gomez, Hegardt and Haro, 1994; Cullingford, Dolphin and Sato, 2002b), MCD (Lee, Kim, Zhao, Cha and Kim, 2004), ACO and catalase. Clearly, with the recent progress of clarifying the physiological idiosyncrasies of lipid metabolism in neonatal piglets at molecular levels, the control of fatty acid oxidation via gene regulation may be an effective tool to improve the utilization of milk fat and lead to improved piglet survivability.

Acknowledgement

We acknowledge support from the US Department of Agriculture CSREES National Research Initiative (grant no. 2007-35206-17897) in preparation of this reivew.

References

Abu-Elheiga L, Matzuk, M.M., Abo-Hashema, K.A. and Wakil, S.J. (2001) Continuous fatty acid oxidation and reduced fat storage in mice lacking acetyl-CoA carboxylase 2. *Science*, **291**, 2613-2616.

Adams, S.H. and Odle, J. (1993) Plasma beta-hydroxybutyrate after octanoate challenge: attenuated ketogenic capacity in neonatal swine. *American Journal of Physiology*, **265**, R761-R765.

Adams, S.H., Lin, X., Yu, X.X., Odle, J. and Drackley, J.K. (1997a) Hepatic fatty acid metabolism in pigs and rats: major differences in endproducts, O2 uptake, and beta-oxidation. *American Journal of Physiology*, **272**, R1641-R1646.

Adams, S.H., Alho, C.S., Asins, G., Hegardt, F.G. and Marrero, P.F. (1997b) Gene expression of mitochondrial 3-hydroxy-3-methylglutaryl-CoA synthase in a poorly ketogenic mammal: effect of starvation during the neonatal period of the piglet. *Biochemical Journal*, **324**, 65-73.

Azain, M.J., Tomkins, T., Sowinski, J.S., Arentson, R.A. and Jewell, D.E. (1996) Effect of supplemental pig milk replacer on litter performance: seasonal variation in response. *Journal of Animal Science*, **74**, 2195-2202.

Barrero, M.J., Alho, C.S., Ortiz, J.A., Hegardt, F.G., Haro, D. and Marrero, P.F. (2001) Low activity of mitochondrial HMG-CoA synthase in liver of starved piglets is due to low levels of protein despite high mRNA levels. *Archive of Biochemistry and Biophysics*, **385**, 364-371.

Battaglia, F.C. and Meschia, G. (1988) Fetal Nutrition. *Annual Review of Nutrition*, **8**, 43-61.

Beiber, L.L., Markwell, M.A.K., Blair, M. and Helmrath, T.A. (1973) Studies on the development of carnitine palmitoyltransferase and fatty acid oxidation in liver mitochondria of neonatal pigs. *Biochimica Et Biophysica Acta*, **326**, 145-154.

Bengtsson, G., Gents, J., Hakkarainen, J., Hellström, R. and Persson, B. (1969) Plasma levels of FFA, glycerol, ß-hydroxybutyrate, and blood glucose during the postnatal development of the pig. *The Journal of Nutrition*, **97**, 311-315.

Benevenga, N.J., Steinman-Goldsworthy, J.K., Crenshaw, T.D. and Odle, J. (1989) Utilization of medium-chain triglycerides by neonatal pigs: 1. Effects on milk consumption and body fuel utilization. *Journal of Animal Science*, **67**, 3331-3339.

Braissant, O., Foufelle, F., Scotto, C., Dauca, M. and Wahli, W. (1996) Differntial expression of peroxisome proliferators-activated receptors (PPARs): tissue distribution of PPAR-α, -ß, and -γ in the adult rat. *Endocrinology*, **137**, 354-366.

Braude, R., Mitchell, K.G., Newport, M.J. and Porter, J.W.G. (1970) Artificial rearing of pigs 1. Effects of frequency and level of feeding on performance and digestion of milk proteins. *British Journal of Nutrition*, **24**, 501-516.

Chatelain, F., Kohl, C., Esser, V., McGarry, J.D. Girard, J, and Pégorier, J.P. (1996) Cyclic AMP and fatty acids increase carnitine palmitoyltransferase I gene transcription in cultured fetal rat hepatocytes. *European Journal of Biochemistry*, **235**, 789-798.

Cheon, Y., Nara, T.Y., Band, M.R., Beever, J.E. Wallig, M.A. and Nakamura, M.T. (2005) Induction of overlapping genes by fasting and a peroxisome proliferator in pigs: evidence of functional PPAR{alpha} in nonproliferating species. *American Journal of Physiology, Regulatory, Integrative and Comparative Physiology*, **288**, R1525-R1535.

Cullingford, T.E., Eagles, D.A. and Sato, H. (2002a) The ketogenic diet upregulates expression of the gene encoding the key ketogenic enzyme mitochondrial 3-hydroxy-3-methylglutaryl-CoA synthase in rat brain. *Epilepsy Research*, **49**, 99-107.

Cullingford, T.E., Dolphin, C.T. and Sato, H. (2002b) The peroxisome proliferators-activated receptor α-selective activator ciprofibrate upregulates expression of genes encoding fatty acid oxidation and ketogenesis enzymes in rat brain. *Neuropharmacology*, **42**, 724-730.

Ding, S.T., Schinckel, A.P., Weber, T.W. and Mersmann, H.J. (2000) Expression of porcine transcription factors and genes related to fatty acid metabolism in different tissues and genetic populations. *Journal of Animal Science*, **78**, 2127-2134.

Duée, P., Pégorier, J., Manoubi, L.E., Herbin, C., Kohl, C. and Girard, J. (1985) Hepatic triglyceride hydrolysis and development of ketogenesis in rabbits. *American Journal of Physiology*, **249**, E478-E484.

Duée, P., Pégorier, J, Quant, P.A., Herbin, C., Kohl, C. and Girard, J. (1994) Hepatic ketogenesis in newborn pigs is limited by low mitochondrial 3-hydroxy-3-methylglutaryl-CoA synthase activity. *Biochemical Journal*, **298**, 207-212

Dyck, J.R., Cheng, J.F., Stanley, W.C., Barr, R., Chandler, M.P., Brown, S., Wallace, D., Arrhenius, T., Harmon, C., Yang, G., Nadzan, A.M. and Lopaschuk, G.D. (2004) Malonyl coenzyme a decarboxylase inhibition protects the ischemic heart by inhibiting fatty acid oxidation and stimulating glucose oxidation. *Circulation Research,* **94**, 78-84

Foster, P.C. and Bailey, E. (1976) Changes in hepatic fatty acid degradation and blood lipid and ketone body content during development of the rat. *Enzyme*, **21**, 397-407.

Girard, J., Ferré, P., Pégorier, J. and Duée, P. (1992) Adaptations of glucose and fatty acid metabolism during the perinatal period and suckling-weaning transition. *Physiology Review*, **72**, 507-563.

Goglia, F., Liverini, G., Lanni, A., Iossa, S. and Barletta, A. (1989a) Effects of 3,5,3'-triiodothyronine (T3) on rat liver peroxisomal compartment during cold exposure. *Experimental Biology*, **48**, 135-140.

Goglia, F., Liverini, G., Lanni, A., Iossa, S. and Barletta, A. (1989b) Morphological and functional modifications of rat liver peroxisomal subpopulations during cold exposure. *Experimental Biology*, **48**, 127-133.

Heo, K. (2000). Nutritional and metabolic assessment of carnitine for young pigs. Ph.D. thesis, North Carolina State University, USA.

Herpin, P., Damon M. and Le Dividich. J. (2002) Development of thermoregulation and neonatal survival in pigs. *Livestock Production Science*, **78**, 25-45.

Herpin, P., Le Dividich, J. and van Os, M. (1992) Contribution of colostral fat to thermogenesis and glucose homostasis in the newborn pig. *Journal of Developmental Physiology*, **17**, 133-141.

Kim, K. (1997) Regulation of mammalian acetyl-Coenzyme A carboxylase. *Annual Review of Nutrition*, **17**, 77-99.

Kliewer, S.A., Xu, H.E., Lambert, M.H. and Willson, T.M. (2001) Peroxisome proliferators-activated receptors: From genes to physiology. *Recent Progress in Hormone Research*, **56**, 239-263.

Lane, M.A., Baldwin, VI, R.L. and Jesse, B.W. (2002) Developmental changes in ketogenic enzyme gene expression during sheep rumen development. *Journal of Animal Science*, **80**, 1538-1544.

Lay, Jr., D.C., Matteri, R.L., Carroll, J.A., Fangman, T.J. and Safranski, T.J. (2002) Preweaning survival in swine. *Journal of Animal Science*, **80**, E74-E86.

Lecce, J. G. (1975) Rearing piglets artificially in a farm environment: A promise unfulfilled. *Journal of Animal Science*, **41**, 659-666.

Le Dividich, J. and Noblet, J. (1984) Effect of colostrums intake on metabolic rate and plasma glucose in the neonatal pig in relation to environmental temperature. *Biology of the Neonate*, **46**, 98-104.

Le Dividich, J. and Sève, B. (2000) Energy requirement of the young piglets. In *The Weaner pig*. Edited by Wiseman J. CAB international, Oxford.

Lee, G.Y., Kim, N.H., Zhao, Z.S., Cha, B.S. and Kim, Y.S. (2004) Peroxisomal-proliferator-activated receptor alpha activates transcription of the rat hepatic malonyl-CoA decarboxylase gene: a key regulation of malonyl-CoA level. *Biochemical Journal*, **378**, 983-990.

Lin, X., and Odle, J. (1995) Regulation of fatty acid oxidation in mitochondria of neonatal pigs via carnitine palmitoyltransferase I. *Journal of Animal Science*, **73** (suppl. 1), 77 (abs.).

Losinger, W.C. (2005) Economoc impacts of mortality rate for sucking pigs in the United States. *Journal of American Veterinary Medical Association*, **227**, 896-902.

Luci, S., Giemsa, B., Kluge, H. and Eder, K. (2007) Clofibrate causes an up-regulation of PPAR-alpha target genes but does not alter expression of SREBP target genes in liver and adipose tissue of pigs. *American Journal of Physiology, Regulatory, Integrative and Comparative Physiology, in press*.

Lyvers Peffer, P.L., Lin, X. and Odle, J. (2005) Hepatic {b}-oxidation and carnitine palmitoyltransferase I in neonatal pigs after dietary treatments of clofibric acid, isoproterenol, and medium-chain triglycerides. *American Journal of Physiology, Regulatory, Integrative and Comparative Physiology,* **288**, R1518-1524.

Mannaerts, M.J. and McCrea, M.R. (1963) Changes in the chemical composition of sow reared piglets during the 1st moth of life. *British Journal of Nutrition,* **17**, 495-513.

Mannaerts, G.P., Debeer, L.J., Thomas, J. and De Schepper, P.J. (1979) Mitochondrial and peroxisomal fatty acid oxidation in liver homogenates and isolated hepatocytes from control and clofibrate-treated rats. *The Journal of Biological Chemistry,* **254**, 4585-4595.

Marion, J. and Le Dividich, J. (1999) Utilization of the energy in sow milk by the piglet. In *Processing of the 7th Biennial Conference of ASPA, 28th November-1st* , Edited by Cranwell, P.D., Adelaide, South Austrialia. Pig Production, VII, p, 254.

Mascaro, C., Acosta, E., Ortiz, J.A., Marrero, P.F., Hegardt, F.G. and Haro, D. (1998) Control of human muscle-type carnitine palmitoyltransferase I gene transcription by peroxisome proliferator-activated receptor. *The Journal of Biological Chemistry,* **273**, 8560-6563.

McGarry, J.D. (2001) Travels with carnitine palmitoyltransferase I: from liver to germ cell with stops in between. *Biochemical Society Transactions,* **29(Pt 2)**, 241-245. Review.

McGarry, J.D. and Foster, D.W. (1980) Regulation of hepatic fatty acid oxidation and ketone body production. *Annual Review of Biochemistry,* **49,** 395-420.

Mersmann, J., Goodman, J., Houk, J.M. and Anderson, S. (1972) Studies of the biochemistry of mitochondria and cell morphology in the neonatal swine hepatocyte. *Journal of Cell Biology,* **53**, 335-347.

Mount, L.E. (1968) The climatic physiology of the pig. In *Monographs of the physiological society,* Edited by Harris, G.W. *et al.* The Williams & Wilkins Co., Baltimore, MD.

NAHMS. (1997) Changes in the U.S. Pork Industry. National Animal Health Monitoring System, Animal and Plant Health Inspection Service, Veterinary Services. United States Department of Agriculture, Washington D.C. pp 10-13.

NAHMS. (2000) Part 1. Reference of swine health and management in the United States. National Animal Health Monitoring System, Animal and Plant Health Inspection Service. Veterinary Services. United States Department of Agriculture, Washington D.C.pp14-16.

Nicot, C., Hegardt, F.G., Woldegiorgis, G., Haro, D. and Marrero, P.F. (2001) Pig liver carnitine palmitoyltransferase I, with low K_m for carnitine and high sensitivity to malonyl-CoA inhibition, is a natural chimera of rat liver and muscle enzymes. *Biochemistry,* **40,** 2260-2266.

Noblet, J. and Le Dividich, J. (1981) Energy metabolism of the newborn pig during the first 24 h of life. *Biology of the Neonate*, **40**, 175-182.

Odle, J. (1997) New insights into the utilization of medium-chain triglycerides by the neonate: observations from a piglet model. *The Journal of Nutrition*, **127**, 1061-1067.

Odle, J., Lin, X., Wieland, T.W. and van Kempen, T.A. (1994) Emulsification and fatty acid chain length affect the kinetics of [^{14}C]-medium-chain triacylglycerol utilization by neonatal piglets. *The Journal of Nutrition*, **124**, 84-93.

Odle, J., Benevenga, N.J. and Crenshaw, T.D. (1991a) Utilization of medium-chain triglycerides by neonatal piglets: Chain length of even- and odd-carbon fatty acids and apparent digestion/absorption and hepatic metabolism. *The Journal of Nutrition*, **121**, 605-614.

Odle, J., Benevenga, N.J. and Crenshaw, T.D. (1991b) Postnatal age and thee metabolism of medium- and long-chain fatty acids by isolated hepatocytes from small-for-gestational-age and appropriate-for-gestational-age piglets. *The Journal of Nutrition,* **121**, 615-621.

Odle, J., Lin, X., van Kempen, T., Drackely, J.K. and Adams, S.H. (1995) Carnitine palmitoyltransferase modulation of hepatic fatty acid metabolism and radio-HPLC evidence for low ketogenesis in neonatal pigs. *The Journal of Nutrition*, **125**, 2541-2549.

Ortiz, J.A., Mallolas, J., Nicot, C., Bofarull, J., Rodrßguez, J.C., Hegardt, F.G., Haro, D. and Marrero, P.F. (1999) Isolation of pig mitochondrial 3-hydroxy-3-methylglutaryl-CoA synthase gene promoter: characterization of a peroxisome proliferator-responsive element. *Biochemical Journal*, **337**, 329-335.

Osmundsen, H., Bremer, J. and Pedersen, J.I. (1991) Metabolic aspects of peroxisomal beta-oxidation. *Biochimica Et Biophysica Acta*, **1085**, 141-58. Review.

Pégorier, J., Duée, P., Assan, R., Peret, J. and Girard, J. (1981) Changes in circulating fuels, pancreatic hormones and liver glycogen concentration in fasting or suckling newborn pigs. *Journal of Developmental Physiology*, **3**, 203-217.

Pégorier, J., Duée, P., Girard, J. and Peret, J. (1983) Metabolic fate of non-esterified fatty acids in isolated hepatocytes from newborn and young pigs. *Biochemical Journal,* **212,** 93-97.

Pégorier J.P., Garcia-Garcia, M.V., Prip-Buus, C., Duee, P.H., Kohl, C. and Girard, J. (1989) Induction of ketogenesis and fatty acid oxidation by glucagon and cyclic AMP in cultured hepatocytes from rabbit fetuses. Evidence for a decreased sensitivity of carnitine palmitoyltransferase I to malonyl-CoA inhibition after glucagon or cyclic AMP treatment. *Biochemical Journal*, **264**, 93-100.

Purdue, P.E. and Lazarow, P.B. (2001) Peroxisome Biogenesis. *Annual Review of Cell and Developmental Biology,* **17,** 701-752.

Quant, P.A. (1994) The role of mitochondrial HMG-CoA synthase in regulation of ketogenesis. *Essays in Biochemistry*, **28**, 13-25.

Reddy, J.K. and Mannaerts, G.P. (1994) Peroxisomal lipid metabolism. *Annual Review of Nutrition*, **14**, 343-70. Review.

Robles-Valdes, C., McGarry, J.D. and Foster, D.W. (1976) Maternal-fetal carnitine relationships and neonatal ketosis in the rat. *The Journal of Biological Chemistry*, **257**, 6007-6012.

Rodriguez, J.C., Gil-Gomez, G., Hegardt, F.G. and Haro, D. (1994) Peroxisome proliferator-activated receptor mediates induction of the mitochondrial 3-hydroxy-3-methylglutaryl-CoA synthase gene by fatty acids. *The Journal of Biological Chemistry*, **269**, 18767-18772.

Rohde Parfet, K.A. and Gonyou, H.W. (1988) Effect of creep partitions on teat-seeking behavior of newborn piglets. *Journal of Animal Science*, **66**, 2165-2173.

Schmidt, I. and Herpin, P. (1998) Carnitine palmitoyltransferase I (CPTI) activity and its regulation by malonyl-CoA are modulated by age and cold exposure in skeletal muscle mitochondria from newborn pigs. *The Journal of Nutrition*, **128**, 886-893.

Schoonjans, K., Staels, B. and Auwerx, J. (1996) The peroxisome proliferators activated receptors (PPARs) and their effects on lipid metabolism and adipocyte differentiation review. *Biochimica Et Biophysica Acta*, **302**, 93-109.

Stanier, M.W., Mount, L.E. and Bligh, J. (1984) In *Energy Balance and Temperature Regulation*. Cambridge University Press, NY.

Sundvold, H., Grindflek, E. and Lien, S. (2001) Tissue distribution of porcine peroxisome proliferator-activated receptor alpha: detection of an alternatively spliced mRNA. *Gene*, **273**, 105-113.

Swiatek, K.R., Kipnis, D.M., Mason, G., Chao, K.L. and Cornblath, M. (1968) Starvation hypoglycemia in newborn pigs. *American Journal of Physiology*, **214**, 400-405.

Tetrick, M.A., Adams, S.H., Odle, J. and Benevenga, N.J. (1995) Contribution of D-(-)-3-Hydroxybutyrate to the energy expenditure of neonatal pigs. *The Journal of Nutrition*, **125**, 264-272.

Thumelin, S., Forestier, M., Girard, J. and Pégorier, J.P. (1993) Developmental changes in mitochondrial 3-hydroxy-3-methylglutaryl-CoA synthase gene expression in rat liver, intestine and kidney. *Biochemical Journal*, **292**, 493-496.

Totland, G.K., Madsen, L., Klementsen, B., Vaagenes, H., Kryvi, H., Frøyland, L., Hexeberg, L. and Berge, R.K. (2000) Proliferation of mitochondria and gene expression of carnitine palmitoyltransferase and fatty acyl-CoA oxidase in rat skeletal muscle, heart, and liver by hypolipidemic fatty acids. *Biology of the Cell*, **92**, 317-329.

Tsukada, H., Mochizuki, Y. and Konishi, T. (1968) Morphogenesis and development of microbodies of hepatocytes of rats during pre- and post-natal growth. *Journal of Cell Biology*, **37**, 231-243.

Tuchscherer, M., Puppe, B., Tuhsccherer, A. and Tiemann U. (2000) Early identificantion of neonates at risk: traits of newborn piglets with respect to survival. *Thenogenology,* **54,** 371-388.

Vamecq, J. and Draye, J.P. (1989) Pathophysiology of peroxisomal b-oxidation. *Essays in Biochemistry,* **24,** 115-225.

van den Bosch, H., Schrakamp, G., Hardeman, D., Zomer, A.W., Wanders, R.J. and Schutgens, R.B. (1993) Ether lipid synthesis and its deficiency in peroxisomal disorders. *Biochimie,* **75,** 183-189. Review.

Vanhove, G.F., van Veldhoven, P.P., Fransen, M., Denis, S., Wanders, R.J.A., Eyssen, H.J. and Mannaerts, J.P. (1993) The CoA esters of 2-methyl-branched chain fatty acids and of the bile acid intermediates di- and trihydroxycoprostanic acids are oxidized by one single peroxisomal branched chain acyl-CoA oxidase in human liver and kidney. *The Journal of Biological Chemistry,* **268,** 10335-10344.

Varley M.A. (1995) Introduction. *In The Neonatal Pig Development and Survival-1995,* pp 1-13. Edited by M.A. Varley, CAB INTERNATIONAL, Wallingford, Oxon OX10 8DE, UK.

Veerkamp, J.H. and van Moerkerk, H.T.B. (1986) Peroxisomal fatty acid oxidation in rat and human tissues. Effect of nutritional state, clofibrate treatment and postnatal development in the rat. *Biochimica Et Biophysica Acta,* **875,** 301-310.

Veum, T.L. and Odle, J. (2001) Feeding neonatal pigs. In *Swine Nutrition, Second Edition.* Edited by Lewis, A.J. and Southern, L. CRC Press. Boca Raton, FL.

Wanders, R.J., Vreken, P., Ferdinandusse, S., Jansen, G.A., Waterham, H.R., van Roermund, C.W. and Van Grunsven, E.G. (2001) Peroxisomal fatty acid alpha- and beta-oxidation in humans: enzymology, peroxisomal metabolite transporters and peroxisomal diseases. *Biochemical Society in Transactions,* **29,** 250-267. Review.

Wolfe, R.G., Maxwell, C.V. and Nelson, E.C. (1978) Effect of age and dietary fat level on fatty acid oxidation in the neonatal pig. *The Journal of Nutrition,* **108,** 1621-1634.

Yu, X.X., Drackley, J.K., Odle, J. and Lin, X. (1997a) Response of hepatic mitochondrial and peroxisomal ß-oxidation to increasing palmitate concentrations in piglets. *Biology of the Neonate,* **72,** 284-292.

Yu, X.X., Drackley, J.K. and J. Odle, J. (1997b) Rates of mitochondrial and peroxisomal ß-oxidation of palmitate change during postnatal development and food deprivation in liver, kidney and heart of pigs. *The Journal of Nutrition,* **127,** 1814-1821.

Yu, X.X., Drackley, J.K. and Odle, J. (1997c) Food deprivation changes peroxisomal ß-oxidation activity but not catalase activity during postnatal development in pig tissues. *The Journal of Nutrition,* **128,** 1114-1121.

Yu, X.X., Odle, J. and Drackley, J.K. (2001) Differntial induction of peroxisomal ß-oxidation enzymes by clofibric acid and aspirin in piglet tissues. *American Journal of Physiology, Regulatory, Integrative and Comparative Physiology,* **281**, R1553-R1561.

12

CONSEQUENCES OF SELECTION FOR LITTER SIZE ON PIGLET DEVELOPMENT

GEORGE FOXCROFT, GUISEPPE BEE[1], WALLY DIXON, MELISSA HAHN, JOHN HARDING[2], JENNY PATTERSON, TED PUTMAN, SABRINA SARMENTO, MIRANDA SMIT, WAI-YUE TSE AND SUSANNA TOWN[3]

Swine Reproduction-Development Program, Swine Research & Technology Centre, University of Alberta, Edmonton, Alberta, T6G 2P5, Canada; [1]Agroscope Liebefeld Posieux Research Station ALP, Posieux, 1725, Switzerland; [2] Department of Large Animal Clinical Sciences, 52 Campus Drive, University of Saskatchewan, Saskatoon, Saskatchewan S7N 5B4, Canada; [3]J.S. Davies Building, University of Adelaide, Roseworthy Campus, South Australia 5371.

Introduction

In a litter-bearing domestic species like the pig, the number of offspring born is an important economic trait, and the components of litter size (ovulation rate, embryonic survival and uterine capacity) responsive to genetic selection are well established (Johnson *et al.*, 1985, 1999). However, as selection for ovulation rate has been associated with selection against early embryonic survival, and because birth weight decreased as litter size increased, Johnson *et al.* (1999) concluded that selection for uterine capacity might be the most productive approach to increasing litter size born in genetic selection programs. Thus, both the developmental competence of the pigs born, as well as the size of the litter, needs critical consideration.

A considerable amount of the variation in growth performance after birth may be pre-programmed during foetal development in the uterus (see Foxcroft and Town, 2004). Furthermore, it is likely that these pre-programmed limitations in growth performance only finally express themselves in the late grower and early finisher stage of production. Furthermore, foetal development may even affect postnatal growth performance in the absence of any associated effects on birth weight (Town, 2004). Therefore, from a practical perspective, sorting pigs by weight at the weaner and grower stages will not resolve inherent variation in growth performance that is still a characteristic of particular pigs or litters. The current chapter covering pre-natal programming, in the context of selection for increased litter size, aims to substantiate the view "that indirect negative effects of intra-uterine crowding on placental development in early pregnancy lead to reprogramming of foetal development, less efficient post-natal growth performance and adverse effects on carcass quality at slaughter".

The effects of prenatal programming on postnatal performance are not limited to effects on muscle development and growth. Epidemiological studies in human infants born with the phenotypic characteristics of intra-uterine growth retardation (IUGR) suggested that these infants had an increased risk of developing cardiovascular disease as adults (Barker *et al.*, 1989, 1990). This, and other epidemiological studies, led to the "Barker hypothesis", linking prenatal programming of the foetus to lifetime health outcomes (see Barker, 1994). The implications of prenatal programming for postnatal health outcomes in the pig are just as real. Harding *et al.* (2006a,b) discussed this in the context of development of the immune system and early postnatal survival. Furthermore, analysis of brain sparing effects that are indicative of IUGR showed that the organs most notably affected in stillborn pigs with low birth weight were the heart, liver and spleen. These and other developmental complications discussed below, undoubtedly underlie the problems of managing low birth weight pigs through the lactation and nursery stages of production.

Recent studies suggest that a lack of understanding of the dynamics of prenatal survival and development in existing mature sow populations is resulting in increasing variance in grow-finish performance. This chapter will explain the biological basis for such effects and the consequences of prenatal programming for postnatal growth performance and carcass quality. The implications of selecting for hyper-prolific sows, and the component traits that collectively determine both the number and quality of pigs born, will then be considered. As background information for considering production practices that might address increased variation in grow-finish performance, the major sources of variance in pig birth weight will be identified and related back to associations with prenatal programming of postnatal development. Finally, there will be a brief review of possible production strategies that might address the concerns raised.

Uterine capacity as the ultimate limitation on litter size and quality

The concept of uterine capacity was established using different experimental approaches to study effects of uterine crowding in the pig. These included uterine ligation, oviduct resection, unilateral hysterectomy and ovariectomy (UHO), super-ovulation and embryo transfer, and led to the conclusion that, when the number of embryos exceeded 14, intrauterine crowding was a limiting factor for litter size born (Dziuk, 1968). Bazer *et al.* (1969a,b) also concluded that increased embryonic loss, associated with a greater number of embryos in the uterus, was due to maternal limitations and not to inherent limitations of the embryo. They suggested that two physiological mechanisms might be involved. Initially, embryo selection might be the result of competition among embryos for some biochemical factor in the uterus necessary for their continued development. However, in later gestation, intrauterine competition for the establishment of adequate surface area for nutrient exchange between foetal and maternal circulations may act to limit litter size.

In the context of variation in development in the uterus, the concept has been advanced that mechanisms promoting competition among embryos in the pre-implantation period will act to reduce within-litter variation in development by selectively removing the least developed embryos (van der Lende *et al.*, 1990). Nevertheless, the more recent results of Père *et al.* (1997) confirm that, even in sows with "normal" ovulation rates, uterine capacity can affect both litter size and the average birth weight of the litter. Furthermore, information from large populations of higher parity commercial sows supports the hypothesis that the dynamics of in-utero development tends to become more variable as sows advance to higher parities (Town *et al.*, 2004a). In turn, it is believed that this creates greater variation in the birth weight characteristics of the litters born and unresolved problems for the appropriate post-weaning management of these litters.

WHEN DOES UTERINE CAPACITY IMPACT FOETAL SURVIVAL AND DEVELOPMENT?

Fenton *et al.* (1970) determined that uterine capacity only becomes a limiting factor for foetal survival after d 25 of gestation. Knight *et al.* (1977) further defined d 30 to 40 of gestation as the critical period when uterine capacity exerts its effects. Subsequent studies in both intact and UHO females support this conclusion (see Vallet, 2000). Vallet *et al.* (2003) suggested that foetal growth rate is less sensitive to intrauterine crowding than placental growth rate and, as in the prolific Meishan female (Ford and Youngs, 1993), within certain limits of uterine capacity an increase in placental efficiency may initially protect the developing foetus from a limitation in placental size. However, in some populations of commercial sows identified in recent studies (Town, 2004), increased placental efficiency did not compensate poor placental development in sows with even relatively modest intra-uterine crowding (15 vs. 9 embryos surviving to d 30) and classic effects of intrauterine growth retardation IUGR were present at d 90 of gestation.

In the context of results from earlier studies of within-litter variation in prenatal development (Adams, 1971; Widdowson, 1971; Hegarty and Allen, 1978; Flecknell *et al.*, 1981), Wooton *et al.* (1983) suggested that the extremes of (IUGR), or "runting", were identified within a discrete sub-set of foetuses. Furthermore, based on data from subsequent studies of the association between within-litter differences in prenatal development and postnatal survival and growth, van der Lende and de Jager (1991) concluded that the lower pre-weaning growth of the runt pigs born could not be entirely explained on the basis of their lower birth weight. They suggested that IUGR, or runting, had a more complex effect on developmental potential. Interestingly, data from the same laboratory led to the suggestion that the extent of IUGR within a litter was associated with specific patterns of embryonic survival (van der Lende *et al.*, 1990), and the larger the litters were in the uterus, the greater the chance that runt foetuses would be present. Furthermore, these data were consistent with the conclusion that within-litter variation in development was already established at the early foetal stage (d 27 to 35) of gestation.

CONSEQUENCES OF CHANGING PATTERNS OF PRE-NATAL LOSS FOR DEVELOPMENTAL POTENTIAL

Pre-implantation embryonic losses are still considered to be the largest proportion of prenatal loss in the pig, with some lesser loss in the post-implantation period that will ultimately reflect uterine capacity (as reviewed by Ashworth and Pickard, 1998). In commercial practice, this generalization probably reflects the situation in gilts in which ovulation rates of 10 to 15, associated with variable embryonic loss, are the primary factors determining litter size at term. Weaned contemporary first parity sows, bred at first post-weaning oestrus, also fall into this category. Although ovulation rate may be higher (18 to 20 ovulations), many sows tend to be in a catabolic state, and this generally decreases embryonic survival to d 30 of gestation (Foxcroft, 1997). However, the dynamics of prenatal loss in existing commercial dam-lines may be changing (Foxcroft, 1997; Figure 12.1). In these populations, several generations of direct selection for litter size have indirectly resulted in an imbalance between the number of conceptuses surviving to the post-implantation period (d 30 of gestation) and uterine capacity.

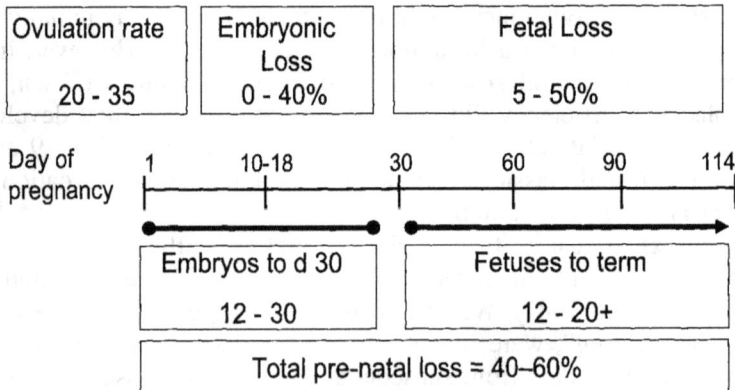

Figure 12.1 Schematic representation of the changing pattern of prenatal loss observed in multiparous sows in recent studies. Ovulation rate recorded as the total number of corpora lutea present at time of embryonic or foetal dissection.

Even in individual gilts with 20 or more ovulations, embryonic survival rate can be 100% at d 28 of gestation (Almeida *et al.*, 2000), whereas average first litter size is still only 10 to 12 piglets. In higher parity females, the situation may be even more extreme, with an increasing proportion of sows having more than 25 ovulations (Figure 12.2). Despite relatively poor embryonic survival to d 30, numbers of conceptuses in the uterus at d 30 (15 or more) still exceeded uterine capacity for normal development. Consistent with the literature reviewed earlier, uterine capacity then exerts its effects and a significant reduction in the number of conceptuses occurs by d 45 to 50 of gestation. An example of these changing trends can be seen in the data in Table 12.1, which represent the same sub-populations of sows in different parity groupings shown in Figure 12.2, but only include data from sows having ovulation rates greater than 25. These changing dynamics of prenatal loss have important implications for placental and foetal development.

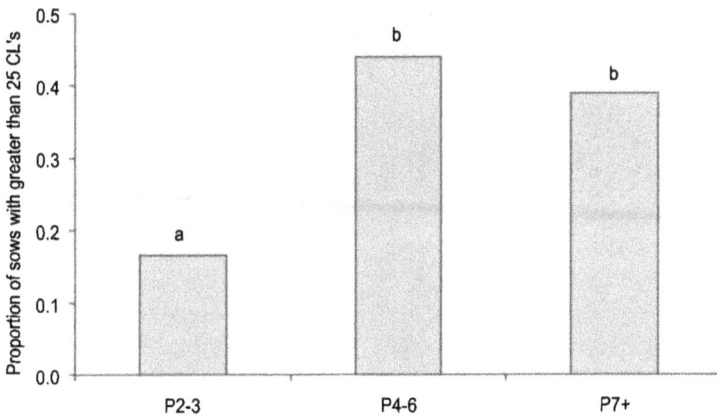

Figure 12.2 Proportion of sows with ovulation rates 25 or higher within parity (P) grouping. (Swine Research & Technology Centre unpublished , 2006) (N = 504)

Table 12.1 OVULATION RATE, VIABLE EMBRYOS AT DAY 30 (D30) AND FOETUSES AT D50 FOR SOWS WITH OVULATION RATES 25 OR HIGHER IN PARITIES 2-3, 4-6 AND 7+. (SWINE RESEARCH & TECHNOLOGY CENTRE UNPUBLISHED, 2006)

Variable	*P2-3*	*P4-6*	*P7+*
Ovulation Rate	27.0	28.3	27.4
Viable Embryos - d30	16.8	16.8	13.3
Viable Foetuses - d50	15.3	11.9	10.5

As has been shown in earlier studies, even modest increases in uterine crowding around d 30 of gestation decreased placental volume (Almeida *et al.*, 2000) and placental weight (Vonnahme *et al.*, 2002). More critically, this restriction in placental weight at d 30 persisted to d 90 (Town *et al*, 2004b; Figure 12.3). Although the size and weight of the embryo was not seen to be affected by crowding up to d44 of gestation, potential impacts on foetal development need careful study. If placental compensatory mechanisms are not adequate, crowding of the uterus in the early post-implantation period of gestation may affect foetal development of surviving conceptuses in a manner analogous to IUGR. This raises important questions for both foetal and postnatal development. In the context of commercial grow-finish performance, a specific interest in effects on the development of foetal muscle fibres, which start to differentiate around d 35 of gestation in the pig, is particularly important. In contrast to situations in which the occurrence of IUGR is limited to a discrete subpopulation of runt foetuses (Royston *et al.*, 1982; Wooton *et al.*, 1983), it was hypothesized that "a changing pattern of embryonic loss that results in uterine crowding in early gestation would produce a more uniform effect on placental development, that would affect the later development of all surviving foetuses" (Foxcroft, 1997; Foxcroft and Town, 2004).

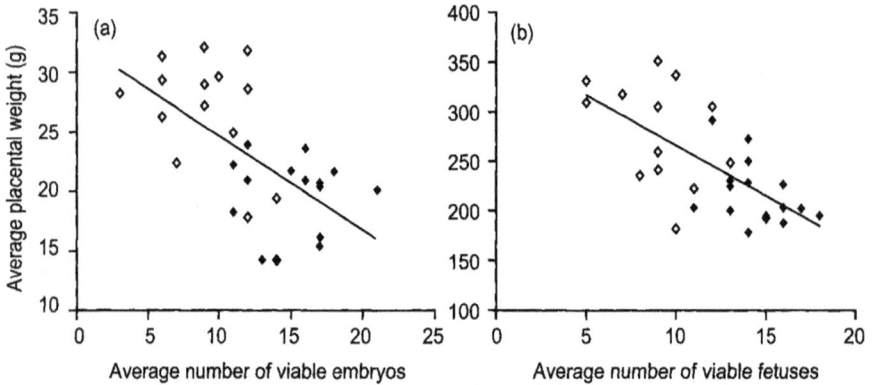

Figure 12.3 Correlation between average placental weight and (a) number of viable embryos at d 30 of gestation ($r = -0.61$; $P = 0.0003$) and (b) number of viable foetuses at day 90 of gestation ($r = -0.67$; $P < 0.0001$) (Control, CTR, ♦; Unilateral oviduct ligated, LIG, ◊ sows) (From Town *et al.*, 2004).

Experimental evidence in support of this hypothesis

Preliminary data from an initial experiment involving analysis of foetal and placental weights at term, and associations with IUGR measured in the neonate, indicated that even when the number of conceptuses in the uterus does not significantly affect birth weight, crowding nevertheless results in measurable IUGR in the foetus (Town, 2004). In another study, unilateral oviduct ligation was used as a surgical approach to vary the number of foetuses developing in the uterus. Even though the uterine crowding observed was not at the level that probably occurs in existing commercial dam-line sows, a higher number of foetuses in the uterus resulted in measurable developmental changes. As shown in Figure 12.3, there were effects of increased numbers of conceptuses and foetuses in the uterus on placental development (Town *et al.*, 2004b). Furthermore, in the same study, among the various measures of IUGR, there were specific effects on the brain:muscle weight ratio, brain to muscle fibre number ratio, estimated total number of secondary muscle fibres, and associated differences in the wet weight and cross sectional area of the semitendinosus muscle (Table 12.2). This provided some of the first evidence that the variation in the number of conceptuses surviving to the post-implantation period will affect not only placental, but also foetal, development.

Table 12.2 MUSCLE FIBRE DEVELOPMENT DATA (MEANS ± SEM) FOR DAY 90 FOETUSES FROM CONTROL (CTR) AND UNILATERALLY OVIDUCT LIGATED (LIG) SOWS (N=28). MUSCLE DATA WERE DERIVED FROM ANALYSIS OF THE SEMITENDINOSUS (ST) MUSCLE. (AFTER TOWN *et al.*, 2004B)

| Parameter | Treatment group | |
	CTR (n=14) "Crowded"	LIG (n=14) "Non-Crowded"
Muscle weight (g)	1.25 ± 0.06a	1.47 ± 0.09b
Muscle CSA (mm2)	47.71 ± 2.85a	58.78 ± 2.65b
Brain: liver weight ratio	1.17 ± 0.04a	0.97 ± 0.04b
Brain: ST muscle weight ratio	10.49 ± 0.43a	9.25 ± 0.33b

Means in rows with different post-scripts a, b were significantly different (P < 0.05)

Placing these experimental results in the context of a discussion of the consequences of selection for increased litter size on piglet development, would lead to the following conclusion: the extent of uterine crowding created experimentally in the studies described above, and a number of comparable studies in both gilts and higher parity sows, is substantially less than the predicted crowding in at least a sub-population of higher parity sows in existing commercial dam-lines (Vonnahme *et al*, 2002; Foxcroft and Town, 2004).

THE BASIS FOR EFFECTS OF INTRA-UTERINE CROWDING ON MUSCLE DEVELOPMENT IN THE PIG

Myogenesis in the pig

A schematic representation of muscle fibre development in the pig is shown in Figure 12.4. Myogenesis is a biphasic phenomenon and involves determination, migration, proliferation, differentiation, and fusion of myoblasts to form myotubes. In the first phase, lasting from d 35 to d 55 of gestation, a primary generation of myotubes (the so-called primary myofibres) develop. In the second phase, which lasts until d 90 of gestation, the formation of a second generation of myotubes (secondary myofibres) occurs. Over 20 secondary myofibres appear around each primary myotube, using them as a scaffold. Considering the fact that hyperplasia in the pig (an increase in myofibre number) ceases by around d 90 of gestation, the number of primary and secondary myofibres formed ultimately determines the total number of myofibres at birth. As reported by Wigmore and Stickland (1983), the number of secondary myofibres, as well as the total number of myofibres, formed is lower at birth in smaller, compared with larger, foetuses. Furthermore, primary myofibres in small foetuses are smaller than in large foetuses. These authors hypothesized that the small size of the primary myofibres may restrict the available surface area for secondary myofibre formation.

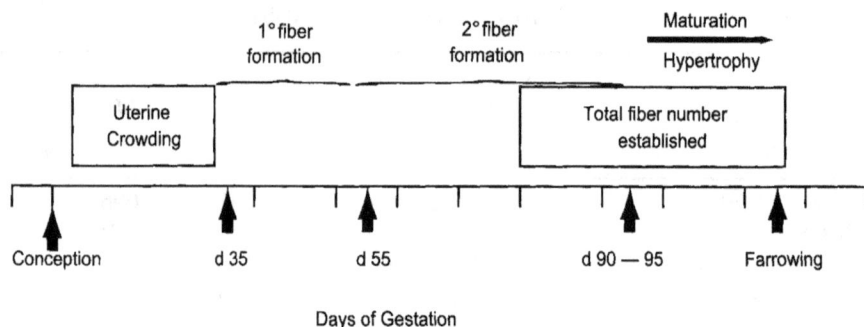

Figure 12.4 Schematic representation of the time-course of muscle fibre development in the pig, indicating a critical window in early pregnancy when crowding effects limit placental development and set in place detrimental effects on foetal development and lifetime growth performance.

As discussed earlier, even the moderate level of intra-uterine crowding established in the experiments of Town *et al.* (2004b) had consequences for the pattern of foetal muscle fibre development. Subsequently, embryonic tissues harvested from these sows were used to study effects on the expression of the myogenic regulatory factors *myogenin* and *myoD*. This study provided direct evidence that crowding at d 30 of gestation can impact the differentiation of muscle fibres through reductions in *myogenin* expression (Tse, 2005). Furthermore, most of the overall litter effect was found to originate from selective effects on *myogenin* expression in the male embryos in the litter.

Birth weight as a source of variation in postnatal growth, and carcass and meat quality

Both birth weight of the individual piglet and the intra-litter variation of birth weight are of considerable economic interest for pork production. Not only survival within the first week after birth, but also postnatal growth in the pre-weaning, nursery and grow-finish stages of production, are impaired in low compared with high birth weight pigs (Quiniou *et al.*, 2002; Figure 12.5).

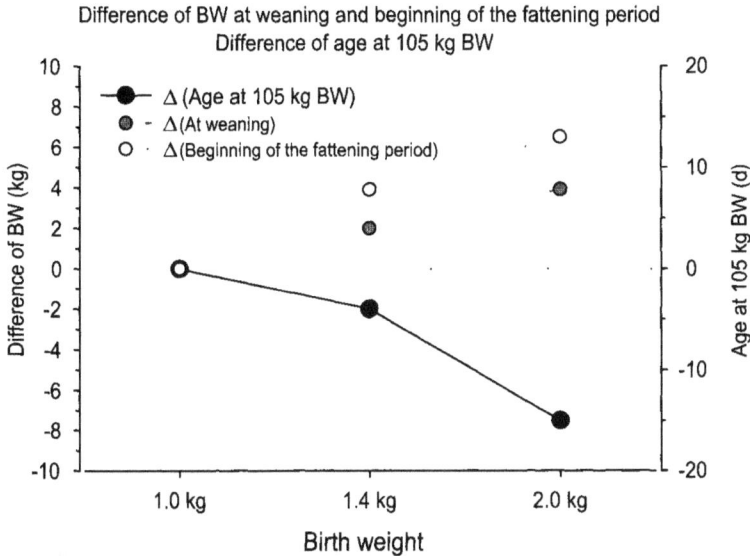

Figure 12.5 Associations between body weight at birth (1.0, 1.4 and 2.0kg) and, 1) the difference in body weight at weaning and at the beginning of the fattening period (left axis), and 2) the difference in age to reach 105kg in body weight. (After Quiniou *et al.*, 2002).

POSTNATAL MUSCLE DEVELOPMENT

The increase in skeletal muscle weight during postnatal growth results from muscle fibre hypertrophy (increase in the size and the length of the individual myofibres). Furthermore, the extent of myofibre hypertrophy and, thus, the capacity of the muscle to grow, depend also on the total number of myofibres within a muscle, which is fixed at birth. It has been shown that myofibre size is inversely correlated with myofibre number, which means that growth rate of the individual myofibre is lower when there are high numbers of myofibres, and higher when there are low numbers of myofibres (Rehfeldt *et al.*, 2000). On the other hand, both number and size of myofibres are positively correlated with the cross-sectional area of the muscle. This raises the question whether hypertrophy or total number of myofibres is more important for lean tissue growth. As reviewed by Rehfeldt *et al.* (2000), it seems that the potential for lean tissue growth depends primarily on the number of the prenatally formed myofibres, because myofibre hypertrophy is limited by genetic and physiological constraints (Figure 12.6). Consequently, impaired postnatal growth can be expected in low birth weight piglets displaying low myofibre numbers.

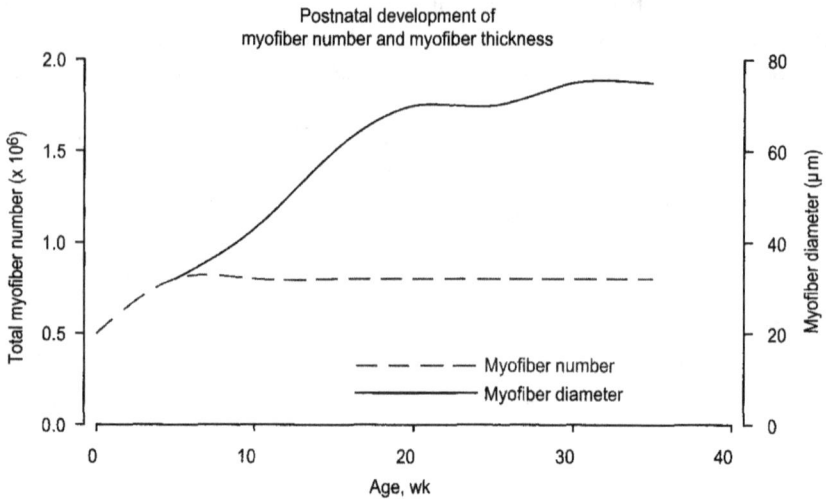

Figure 12.6 Representation of the relative time points postnatally at which the number of myofibres (broken line) and myofibre size (unbroken line) cease to contribute to the increase in muscle mass. (After Rehfeld *et al.* 2000).

EFFECTS OF LOW BIRTH WEIGHT ON CARCASS AND MEAT QUALITY

Recent results from various experiments demonstrate the close relationship between birth weight, carcass characteristics and meat quality traits. Rehfeldt *et al.* (2004) determined that, at birth, the lightest piglets exhibited the lowest proportion of muscle tissue, total protein, total fat, and the lowest semitendinosus muscle weight and total number of myofibres, whereas the proportions of internal organs, skin, bone, and total water were highest, compared to their heavier littermates. In finishing pigs slaughtered at fixed age of 182 days, the pigs of low birth weight were lighter, had lower meat contents and loin area was smaller compared to pigs of high birth weight, whereas the proportion of omental fat tended to be higher (Table 12.3). The pigs of low birth weight exhibited the lowest myofibre numbers, the largest myofibre size, and the highest proportion of abnormal "giant" myofibres in both muscles under investigation. With respect to meat quality, higher drip losses were determined in the longissimus muscle of low birth weight pigs. Similar effects of relative birth weight within litters on postnatal growth performance were reported by Nissen *et al.* (2004) and Gondret *et al.* (2004). Furthermore, Bee (2004) reported larger myofibres and fatter carcasses in low birth weight pigs slaughtered at a fixed weight of 105 kg. Finally, Gondret *et al.* (2006) reported that, compared to heavy birth weight pigs, low birth weight pigs required an extra 12 days to reach the same slaughter weight of 112 kg(Table 12.4). Not only was growth rate impaired, but feed conversion ratio was also inferior in low birth weight pigs. Accordingly, low birth weight pigs exhibited a fatter carcass, associated

Table 12.3 ASSOCIATION BETWEEN HIGH AND LOW BIRTH WEIGHTS IN A LITTER
AND CHARACTERISTICS OF POSTNATAL GROWTH, CARCASS CHARACTERISTICS,
AND PORK QUALITY TRAITS. (REHFELD *et al.*, 2004)

	Birth weight		
	Low (0.9 kg)		*High* (1.8 kg)
Overall ADG, kg/d	0.582	<	0.641
Live Weight, kg	106.1	<	116.0
Hot carcass weight, kg	84.2	<	92.5
Drip loss, %	6.6	>	4.5
Myofibre area, µm²	3900	>	3200
Myofibre number x 1000	900	<	1200
"Giant" myofibres, %	0.44	>	0.07

Table 12.4 ASSOCIATION BETWEEN HIGH AND LOW BIRTH WEIGHTS IN A LITTER
AND CHARACTERISTICS OF POSTNATAL GROWTH, CARCASS CHARACTERISTICS,
AND PORK QUALITY TRAITS. (GONDRET *et al.*, 2006)

	Birth weight		
	Low (0.9 kg)		*High* (1.8 kg)
Overall ADG, kg/d	0.650	<	0.690
Hot carcass weight, kg	90.2	=	89.5
Lean meat content, %	61.1	<	63.0
Backfat, %	6.7	>	5.2
Fatty acid synthase, nmol/min	269.0	>	214.0
Adipocyte diameter, µm	64.9	>	57.2
Tenderness score (10 point scale)	4.0	<	4.7

with markedly higher activity of enzymes involved in lipogenesis, such as
fatty acid synthase and malic enzyme. Again, the total myofibre number was
lower in the semitendinosus muscle, and the myofibres were larger in both the
semitendinosus and longissimus muscles of low, compared to high, birth weight
pigs. Of great importance with respect to consumer's satisfaction with pork,
was the finding that the low birth weight pigs exhibited a lower score for loin
meat tenderness compared with high birth weight pigs.

Collectively, available results indicate that pigs of low birth weight develop
lower carcass and meat quality. There is increasing evidence that this is related
to the low number of myofibres that undergo accelerated hypertrophy during
postnatal growth. A possible explanation for the differences in adipose tissue
accretion observed between low and high birth weight pigs, is that in low birth

weight pigs the increase in myofibre size is faster because of the low myofibre number and the plateau of myofibre growth is attained earlier compared to high birth weight pigs (Rehfeld and Khun, 2005).   Consequently, in low birth weight pigs, nutritional energy which can no longer be used for muscle accretion is used for lipogenesis instead.

Implications of selection of hyper-prolific sows for increasing litter size born

Given the impact of an imbalance between ovulation rate, embryonic survival, and uterine capacity on foetal and post-natal development, the reproductive characteristics of prolific dam-lines need careful consideration. Although the primary goal of increasing the number of pigs born per litter may be achieved, data from some of these prolific dam-lines lead us to suggest that many of the adverse pre-natal programming effects associated with inadvertent crowding of foetuses in the uterus may be prevalent in the mature sows in these populations.   Over the last decade, selection for improved prolificacy has resulted in an increase of litter size at birth in most breeding populations.   In France for example, total litter size born increased from 10.9 in 1992 to a mean litter size of 12.2 piglets in 2001 (Gondret *et al.*, 2005).   The selection for sow's ability to give birth to a higher number of piglets has led to an increased within-litter variation in piglet birth, as well as to an overall decrease in birth weight.   A possible cause for these observations is the increased competition among littermates for maternal nutrients in utero, because fetal weight and birth weight have been shown to be inversely related with litter size.

A consideration of the proportion of live-born vs. dead-born pigs within the litters of one population of hyper-prolific French sows (Table 12.5) suggests that the growth potential of the live-born pigs that survive to weaning will have been seriously compromised by intra-uterine competition with an increasing number of foetuses born dead.   Ongoing studies of the reproductive characteristics of a similar line of French hyper-prolific sows being imported into Canada (Harding, J.R., personal communication) confirm that, in the higher parity sows in this population, high ovulation rates can be associated with substantial crowding of viable conceptuses at d 30 of gestation, with obvious implications for deleterious effects of pre-natal programming on post-natal performance.   Furthermore, the overall distribution of birth weights in these prolific sows shows the same shift to a lower median birth weight and an increase in variance in birth weights discussed above (Duggan, M., personal communication).

A better understanding of the characteristics of specific hyper-prolific dam-lines is needed.   This information, and an increasing focus on the need to maximize total net revenues per sow in terms of the value of saleable pork

Table 12.5 PRODUCTION DATA RECORDED FOR INDIVIDUAL HYPERPROLIFIC WHITE-TYPE SOWS FROM COMMERCIAL UNITS IN BRITTANY, FRANCE.[a]

Sow parity	Total pigs born	Pigs born dead	Pigs born live	Adjusted litter size 48 h after farrowing
7	*20*	*6*	*14*	*12*
2	15	2	13	13
5	*19*	*5*	*14*	*11*
2	15	1	14	11
9	*14*	*1*	*13*	*12*
5	*13*	*0*	*13*	*12*
4	19	1	18	13
2	12	0	12	12
5	*13*	*1*	*12*	*10*
5	*18*	*0*	*18*	*11*
4	16	1	15	12
1	10	2	8	12
4	16	0	16	12
5	*18*	*3*	*15*	*11*
8	*22*	*5*	*17*	*11*
5	*13*	*7*	*6*	*12*

[a]Individual higher parity sows (***data shown in bolded italics***) with high ovulation rates tend to show both an increase in total and dead born pigs per litter. Data are from personal communication (Leveneau, P.).

products relative to the input costs involved per kg of pork sold, should allow the most commercially acceptable terminal-line dams to be developed in the future. Ultimately, selection of sows with increased uterine capacity offers the best opportunity for increasing the number of pigs born per litter without compromising the post-natal growth performance of these pigs. Recently, Rosendo *et al.* (2007) selected two lines for six generations on high ovulation rate at puberty (OR-line), or high prenatal survival corrected for ovulation rate in the first two parities (PS-line). No significant difference was found for prenatal survival, or prenatal survival corrected for ovulation rate. However, although the total number of piglets born did not change in the OR-line, numbers born were significantly improved in the PS-line (0.24 ± 0.11 piglets per generation). Perhaps at a population level, selection for litter size also needs to include data from multiparous sows, with average litter birth weight being used, in addition to litter size born live, as a means of identifying sows in which the dynamics of pre-natal loss does not result in detrimental pre-natal programming effects on post-natal growth potential.

Sources of variance in postnatal growth performance

As described above, a comparison between the largest and smallest pigs within a litter has been the most frequently used experimental paradigm to study impacts of birth weight on postnatal growth performance. However, as discussed earlier, limitations in functional uterine capacity in hyper-prolific sows is predicted to result in prenatal programming effects on entire litters. If this assumption is correct, then the origins of increasing variance in postnatal growth performance needs to be clarified as the basis for developing selection and production strategies that effectively address the problem.

As an initial approach to determining the origins of increased variation in litter birth weight and postnatal growth performance, a retrospective study of production data from a large breeding nucleus population was conducted (Smit, 2007). One of the goals of this study was to characterize litters that had been subjected to prenatal programming in a dataset with information on litter phenotype after birth, but no information about ovulation rates. Based on earlier discussions, intra-uterine crowding is predicted to have an effect on birth weight; however, birth weight is also influenced by litter size. As shown in Figure 12.7, both the mean and the variance in birth weight decreases in bigger litters. This suggests that the birth weight of most pigs born in litters >15, irrespective of their litter of origin, is relatively low, because limited uterine capacity is unable to support a higher birth weight in these situations. Furthermore, there appears to be a lower limit of average birth weight which is more or less independent of litter size.

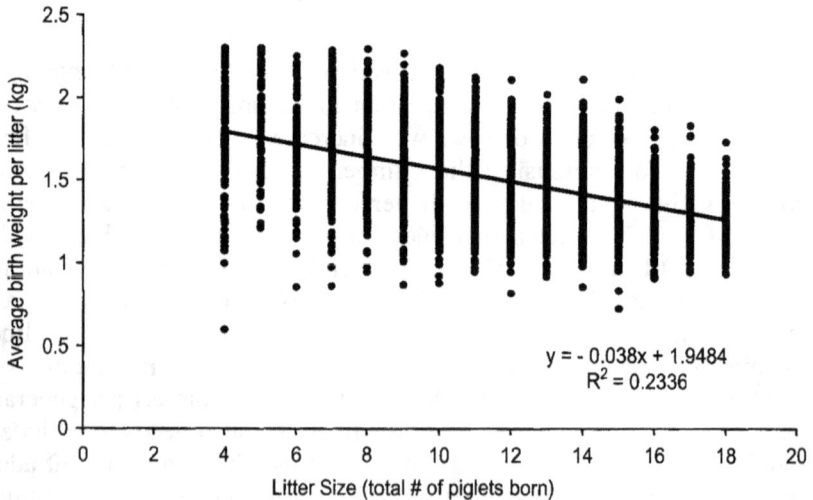

Figure 12.7 Correlation between the total number of pigs born and average birth weight of the litter (N = 5290). (From Smit, 2007)

Because it can be assumed that most, if not all, litters >15 have suffered from intra-uterine crowding, this population of litters is not likely to show the greatest variation in average birth weight. At the other extreme, litters of <10 should not have suffered from extreme intra-uterine crowding in early gestation. Therefore, the greatest likelihood of finding variation in average litter birth weight that might be associated with different patterns of prenatal survival appeared to be in litters between 10 and 15 total born. It was assumed that intra-uterine crowding negatively affects birth weight, and the characteristics of litters of between 10 and 15 pig, with low and high average birth weight, were compared. For each litter size in this range, the average birth weight was calculated and litters within an average birth weight plus or minus 0.2 kg of the average were excluded from the analysis. This resulted in a new dataset with data of 1,094 litters, which were classified as having "High" or "Low" average birth weight, as shown in Figure 12.8.

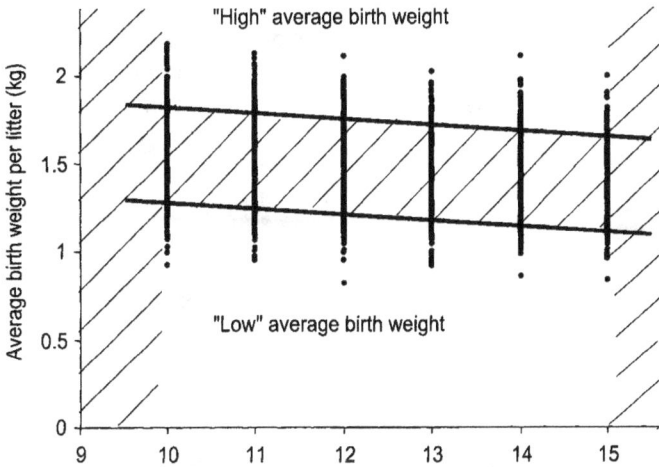

Figure 12.8 Litters classified as having a "High" (average litter birth weight > mean birth weight + 0.2 kg for a particular litter size born) or "Low" (average litter birth weight < mean birth weight - 0.2 kg for a particular litter size born) average birth weight within the litter size range from 10 to 15 total pig born. (N = 1,094) (After Smit, 2007).

The effects of High or Low average birth weight on the total number of piglets born per litter, number of piglets born alive and dead, the number of piglets weaned, the sex-ratio of litters and the variance in within-litter birth weight were then determined (Table 12.6). In litters with a High average birth weight, significantly ($P < 0.001$) more pigs were born alive, less pigs were born dead, and more pigs survived to weaning, compared to the Low average birth weight litters. The standard deviation of birth weight was also significantly higher in the High birth weight compared to the Low birth weight litters (0.287 vs. 0.272 respectively, $P = 0.01$). There was also a significant increase in within-litter variance in birth weight with increasing parity of the sow.

Table 12.6 CHARACTERISTICS OF LITTERS CLASSIFIED AS HAVING A HIGHER THAN AVERAGE (HIGH) AND LOWER THAN AVERAGE (LOW) BIRTH WEIGHT, AND BORN IN LITTER SIZES OF BETWEEN 10 AND 15 TOTAL PIGS BORN. (AFTER SMIT, 2007).

	High	*Low*	*P-Value*
Average birth weight (kg)	1.8 ± 0.01	1.2 ± 0.01	< 0.001
Total born	12.3 ± 0.08	12.3 ± 0.07	0.91
Born alive	11.7 ± 0.09	11.0 ± 0.09	< 0.001
Born dead	0.6 ± 0.07	1.2 ± 0.06	< 0.001
Weaned	10.8 ± 0.10	9.4 ± 0.10	< 0.001

Usually, litters with a low average piglet birth weight are litters with a high number of piglets born in total, due to the negative effect of litter size on birth weight. However, when only litters between 10 and 15 piglets born were taken into account, the impact of numbers born on average litter birth weight is relatively small (<40g for each additional pig born between 10 and 15). In contrast, the difference in mean birth weight between High and Low birth weight litters in the 10 to 15 total born range was 590g, and each of these High and Low subsets represented about 15% of the total population of litters analyzed. Clearly, some factor other than total born per litter is driving these substantial differences in average litter birth weight. The observation that the Low average birth weight group had more pigs born dead and less piglets weaned is consistent with the notion that these litters have been subjected to prenatal programming in utero. The lower within-litter standard deviation of birth weight in the Low average birth weight group may also be a consequence of the prenatal loss of the weaker and smaller pigs, thus already reducing the variation in litter birth weight. In contrast, in litters not subjected to extremes of intra-uterine crowding, pigs across a wider range of birth weights have the opportunity to survive to term and this would explain the higher variance in birth weights in the High average birth weight litters.

This retrospective study of litter birth weight appears to identify between-litter variance in birth weight as the greatest potential contributor to variation in postnatal growth performance. However, the inference that low average birth weight is likely the result of intra-uterine crowding and would be associated with prenatal programming needs verification. Therefore, in a subsequent study using the population of sows for which the dynamics of prenatal loss has already been established (shown earlier in Figure 12.1 and Table 12.2), a more extensive set of phenotypic data were collected from some 600 litters. In this study, the data included parity of the sow, previous breeding history, records of total pigs born live and dead, individual birth weight of all pigs born, sex ratio of the litters and estimates of wet placental weight. However, in addition: 1) Necropsy was performed on a subset of still-born pigs that fell within the mid-weight range for their respective litters, and data on organ weights were used to estimate brain sparing effects as a measure of prenatal programming; 2) Two male and two female offspring in the mid-weight range of each litter were identified at birth with ear tags in the hope of capturing growth performance data in commercial finishing barns at a later date.

The variation in average litter birth weight in this study followed the same trends as in the study of Smit (2007). Having confirmed that the birth weight of the subset of still-born pigs necropsied fell within the mid weight range for their respective litters (Figure 12.9), it was then possible to demonstrate that Low average birth weight litters carried all the negative phenotypic characteristics associated with IUGR (Figure 12.10). These data provide further support for the suggestion that one of the major causes of variation in postnatal growth performance is between-litter variation in average birth weight. Furthermore, the lower average birth weight of litters from mature sows in the mid-range of litter sizes (10 to 15) are likely the result of intra-uterine crowding, as these litters show all the benchmarks of IUGR. Linking back to the extensive data on the impact of birth weight on postnatal growth performance reviewed earlier, the postnatal growth potential of low birth weight litters should be a major concern for the pork industry.

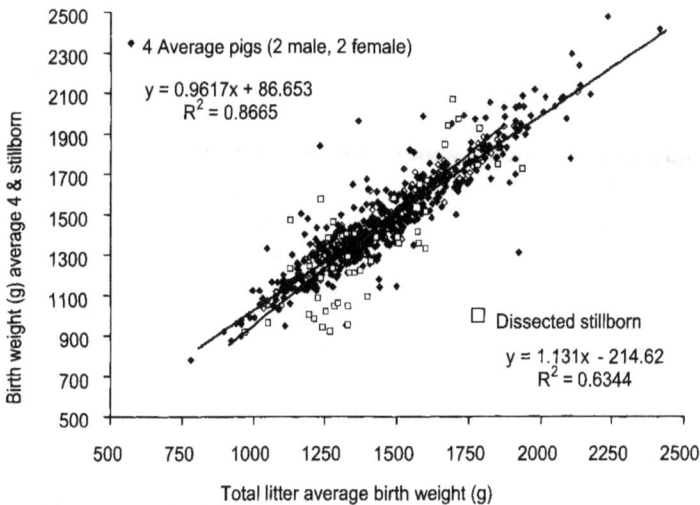

Figure 12.9 Relationship between the average litter birth weight of the whole litter (Total litter average birth weight) and 1) the average birth weight of the four pigs (2 males, 2 females) selected to represent the mean litter and tagged for later estimates of postnatal growth and 2) the weight of stillborn pigs from the same litters used for necropsy to determine the extent of prenatal programming of foetal development. (Swine Research & Technology Centre unpublished data, 2007).

Implications for better production systems

If pre-natal development can have measurable effects on post-natal variation in growth performance, what are the possible practical resolutions to this problem? It is certainly appropriate to question whether the continued selection

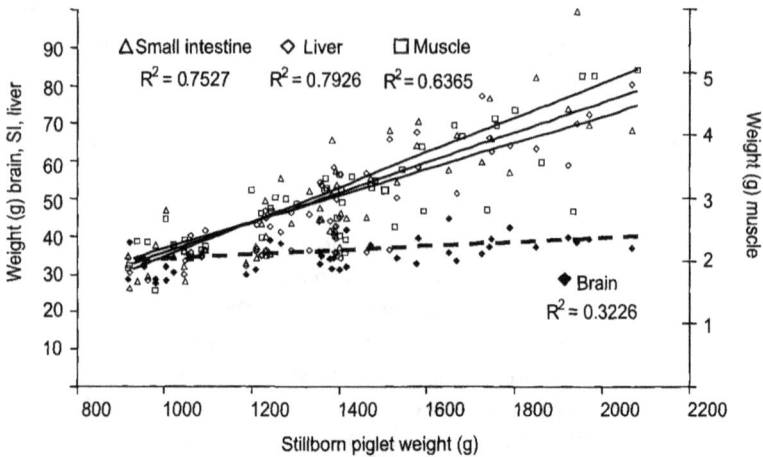

Figure 12.10 Relationships between weight of still born pigs used for necropsy and the weight of the brain, liver, small intestine and semitendinosus muscle. (Swine Research & Technology Centre unpublished data, 2007).

for increased litter size born is an appropriate strategy for the pork industry. Feed costs for terminal line progeny continue to rise and the livestock industry is increasingly in competition with other industries that seek to divert the same feedstock to other industrial processes, with methanol production being the most topical example. The efficiency of feed utilization, and minimal environmental impacts of food-animal production, will become increasingly important issues in our ability to sustain pork production in existing pork producing countries. The net efficiency with which we can produce a kilogram of high quality pork product is, therefore, critical.

SEGREGATED PRODUCTION SYSTEMS

As intensive pork production systems continue to evolve, increasing attention is being paid to the concept of segregated management systems. The reasons for adopting segregated production flows vary, but the underlying principle remains the same – the net advantages that come from managing particular populations of the pork production chain as a sub-population. Segregation may be a spatial concept, in a geographic sense, to improve the control of disease transmission at different levels of the production pyramid. Increasingly, segregation involves separation of sub-populations on the basis of their susceptibility to disease challenges compared to say more mature animals, or because this segregation allows specialized management to be applied in a cost-effective way to these segregated populations. Segregation in this instance can be on-site within a farm, or even within-barn, depending on the goals and situation.

As reviewed by Moore (2005), the origins of segregated parity management systems vary, and have initially been directed to improving the management of the gilt and first litter sow. However, this trend has also simultaneously recognized the problems of co-mingling the progeny of different parity sows, and the advantages to be gained from adopting segregated nursery systems for at least the progeny of parity 1 sows, compared to the progeny of higher parity females. These concepts were expanded by Deen (2005), who discussed the risk factors associated with offspring born in the lower percentile of birth weights per se. By definition, the proportion of offspring from gilts litters falling into these lower weight percentiles is higher. However, the progeny of higher parity, hyper-prolific, sows are also at increased risk of producing litters that have been exposed to adverse prenatal programming. In the extreme situations, developmental limitations will be associated with low average litter birth weights, and at least this population of litters could be designated to segregated production flows at the nursery and grow-finish stages. Therefore, there is good reason to think that segregated management of these offspring at the nursery level will bring overall improvements to a production system.

Overall conclusions

The current chapter hopefully provides an understanding of the very complex interactions that determine the development of a market pig from conception to consumption. In the new era of epigenetic regulation of pre-natal development, and expanding information on the mechanisms controlling various levels of IUGR, the profound influence that the environment of the sow can have on the phenotypic characteristics of her off-spring are becoming evident. Clearly, simple selection of genetically superior sows and terminal line boars will determine the potential to produce a desired genotype in their terminal line off-spring. However, inappropriate management may interact with this genetic potential to produce a very different outcome. In this review, the profound effects of sow maturity, interacting with the component biological traits that result from selection of hyper-prolific females, have been considered. Observations from controlled experiments and from the analysis of the reproductive and developmental characteristics of existing commercial dam-line sows indicate the diversity of postnatal outcomes that are possible. In part, these studies start to provide a better understanding of the reported benefits of segregated parity management. Innovative approaches to addressing the problems, as well as the opportunities, presented by pre-natal programming of post-natal performance will likely be the benchmark of the most profitable pork production systems in the next decade. In particular, these approaches will need to address the possible conflict between continued selection for hyper-prolificacy and increased variance in post-natal growth performance.

Acknowledgements

The financial support of the funding partners supporting the research of the Swine Reproduction-Development Program at the Swine Research & Technology Centre, University of Alberta is gratefully acknowledged, as is the financial support of Sask Pork and the Agricultural Development Fund of Saskatchewan for the collaborative research based at the Prairie Swine Centre Inc.and conducted in collaboration with Dr. John Harding at the Western College of Veterinary Medicine, Saskatoon, Canada. The support of Intervet (USA) Inc., PIC Technical Services (N. America) and The Maschoff Farms in conducting ongoing collaborative studies in commercial sow farms is also greatly appreciated. Those private and corporate sponsors that have supported the establishment of the Swine Research & Technology Centre as one of the premier swine research facilities worldwide provided the operational framework for much of the research presented in this review. Finally, I wish to acknowledge the input of my colleagues in the Swine Reproduction-Development Program at the University of Alberta; it has been a privilege to work with this group and this review reflects the collective wisdom and efforts of the SRDP team. George Foxcroft presently holds a Canada Research Chair in Swine Reproductive Physiology at the University of Alberta.

References

Adams, P. H. (1971) Intra-uterine growth retardation in the pig. II. Development of the skeleton. Biol. Neonate 19: 341-353.

Almeida, F. R., R. N. Kirkwood, F. X. Aherne and G. R. Foxcroft. (2000) Consequences of different patterns of feed intake during the estrous cycle in gilts on subsequent fertility. J. Anim. Sci. 78: 1556-1563.

Ashworth, C. and A. Pickard. (1998) Embryo survival and prolifacy. Page 303 in Progress in Pig Science. J. Wiseman, M. Varley and J. Chadwick, eds. Nottingham Univ. Press, Nottingham, UK.

Barker, D.J.P. Mothers, Babies and Diseases in Later Life. (1994) London BMJ Publishing.

Barker, D.J.P., Bull, A.R., Osmond, C. and Simmonds, S.J. (1990) Foetal and placental size and risk of hypertension in adult life. Brit Med J. 301:259-62.

Barker, D.J.P., Winter, P.D., Osmond, C., Margetts, B. and Simmonds, S.J. (1989) Weight in infancy and death from ishaemic heart disease. Lancet. ii:577-580.

Bazer, F. W., A. J. Clawson, O. W. Robison and L. C. Ulberg. (1969a) Uterine capacity in gilts. J. Reprod. Fertil. 18: 121-124.

Bazer, F. W., O. W. Robison, A. J. Clawson and L. C. Ulberg. (1969b) Uterine capacity at two stages of gestation in gilts following embryo superinduction. J. Anim. Sci. 29: 30-34.

Bee, G. (2004) Effect of early gestation feeding, birth weight, and gender of progeny on muscle fibre characteristics of pigs at slaughter. J. Anim Sci 82:826-836.

Deen, J. (2005) Why segregation works and what it means economically. Proceeding of Pre-conference Seminar #12 on Parity Segregation: Application in the industry., AASV Annual Meeting, pp5-8.

Dziuk, P. J. (1968) Effect of number of embryos and uterine space on embryo survival in the pig. J. Anim. Sci. 27: 673-676.

Fenton, F. R., F. W. Bazer, O. W. Robison and L. C. Ulberg. (1970) Effect of quantity of uterus on uterine capacity in gilts. J. Anim. Sci. 31: 104-106.

Flecknell, P. A., R. Wootton, M. John and J. P. Royston. (1981) Pathological features of intra-uterine growth retardation in the piglet: Differential effects on organ weights. Diagn. Histopathol. 4: 295-298.

Ford, S. P. and C. R. Youngs. (1993) Early embryonic development in prolific meishan pigs. J. Reprod. Fertil. Suppl. 48: 271-278.

Foxcroft, G. R. (1997) Mechanisms mediating nutritional effects on embryonic survival in pigs. J. Reprod. Fertil. Suppl. 52: 47-61.

Foxcroft, G. R. and S. Town. (2004) Prenatal programming of postnatal performance - the unseen cause of variance. Adv. in Pork Prod. 15: 269-279.

Gondret, F., Lefraucheur, L., Louveau, I., Lebret, B., Pichodo, X. and Le Cozler, Y. (2004) Influence of piglet birth weight on postnatal growth performance, tissue lipogenic capacity and muscle histological traits at market weight. Livestock Prod Sci: 93, 137-146.

Gondret, F., L. Lefaucheur, I. Louveau, B. Lebret, X. Pichodo and Y. Le Cozler. (2005) Influence of piglet birth weight on postnatal growth performance, tissue lipogenic capacity and muscle histological traits at market weight. Livest. Prod. Sci. 93:137-146.

Gondret, F., L. Lefaucheur, H. Juin, I. Louveau and B. Lebret. (2006) Low birth weight is associated with enlarged muscle fibre area and impaired meat tenderness of the longissimus muscle in pigs. J. Anim. Sci. 84:93-103.

Harding, J.C., Auckland, C., Patterson, J. and Foxcroft G.R. (2006a) Hidden ramifications of attaining 30 pigs per sow per year induced by adverse foetal programming. Proceedings of the 37th Annual Conference of the American Association of Swine Veterinarians; Pre-Conference Symposium on "Sow Productivity and Reproduction", pp1-10.

Harding, J.C, Auckland, C., Patterson, J. and Foxcroft, G.R. (2006b) Prenatal programming of post-natal health and survival. Proceedings of Leman Reproductive Workshop: Achieving and exceeding sow production targets. College of Veterinary Medicine, Univ. Minnesota, pp73-82.

Hegarty, P. V. and C. E. Allen. (1978) Effect of pre-natal runting on the postnatal development of skeletal muscles in swine and rats. J. Anim. Sci. 46: 1634-1640.

Johnson, R. K., D. R. Zimmerman, W. R. Lamberson and S. Sasaki. (1985) Influencing prolificacy of sows by selection for physiological factors. J. Reprod. Fertil. Suppl. 33: 139-149.

Johnson, R. K., M. K. Nielsen and D. S. Casey. (1999) Responses in ovulation rate, embryonal survival, and litter traits in swine to 14 generations of selection to increase litter size. J. Anim. Sci. 77: 541-557.

Knight, J. W., F. W. Bazer, W. W. Thatcher, D. E. Franke and H. D. Wallace. (1977) Conceptus development in intact and unilaterally hysterectomized-ovariectomized gilts: Interrelations among hormonal status, placental development, foetal fluids and foetal growth. J. Anim. Sci. 44: 620-637.

Moore, C. (2005) The beginnings of parity segregation, what we have learned, and how it will evolve. Proceeding of Pre-conference Seminar #12 on Parity Segregation: Application in the industry, AASV Annual Meeting, pp1-4.

Nissen, P. M., P. F. Jorgensen and N. Oksbjerg. (2004) Within-litter variation in muscle fibre characteristics, pig performance, and meat quality traits. J Anim Sci 82: 414-421.

Patterson, J., A. Wellen, M. Hahn, M. Smit, A. Pasternak, J. Lowe, N. Williams and G. Foxcroft. 2007. Responses to delayed estrus after weaning in sows using oral progestagen treatment. Proceeding Midwest ASAS Ann. Conf.,

Pere, M. C., J. Y. Dourmad and M. Etienne. (1997) Effect of number of pig embryos in the uterus on their survival and development and on maternal metabolism. J. Anim. Sci. 75: 1337-1342.

Quiniou, N., J. Dagorn and D. Gaudré. (2002) Variation of piglets' birth weight and consequences on subsequent performance. Livest. Prod. Sci. 78:63-70.

Rehfeldt, C., I. Fiedler, G. Dietl and K. Ender. (2000) Myogenesis and postnatal skeletal musle cell growth as influenced by selection. Livest. Prod. Sci. 66:177-188.

Rehfeldt, C., G. Kuhn, I. Fiedler and K. Ender. (2004) Muscle fibre characteristics are important in the relationship between birth weight and carcass quality. J. Anim. Sci. (Suppl. 1) 82:250.

Rehfeldt, C. and Kuhn, G. (2006) Consequences of birth weight for postnatal growth performance and carcass quality in pigs as related to myogenesis. J. Anim. Sci.:84 (E-Suppl.) E113-E123.

Rosendo A., Druet T., Gogué J. and Bidanel J.P. (2007) Direct responses to six generations of selection for ovulation rate of prenatal survival in Large White pigs. J. Anim. Sci. 85:356-364.

Royston, J. P., P. A. Flecknell and R. Wootton. (1982) New evidence that the intra-uterine growth-retarded piglet is a member of a discrete subpopulation. Biol. of the Neonate 42: 100-104.

Smit, M.N. (2007) Study of fetal programming, sex ratio and maternal parity in a multigenerational field dataset. MSc Minor Thesis, Wageningen University, The Netherlands. 43pp.

Town, S. (2004) Patterns of prenatal loss: Implications for placental and foetal development. PhD, Univ. Alberta, Edmonton, Alberta, Canada.

Town, S., J. Patterson, C. Pereira, G. Gourly and G. R. Foxcroft. (2004a) Embryonic and foetal development in a commercial dam-line genotype. Anim. Reprod. Sci. 85: 301-316.

Town, S., C. Putman, J. Turchinsky, W. Dixon and G. R. Foxcroft. (2004b) Number of conceptuses in utero affects porcine foetal muscle development. Reprod. 128: 443-454.

Tse, W-Y. (2005) Consequences of uterine crowding during early gestation on myogenesisin the pig. MSc Thesis, Univ. Alberta, Edmonton, Alberta, Canada.

Vallet, J. (2000) Foetal erythropoiesis and other factors whcih influence uterine capacity in swine. J. Appl. Anim. Res. 17: 1-26.

Vallet, J. L., H. G. Klemcke, R. K. Christenson and P. L. Pearson. (2003) The effect of breed and intrauterine crowding on foetal erythropoiesis on day 35 of gestation in swine. J. Anim. Sci. 81: 2352-2356.

van der Lende, T., W. Hazeleger and D. de Jager. (1990) Weight distribution within litters at the early foetal stage and at birth in relation to embryonic mortality in the pig. Livest. Prod. Sci. 26: 53-65.

van der Lende, T. and D. de Jager. (1991) Death risk and preweaning growth rate of piglets in relation to the within-litter weight distribution at birth. Livest. Prod. Sci. 28: 73-84.

Vonnahme, K. A., M. E. Wilson, G. R. Foxcroft and S. P. Ford. (2002) Impacts on conceptus survival in a commercial swine herd. J. Anim. Sci. 80: 553-559.

Widdowson, E. M. (1971) Intra-uterine growth retardation in the pig. I. Organ size and cellular development at birth and after growth to maturity. Biol. Neonate 19: 329-340.

Wigmore, P. M. and N. C. Stickland. (1983) Muscle development in large and small pig foetuses. J. Anat. 137 (Pt 2):235-245.

Wootton, R., P. A. Flecknell, J. P. Royston and M. John. (1983) Intrauterine growth retardation detected in several species by non-normal birthweight distributions. J. Reprod. Fertil. 69: 659-663.

13

ARTIFICIAL INSEMINATION AND EMBRYO TRANSFER IN THE MODERN PIG INDUSTRY – AN APPLIED COMMERCIAL APPROACH

MERITXELL DONADEU
PIC Europe. 2 Kingston Business Park, Kingston Bagpuize, Oxfordshire OX13 5FE, UK

Assisted reproductive technologies (ART) play a very important role in modern livestock production. The modern pig industry uses several technologies which are available such as artificial insemination (AI), frozen semen, semen sexing and embryo transfer (ET). Their level of implementation at field level varies considerably, with artificial insemination being by far the most popular technique amongst them. This chapter aims to give an overview of the recent application of these technologies and the progress made in recent years within Europe.

Artificial insemination

Artificial insemination (AI) is a useful tool to achieve faster genetic improvement: the best boars can be used more efficiently as more inseminations can be carried out with an AI boar than with a natural service (NS) boar. A boar used for natural service can sire approximately 600 piglets per year, whilst a boar used for artificial insemination can sire up to 10,000 piglets per year (Almond, Britt, Flowers, Glossop, Levis, Morrow and See, 1998).

AI also facilitates genetic dissemination if the boars are located at a stud, since their semen is available for different farms at the same time. From the farmers' point of view, AI gives access to boars that otherwise might not be accessible to them and it might also help to increase uniformity as fewer sires will be needed to produce the same number of piglets.

The use of AI within Europe has increased dramatically over the last two decades. Information from the Swine Qualivet Meeting in 2006 indicated that the use of AI in Europe in many countries is around and, in some instances, even above 90% (Table 13.1).

Semen production can be set up in two distinct ways:

1   On-farm AI (also known as "Do it yourself" or "DIY" boars): the boars

are located at the farm itself and they only produce semen for the farm or groups of farms owned by the same producer. Usually the local legislation does not allow the semen produced within a farm to be sold to third parties.

2   Commercial boar studs or AI Centres: the boars are located in commercial studs, housing only boars destined to produce commercially sold semen. Their purpose is to produce, sell, and distribute semen as a commercial operation.

Table 13.1 ARTIFICIAL INSEMINATION (AI) PROPORTION FOR SEVERAL EUROPEAN COUNTRIES; INCLUDES PURCHASED COMMERCIALLY PRODUCED SEMEN AND ON FARM COLLECTION. Source: Qualivet Swine Meeting, Sweden. 2006

	AI percentage
Belgium	90%
Denmark	97%
Germany	80%
Netherlands	>95%
Norway	92%
Sweden	90%
Switzerland	65%
UK	75-80%

ON-FARM AI

Within the overall European pig AI industry, on-farm AI is not the preferred option, but it is still popular in some countries or regions by tradition, for example in Italy. It is also sometimes preferred for health reasons when the health status of the commercial studs is not acceptable for the farmer (for example when the studs have animals positive for porcine reproductive and respiratory syndrome virus - PRRSv - and the recipient farm is PRRSv negative); or when the farmers do not have confidence in the health status of the commercial stud. In addition on-farm AI could be used for logistical reasons, when the commercial studs are too far from the farm and the semen transport is not viable or practical.

As a system it has some limitations. The genetic dissemination is not as efficient as the best boars will probably be located at commercial studs and there is less flexibility to change to different products if the market changes. The technology used for semen collection and processing is usually more limited when compared to a commercial stud as it is not economically viable to use some of the most expensive technologies at farm level. Typically, at commercial studs, the quality control at the critical steps will be more detailed and complete. On such dedicated sites the staff will usually be more specialised as they will not have to perform other farm tasks; for example at farm level it is not unusual that the person collecting and processing the semen, is the same person who will perform the heat detection and inseminate the sows. With generally few boars being required for an on farm stud, even on a large pig breeding unit it is likely that excess boars will be held. This can, however, very quickly turn to shortage with no on farm alternative supply being available if semen defects or boar issues such as lameness occur.

Nevertheless, there are farms that achieve good reproductive performance because they are very professional and hygienic when collecting and processing semen.

COMMERCIAL BOAR STUDS

The structure of the commercial boar stud industry varies between different European countries. In some countries the commercial studs tend to be owned by Co-ops (e.g. Netherlands, Germany), in other locations they are commercial institutions in their own right (e.g. Spain), or they might be owned by breeding companies (UK) or AI companies (France). The nature of the set up influences whether the semen doses are positioned as a commodity or as a genetic product.

RECENT DEVELOPMENTS AT STUD LEVEL, BOAR MANAGEMENT AND SEMEN PROCESSING

Some of the recent advances in male pig AI technologies were reviewed by Donadeu (2004). A summary and an update on those developments include:

Housing and environment

The latest tendency is to house the boars in deep sawdust instead of solid floors with straw or other types of flooring. The deep sawdust has several advantages: it keeps the boars clean and dry, and does not require too much labour. The bedding needs to be changed only once or twice per year. There are not many disadvantages, but the animals need to be dewormed before entry and checked for parasites periodically.

Welfare

Current European welfare legislation requires a living area of at least 6m^2 per adult boar. Some welfare organisations are recommending increasing the requirement to 7.5m^2. Research carried out (Hoy and Rohrmann, 2004) showed that the boar's pen size had low influence on the boar behaviour when using pens measuring at least 4.9 m^2 up to 7.8 m^2.

Health

The risk of disease introduction into a farm via semen can be seen from two points of view. On the one hand, there are fewer diseases that can be spread via semen when compared with diseases potentially spread by the introduction of live animals. On the other hand, the frequency of semen introduction is high (usually a minimum once per week) so that increases the risk. The impact of a disease breakdown in a boar stud could be very large as many farms could be potentially affected. For example a 250 place boar stud would supply sufficient semen for approximately 75,000 sows in commercial pig production. Therefore, it is very important to protect the health status of the stud. In recent years the industry seems more aware of the health risk, and more emphasis is made in this area:

•   Polymerase Chain Reaction (PCR) diagnostic tests have been developed and improved in recent years, and these developments can be applied to the boar studs. From the traditional PCRs there are now available nested and multiplex PCRs. The tests can be used in semen, in serum, or in both, depending on the type of test and the validation done. At boar studs, PCRs are used mainly to monitor semen for the presence of PRRSv. There are several PCRs available in the market to test PRRSv in semen, but the RT-nested PCRs are the tests with the highest sensitivity. Is not uncommon to pool semen samples for the PCR test, but the number of samples per pool has to be related to the sensitivity of the specific PCR used.

•   Semen and blood sampling: testing semen has the advantage that the animals do not need to be restrained to obtain the samples, but it has the disadvantage that the results might not be returned in time before the semen is used (based on the current extenders and practices, where most farmers prefer to use the semen as soon as possible after collection).

Based on the facts that most viruses will appear in blood some days before they appear in semen, that some of the PCRs do not perform as well in semen, and that the main concern is probably the transmission of PRRSv via semen, new blood sampling techniques for the boars are becoming more common (Reicks, 2005). This blood testing technique involves sampling from visible veins from the ear or the back legs using swabs, capillary tubes, syringes, Vaccutainers and Eppendorf tubes, while the boars are mounting the dummy. Although this sounds disruptive, the majority of the boars show no concern and they don't seem affected with the testing during mounting.

Blood testing can be very useful when monitoring PRRSv negative studs for recent infections. But when monitoring PRRSv in a stud with PRRS serologically positive boars that is trying to avoid the production of PRRSv positive semen, probably the best option is still to test the semen, as the blood sampling techniques may not always predict the presence of the PRRS virus in the semen.

- Semen washing: semen can be contaminated with viruses during the sperm production process, but it can also be contaminated from the urogenital tract, preputial sac and environment. Numerous viruses have been isolated from semen and several viruses have been proved to be transmitted via AI (i.e. Aujeszky's disease virus, Classical swine fever virus and PRRSv). So far, techniques or treatments are not commercially available for virus control in boar semen. Semen washing techniques are already used in humans to reduce the risk of HIV transmission in HIV-discordant couples wishing to have children (Kim, Johnson, Barton, Nelson, Sontag, Smith, Gotch and Gilmour, 1999). An adapted version of the semen washing technique has been proved to eliminate PRRSv from boar semen (Morfeld, White, Mills, Krisher, Mellencamp and Loskutoff, 2005). These types of techniques have the potential to remove other known and unknown emerging viruses. PIC and the Henry Doorly Zoo (Omaha, NE, USA) have performed a small scale trial (data not published) in where 8 boars were infected with PRRSv. Ninety PRRSv PCR-positive ejaculates were treated using this protocol and the PCR testing of the end product demonstrated a 0.957 success rate in not detecting the PRRSv by PCR after the treatment. A fertility study was also done using PRRSv negative washed semen, to see if the processing would affect semen quality and farm performance. Seventy five (75) inseminations were done with washed semen, and 47 inseminations were done with normal processed semen (split ejaculates) in 2 farms. Farrowing rate was over 95% for both groups and litter size was over 12 for both groups. No statistically significant differences were found between the two groups. These are only preliminary results, and the technique is not yet ready for commercial use.

Semen collection equipment

Until recently semen collections were done manually. In some Eastern European countries artificial vaginas and semi-automatic collection systems were used, but generally those systems were abandoned 15-20 years ago. During the last 6 years, automatic collecting machines like the Collectis® (produced by Genes Diffusion) have appeared in the market. Several of these machines are arranged around a central pit that looks like a milking parlour. The collection pens have a stainless steel dummy and an artificial vagina. An electro-pneumatic regulation unit drives the automatic collection of the boar. With this machine, one employee can manage up to 3 collection pens – that means that the same number of collections per hour can be obtained with half the number of employees. Semi-automatic collection methods are also recently available, like the AutoMate® produced by Minitube. The AutoMate® consists of a dummy with a movable slide on the bottom side of the dummy with a holder for the semen collection cup and a clamp for the artificial cervix. After the semen collection, the boar is able to leave the dummy without any assistance. All these systems improve the speed and efficiency of the boar collection process.

Semen processing

- Concentration: sperm concentration is evaluated today mainly with spectrophotometers. These have advantages over haemocytometers in their speed, but they also have sources of error like the coloration of the seminal plasma. Some of the newer Computer Assisted Sperm Analyser systems (CASA) even though they were not originally designed to perform sperm counts, have improved their counting mechanisms and some, but not all, are reliable enough to be commercially used. A newer system is the NucleoCounter® (produced by ChemoMetec A/S). This system uses fluorescent stains to count seminal DNA, and it is a very accurate and precise system, simple and fast, but both the CASA systems and the NucleoCounter® are still relatively expensive for small studs.

- Motility evaluation: in order to evaluate motility and motility characteristics in a more objective way, computer assisted sperm analysis (CASA) systems have been developed. They are still dependent though on the programming and the operator. There are several machines and manufacturers, for example the Hobson-Tracker, Hamilton Thorne (produces different models like IVOS and CEROS), Spermvision™. The evaluation is not done in the same way in the different machines; some will follow the sperm for as long as programmed, others will analyze movement in one second or in a predetermined number of fields. The movement values between the different machines are not comparable because the calculations are different. The trends are the same, but not necessarily the values. The CASA systems have been improved in recent years and they have become more affordable, even if they are still relatively expensive; some of them are also accurate for measuring concentration, some of them are also good to evaluate morphology and some have the capability of analyzing fluorescent stains and measure viability.

- Fertility evaluation: currently it is possible to evaluate several components of sperm quality like morphology and functionality of the plasma membrane. However, the aspects we are able to evaluate might not necessarily be linked to reproductive performance if that component is not the limiting factor. Advances have been made to improve the evaluation of the individual parameters, but there is not a single test that is directly related to fertility and that is practical and commercially viable.

Quality control programmes

The quality control programmes have improved greatly in recent years. From being basic, they have developed into complex programmes that target the critical control points and the quality of the final product. Usually the quality control programmes are designed and managed by each centre or institution, but countries like UK, have developed an industry standard (BPEX standards). This BPEX standard is quite unique in that a group of breeding companies have agreed on a common quality standard.

Extenders for fresh semen

There are short (preserving the semen up to 3 days post collection), medium (up to 5-6 days) and long term extenders (9-8 and even up to 10 days) available in the market. The long term extenders have improved, and they have the advantage that facilitate the testing of the semen (for example PCR results can be available before the semen is shipped) and also facilitate the organisation of the work at the stud, but they are more expensive and there is still a general perception that "fresher" is better.

Semen containers

The semen doses are still usually packed in bags or tubes. Newer systems like the Gedis® (a catheter containing the semen along the length of the catheter in a collapsible membrane) have appeared, but European wide they are not as popular as the tubes or bags. The newest bags and tubes are lighter, more flexible, and can be coloured. Improvements have been made in the openings of the bags and tubes, but the basic concept has not changed significantly.

Semen packing machines

The speed of the packing machines continues to improve, and nowadays there are machines that can fill up to 2100 tubes per hour.

INSEMINATION TECHNIQUES

Catheters have come a long way since early AI days. The original "Melrose" re-usable catheters have mostly been replaced by disposable catheters and there are only few small farmers using them today in Europe. The disposable catheters usually have a foam tip, which either has a spiral mimicking the boar penis, or a foam tip. They are usually available in sow or gilt sizes.

The traditional insemination technique has been to deposit the semen in the caudal segment of the cervix, after the catheter has been locked at the entrance of the cervix (intra cervical technique). In recent years, techniques and catheters have been developed for intra uterine insemination. The intra-uterine insemination can be done depositing the semen in the uterine body or in the uterine horn.

a)   Uterine body (IUBI): the semen is deposited in the uterine body (approx 20 cm further than with the cervical technique). A special catheter is needed; it looks similar to the traditional disposable catheter but it has a 15-20 cm extension than passes the cervix after careful pressure is applied once it has been locked at the entrance of the cervix. Using doses of 1000×10^6 sperm, some trials have demonstrated similar farrowing rates (FR) and litter size (LS) with the IUBI technique, compared to using 3000×10^6 sperm doses with the standard AI technique (Watson and Behan 2002); however some other studies have demonstrated significantly lower litter size using the IUBI technique (Rozeboom, Reicks and Wilson, 2004).

b)   Uterine horn or deep intrauterine insemination (DIUI): the semen is deposited using a flexible catheter (working length of 1.8 m, 4 mm in diameter) inserted up to about the middle or the upper first third of one uterine horn. With DIUI using 150 million of sperm, farrowing rates are similar to standard AI groups, but litter size decreases in 1-2 piglets due to unilateral fertilisation. This effect can be avoided by increasing the number of sperm to 600×10^6 per dose (Martinez, Vazquez, Roca, Cuello, Gil, Parrilla and Vazquez, 2002; Vazquez, Roca, Gil, Cuello, Parrilla, Vazquez and Martinez, 2006).

Results in the field with the deep insemination in the uterine body are comparable to the results using the traditional catheter (depending on the dose used), but the technique has not really taken off. The technique is a bit more laborious, requires more specialized labour and attention to avoid tissue damage, and it is not always physically possible in gilts as the individuals cervix sometimes proves too tight.

The main benefit of the deep intrauterine insemination resides in the decreased number of sperm that could be used in an insemination dose. At the moment this benefit is not compensated by the extra effort. In the mid and late 90's several models of deep AI catheters were developed, good results were available and many farmers tried the technique, but the big majority of them returned to the traditional catheters.

Insemination in the uterine horn is a more recent development, and there are some catheters available in the market. Because the semen is deposited in the middle of the horn, this avoids most of the phagocytosis (responsible for removing up to 0.80 of the sperm). The other potential benefit is that the dose could be inseminated faster as the reflux is minimum, but it is debatable if the sow will still need some degree of time and stimulation to obtain good results. This catheter will need better trained people than for the deep uterine body AI, so the same limitations apply. The deep uterine body insemination seems to have more benefits when using lower doses of "specialized" semen, like frozen semen or sexed semen. It might be that this is the future of this AI technique.

Intraoviductal insemination has also been done, but it requires a laparoscopic procedure, so again it is more interesting when using it with specialized semen like sexed semen (Vazquez *et al*, 2006).

FROZEN SEMEN

The advantage of frozen semen is that it has a longer preservation period allowing transport to remote places or places of difficult access. It also has health advantages because it allows the possibility of testing boars and semen for pathogens, as well as increased observation period for the boars before the semen is released for use. It is also useful as a genetic bank.

Unfortunately, the advances in this area have been very slow. The technique has improved, and better performance can be obtained, but it is still a very laborious technique at the stud (the semen needs to be centrifuged in order to concentrate the semen). Also, it is expensive, and the breeding performance produced is not at the same level as fresh semen (variable decrease in farrowing rate and litter size when compared on farm with fresh semen). The main reason for the variable results is that during the freezing-thawing process the sperm function and structural integrity are compromised. Consequently some sperm show a response that resembles capacitation (Watson and Green, 1999) and a reduction in viability. It is also associated to an inadequate transport of the thawed sperm in the sow genital tract and that would explain the improved results using deep intrauterine insemination (Roca, Carvajal, Lucas, Vazquez and Martinez, 2003). At the commercial farm level there have also been challenges in implementing the technique, as the staff are not used to dealing with liquid nitrogen, and also operating to a very strict protocol where the temperatures and timing are critical to minimise thawing damage.

SEXED SEMEN

It could be highly beneficial for commercial producers and for breeding companies to have specific systems that produce mainly gilts or mainly boars.

In order to achieve the desired sex of offspring, the semen has to be sexed and the X-chromosome has to be separated from the Y-chromosome.

Different technologies have been tried: sperm head volume, swimming velocity, the Y-linked histocompatibility antigen (H-Y antigen) expressed on the surface of male somatic cells, plasma membrane proteins, time of insemination and DNA content (Levis, 2002). Using 2D electrophoresis over 100 proteins of the sperm surface have been isolated and characterized with no differences found between X and Y. More recent research is attempting to develop mass sorting techniques based on markers of chromosome specific loci on the X and Y chromosomes (Wilson, 2006).

The method that has been repeatedly proven to be effective is the Beltsville Sperm Sexing Technology (BSST). This technique is based in that the Y chromosome is smaller and carries 3.6% less DNA than the X chromosome. Currently, X and Y sperm can be produced with a purity of 0.90 or greater at a rate of 15 to 20 million X and Y sperm per hour (Johnson, Rath, Vazquez, Maxwell and Dobrinsky, 2003). This technology is used in livestock, laboratory animals, zoo animals and in humans with a success rate of 0.90 to 0.95 (Johnson *et al*, 2003).

Sexing semen is not a routine practice yet and improvements need to be made to the current technology as they are too slow for the large number of semen cells needed for a boar dose compared to that of the other species where the technique is successfully used. New insemination techniques (e.g. deep intrauterine) where lower amounts of sperm are needed will help to make this type of techniques more viable.

Embryo transfer

Embryo transfer is a technique with good potential and some key advantages, but it is still not a commercial reality. Some of the challenges that the technique involves are the surgical components of the technique and the low viability of the frozen embryos. Advances have been made in both areas.

Embryo transfer can be used to move genes in a safer way from the health point of view when compared to using semen or introducing live animals. The embryo is wrapped in a protective layer (zona pellucida) until the blastocyst stage, and viruses might bind to the zona pellucida. However if the zona pellucida is intact, there is the possibility of washing the embryos with enzymes like trypsin or hyaluronidase to remove the potentially attached viruses. The hyaluronidase treatment has proved to be more effective than the trypsine one. Hyaluronidase treatment has proved effective to decontaminate porcine embryos from viruses like the PRRSv, Porcine Circovirus type 2, Porcine Parvovirus and Encephalomyocarditis virus (Bureau, Dea and Sirard, 2005; Trincado, Roycewicz, Dee, Joo and Pijoan, 2006).

Embryo transfer has been a surgical technique for many years, but recently the non-surgical embryo transfer technique (nsET) has been developed. It still requires some degree of intervention in the donor animals, but for the recipient animal it is a non surgical procedure, using a special flexible catheter. The embryos are deposited in the cranial end of the uterine horns instead of the uterine body as it has been demonstrated better farrowing rate and litter size if the embryos are deposited in the horns (Wallenhorst and Holtz, 1999). Recent results using the nsET at farm level showed a farrowing rate of 71.4% and litter size of 6.9 piglets (Martinez, 2006).

This non-surgical technique is more promising, and even if it is still not financially viable for commercial pig production, it is closer to fruition for genetic breeding programmes where the movement of very valuable genetics plays a very important role.

Embryo freezing has proved difficult as the pig embryos are very sensitive to low temperatures. The freezing techniques have improved thanks to systems like vitrification that avoid the formation of ice crystals. At the same time, vitrification has improved by using "open pulled straws" (OPS) that allow a faster cooling and allow a decreased use of cryoprotectors which are toxic for the embryo. One of the issues is that the embryos frozen in peri-hatching stage have demonstrated a better in-vitro development, but for sanitary reasons, the embryos with an intact zona pellucida are preferred. Improvements in the vitrification area have also been done by simplifying the thawing process from 3 steps to 1 step (Martinez, Vazquez, Roca, Cuello, Gil, Parrilla and Vazquez, 2006).

Again, embryo freezing is not used routinely in commercial pig production, but there are banks of pig embryos maintained by pig breeding companies, or by companies or societies wanting to maintain certain breeds.

Future developments and conclusions

Important advances have been made in assisted reproductive technologies, especially in the area of artificial insemination and embryo transfer. Artificial insemination techniques have been improved at its different stages, from semen collection to semen deposition. Embryo transfer techniques have improved, and the availability of non surgical techniques has been a big step, though it is still far from reality for the commercial pig producer.

However, there is still plenty of room for improvement in the pig AI industry. Developments are needed in several areas, from both the pig producers' and the semen producers' point of view. Some of the areas where improvements would be of benefit are:

- Use of environmental friendly disposable material for example for catheters and bags. The amount of disposable material used in a farm can be considerable, and in some countries the disposal of this type of products is taxed or expensive. It will be of benefit for the environment and for the farmer to have environmental friendly material.

- Practical and inexpensive methods to evaluate semen fertility. The simple methods that are available today are not directly related to fertility.
- Methods to control viruses in semen. Tests for some viruses are becoming available, but they are still expensive, they take time, and they are only testing for certain known pathogens. The option to "remove" all types of virus from the semen would be of benefit for the industry but current techniques are not quite at a level where they would be able to be used commercially.
- Improvements in frozen semen techniques which make it more practical, reliable and economical would be beneficial.
- Tools that facilitate heat detection. In most European countries staffing is an issue. Generally staff turnover has increased and obtaining trained staff has become more difficult. Therefore, a tool or easy technique that could facilitate relatively inexperienced staff to do a better job at heat detection and thus optimise timing of insemination would be useful.
- Tools that could allow more flexibility with the timing of insemination. As already mentioned staffing can be a problem, so tools that can decrease labour or would compensate for less experienced staff would be beneficial. This could be from encapsulated semen that is released over a period of time while still fertile, to tools that allow better communication between the eggs and the sperm or induce ovulation.

References

Almond G., Britt J., Flowers W., Glossop C., Levis D., Morrow M. and See T. (1998) *The Swine AI Book: Second Edition*, pp 20. Publisher Morgan Morrow, Editor: Ruth Cronje.

Bureau M., Dea S. and Sirard M-A. (2005) Evaluation of virus decontamination techniques for porcine embryos produced in vitro. *Theriogenology, 63*, 2343-2355.

BPEX standard for porcine semen quality in AI centres: http://www.bpex.org/ technical/ general/pdf/BPEX_standard_for_Porcine_Semen_Quality-19Feb07.pdf

Donadeu, M. (2004) Advances in male swine artificial insemination (AI) techniques. *The Pig Journal, 54*, 110-122.

Johnson, L.A., Rath D., Vazquez J.M., Maxwell W.M.C. and Dobrinsky J.R. (2003) Pre-selection of sex in swine for production of offspring: an updated on the process and application. *Abstracts 5th International Conference on Boar Semen Preservation, Doorweth, The Netherlands,*

24-27 *August 2003*, VI-060.

Hoy S. and Rohrmann S. (2004) Long term survey on boar housing and behaviour in AI Centres. *Proceedings 16th Meeting of AI vets 2004 in Berlin.*

Kim L.U., Johnson M.R., Barton S., Nelson M.R., Sontag G., Smith J.R., Gotch F.M. and Gilmour J.W. (1999) Evaluation of sperm washing as a potential method of reducing HIV transmission in HIV-discordant couples wishing to have children. *AIDS,* **13**, 45-651.

Levis, D.G. (2002). New Reproductive Technologies for the AI Industry. www.nsif.com/conferences/2002/NewReproductiveTechnologies AIIndustry.htm

Martinez, E.A., Vazquez, J.M., Roca, J., Lucas, X., Gil, M.A., Parrilla, I., Vazquez, J.L. and Day, B.N. (2002) Minimum number of spermatozoa required for normal fertility after deep intrauterine insemination in non-sedated sows. *Reproduction,* **123**, 163-170.

Martinez, E.A., Vazquez, J.M., Roca, J., Cuello, C., Gil, M.A., Parrilla, I. and Vazquez, J.L. (2006). Tecnologías reproductivas con una aplicación potencial a corto plazo en el ganado porcino (II). *Avances,* **III**, 27-42.

Morfeld, K.A., White B., Mills, G., Krisher , R., Mellencamp, M.A. and Loskutoff, N.M. (2005) A novel method for eliminating porcine reproductive and respiratory syndrome virus (PRRSV) from boar semen and its effects on embryo development. *Proceedings of the annual conference of the international embryo transfer society,* 9-11 January 2005.

Reicks, D.L. (2005) An overview of blood collection strategies for boar studs. *Proceedings 2005 Allen D Leman Swine Conference,* 54-55.

Roca, J., Carvajal, G., Lucas, X., Vazquez, J.M. and Martinez, E.A. (2003). Fertility of weaned sows after deep intrauterine insemination with a reduced number of frozen-thawed spermatozoa. *Theriogenology,* **60**, 77-87.

Rozeboom, K.J., Reicks, D.L. and Wilson, M.E. (2004) The reproductive performance and factors affecting on-farm application of low-dose intrauterine deposit of semen in sows. *J Anim Sci,* **82**, 2164,2168.

Trincado, C., Roycewicz, J., Dee, S., Joo, H.S. and Pijoan, C. (2006) Evaluation and viability testing of two washing protocols for porcine oocytes as a model for embryo transfer. *Proceedings of the 19th IPVS Congress, Copenhagen, Denmark,* **2**, 543.

Vazquez, J.M., Roca, J., Gil, M.A., Cuello, C., Parrilla, I., Vazquez, J.L. and Martinez, E.A. (2006) Factors affecting the success of the deep intrauterine insemination. *Proceedings of AI Vets 18th Meeting, Boras, Sweden.* October 2006. Section 18.

Wallenhorst, S. and Holtz, W. (1999) Transfer of pig embryos to different uterine sites. *J Anim Sci,* **77**, 2327-2329.

Watson, P.F. and Green, C.E. (1999). Cooling and capacitation of boar sperm:

what do they have in common? *Abstracts IV International Conference on boar semen preservation, Beltsville, Maryland.* O7

Watson, P.F and Behan, J.R. (2002) Intrauterine insemination of sows with reduced sperm numbers: results of a commercially based field trial. *Theriogenology,* **57,** 1683-1693.

Wilson, M. (2006) Reproductive innovations for swine production: future impacts of gender pre-selection, embryo transfer and cloning. *Proceedings London Swine Conference April 2006,* 169-180. [ax]

14

FUTURE BREEDING POLICIES FROM AN APPLIED PERSPECTIVE

THOMAS A. RATHJE
Chief Technical Officer Danbred North America Columbus, NE USA

Introduction

The breeding policies of seedstock organizations have historically been designed to improve the economic performance of the production systems they serve. This business objective will remain in the future. However, dynamic change within the pork production industry across the globe will result in new economic realities that must be addressed if breeding stock suppliers are to remain competitive.

The objective of the current chapter is to discuss some of the most dramatic changes in the pig industry and how those changes will impact future animal improvement programs. The topics reviewed reflect personal opinions and will not be exhaustive. Any omitted topics and/or research on the various subjects discussed is not meant in any way to undervalue the importance of any industry trend or quality of research. Within this chapter, references are made to review and/or cite key articles that serve to illustrate a concept and serve as a starting point for the individual interested in more detail on a particular topic.

Industry trends affecting breeding policies

The amount of change in economic value that comes about through genetic progress is equivalent to roughly 1-3% of the cost of production ($1-3 U.S.), a relatively small change. In addition, due to genetic lag, it takes several years for the selection decisions made today to have an impact at the level of commercial production. Thus, genetic programs have a unique challenge in that genetic progress must be in traits that will be valuable to the industry several years in the future. Foresight is required to anticipate the direction of the industry in order to have an effective breeding program today. Various trends are considered below

Global consolidation. The production of pork world-wide has been undergoing a trend of consolidation for well over two decades. Production in almost all countries around the world is concentrated in the hands of fewer decision-makers who specialize in producing pork. This trend is perhaps most notable in the United States (Table 14.1, Plain *et al.*, 2006). However, examples from Europe show a similar pattern (Table 14.2, European Commission, Eurostat, 2005).

Table 14.1 NUMBER OF U.S. PIG OPERATIONS AND AVERAGE INVENTORY BY YEAR[1]

Year	# Operations, U.S.	Avg. inventory, U.S.
1965	1,057,570	47.8
1970	871,200	77.2
1975	661,700	74.5
1980	666,550	96.7
1985	388,570	134.6
1990	268,140	202.9
1995	168,450	345.5
2000	85,760	689.6

[1]U.S. Pork Board Fact Sheet PIG 15-01-01

Table 14.2 NUMBER OF E.U. PIG OPERATIONS AND AVERAGE INVENTORY BY YEAR[2]

Year	# Operations, E.U.	Avg. inventory, E.U.
1997	1,152,000	105.9
1999	1,014,000	123.3
2001	744,000	166.1
2003	644,000	189.3

[2]Source: European Commission, Eurostat. http://ec.europa.eu/agriculture/agrista/2005/table_en/35311.pdf, accessed 9 April, 2007.

Changes to the structure of the industry. The impact of consolidation is largely seen in how the remaining pork producers approach their business. A traditional industry of diversified farming operations is being replaced, in many parts of the world, by specialized pork producers taking advantage of economies of scale and the efficiency of specialized labor (Plain *et al.*, 2006). Moreover, these businesses increasingly have the volume to contribute a consistent flow of pork to support a brand. Thus, many production businesses are becoming an integral part of pork value chains that leverage product characteristics contributed by each segment of the chain to create differentiated products.

Consumer influence. The emphasis on delivering value as part of a food system connects the pork producer to the consumer in ways not seen before this point in history. To remain competitive, food systems increasingly move information upstream to force differentiation into a traditionally commodity product. Through application of supply-chain management tools to food systems, driven predominantly by retailers, consumers now have more power to influence the type of product they want to buy. The pork production chain has no choice but to respond.

The combination of these trends impacts a genetic program by influencing what is economically important to the user of seedstock. Several important new areas are emerging that must be considered in the development of breeding policy:

- Traditionally, breeding policies seek to improve production efficiency. One with reasonable knowledge of the economic drivers of successful pork production can quickly and easily develop a list of important economic traits, and their underlying component traits, that improve the efficiency of producing a kilogram of pork:

    - Increased pigs marketed per sow per year
        - Number of pigs born alive
        - Number of pigs weaned
    - Increased Average Daily Gain in weaner and finishing phases
    - Improved efficiency of converting feedstuff to pork
    - Improved carcass quality (lean content, carcass yield)

- Improvement in these traits will continue to be important to increasingly large, economically driven, pork production systems. Quantitative selection will continue to be the mainstay of genetic programs. However, new biotechnology tools, such as marker-assisted selection may help to improve the rate of genetic progress. Genetic suppliers must actively identify and adopt such tools to remain competitive.

- The integration of the pork value chain is beginning to influence the traits important to include in breeding goals. Perhaps most notable are consumer demand for differentiated product with regard to quality and standards for production. Supply chains are differentiating product with high intramuscular fat, darker colour, higher pH and improved tenderness to capture additional margin by satisfying markets that are willing to pay a premium for these characteristics. Improving these traits doesn't improve efficiency, the traditional goal of breeding policies, and may in fact be antagonistic in some cases. Nevertheless, there appears to be an economic value for eating quality driven from the consumer. However, it has been difficult for breeding stock suppliers to implement selection for improved meat quality due to the lack of clear direction from the value chain regarding economic values for the traits involved.

- Consolidation of the industry typically results in construction of larger production systems that are staffed by salaried, non-owner, employees. Attention to detail, animal husbandry, and animal health are large challenges for these systems. These systems demand animals that are easy to care for and have the ability to thrive in modern facilities. As larger systems begin to move toward loose housing of sows, these problems will likely increase (e.g. Smithfield Foods, www.smithfieldfoods.com/Investor/Press/press_view.asp?ID=394, 2007). Genetic programs can impact animal welfare a variety of ways by improving longevity and its underlying component traits.

- Finally, larger populations of animals housed within the same air space present unique health challenges to animals due to the difficulty of maintaining a stable health status throughout the population, especially for populations affected by the PRRSv. However, PRRSv is not the only pathogen of concern to pork producers. In addition, consumer pressure for rearing animals without the use of antibiotics creates new and unique challenges. Genetically improving general disease resistance would result in animals that continue to have acceptable performance despite an outbreak of disease.

In summary, future breeding policies will be affected by pressure to improve the rate of genetic gain through the use of biotechnology. Non-traditional traits, including longevity (and its component traits), pork quality and disease resistance are new and uncharted territory for genetic improvement programs. These latter traits, especially disease resistance, must be part of any program that will remain successful well into the future. Selection for production efficiency, the impact of biotechnology and improvement in sow longevity and disease resistance will be discussed in more detail, below.

Biotechnology (Is Marker-Assisted Selection Dead?)

Several recent reviews (Rothschild *et al.*, 2007; Tuggle *et al.*, 2007; Chen *et al.*, 2007; Feng-Xing *et al.*, 2007; Lunney, J.K., 2007; and, Green *et al.*, 2007) present excellent detail on the current state of affairs for the application of biotechnology to pig genetics research and serve as an entry into the subject. The use of biotechnology to improve the performance of animals has relied largely on the concept of identifying allelic variation in genes that impact the differences in phenotype among pigs.

As outlined by Rothschild *et al.* (2007), over the course of the last 13 years, 110 papers have identified 1675 quantitative trait loci (QTL) in the pig. These loci have been shown to affect exterior traits (e.g. coat colour), pig health, meat and carcass quality, performance and reproduction with the greatest number of identified loci shown to affect meat quality. Relatively few of these loci have been pursued to the point that the causative mutation has been identified. The effects of many QTL have been demonstrated to differ among breed backgrounds (Buske *et al.*, 2006). Consequently, use of most QTL for pig improvement requires (re)estimation of effects within any given population which has largely inhibited widespread adaptation.

Feng-Xing *et al.* (2007) discussed in detail the identification of linkage disequilibrium between genetic markers at two or more loci and the use of such markers to localize genes affecting a trait of interest. The authors suggest that marker spacing across the genome of 0.1 to 1 centiMorgan to detect association between markers and QTL, considerably less than the tens of centiMorgan spacing most microsattelite studies have traditionally utilized. An increasing number of single nucleotide polymorphisms (SNPs) are now available. Combined with decreasing genotyping costs, it is now more feasible to collect data from high density placement of genetic markers. Identification of multiple regions of the genome controlling genetic variation could perhaps be used to improve the accuracy of selection, even without knowledge of the causal mutation.

Dekkers and Chakraborty (2001) are one of numerous authors who have modeled the impact of information on an underlying genotype on response to selection. In most cases, response to selection was improved by less than 5% over traditional selection using BLUP for prediction of breeding values for traits of interest. The exception to this result was for traits of low heritability (e.g. litter size, survival) and alleles present at very low frequency. For these conditions, response was up to 55% greater for a gene having an effect of one phenotypic standard deviation.

Two key observations from Dekkers and Chakraborty (2001) are, first, that the gene effect modeled is large (one standard deviation). Genes of smaller effect will not achieve the same level of contribution to response. Numerous genes affect any given trait. While examples of genes with large effect do exist (e.g., HAL1843, rendement napole), most genes affecting quantitative traits are likely to be small and moderate in effect resulting in little improvement in response for any one locus considered alone. Second, there was a significantly larger effect of molecular information on traits of low heritability. Traits of low heritability are more difficult to improve through traditional quantitative selection and thus, the application of marker-assisted selection might be better applied to traits of lower heritability. The same can be said for traits that are difficult to measure, particularly on the live pig (e.g. meat quality, disease susceptibility) where genes of known effect can be selected for increased frequency in nucleus populations.

Nejati-Javeremi *et al.* (1997) demonstrated the use of genetic markers to estimate the allelic relationship among animals. As contrasted with the traditional relationship matrix used in BLUP evaluation, the allelic relationship does not simply assume an average sharing of alleles among related individuals (e.g. a 50% relationship among full-sibs), but instead uses knowledge of allelic variation to produce an estimate closer to the true number of alleles shared in common between related individuals. Their model demonstrated improved response to selection due to a reduction in prediction error (higher correlation between predicted and true breeding value). Such an approach could be widely implemented to improve response, without requiring knowledge of causative mutation. However, significantly reduce genotyping costs to generate marker phenotypes would be required.

This all questions the future of marker assisted selection. As Dekkers (2004) points out, there is a feeling of 'cautious optimism' for application of genetic marker approaches to improvement programs. The paradigm of identifying causal mutations will continue, but may be significantly augmented in application to breeding programs by leveraging marker information to improve the accuracy of predicting the breeding value of individuals undergoing selection. Such an approach will require extremely inexpensive genotyping costs, substantial information systems and powerful computing resources, all of which continue to develop at a rapid pace. Integration to a breeding program will simply require a positive benefit:cost ratio, supported by the customer base of a breeding organization, in order to be implemented on a large scale.

It is most interesting that to date, to the knowledge of this author, no controlled study (a comparison of lines originating from the same base population, one undergoing BLUP selection, the other BLUP-MAS selection) has ever been published essentially to provide biological proof that the use of marker-assisted selection actually improves response over the long-term. Only simulation studies have been published.

Longevity of the sow

Sow longevity impacts the efficiency of pig production systems primarily by its association with the number of pigs produced per lifetime. In addition, high replacement rates within breeding herds increase the risk of introducing disease. Finally, it is an animal welfare issue to produce sows from breeding systems that are not capable of withstanding the rigors of modern production systems (Serenius and Stalder, 2006). This latter issue is very difficult to assign an economic value to, but will become increasingly important as production systems return to loose housing designs driven by consumer demand.

Rodriquez-Zas *et al.* (2003) examined the profitability of sow enterprises by examining the production records of 148,568 sows from 32 herds in central Illinois (USA). In this data set, the average replacement rate was 59.8%, the average culling rate, 41.6% and death rate, 9.7%. The genetics of the sow affected the length of time in the herd by up to one parity with differences noted in removal rate between lines. Interestingly, the effect of longevity on profit was mixed. At a zero discount rate, sows that remained in the herd longer produced greater profit because the cost of a replacement female is recouped during the early parities. However, as discount rate increased, the difference between longer and shorter herd life were reduced. This is because the value of future profit is reduced by the level of the discount rate.

At low levels of profit per litter and increasing discount rates, the value of longevity decreased significantly. For example, modeling $50 profit per litter and a 10% discount rate showed that the advantage was $16.37 for the best compared to worst longevity line. However, modeling $10 profit per litter reversed the comparison. These results demonstrate that profit per litter affects the value of longevity and longevity alone cannot recoup the cost of a replacement animal in periods of low profitability (Rodriquez-Zas *et al*, 2003).

Several key traits have been shown to be associated with longevity in sows. Serenius *et al*. (2006) examined several genetic lines evaluated by the U.S.-based National Pork Board. Within lines, sows that had low feed intake and high backfat loss during lactation had a shorter productive lifetime. Energy balance, the ability of a sow to consume the energy required to meet the metabolic requirements for piglet development, lactation, growth and maintenance, is a primary indicator of length of reproductive life.

Heritability estimates for longevity (length of productive life) have ranged between 0.02 and 0.25 (Serenius and Stalder, 2006). This range of heritabilities indicates selection for longevity would be effective. Reproductive failure showed the strongest relationship with longevity. However, reproductive failure (failure to return to estrus) may be a by-product of energy balance.

Leg conformation is the second greatest reason for culling (Serenius and Stalder, 2006). Correlations between longevity and leg conformation range from 0 to 0.36 (average of 0.21), suggesting improved conformation scores will increase longevity. Heritability estimates range from zero to quite large (0.4).

Sow longevity is a complex trait that is primarily impacted by involuntary culling for reproductive failure and soundness. Research has demonstrated associations between various traits and longevity, but has not provided clear answers to the underlying relationships among the component traits, particularly for reproduction. Therefore, a logical approach to improving longevity would be to select for improved skeletal structure (front and hind legs, body) as well as length of reproductive life. Breeding companies should also research gilt development protocols that produce a replacement female of sufficient sexual maturity and body mass to enter the production herd.

The Danish National breeding program, until recently, approached improvement in longevity by including conformation in the selection objective (policy). Selection for improved structure score has taken place since 1995.

Using a one to five scale, trained technicians score each nucleus candidate at the end of the testing period for front and rear legs and body conformation. A combined score is included in the genetic analysis as a single trait with an estimated heritability of 0.12. The genetic trend indicates improvement of 0.03 to 0.06 conformation points per year (2006 Annual Report, National Committee for Pigs, Copenhagen, Denmark).

Recently, the Danish program has added a partial measure of longevity to the genetic analysis. Longevity is difficult to measure in the nucleus herd itself because most sows are culled early in their reproductive life for genetic merit and never have the opportunity to fully express their phenotype for longevity. Consequently, to measure longevity, it is important to use a data source outside the nucleus where culling is unaffected by genetic index. In addition, the data source must have a sufficient amount of data, the trait must be easy to measure and the structure of the data must be useful for estimation of breeding values. This source of data in the Danish program is the multiplication herds wherein sows are not typically culled for their index during the early parities, use nucleus level sires that create connectedness within the data and provide ample records (7000 Yorkshire and 15,000 Landrace sows per year).

Nielsen, B. and Vernersen, A. (personal communication) examined two 'longevity' traits. The first measure of longevity was defined as the probability that a female that was mated for her first litter that was also mated to produce her second litter (LT1, Figure 14.1). The second was the probability that a female that produced her first litter that was mated to produce her second litter (LT2, Figure 14.2). For LT1, 0.85 of the gilts mated for their first litter had their first litter and 0.75 were remated for their second litter. For LT2, 0.88 of gilts having their first litter were mated for the second.

Heritability estimates for this trait average 0.17 across the two traits for Yorkshire and Landrace, a moderate heritability estimate (Table 14.3). Genetic correlations of LT1 and LT2 with 'fitness' (defined as the structure score given to animals off-test) ranged from 0.14 to 0.53 and with 'growth' (defined as the growth rate during the testing period) ranged from 0.035 to -0.288. These results indicate that genetic progress in the trait is possible, progress in longevity will be positively associated with fitness and somewhat negatively associated with growth rate. Nielsen, B. (personal communication) also indicates the highest level of genetic variation is obtained when the probability is nearest 0.5 for the trait. Therefore, LT1, with an incidence rate of 0.75 is preferred as a higher rate of genetic gain can be expected.

Measuring longevity as success from first to second mating certainly does not capture all variation expected for longevity. However, this trait captures a significant portion of that variation, measuring the ability of a sow to transition from a gilt to mating for her second parity, perhaps the highest risk time for culling. Additional information required is a better understanding of the correlation between LT1 and 'true' longevity, the true economic value of longevity and how to include censored information from third parity and beyond in the genetic analysis.

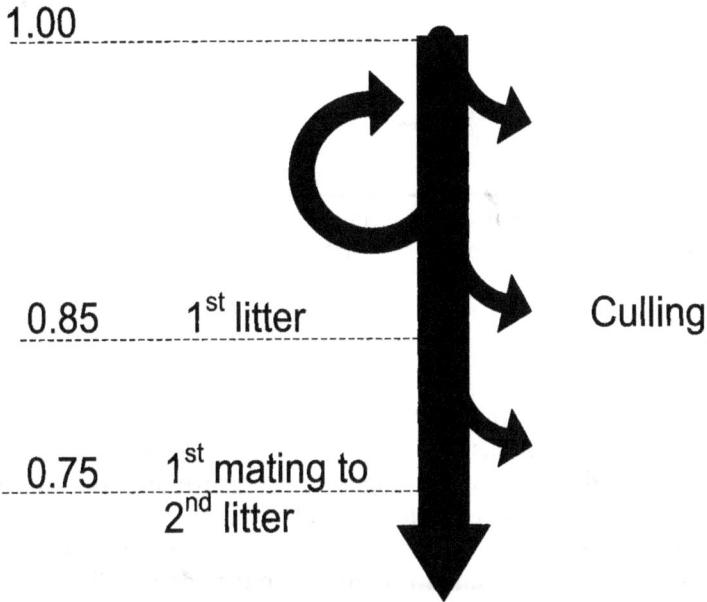

Figure 14.1 Proportion of gilts mated for first litter that are also mated for their second litter (LT1)

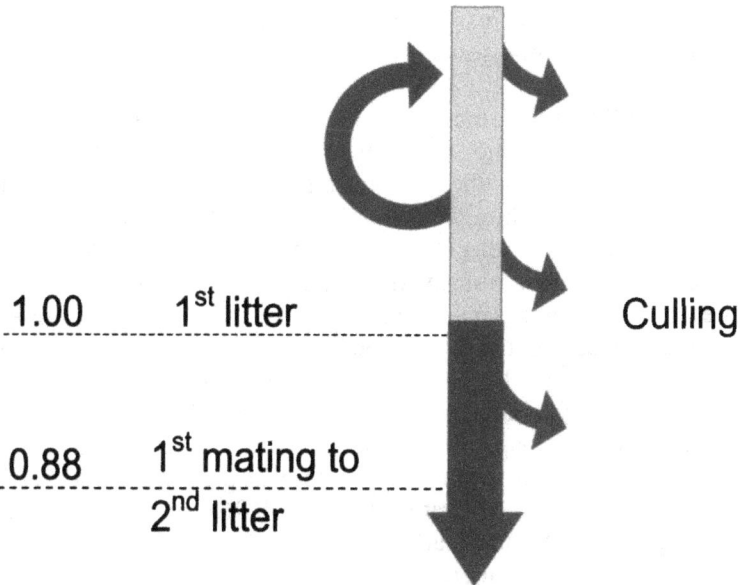

Figure 14.2 Proportion of sows having their first litter that were remated for their second litter (LT2)

Table 14.3 HERITABILITY AND GENETIC CORRELATION ESTIMATES FOR LT1 AND LT2[1].

Breed	Trait	Heritability	Fitness score	Growth
Yorkshire	LT1	0.16	0.28	-0.21
Yorkshire	LT2	0.23	0.36	0.04
Landrace	LT1	0.17	0.14	-0.29
Landrace	LT2	0.14	0.53	0.12

[1]LT1 is the probability of first mating to second litter if the first mating for first litter is recorded. LT2 is the probability of the first mating to second litter if the first litter is recorded.

Disease susceptibility

Consumer interest in restricting the use of antibiotics, increased density of production animal populations and animal welfare concerns combine to increase our interest in producing animals capable of performing and thriving in the presence of natural pathogens with minimal intervention. Biotechnology has provided tools to aid in unraveling the genetics of disease resistance and several recent studies indicate substantial progress in our understanding of the impact of genetic variation on susceptibility to pathogens. Danbred North America has taken a lead role in understanding genetic variation for resistance to Porcine Reproductive and Respiratory Syndrome (PRRS), a leading cause of economic losses to swine producers across the globe. The PRRS virus (PRRSv) typically infects the macrophage cells resident in various tissues, particularly pulmonary alveolar macrophage.

Petry *et al.* (2005) performed the first controlled experiment designed to investigate within line genetic variation for susceptibility to the PRRS virus. Two lines were examined for their response to infection with PRRSv strain 97-7985, a commercial Hampshire by Duroc line (HD, Danbred), and the Nebraska Index line (NEI), a closed selection line at the University of Nebraska (USA) that has been selected for improved reproduction for 20 generations. Two pigs from each of 50 litters per line were sampled with one pig from each litter serving as an uninfected control. Uninfected pigs from the HD line grew faster (0.67 kg) from day 0 to 14 post-infection and had higher average rectal temperature (0.34 C) compared to NEI pigs. Infected pigs from the NEI line grew 0.34 kg faster and had lower body temperatures (-0.54 C) than the HD line. Viremia was higher in HD compared to NEI pigs at days 4, 7, 10 and 14 days post- infection. Elisa S/P ratios for the PRRSv were also higher in HD versus NEI pigs and HD pigs had higher lung lesion scores. These results demonstrate difference in underlying genetic variation.

Petry *et al.* (2007, submitted), using the data set from the previously described experiment, conducted a principle component analysis and identified two unique groups of pigs within both the HD and NEI populations. Within each line, Low (L) responders existed which had high average daily gain, low levels of viremia and few lung lesions. High (H) responders had low daily gain, high levels of viremia and many lesions. Expression of 11 innate and T helper 1 immune markers were evaluated. All genes except one showed significant up-regulation following infection. Expression was higher in the HD line compared with the NEI line. There was a significant down-regulation in L pigs, relative to controls, in lung and bronchial lymph nodes. Serum level of cytokines affirmed these differences. Most notable was the fact that high pre-infection levels of interleukin-8 were associated with PRRSv resistance in the L pigs and, following infection, lower levels of interferon-gamma expression were associated with the resistant pigs. Further validation of this model is underway.

Vincent *et al.* (2006), following Petry *et al.* (2005), examined the susceptibility of two diverse commercial lines of pigs to the PRRSv. Two lines were characterized by the percentage of macrophage cells infected in vitro when exposed to PRRSv. A Large White population was classified as 'High' (meaning that a higher percentage of macrophage were infected compared to the 'Low' line) and pigs derived from the Duroc and Pietran breeds were classified as 'Low'. Pigs from each line were exposed to the PRRSv (strain VR-2385) at six weeks of age and necropsied at 10 or 21 days after infection to examine response to infection in vivo.

It was interesting to note that the Low line showed greater severity of clinical signs early, during the acute phase of the infection. The Low line had higher lung lesion scores, virus level in serum and number of febrile days. However, 21 d post-exposure the lines reversed with the High line having higher levels of viremia in the blood than the Low. These authors (Vincent *et al.*, 2006) conclude that obviously the host:virus interaction is complex. A line of pigs showing lower rates of infected macrophage cells indeed had a higher incidence of undesirable symptoms during the acute phase. However, these pigs appeared to clear the virus more quickly because the level of viremia was lower in the High group 21 days after infection. Genetic variation in response was clearly demonstrated however, the underlying mechanism for these differences remains to be understood.

PRRSv is not the only pathogen causing major economic losses in the Danish and North American swine industry. Other areas of emphasis in the Danish National Program include resistance to Salmonella, E.coli. and general resistance to pneumonia. (2006 Annual Report, National Committee for Pigs, Copenhagen, Denmark). The projects are discussed below.

The Pneumonia Project was developed with the objective of identifying variation in the incidence of lung disease among progeny of various sires. Boars were mated to sows in a common commercial herd. Pigs were finished at three separate finishing sites with varying levels of mycoplasma prevalence. Table 14.4 lists the average incidence rate of various types of lung disease identified among the progeny of the sires. Early evidence from over 13,000 progeny of 62 sires suggests fairly substantial variation exists among sires for general resistance to lung infection. This project will serve as the basis for genetic marker identification research in the near-term as well as estimation of genetic parameters for resistance to lung disease (e.g. heritabilties and genetic correlation with production traits).

Table 14.4 FREQUENCY OF SURVIVAL AND LUNG LESIONS BY DUROC SIRE

| Sire | Survival | | At slaughter | |
	Weaners	Finishers	Mycoplasma	Actinobacillus
A[a]	0.95	0.98	0.45	0.39
B[a]	0.95	0.97	0.14	0.54
C	0.97	0.97	0.18	0.54
D	0.97	0.98	0.25	0.54
E	0.96	0.95	0.36	0.60
F	0.93	0.98	0.40	0.59
G	0.96	0.97	0.46	0.55
H	0.95	0.93	0.46	0.53
I	0.95	0.95	0.38	0.57
J	0.96	0.93	0.32	0.64
K	0.97	0.90	0.34	0.62
L	0.98	0.94	0.30	0.47
M	0.93	0.94	0.50	0.60
N	0.92	0.92	0.43	0.62

[a]Sire with few litters

Salmonella resistance is important for the safety of meat products as it can impact human health. Current production and packing plant practices help protect against transmission of the disease; however, it would be desirable to identify pigs that are resistant to the pathogen. A two-phase trial was developed to examine if resistance to Salmonella is heritable. Within the Danish pig population, approximately 0.07 of animals exposed to Salmonella do not develop antibodies against it. For this study, a population of 600 pigs was inoculated with Salmonella resulting 21 gilts that did not show any antibody response. These gilts were retained and mated to 'resistant' boars (identified from the same population) producing approximately 200 progeny. The second generation was inoculated with the same pathogen, resulting in approximately 0.08 showing 'resistance', the same percentage as the parent generation. A simple pattern of inheritance could not be identified from these results and further research is underway to identify if genetic variation does indeed exist for resistance to Salmonella, albeit in a more complicated pattern involving multiple genes.

A major cause of scours in young pigs is the E. coli. bacterium in its various forms. In the United States, the F18 and F4 (K88) forms of the bacteria are the predominant strains with a trend toward more increasing incidence of F18 infection (Vansickle, 2001). In Denmark, the focus has been on the F4 form as the dominant source of infection. Jorgensen, C.B. *et al.* (2003) identified the chromosomal region containing candidate genes for resistance to E.coli F4 ab/ac. Subsequently the gene containing the causative mutation has been identified with two forms, one conferring resistance and the alternative resulting in susceptibility. Resistance is recessive to susceptibility, thus, two copies of the resistance allele are required for an animal. Currently, an ongoing effort (F4 Project) is in place to increase the frequency of resistance within the various Danbred nucleus lines.

In 2003, before initiation of the F4 Project, the frequency of resistance alleles in the Landrace breed was 0.01, Yorkshire approximately 0.20, Duroc approximately 0.88 and Hampshire at 1.00. As of August, 2006 the number of Yorkshire boars testing positive for resistance has risen to 0.75 and Landrace boars to 0.35. All Duroc boars used in the nucleus herds since April 2006 are homozygous for the resistance allele.

Frydendahl *et al.* (2003) and Meijerink *et al.* (2000) discovered and developed a test to identify pigs resistant to the F18 strain of E. coli., which causes Edema disease in weaned pigs. PIC, in 1999, established a program call "EdemaGard" to deliver customers grandparent dam-line boars and gilts and terminal boars resistant to E. coli F18 (Van der Steen, H.A.M. *et al.*, 2005). Like the allele for F4, the F18 resistance is conferred from a recessive allele that only eliminates susceptibility when present in two copies in the genome. Vansickle (2001) published a report summarizing a field result in a 1200 sow farrow-to-finish operation in Minnesota (USA). Death loss in the nursery suddenly increased to 0.08-0.12 due to F18 toxigenic E. coli. Treatment with antibiotic did not decrease death loss. Following introduction of resistant terminal boars, the death loss due to the pathogen reduced to 0.01 in the nursery simply by increasing the frequency of resistance among the terminal pigs to 0.31 following the initial use of a resistant boar. Gradually, the frequency of resistance in the dam line would increase. However, this field result demonstrates not only the effectiveness of genetic resistance, but also how simply increasing the frequency of resistant animals within a population can be used to minimize the pathogenic affects of the disease on a population of animals.

Improving the resistance of pigs to pathogens present in their environment provides what is perhaps the best use of the tools of biotechnology. Identifying causative mutations that confer resistance can be used to select animals to be retained as parents without exposing nucleus populations to the pathogen itself.

Such traits are difficult to measure in a pig housed within a high-health environment. The use of genetic markers for these traits will continue to be a focal area of research in pig breeding.

Summary and a few comments on improving production efficiency

Falconer (1989) defines genetic progress in any single trait under selection with the following formula:

$$Gain / year = [(r_{bv,ebv} * i * \sigma_{bv})/t]$$

Wherein, $r_{bv,ebv}$ is the accuracy with which we rank our animals. This is impacted by the heritability for a given trait (generally beyond our control), and by the amount and quality of information available to us (within our control); i is the selection intensity we are able to achieve. The smaller the number of animals we retain, the greater the genetic potential of the selected group will differ from the average pig; σ_{bv} is a measure if the inherent genetic variation present in the population; and, finally, t, is generation interval. Generation interval is the average age of an animal when its replacement is born.

In all its simplicity, the formula above remains the key driver to success in operating a successful genetic program. To improve production efficiency over the long term, a well-disciplined trait recording program within a large nucleus population, with a low generation interval, that allows for the top ranking animals to become parents day-in and day-out, is an absolute necessity. Worldwide, those programs that have done the best job of implementing the fundamentals are currently gaining and holding market share. This is mainly the result of sustained superior performance in their product lines. Companies that short-cut the investment required (e.g. maintaining small nucleus herds) will not maintain a sustainable competitive advantage.

Improvements in efficiency will continue to come about through selection for increased litter size, improved pig quality, higher daily gain, better feed efficiency and improved yield of lean, high-quality, pork. Additional traits, such as sow longevity and meat quality, have and will continue to enter the selection objective of successful breeding stock suppliers.

Biotechnology will not overcome the need to practice the fundamentals daily. Instead, the tools of molecular biology will result in enhanced accuracy of selection within already disciplined nucleus program. In addition, biotechnology will have positive impact on being able to identify and select animals that are more resistant to common pathogens and may provide a means to control and manage the outbreak of disease within large commercial operations.

Future success for breeding stock suppliers will depend upon

✓ Maintenance of *large* nucleus populations that minimize inbreeding and maximize genetic progress.
✓ Development of consistent trait recording schemes and maintenance of recording discipline in large systems across numerous nucleus herds. * World-class production at the nucleus level.
✓ World-class health status at the nucleus level.
✓ Addition of new traits to the breeding policy well ahead of when the industry will demand improved performance in those traits.
✓ Active investment in biotechnology as a tool to improve response within the existing framework of selection schemes with the caveat that biotechnology is not a replacement for fundamental quantitative selection schemes.
✓ Developing an understanding of the genetic variation underlying disease susceptibility and use of molecular tools to increase the frequency of resistant animals in nucleus populations.

Bibliography

Buske, B., I. Sternstein and G. Brockman. 2006. QTL and candidate genes for fecundity in sows. *Anim. Reprod. Sci.* **95**(3-4):167.

Chen, K., T. Baxter, W.M. Muir, M.A. Groenen and L.B. Schook. 2007. Genetic resources, genome mapping and evolutionary genomics of the pig (Sus scrofa). *Int. J. Biol. Sci.* **3**:153.

Dekkers, J.C.M. 2004. Commercial application of marker- and gene-assisted selection in livestock: Strategies and lessons. *J. Anim. Sci.* **82**(E. Suppl.):E313.

Dekkers, J.C.M. and R. Chakraborty. 2001. Potential gain from optimizing multigeneration selection on an identified quantitative trait locus. *J. Anim. Sci.* **79**:2975.

Du, F., A.C. Clutter and M.M. Lohuis. 2007. Characterizing linkage disequilibrium in pig populations. *Int. J. Biol. Sci.* **3**:166.

Eurostat. 2005. http://ec.europa.eu/agriculture/agrista/2005/table_en/35311.pdf, accessed 9 April, 2007. European Commission, Eurostat.

Falconer, D.S. 1989. Selection: I The response and its prediction. In: Introduction to Quantitative Genetics, pp. 187-203. Longman Scientific & Technical, Essex, England, U.K.

Frydendahl, K., T.K. Jensen, J.S. Andersen, M. Fredholm and G. Evans. 2003. Association between the porcine Escherichia coli F18 receptor genotype and phenotype and susceptibility to colonisation and postweaning diarrhoea caused by E. coli O138:F18. *Vet. Microbiol.* **93**:39.

Green, R.D., M.A. Qureshi, J.A. Long, P.J. Burfening and D.L. Hamernik. 2007. Identifying the future needs for long-term USDA efforts in agricultural animal genomics. *Int. J. Biol. Sci.* **3**:185.

Lunney, J.K. 2007. Advances in swine biomedical model genomics. *Int. J. Biol. Sci.* **3**:179.

Meijerink, E., S. Neuenschwander, R. Fries, A. Dinter, H.U. Bertschinger, G. Stranzinger and P. Voegeli. 2000. A DNA polymorphism influencing alpha(1,2) fucosyltransferase activity of the pig FUT1 enzyme determines susceptibility of small intestinal epithelium to Escherichia coli F18 adhesion. *Immunogenetics* **52**:129.

National Committee for Pigs. 2006 Annual Report. Danske Slagterier, Copenhagen, Denmark.

Nejati-Javaremi, A., C. Smith and J.P. Gibson. 1997. Effect of total allelic relationship on accuracy of evaluation and response to selection. *J. Anim. Sci.* **75**:1738.

Petry, D.B., J. Lunney, P. Boyd, D. Kuhar, E. Blankenship and R.K. Johnson. 2007. Differential immunity in pigs with high and low responses to porcine reproductive and respiratory syndrome virus (PRRSv) infection. *J. Anim. Sci.* (submitted).

Petry, D.B., J.W. Holl, J.S. Weber, A.R. Doster, F.A. Osorio and R.K. Johnson. 2005. Biological responses to porcine respiratory and reproductive syndrome virus in pigs of two genetic populations. *J. Anim. Sci.* **83**:1494.

Plain, R., J. Lawrence and G. Grimes. 2006. The structure of the U.S. pork industry. National Pork Board publication PIG 15-01-01. Des Moines, Iowa, U.S.A.

Rodriguez-Zas, S.L., B.R. Southey, R.V. Knox, J.F. Connor, J.F. Lowe and B.J. Roskamp. 2003. Bioeconomic evaluation of sow longevity and profitability. *J. Anim. Sci.* **81**:2915.

Rothschild, M.F., Z. Hu and Z. Jiang. 2007. Advances in QTL mapping in pigs. *Int. J. Biol. Sci.* **3**:192.

Serenius, T. and K.J. Stalder. 2006. Selection for sow longevity. *J. Anim. Sci.* **84**(E. Suppl.):E166.

Serenius, T., K.J. Stalder, T.J. Baas, J.W. Mabry, R.N. Goodwin, R.K. Johnson, O.W. Robison, M. Tokach and R.K. Miller. 2006. National Pork Producers Council Maternal Line National Genetic Evaluation Program: A comparison of sow longevity and trait associations with sow longevity. *J. Anim. Sci.* **84**:2590.

Smithfield Foods. 2007. http://www.smithfieldfoods.com/Investor/Press/press_view.asp?ID=394. Accessed 9 April 2007. Smithfield Foods Corporation, Virginia, U.S.A.

Tuggle, C.K., Y. Wang and O. Couture. 2007. Advances in swine transcriptomics. *Int. J. Biol. Sci.* **3**:132.

Van der Steen, H.A.M., G.F.W. Prall and G.S. Plastow. 2005. Application of genomics to the pork industry. *J. Anim. Sci.* **83**(E. Suppl.):E1.

Vansickle. J. 2001. Genetically resistant line stops E. coli cold. In: National Hog Farmer. Prism Business Media. 15 November 2001.

Vincent, A.L., B.J. Thacker, P.G. Halbur, M.F. Rothschild and E.L. Thacker. 2006. An investigation of susceptibility to porcine reproductive and respiratory syndrome virus between two genetically diverse commercial lines of pigs. *J. Anim. Sci.* **84**:49.

15

REPRODUCTIVE BIOTECHNOLOGY IN PIGS: WHAT WILL REMAIN?

HERIBERTO RODRIGUEZ-MARTINEZ
Division of Reproduction, Department of Clinical Sciences, Faculty of Veterinary Medicine and Animal Science, Swedish University of Agricultural Sciences (SLU), Uppsala, Sweden

Introduction

Pigs are among the firstly domesticated animals, somewhere around 9,000 years ago, and are now represented by nearly 500 breeds after being established by migration or transport by mankind around the globe. Currently most pigs are raised mainly for human consumption of meat and other co- products. About 0.40 of the red meat consumed worldwide is provided by pigs, and a projection by FAO (http://apps.fao.org/cgi-bin/), based up on the tendency for increased pig production in the developing countries over the past decade (Steinfeldt *et al.*, 2006), foresees an increased demand for pig products of over 25% of current production by the year 2020 (Delgado *et al.*, 1999). Pig production requires efficient fast growth, reduced feed intake, high carcass merit and meat quality. Obviously, there must also be high levels of reproductive success among breeding animals as well as disease resistance and increased survivability in piglets and growing pigs. Public concerns, such as those linked to BSE, have damaged beef production in Europe and other countries, and have certainly promoted pig production along with an increase in production in developing countries (Steinfeldt *et al.*, 2006). Pig consumers also demand the removal of antibiotic growth promoters from the feed, and the altering of facilities to promote better animal welfare. In addition, concerns regarding the environmental impact of pig wastes have prompted producers to consider ways to reduce feed wastage and improve feed efficiency. Therefore, implementation of new production techniques, based on advances in disease control, nutrition and management, and the promotion of desired genetics play a major role in developing this agricultural sector which, by definition, is not providing large benefits to producers unless large volumes of a product, valuable to the ultimate consumer, is obtained at the lowest possible cost. Consumers, however, represent a large grouping. Most want pigs for their meat; others want herd sires or gilts for replacement; while the smallest group wants them for research and/or potential industrial use. It is obvious that numbers of delivered pigs with certain phenotypic characteristics, such as high feed-conversion rate, growth rate, desirable meat quality (whatever the definition would be in that particular market) and prolificacy are leading the target set -and reached - by producers. Such achievements are not only based on cheap production of large numbers of pigs; they rely on the application of genetic selection of sires and dams and on reproductive techniques that would ensure the most effective propagation of their genes to the general breeding and terminal population. Stud boars are generally selected for their genetic potential to produce litters that grow quickly, efficiently and have commercially-attractive carcass types. Breeding dams are selected for fertility, prolificacy, good motherhood and longevity, so that replacement figures are kept at minimum.

Application of reproductive biotechnologies, such as artificial insemination (AI), has contributed enormously to the propagation of the genetic material of selected stud boars and, probably, will continue as boars shall continuously replace the best ones currently used (Gerrits *et al.*, 2005; Robinson and Buhr, 2005). Unfortunately, some problems arise. Firstly, traits related to boar fertility are of low heritability (h^2 ~0.01-0.06); secondly, these traits are strongly affected by genetic and environmental effects of the boar itself, the dam and the offspring and, thirdly, female fertility has a greater impact on the reproductive performance realized in a commercial herd than boar fertility. Therefore, additional efforts have to be made to elucidate genetic components controlling female fertility.

Recently, the role of the pig has expanded -albeit in most cases only potentially- from being a source of food to serve as important animal model for basic and applied research for human health. The advances in genome research (see Womack, 2005, for review) and molecular technologies seen recently has made it possible to design transgenic pigs carrying gene constructs for improved commercial traits (growth, prolificacy, etc), commercial products (specific proteins, enzymes, antibiotics, etc) or even modified cells/tissues/ organs capable of being xeno-transplanted (Rothschild, 2004). Achievements in molecular genetics (functional genomics, transcriptomics and proteomics among others; Rothschild, 2003; Womack, 2005), gene manipulation (transgene constructs) and reproductive biotechnologies (e.g., somatic cell nuclear transfer for cloning (SCNT), transgenesis, *in vitro* embryo production, cryopreservation, etc) are today able to develop these pigs, even when their efficiency and costs are still very high (Prather *et al.*, 2004; Vajta and Gjerris, 2006). In addition once initial animals are produced, there are subsequent propagation techniques for these, usually extremely expensive individuals, such as testicular transplants, intra-cytoplasmic sperm injection (ICSI), *in vitro* fertilisation, embryo-, foetal- or even adult somatic cloning, followed by embryo transfer (ET) of the resulting embryos (Vajta *et al.*, 2007). However, genetic propagation for commercial application is constrained by (i) the assured retention of the transferred gene into the breeding population and (ii) the economic feasibility of their incorporation into classical, well-controlled, breeding schemes; thus far it is very expensive. Further constraint is exerted by public concerns on the production and dissemination of genetically modified organisms (GMOs), for matters of animal welfare and also for their possibility of reaching the food chain, something that is considered -for parts of the globe- unacceptable or even banned. However, use of transgenic pigs for non-food purposes, especially for industrial and, particularly, medical (pharmaceutical) application, seems more acceptable. The control of this area has been - or shall be – the domain of only a few, financially strong, companies. In any case, there shall be a growing need for suitable reproductive technologies that could propagate permanently the desired genotypes, obtained by gene manipulation or by directed breeding.

Propagation techniques, such as AI or embryo transfer (ET), have benefited largely from research in cell biology (mostly for *in vitro* handling of gametes and embryos) as well as animal physiology (for semen or embryo deposition strategies) and, although the transfer of the best genetic material is still undertaken by the well-established AI with "fresh", liquid semen, major developments have been made in the cryopreservation of embryos and spermatozoa, as well as in the techniques for non-surgical ET (nsET) or the deposition of low sperm numbers (intra- and deep-intrauterine AI, diu-AI), two strongly emerging techniques, that would probably be established as routine (see a recent review by Martinez *et al.*, 2005). While the former enables the deposition of vitrified-rewarmed embryos (transgenic or not), the latter facilitates the use of frozen-thawed spermatozoa or of sperm doses containing "sexed-semen", using high-output flow-cytometry sorting of spermatozoa for chromosomal sex (Garner, 2006). Other biotechnologies, such as *in vitro* production of embryos (IVP), and related techniques such as intracytoplasmic sperm injection (ICSI), cloning, isolation of stem cells, transgenesis, etc., are not yet used in pig production, despite their potential to incorporate the results of ongoing research in genomics, transcriptomics and proteomics into pig breeding (Prather *et al.*, 2004).

The present review describes the state-of-the-art regarding established and emerging reproductive biotechnologies in pigs, attempting a critical appraisal of their value for future breeding and selection, research applications, design of disease models or bioreactors, and for their use in commercial production. It hopes to avoid reiteration of the large volume of literature available elsewhere, including companion papers in these proceedings.

Artificial insemination: the most successful reproductive technique

As a reproductive biotechnology, pig AI has more than 80 years of documented success (Serdiuk, 1970), starting in the Soviet Union in the 1920s owing to the initial stimulus from Ivanov and continuing, efforts by Milovanov in the next decade, designing extenders for the handling of boar semen at room temperature (Milovanov, 1962). The use of boar semen preserved for AI has increased more than 3 times over the past 20 years (in Europe >0.90 of the females are bred via AI with liquid semen), with >0.99 of the approximately 30 million of registered first AIs in the world achieved with liquid semen while the remaining 0.01 with frozen-thawed semen (Wagner and Thibier, 2000). Most AIs concern breeding of terminal pigs, with semen either collected, processed and inseminated on-farm (do-it-yourself) or commercially supplied by specialised boar centres. Genetic improvement also benefits from the use of AI as propagation technique, where the semen from selected (either commercially, national or international breeding programmes) boars is distributed to breeding sows as liquid- or cryopreserved semen. The former is used for genetic improvement at national or regional level while the second concerns international trade, avoiding movement of livestock.

USE OF AI WITH LIQUID-STORED BOAR SEMEN: SOMETHING NEW?

This is probably not novel with the exception that the market has a plethora of extenders, one promising longer sperm survival than the other, despite most producers still inseminating (whenever possible) on the first, second or third day following collection (Levis, 2000). When semen delivery is constrained by distances or distribution problems, semen is shipped and stored at moderately reduced temperatures (i.e., 16-20°C), most often for up to 5 days before AI, although there are extenders available for much longer storage (Johnson *et al.*, 2000). In this respect, the most common extenders, including those that have been around he longest, held boar spermatozoa within expected ranges of sperm viability and potential fertilizing capacity (De Ambrogi *et al.*, 2006). If fertility and prolificacy are kept high (e.g., not lower than that obtained with controlled natural mating) by 2 AIs per oestrus, even with less spermatozoa than those used at present (ranging 2.5-4x10^9 spermatozoa/dose), the use of liquid semen-AI -done within 3 days post-collection- is here to stay. The lower the number of spermatozoa included per dose and the shorter the storage prior to AI, the higher the number of usable AI-doses produced from one ejaculate, and thus the improved revenues. How low sperm numbers per dose can be reached whilst still maintaining fertility/prolificacy is most likely to be boar-dependent (as it is sire-dependent in bovine AI) rather than extender-dependent, but the answer can only be provided after well-designed and controlled studies by the pig industry.

CRYOPRESERVATION OF SEMEN: ANY BREAKTHROUGH?

This again is probably not likely. Despite the impressive amount of research undertaken in the subject since the first successful inseminations with frozen-thawed boar semen in the early 1970s, current commercially-available semen cryopreservation techniques are still sub-optimal, technically demanding and basically limited to research, genetic banking or the export of semen for selected nuclei lines (see Johnson *et al.*, 2000 for review). The major constraint is not only the inherent difficulties in freezing pig spermatozoa (Holt, 2000), but the sire-dependent cryosurvivability in the current procedures (Eriksson *et al.*, 2002; Holt *et al.*, 2005; Gil *et al.*, 2005; Waterhouse *et al.*, 2005; Hernandez *et al.*, 2005; 2007; Roca *et al.*, 2006a), which most often yields thawed spermatozoa with a shortened life span, a drawback that has been compensated by the AI of excessive sperm numbers (at least $5x10^9$ spermatozoa/AI-dose). However, even with these huge sperm numbers, fertility (as farrowing rates) and prolificacy (as litter size) are still lower than for liquid semen (around 10-30 % lower farrowing rates, and 1-3 less piglets), indicating that other factors are limiting, such as the timing of insemination in relation to ovulation (Bolarín *et al.*, 2006; Wongtawan *et al.*, 2006). Moreover, the procedure of freezing is expensive, both in terms of labour and laboratory equipment costs, and time-consuming (rev by Roca *et al.*, 2006b). Finally there is a lack of reliable laboratory tests for the accurate assessment of semen quality *in vitro*, that limits the capacity to monitor properly the methods used to freeze-thaw boar semen and, particularly, its relationship to AI-fertility (Rodriguez-Martinez, 2007).

Therefore, it seems unlikely that deep frozen semen will replace the use of fresh semen on an extensive basis even if the fertility levels were similar. It is, simply, too expensive. However, having a reliable cryopreservation method for boar semen would (i) allow selection of genetics from all over the world, (ii) enable planned, essential AIs at the top of the breeding pyramid, thus (iii) facilitate preservation of top quality genetic lines for ongoing or future breeding programmes and (iv) offer an extra health safeguard, by allowing completion of any health test specified by a country or breeding organisation before use. Thus, since problems are there to be solved, attempts are made, worldwide, to design novel cryopreservation methods and containers (Eriksson and Rodriguez-Martinez, 2000; Kumar *et al.*, 2003; Saravia *et al.*, 2005; Woelders *et al.*, 2006; Ekwall *et al.*, 2007) attempting to increase cryosurvival (Peña *et al.*, 2003, 2004a-b, 2005, 2006; Roca *et al.*, 2004), and the number of AI-doses per processed ejaculate (Bussiere *et al.*, 2000; Saravia *et al.*, 2005). Fertility post-AI is nowadays substantially better, closer to AI with liquid semen (Eriksson *et al.*, 2002), even using lower sperm numbers, albeit requiring alternative sites of sperm deposition, such as deeply intra-utero (see below) (Bathgate *et al.*, 2005; Roca *et al.*, 2006b).

IS GENDER SELECTION BY THE USE OF AI OF SPERMATOZOA SEPARATED FOR CHROMOSOMAL SEX REALISTIC?

Gender selection is, in livestock production, highly desirable (Seidel, 2003). For pigs, it would allow the production of either male or female crossbred lines, or ameliorate the imminent banning of male piglet castration in Europe. Gender selection has been tested in cattle using ET of gender-diagnosed embryos and it is also possible using AI-doses where either X- or Y-chromosome-bearing spermatozoa have been enriched to >0.95, resulting, with commercial studies, in the birth of >30,000 calves of the desired sex, with >0.95 success (Garner, 2006; Schenk and Seidel, 2007). The latter has been possible with the Beltsville Sperm Sexing Technology, which uses high speed flow cytometry sorting of DNA-stained spermatozoa (including those of pigs) based on the difference in size (and thus emitted fluorescence to a laser beam exposure) between the sex chromosomes (rev by Johnson *et al.*, 2005). A series of problems accompanies the commercial application of sex-sorted boar semen for AI; initially, the well-known low survival of the sorted spermatozoa, a matter that affects spermatozoa from all species related to the high pressure and to the extreme sperm extension applied during the process (Suh *et al.*, 2005), conveying detrimental effects of the absence of seminal plasma components (Maxwell and Johnson, 1999; Caballero *et al.*, 2004; 2006). Secondly, boar spermatozoa can only be routinely separated at ~15 million spermatozoa-bearing X- or Y-chromosomes per hour (Johnson *et al.*, 2005), a speed considered too slow for the production of doses for standard pig AI. In other words, in contrast to what is seen in cattle, sex-sorted boar semen is not commercially available for conventional, cervical AI.

The above-mentioned drawbacks of the current technology have been compensated by the use of additives to the sperm-media (mostly in the form of seminal plasma) and the growing application of deep intra-uterine AI (Johnson *et al.*, 2005). Although piglets have been successfully born after embryo transfer of IVF- or ICSI-derived embryos using sex-sorted boar spermatozoa (Abeydeera *et al.*, 1998; Rath, 2002; Probst and Rath, 2003), or after deep intra-uterine AI (Vazquez *et al.*, 2003; Rath *et al.*, 2003; Grossfeld *et al.*, 2005), the commercial application of sex-sorted boar semen is very limited, since the numbers of spermatozoa required for the application of diu-AI are still large (50-70 million) calling for even more complicated techniques such as laparoscopic intra-uterine or intra-oviduct-AI (Vazquez *et al.*, 2005). Freezing of sex-sorted boar spermatozoa has been tested and proven usable for IVF, the embryos obtained capable of establishing early pregnancies (but not carried to term) after nsET (Bathgate *et al.*, 2007). Obviously, owing to the enormous impact that sperm-mediated gender selection would have in pig production, it is hoped that further developments of the flow cytometry sex-sorting technique or alternative methods will emerge, albeit still probable far-removed from commercial use.

DOES INTRAUTERINE AI HAVE A FUTURE IN COMMERCIAL PIG BREEDING?

Commercial cervical pig AI implies the deposition of 2.5–4x10^9 "fresh" spermatozoa per AI-dose, usually extended to 70-100 mL, with 1-3 AIs per oestrus. Such sperm numbers/breeding implies an ejaculate yield of ~20-25 AI-doses. Popular boars, in particular those of high genetic merit, should often yield higher numbers of AI-doses, a matter that can only be solved by decreasing the sperm number/dose or inseminating only once per oestrus. The latter functions well with particular boars and often with the majority of sires, provided the AI to ovulation interval is optimal (usually less than 8h; Waberski *et al.,* 1994; Soede *et al.,* 1995). Such an interval can be attempted by hormonal induction of oestrus and ovulation, a procedure not recommended in commercial pig production. With non-manipulated females, the duration of the standing oestrous period can provide an estimate of the expected moment of spontaneous ovulation, since ovulation most often takes place when two-thirds of oestrous time has passed (Soede *et al.,* 2000). Although the duration of standing oestrus is highly consistent within females over several oestrous cycles and even within a farm, it varies considerably among farms (Steverink *et al.,* 1999), implying that the recordings of standing oestrus (at least twice daily) for each individual female are crucial before this procedure for AI within an optimal interval to spontaneous ovulation is attempted. Another alternative is the use of intra-uterine AI. Two procedures have been adopted for commercial use, either depositing the spermatozoa into the uterine body (the so-called post-cervical-AI, Watson and Behan, 2002) or deeper into one uterine horn (diu-AI, Martinez *et al.* 2001, 2002). Using the post-cervical-AI, numbers of 1×10^9 liquid-extended spermatozoa have been considered the threshold for this procedure (Levis *et al.,* 2002; Watson and Behan, 2002), with significant decreases in fertility (prolificacy in particular) if lower than 1×10^9 (Rozeboom *et al.,* 2004). The AI-catheter can be used both for sows or gilts. It is important to remember that there is a significant sire effect, with some sires sustaining sperm extension with maintained fertility. Such a procedure, together with competent oestrus detection in a well-managed farm, allows for AI-doses down to 1.5×10^9 spermatozoa, conventionally deposited, a figure more and more commonly seen in practice. The mechanical stimulation of the uterus might induce myometrial contractions and promote sperm transport (Rodriguez-Martinez *et al.* 2005), effects that could mimic what constituents of the seminal plasma (Waberski *et al.,* 1997) or the boar (Langendijk *et al.,* 2005) might physiologically induce during mating.

The diu-AI can be routinely performed in non-sedated sows in standing oestrus using a 180 cm-long flexible catheter that can be readily introduced into one uterine horn to deposit an AI-dose 8-55 cm from the tip of the horn (Martinez *et al.,* 2002). Both farrowing rates and litter size were comparable to conventional cervical-AI in weaned sows, whose oestrous and ovulation were hormonally-induced, even lowering both the volume of the AI-dose and the number of *liquid-extended spermatozoa* by 20 fold (Martinez *et al.,* 20001, 2002), thus suggesting the mere introduction of the catheter induces an effective sperm transport by the uterus to warrant bilateral fertilisation (Martinez *et al.,* 2005). Similar doses, diu-AI deposited during spontaneous oestrus in weaned sows, yielded comparable farrowing rates, but smaller litter sizes than controls (conventional AI), most likely owing to a longer interval onset of oestrus-to-ovulation in some of the inseminated sows, which resulted in less oocytes fertilised (see Martinez *et al.*, 2005). Although the use of diu-AI for liquid semen is probably going to be unlikely under commercial pig production, the technique becomes more interesting when cryopreserved or sex-sorted boar spermatozoa are considered, related primarily to their compromised lifespan (see above). Cryopreservation causes death of a proportion of spermatozoa and, among those surviving the procedure, a high proportion have lost fertilizing ability (Watson, 1996), owing to an irreversible deterioration of the plasma membrane, not related to a capacitation-like process (Guthrie and Welch, 2005; Saravia *et al.*, 2007). At present, deposition of $0.5-1 \times 10^9$ total *frozen-thawed boar spermatozoa* in low volumes (0.5-10mL) by diu-AI could yield levels of fertility close to controls (AI with liquid semen; Roca *et al.*, 2003). However, both in this study and follow-up trials (Bolarín *et al.*, 2006) and in programmes where low volumes (0.5 mL) of highly concentrated (1×10^9) boar spermatozoa were inseminated without post-thaw re-extension (Wongtawan *et al.*, 2006), it was clear that frozen-thawed boar spermatozoa require an AI-to-ovulation interval not longer than 8 hours, thus making peri-ovulatory AI a pre-requisite to obtain the highest possible fertility. As noted above, deep intrauterine-AI can also be used to deposit *sex-sorted boar spermatozoa*, and obtain piglets (Vazquez *et al.*, 2003; Rath *et al.*, 2003; Grossfeld *et al.*, 2005) but using numbers as high as $50-70 \times 10^6$ spermatozoa, which implies a sorting working day per sow.

The counterpart: non-surgical embryo transfer (nsET)

Despite ET being, initially, considered as a biotechnology not suited to commercial pig production, the pig industry has shown considerable interest, since it would allow dissemination of genetic resources with minimal risk of disease transmission and reduced transportation costs, compared to the movement of livestock (Wrathall and Sutmoller, 1998). Recent outbreaks of foot and mouth disease in Europe vividly illustrated the importance of embryo preservation and ET for genetic rescue in cases of loss of breeding units. The same would apply for the re-inclusion of genetic lines that have been, for any reason, separated from current production. Pig ET is, moreover, essential for the application of other reproductive biotechnologies such as embryo production *in vivo* or *in vitro*. However, the estimated numbers of transferred pig embryos during 2003 were merely ~20,000, most of them transferred as "fresh" embryos, developed *in vivo* (Thibier, 2004). This low applicability of ET is attributable to the needs of surgical collection procedures which, together with the need for hormonally manipulating oestrus and ovulation, are two concerns for animal welfare. In addition, proper preservation of pig embryos using slow-cooling methods has been difficult, since pig embryos are very sensitive to cooling.

FROM THE DEVELOPMENT TO THE TRANSFERENCE OF THE EMBRYOS

As stated above, most embryos used for ET are developed *in vivo*, following AI after a spontaneous oestrus or an oestrus where ovulation has been hormonally-induced/synchronised. In the latter case, the females acting as "donors" of embryos (morulae, un-hatched blastocysts) can either be pre-pubertal pigs, gilts, or sows. The former are usually treated with progestagens, followed by an injection of eCG 24 h after progestagen treatment is finished, while sows are treated with eCG at weaning. In gilts (and often also in sows), synchronisation of ovulation usually begins by the recording of the onset of the first oestrus followed, 15-days later, by two injections (8h-apart) of the antiluteolytic cloprostenol, followed 22 h thereafter by a single administration of eCG. Ovulation is induced 72 to 80 h later by administration of hCG. Two AIs are performed during detected oestrus, alternatively 30-40h after injection of hCG, independently of the time of onset of oestrus. Ovulation rate, numbers of collectable and transferable embryos vary largely, depending on eCG-dose, the genotype and the category of female considered. The collection of *in vivo*-developed embryos is done either post-mortem or surgically (under narcosis) by flushing of the oviducts and the upper segment (20-25 cm) of the uterine horns. Collection rate is often high (>0.80). Repeated synchronisation and surgical flushing are not advisable, mainly through issues of animal welfare, but also due to the establishment of post-surgical adhesions. Alternative collection methods, such as laparoscopy or non-surgical embryo flushing have never passed the experimental stage. As already pointed out, *in vitro* embryos are rarely used for commercial ET. The collection, cryopreservation, transport and commercial ET of *in vivo* developed pig embryos are restricted to those with an intact zona pellucida (ZP), e.g., morula and un-hatched blastocyst (International Embryo Transfer Society, IETS; Stringfellow, 1998).

EMBRYO CRYO-PRESERVATION FOR IN VIVO-DEVELOPED AND IN VITRO-PRODUCED EMBRYOS

Early stage pig embryos are rich in cytoplasmic lipids, and very sensitive to temperatures below 15°C (Wilmut, 1972), sensitivity decreasing -along with the amount of lipids, with development, towards peri-hatching blastocysts (Niimura and Ishida, 1980). Offspring has been obtained after ET of slow-frozen and thawed 2-4 cell pig embryos where these cytoplasmic lipids were removed *in vitro* (de-lipation) before cooling (Nagashima *et al.,* 1995). Recently vitrification, which enables rapid cooling by direct plunging in liquid nitrogen (LN_2) without ice being formed either in the intra- nor the extracellular medium due to a very high concentration of cryoprotectant (Rall and Fahy, 1985) appears a better alternative for long-term storage of pig embryos. Vitrification of *in vivo*-developed, ZP-intact pig embryos, where lipids were polarized by centrifugation of the blastomeres, by delipation and/or treatment with cytochalasin for cytoskeleton stabilization, has resulted, after rewarming and ET, in piglets (Dobrinsky *et al.*, 1997; 2001; Kobayashi *et al.*, 1998; Berthelot *et al.*, 2000, 2003; Cameron *et al.*, 2000; Dobrinsky *et al.*, 2000). Recently, piglets were even obtained following vitrification of delipated 4-8 cell stages of IVP, parthenogenetic embryos and ET (Nagashima *et al.*, 2007). Vitrification, usually done within 0.25 mL plastics-straws, has yielded better embryo survival post-warming when Open Pulled Straws (OPS; Vajta *et al.*, 1997), which increases the cooling rate achievable in 0.25 mL straws (2,500°C/min) by almost 8-fold (Cuello *et al.*, 2004a-b), were used, again resulting in piglets born (Berthelot *et al.*, 2000; 2001). Higher cooling-rates (>20,000°C/min) can nowadays be reached using using cryo-loops (Lane *et al.*, 1999) or with straws with a smaller inner diameter and wall thickness (the Superfine Open Pulled Straws: SOPS; Isachenko *et al.*, 2003), and by applying negative pressure to decrease the LN_2-temperature down to -210 °C (Vit-Master®), which allows for the use of lower concentrations of toxic cryoprotectant. Despite peri-hatching, blastocyst stage embryos are those better sustaining vitrification and warming with continued *in vitro* development (Dobrinsky, 2001), this particular embryo stage can not be commercially used since there is no ZP.

Nevertheless, vitrification of untreated morulae and blastocysts has resulted in high survival rates after warming (Berthelot *et al.*, 2003), especially when re-warming after SOPS is done in one stage (direct warming, a very practical solution for ET, Cuello *et al.*, 2004b), yielding live litters (Cuello *et al.*, 2005). These latter developments indicate that the procedure is now reaching maturity for commercial application (Cameron *et al.*, 2004; Martinat-Botté *et al.*, 2006). Cryopreservation of *in vitro*-produced (IVP) pig embryos -owing to differences in the cytoskeleton and the distribution of the lipid deposits- has, until recently (Esaki *et al.*, 2004), been considered as more difficult than for *in vivo*-developed, but the birth of piglets resulting from ET of IVP, transgenic pig embryos, has modified this view opening for the commercialization of highly valuable, modified genetic material (Li *et al.*, 2006).

NON-SURGICAL EMBRYO TRANSFER (nsET)

Almost 40 years ago, pregnancies could be established in pigs by a non-invasive procedure of embryo transfer (nsET, Polge and Day, 1968). However, and despite interest shown by the pig industry for this technology, it took almost 30 years before nsET could be considered as practical to place early, non-hatched embryos in the uterus, owing to the anatomical features of the female internal reproductive tract. The cervical folds and the lack of suitable instrumentation to gain access to the uterus were the first obstacle, a matter first solved in sows, with deposition of early embryos in the uterine body in un-sedated sows, 0.33 of them farrowing a mean of 6.7 piglets / litter (Hazeleger and Kemp, 1994). Fertility improved over the years (Li *et al.*, 1996; Yonemura *et al.*, 1996) to peak with 0.59 pregnancy rate and a mean of 11 30-day-old foetuses using a newly designed instrument (Hazeleger *et al.*, 2000). In a large field trial using gilts, lower farrowing rates (0.41) and fewer piglets born (mean 7.2) were obtained by the same procedure for nsET (Ducro-Steverink *et al.*, 2004), lower rates which could be due to the type of recipients, the reflux of embryos, the lack of intrauterine transport or the unsuitability of deposition in the uterine body (Hazeleger and Kemp, 2001). In the pig, once the embryos reach the uterus, they remain in the tip of the horn until Day 6, to progress thereafter through the uterine horn during Days 7 and 8 (Dziuk, 1985).

Therefore, it is logical to consider that the deposition of morulae or un-hatched blastocysts mimics better the physiological events if the ET is done in the anterior third of the uterine horn than in the uterine body, a matter already seen after surgical ET in the pig (Wallenhorst and Holtz, 1999). This rationale led to the adaptation of the catheter used for diu-AI to allow nsET deep (diu-nsET) into one uterine horn of un-restrained, non-sedated gilts and sows at days 4-6 of the oestrous cycle, obtaining –with *in vivo*-developed fresh embryos - a 0.714 farrowing rate and 6.9 piglets born (Martinez *et al.*, 2004). Shortly thereafter, the first litters were born after diu-nsET of *in vivo*-developed vitrified embryos (Cuello *et al.*, 2005), making - provided follow-up experiments show similar or better results in the field - the use of diu-nET a good tool for commercial pig breeding. In addition, IVP-embryos have been deposited via diu-nsET but only established early pregnancies (Bathgate *et al.*, 2007). Further development would increase the total numbers of nsET (steadily maintained at ~2,000 worldwide) as well as in total effectiveness (today ~0.30 of transferred embryos develop into live offspring) (Cameron *et al.*, 2006) by unravelling the endometrial response of recipient females to diu-nsET.

In vitro embryo production: a lagging technology?

Production of pig embryos by *in vitro* maturation (IVM), *in vitro* fertilisation (IVF) and *in vitro* culture (IVC) is, despite intensive research and demands for ET or delivery of transgenic pigs via somatic cloning, lagging behind other species such as cattle, where the technology is commercially applied (>100,000 IVP-embryos transferred worldwide in 2003). In contrast, of the ~30,000 pig embryos reported, only 1/3 were through IVP. Of the total, ~20,000 embryos were transferred "fresh", of which ~1,500 were IVP and ~2,000 more were cloned embryos, the majority of these ETs performed in Asia (Thibier, 2004). Most embryos are still being developed *iv vivo*, which confirms the lack of interest in using IVP-embryos for pig breeding, despite surgical collection of embryos from "donor" pigs being time-consuming, expensive and controversial. However, numbers of IVP-embryos is growing, a reflection of the better techniques for IVM and IVC that, following intensive research over the past 7-8 years, improved the ability not only to fertilise oocytes but also to produce blastocysts. Ten years ago, piglets were only obtained after the sET of 2-4 cell stage-IVP embryos, while today piglets can be produced after sET of IVP-bastocysts, and even by nsET (see Kikuchi, 2004). However, behind these positive developments, the reality is that current IVM-IVF-IVC systems need improvement if success with transgenics is to increase significantly.

Basic research is needed to unravel the peculiarities of porcine gametes and the developing embryo. Porcine oocytes and embryos are fragile, particularly sensitive to the physical and chemical particularities of the work *in vitro*, probably related to the copious amounts of cytoplasmic lipid deposits in the oocyte and the early blastomeres. Although their role is unknown, there should be a reason for such deposits, which usually diminish by the blastocyst stage. An option is the need for energy reserves for early development, before hatching occurs and the embryo communicates directly with the uterine environment. However, this speculation does not explain why other species lack these reserves, despite having a similar chronology of development. Dislocation or even removal of the lipid deposits by ultracentrifugation and de-lipation do not impair *in vitro* development, and the procedure is even considered "beneficial" for cryopreservation (Nagashima *et al.,* 1995). Despite these constraints, it is obvious that animal welfare issues, practicality and costs direct the use of embryos for research and transgenic applications towards IVP.

IN VITRO EMBRYO PRODUCTION: HINDERS AND SUCCESSES

The greatest hurdles to successful large-scale IVP or pig embryos are: (i) inefficient oocyte maturation (IVM) and fertilisation techniques, which leads to no-fertilisation or abnormal (often polyspermic) fertilisation or, poor developmental capacity of the IVF-oocytes and, (ii) sub-optimal embryo culture conditions (IVC). Selection of oocytes entering the *in vitro* production system is more difficult in pigs than in other species, owing to the variation within the so-called medium-size follicles, which in turn relate to meiotic progression, cytoplasmic maturation and subsequent competence to develop. Therefore, a period of pre-maturation culture is recommended, since it does not complicate the routine of the system, and can be done prior to or during the process of selection of the cumulus-oocyte complexes (COCs; Abeydeera, 2002). Alternatively, the use of COCs from large follicles (>5mm) rather than from small follicles from offal ovaries from sows rather than from pre-pubertal gilts, is preferred to obtain high oocyte development rates after IVF (Marchal *et al.,* 2002). Oocyte IVM is initiated after COCs are selected, usually those oocytes with >3 layers of surrounding cumulus cells that are considered essential to support the two interdependent processes of nuclear and cytoplasmic maturation. Media for IVM vary among laboratories with an increasing tendency to use chemically-defined ones. Additives are commonly used, ranging from hormones (mainly FSH and leptin; Craig *et al.*, 2005), growth factors, vitamins, energy substrates, exogenous glutathione etc to some undefined fluids, such as porcine follicular fluid. Follicular fluid, rich in hyaluronan, has a protective role against oxidative stress of the maturing oocytes which, rich in cytoplasmic lipids, should be an easy target for oxidation (Bing *et al.*, 2001; 2002; Tatemoto *et al.*, 2004; Nagai *et al.*, 2006).

With this reasoning, a reduction in partial pressure of oxygen (from 20% to 5%) during IVM would be beneficial, but these effects are still a matter of controversy among researchers (Kikuchi *et al.*, 2002; Karja *et al.*, 2004; Park *et al.*, 2005). The completion of nuclear maturation -an apparently easily achieved process with the current protocols- can be determined comparatively easily in a proportion of IVM-oocytes, while cytoplasmic maturation is not that easy to assess unless invasive methods are used (EM or glutathione content measurement). Expansion of the cumulus cells is a rather safe diagnostic tool for nuclear maturation, but not necessarily for cytoplasmic maturation. Absence occurrence of polyspermy has historically been a problem in porcine IVF, where different sperm sources (epididymal vs ejaculated), and handling (fresh- or frozen-thawed) were used. Epididymal spermatozoa, either fresh- or frozen-thawed, seem advantageous since they lack exposure to seminal plasma, but it narrows male availability/diversity. Therefore, ejaculated spermatozoa are mainly used, often frozen-thawed.

Absence of fertilisation (absence of penetration) is probably related to the spermatozoa used, that must be quality-checked before use. Since some of the preparations for IVF include washing and centrifugation or even sperm cleansing by density gradients, the spermatozoa used are often of good quality (high proportion of motile spermatozoa). Polyspermy should, under optimal conditions (e.g., >0.80 of penetration rate), occur in less than 0.10 of the IVF-oocytes. However, polyspermy is common in pig-IVF, often related to insufficient oocyte maturation, to excessive numbers of spermatozoa during IVF and even to too high sperm quality. Sperm numbers and their motility can be down-manipulated, but may risk low penetration rates. The best approach is to ensure IVM conditions are optimal, providing cytoplasmic maturation is reached (Nagai et al., 2006). Inclusion of oviductal components, such as hyaluronan, increases monospermic fertilisation (Suzuki *et al.*, 2000), especially in presence of cumulus cells (Suzuki *et al.*, 2002). In addition, other factors such as theophylline, adenosine and cysteine appear beneficial to the outcome (Yoshioka *et al.*, 2003). Embryo culture has changed in recent years, avoiding oxidative insults on the zygotes caused by Reactive Oxygen Substances (ROS), either eliminating dead cells (spermatozoa and cumulus cells) as quickly as possible or by addition of antioxidants (Kitagawa *et al.*, 2004). As well, a low O_2 concentration (5%-7%, Karja *et al.*, 2004) improved IVC, by decreasing cytoplasmatic H_2O_2 content (Kitanawa *et al.*, 2004). Currently most IVC-media are almost completely chemically defined even when beneficial additives are substances present in the oviduct, such as hyaluronan (Yoshioka *et al.*, 2004). However, and probably related to the situation *in vivo* where development beyond 4 cell-stage only occurs when the embryos enter the uterine lumen, addition of serum still has a beneficial effect in the formation and hatching of porcine blastocysts *in vitro* (Yoshioka *et al.*, 2005). Modified culture systems, such as the Well-of-the-Well (WOW, Vajta *et al.*, 2007) are being increasingly used for IVC of individual embryos. Taken together, the improvements of the different stages of IVP, has increased the proportion of blastocyst development *in vitro* over the years from <0.50 to >0.80 (Vajta *et al.*, 2007). However, only few of these embryos develop after ET to term (<0.10), implying that porcine embryo-IVP is still sub-standard. An excellent review, up-dating pig IVP-procedures, has been recently published on-line (Nagai *et al.*, 2006).

WHAT KIND OF PIGLETS DO WE WANT TO PRODUCE IN VITRO?

Since oocytes for IVP of pig embryos almost always derive from offal ovaries, they are probably not appropriate for breeding, unless ovaries came from sows with high genetic value. Apart from such very occasional situations, most IVP-embryos are produced for research or for production of genetically-modified pigs, the latter through an increased use of somatic cloning (Li *et al.*, 2006). The genetic complement of pigs, as for any other animal, undergoes continuous alteration by mutation, natural selection and genetic drift. The artificial modification of the genome can, by the correction of a specific natural mutation or the inactivation of an endogenous gene, create a so-called Knock-Out (KO) pig, in which the ability to replicate, transduce and translate information of a gene is lost (e.g., "loss of function"; Phelps *et al.*, 2003; Lai and Prather, 2004). The insertion of an exogenous transgene construct (through which it may be responsible to translate a new, foreign protein; e.g., "gain-of-function") into the nuclear genome of the developing embryo shall, ultimately, produce a transgenic piglet (Paterson *et al.*, 2003).

PRODUCTION OF GENETICALLY-MODIFIED PIGS: WHICH METHOD IS MOST EFFICIENT?

The success of DNA transfer while attempting production of genetically-modified pigs relies on its accurate insertion into the host genome, so that expression of the inserted DNA is effective and high. The major problem in the production of genetically-modified pigs is still the method used for transferring the DNA construct, trying to avoid its random insertion into the host genome and thus resulting in variable, or lack of expression (Niemann *et al.*, 2003). Among these we find: (i) the micro-injection of the transgene construct into the pronucleus of a zygote (Wall, 2002), a rather inefficient

(<0.04), costly and time-consuming method that requires high operator skills and where integration to the genome occurs at random and of a variable number of copies (Müller *et al.*, 1992). The first transgenic pigs were obtained via micro-injection 20 years ago (Hammer *et al.*, 1985) and the method is still used (Niemann *et al.*, 2005). The introduction of exogenous DNA into embryos can also be done by (ii) DNA transfer using viral-based constructs as vectors. Two types of retroviral vectors have been developed for producing transgenic animals; either derived from the genome of simple prototypic retrovirus or, the genome of more complex retroviruses (e.g., lentiviruses); the latter most often used since they are larger and can be actively transported into the interface nucleus, allowing deliverance of transgenes to non-dividing cells (rev by Niemann *et al.*, 2005; Robl *et al.*, 2007). Use of these lentiviral vectors have allowed for the efficient (up to 0.31) production of transgenic pigs, most of them (>0.90) expressing high levels of an equine infectious anemia virus (EIAV) transgene (Whitelaw *et al.*, 2004). Despite this efficiency, the lentivirus has a genome limitation, and there is an inherent risk that the random insertion can mutate the endogenous genome.

Based on a discovery by Brackett *et al.* (1971) that mammalian spermatozoa has an intrinsic ability to bind and internalise exogenous DNA, and to transfer it into the ova during IVF, the (iii) Sperm-Mediated Gene Transfer (SMGT) has resumed as a practical method to produce transgenic animals, without requiring embryo handling or expensive equipment (Lavitrano *et al.*, 1989). Although the mechanism governing foreign DNA integration is not well understood, SMGT has proven highly efficient in integrating the transgene into the genome of the pig embryo (0.57-0.80, based on ~200 generated pigs; Lavitrano *et al.*, 2006). Transgenic pig lines have been reported as being produced by SMGT since 1997 for human decay accelerating factor (hDAF, which helps to overcome the first rejection barrier in pig-to-primate transplantation models; Lavitrano *et al.*, 1997; 2002) as well as for multigene transgenic pigs, following the simultaneous introduction of 3 reporter genes (EGFP, EBFP and DsRed2; Webster *et al.*, 2005). SMGT efficiency has recently increased by ICSI of spermatozoa coated with DNA (Kurome *et al.*, 2006). However, introduction of some constructs can be impaired if sperm overload with the foreign DNA occurs (Lavitrano *et al.*, 2003).

Finally, transgenic pigs have been produced using (iv) the Somatic Cell Nuclear Transfer (SCNT) or "somatic cloning", a 3-step method which includes the enucleation of oocytes, the insertion of the donor cells (or nuclei) and the activation of the reconstructed embryo, which are subsequently cultured *in vitro* until ET is performed to a recipient mother. Park *et al.* (2001) inserted genetically modified somatic cells (by transfection of a transgene during *in vitro* culture) into an enucleated oocyte by nuclear transfer (NT) and created the first transgenic pigs. Since the success of the DNA-transfer made in the donor cells can be assessed *in vitro*, only those cells with correct integration and maximal expression are selected for NT, ensuring all offspring will be transgenic (Fujimura *et al.*, 2004). Moreover, use of targeted DNA insertion and NT, allows for multiple modifications in a single cell line (Robl *et al.*, 2007), an approach used to produce heterozygous $\alpha(1,3)$-galactosyltransferase (GT) gene KO-piglets which also express both hDAF and GnT-III (e.g., a heterozygous knockout GT-gene and two transgenes, Takahagi *et al.*, 2005). Using NT, existing transgenic animals can be multiplied without going through the germ line with its recombination events, thus allowing the transgenic trait to be found identical in the cloned offspring, expressing the transgene as much as their nuclear donor cells (Fujimura *et al.*, 2004).

Combinations of methods have also been tested, such as using frozen boar spermatozoa coated with a bicistronic gene constituted of the human albumin (hALB) and enhanced green fluorescent protein (EGFP) genes that, injected by ICSI into pig enucleated oocytes produced embryos which, after ET, resulted in pregnancies and the birth of live piglets, of which one (0.30) was transgenic. Cells lines were established from kidney, lung and muscle tissues from this founder piglet and used as donor nuclei for SCNT. The ET of the transgenic clones resulted in the birth of 6 live transgenic clones (Kurome *et al.*, 2006). Another possibility to be considered is the genetic manipulation of isolated germ line stem cells (or even embryonic germs cells, Kerr *et al.*, 2006) whose subsequent transplantation to testes would result in the production of transgenic spermatozoa (Dobrinski, 2006). Transgenesis through the male germ line has tremendous potential in pigs, a species where embryonic stem cell (ESC)-technology is yet undeveloped (although some ECS-lines have been recently obtained from pig parthenotes, Brevini *et al.*, 2007; see below), and other technologies like SCNT are resisted due to excessive manipulation. However, if immortal ESC-lines would be made commercially available for use in NT, SCNT would become the most effective method to produce transgenic pigs, since multiple modifications could be rapidly made *in vitro* into the ESC´s, allowing the introduction of genetic changes in the germ line.

PRODUCTION OF GENETICALLY-MODIFIED PIGS: WHAT FOR?

The first transgenic livestock, including pigs, were reported some 20 year ago (Hammer *et al.*, 1985) using microinjection to introduce the transgene into the embryo. Since then, transfers of foreign genes into pig embryos were undertaken using the same methodology in order to modify carcass composition by introduction of a transgene that directed the expression of human insulin-like growth factor-I (hIGF-I) specifically to striated muscle in order to obtain a leaner pig with increased growth rate. The transgene was successfully passed from the founders to the first generation, who significantly improved several carcass characteristics (particularly lower rates of fat accretion than controls), without showing gross abnormalities, phenotypic alterations or health-related problems (Pursel *et al.*, 2004). Using SNCT, pigs with fat rich in omega-3 fatty acids, a component supposedly healthy for both humans and also for pigs, have also been generated (Lai *et al.*, 2006). Sows can produce about one kg of milk per piglet/day and >0.40 of the growth rate of a developing piglet depends on this milk. Transgenic pigs were produced to express bovine α-lactalbumin (α-lac) in their milk, which resulted in 20-50% higher milk yield on days 3-9 post-partum and improved lactose content during early lactation (first parity). Such change resulted in the suckling piglets gaining more weight at 7 and 21 days post-partum than controls (Noble *et al.*, 2002). Phytase-transgenic Yorkshire pigs (Golovan *et al.*, 2001) have been produced using a bacterial transgene directed to the salivary glands. These transgenic pigs, trademarked as the "Enviropig™" (Ontario Pork, Canada), can digest feed phytate (a non-digestible form of phosphorous present in cereal diets that pigs can not digest unless exogenous phytase enzyme is added to the feed) more efficiently, substantially decreasing their faecal excretion of phosphorous into the environment. The latter ameliorates potential phosphorous pollution of local surface and ground water which increases algae and plant growth in fresh water systems such as ponds, rivers or lakes. Since an important aspect of gene transfer for transgenesis is the improvement of disease resistance, attempts were made to produce pigs that could be resistant to influenza virus infection (Müller *et al.*, 1992), a topical subject in times where avian flu is of major importance. Transgenic pigs were also produced as bioreactors for human proteins. An example is the production of human erythropoietin (hEPO) by SCNT (Lee *et al.*, 2005).

Generation of genetically-modified pigs for xenotransplantation has been a major topic since the 1990s, owing to the assumption that the pig is, anatomically and physiologically, not too different from humans. This assumption, together with known facts such as their relatively short reproductive cycles and generation intervals, the large litters delivered and the rapid growth rate shown by young pigs, added to the relatively low cost for maintenance of this domestic species, stimulated the research community to attempt producing pigs that could overcome the incompatibility of "donor" cells, tissues or organs when transplanted to humans (see reviews by Piedrahita and Mir, 2004; Zhu *et al.*, 2007). A xenograft is usually affected by several immunologically-mediated rejection mechanisms, which often overlap: the hyperacute rejection (HAR, which kills the graft in minutes or hours post-transplant), the acute vascular rejection (mixed with the above), the delayed (cellular) rejection of the receptor´s immune system and, for those xenografts that still survive, chronic rejection.

Most xenografts are usually destroyed by the first two above mechanisms, where the presence of foreign antigens in the cell surface of the donor tissue induces a rapid binding of antibodies and the activation of the complement cascade which, destroying the endothelia, causes occlusion of the vascular bed and hypoxia of the graft. In view of this chain of events, research has focused on two possibilities to prevent the hyperacute rejection: to interfere with the reaction and to eliminate the antigen. As a result, pigs transgenic for a series of factors related to HAR, mainly suppressive of the complement activation, have been produced since the early 1990s (Bucher *et al.,* 2005; Zhu *et al.,* 2007; Liu *et al.* 2007). Regarding antigen/s, the α-1,3-galactose (α-1,3Gal) epitopes (Galα-1, Galß-1, 4GlcNAc-R), present in most mammals (including pigs) but absent in New World monkeys or humans, has been considered as the most important inducer of the rejection processes following xenotransplantation. This carbohydrate structure, is synthesized by the enzyme α-1,3-galactosyltransferase (α-1, 2GT or GGTA1), whose gene has been focus for the production of Gal-knockout pigs, in the belief that the complete removal of the carbohydrate xenoantigen α-1,3Gal from the surface of the pig cells is the critical step toward the success of xenotransplantation. Thus, the gene encoding pig α-1,3-galactosyltransferase has been disrupted by homologous recombination (Tai *et al.,* 2007; Zhong, 2007), and KO-pigs have been generated via SCNT whose tissues and organs resulted in survivals following pig-to-baboon xenotransplants (Tseng *et al.,* 2005).

The α-1,3Gal sugar seems also the main antigen for the acute vascular rejection (and the following delayed cellular rejection), which occurs within a few days to weeks post-transplant, a reaction that can be delayed by immunosuppresive treatment but also by the use of organs from hDAF-transgenic pigs (Garcia *et al.,* 2004). Despite all these achievements using KO-pigs, recent concerns have been raised that the procedure might have created more problems than solutions, such as the exposure of the N-acetyllactosamine epitope that can bind natural antibodies. (Milland *et al.,* 2005). Surprisingly, Galα(1,3)Gal has been found still present in α1,3-galactosyltransferase KO-animals! (Milland *et al.,* 2005) probably due to the presence of other cover-up enzymes to synthesize Galα(1,3)Gal (Milland *et al.,* 2006); once again proving that nature has always a back-up system available....

Moreover, besides rejection or even inappropiate function of the the xenografts (after all, there are cell-signalling and physiological differences between pigs and humans, Bucher *et al.,* 2005), the greatest danger of using xenotransplantation is the possible transmission of zoonoses to the human recipient, such as of porcine endogenous retroviruses (PERV), retroviral sequences present in the pig genome which -in some *in vitro* experiments- have infected human cell lines (Patience *et al.,* 1997). With the observation that some individual pigs or even some minipig breed lines are free from PERV (Lu *et al.,* 2004), these animals are currently being used for xenotransplantation experiments. Unfortunately, other viruses such as the porcine cytomegalovirus and porcine lymphotropic herpesvirus can be cross-species transmitted (Fishman and Patience, 2004). However, since porcine cells, tissues and bioproducts (such as Langerhans islet cells, insulin, neurons, skin, extracorporeal liver, heart valves, etc) have been applied for decades to humans without evidence of infection (Paradis *et al.,* 1999) there is still controversy on this issue (Martin *et al.,* 2006; Bisset *et al.,* 2007). Such doubts, as well as an increasingly negative public opinion and the reluctance of decision-makers to address this issue have, basically, stopped the once-considered forthcoming explosion of xenotransplantation. Use of genetically-modified pigs, designed as models for human diseases is, apparently, less controversial and they are found acceptable for longitudinal studies in infections where one- or few genes cause well-defined pathologies (e.g., cystic fibrosis, retinitis pigmentosa, neurodegenerative diseases - Parkinson, Alzheimer -, skin alterations (psoriasis, epidermolysis bullosa), some forms of diabetes mellitus, arteriosclerosis, conformational diseases, breast cancer etc).

It is important to note that none of the above described genetically-modified pigs have been approved for dissemination to pig producers nor have they reached the food chain, pending approval by regulatory agencies whose requirements to cope with these new production methods are still evolving. Obviously, insertion of porcine genes for over-expression, instead of the genes from other species, would be the only way these agencies and the public would consider their entry into production and the market. Last but not least, their eventual use needs, besides the acceptance of the public- and regulation bodies, to be profitable, based on costs and expected revenues (Pratt *et al.,* 2006). The technology, particularly when somatic cloning (SCNT) has been used, has become more efficient, making profits larger for some applications (biomedical in particular) than for others (i.e. carcass modifications).

Somatic cloning: advantages and problems

The first piglet born after NT was produced using blastomeres of 4-cell stage embryos as donors (Prather *et al.,* 1989) and, thereafter. SCNT has enabled production of piglets using somatic cells from foetal and adult porcine tissues, using either *in vivo-* or *in vitro-*matured oocytes, to where the donors were transferred by electrofusion or injection (Onishi *et al.,* 2000; Polejaeva *et al.,* 2000; Betthauser *et al.,* 2000). Somatic cloning of pigs can be used to preserve genetic resources and the rescue of species, when no oocytes or spermatozoa are available to produce embryos *in vivo* or *in vitro*. It has also been linked to several potential application areas, from embryological research (including differentiation and senescence) to biomedicine (production of animal models for human diseases, xenotransplantation, or bio-reactors) or the improvement of pig conformation and health. However, it is particularly interesting for the large-scale production of transgenic founder pigs (see above) which, being mostly homozygous for the respective transgene, can generate transgenic lines by conventional breeding, avoiding time-consuming back-crossing. The frequency of developmental anomalies from established pregnancies with NT-cloned pig embryos is lower than that in cattle or sheep (Santos and Dean, 2004) although decreased survival, infections and cardiac problems have been seen (Carter *et al.,* 2002) together with syndromes specific to pigs, such as contracted tendons and enlarged tongue and kidneys (Phelps *et al.,* 2003; Prather *et al.*, 2004). On the other hand, as seen in other species, the offspring of the founder clones are almost always healthy (Martin *et al.*, 2004) likely because any abnormalities related to epigenetic alterations seem to be erased when cell nuclei go through the germ line gametogenesis.

However, the overall efficiency of porcine SCNT is still very low (<0.02), despite the increasing number of attempts done to produce live piglets. Few changes have been done on the original SCNT method *per se*, with the exception of the overall improvement of the IVM and IVC procedures, the activation of the oocyte, or the zona-free handling of cytoplasts where micromanipulation is waived for enucleation and fusion, gathered under the name "handmade cloning" (HMC; see review by Vajta *et al.*, 2007). More than 0.90 of the cells used as donors are foetal-derived, 0.80 of the recipient oocytes are *in vitro*-matured (mostly from sow ovaries), the embryos obtained are transferred at 1- to 4-8 cell stages in large numbers per recipient pig (50-150 or even larger numbers) leading to pregnancy rates as high as 1.00. Despite all these improvements, less than 200 viable cloned piglets have been produced, the largest litter being 9 piglets (Vajta *et al.*, 2007), thus calling for further research on the way these NT-produced embryos develop following ET.

Isolation of stem cells: any breakthrough yet?

Stem cells are characterised as either totipotent, pluripotent, multipotent or monopotent depending on the number of distinct cell types they can give rise to, following differentiation. Examples of totipotent stem cells are the first few blastomeres in the course of early embryo development (which can differentiate in any cell type), pluripotent stem cells are those contained in the inner cell mass (ICM) of the blastocyst (which can differentiate into the primitive cell layers and the cell lines of the organism); while those multi/monopotent can be found among adult (tissue/organ specific) stem cells, which have the capacity to renew themselves (self-renewal) as well as the ability to generate differentiated cells. Stem cells, with either pluri-, multi- or monopotentiality, can be isolated from embryos or adult tissues. Isolated ICM-cells give rise to the so-called embryonic-stem cells (ESC) which can be maintained *in vitro* (ESC lines) and, under the influence of different factors and conditions can also *in vitro* differentiate in certain tissue direction. The two cell types, ICM and ECS are, however, not equivalent, since the ICM only exist transiently in the embryo and does not act as a stem cell compartment *in vivo* while the ECSs form a stable cell line *in vitro*. ECS are, therefore, to be considered the result of a selection and adaptation process to the culture environment, an artefact, actually. Despite a lot of effort expended in obtaining ECSs from pigs, not until recently (Brevini *et al.*, 2005) have ECS lines been etablished. The cell lines have been stable, undifferentiated for more than 2 years in culture, but could be induced to spontaneous differentiation (Brevini *et al.*, 2007). Recently, multipotent self-renewing stem cells have been established using peripheral blood mononuclear cells from adult green-fluorescent protein transgenic pigs. Under specific conditions, the cells have shown ability to differentiate into angio-, osteo-, adipo- and neurogenic phenotypes and have been designated blood-derived multipotent adult progenitor cells (PDB-MAPCs) and, given their proven longevity as primordial cells, they appear as a given alternative to embryonic stem cells for research and eventual cell therapy (Price *et al.*, 2006).

Would reproductive biotechnologies benefit of current advances in pig genomics?

Human-driven selection of pigs for desirable traits (or against undesirable traits) has been practiced since domestication started, to increase their health and usefulness, ultimately increasing production profits. Being a pluriparous species, profit is obviously related to increases in litter size, even moderate ones. Selection programmes have moved from being largely based on phenotypical traits (which are laborious, expensive and especially time-consuming) towards use of Marker Assisted Selection (MAS) which, used together with traditional methods for selection, have accelerated selection for various traits such as fertility. Implementation of MAS requires the identification of genetic markers such as the Quantitative Trait Loci (QTLs) which associates the region of a chromosome with a phenotypical performance trait. QTLs have been mostly been determined using genome scans (with anonymous DNA markers such as microsatellites) but, because at least 3 generations are required, QTL-analyses are time-consuming and costly (Buske *et al.*, 2006).

Recently, when recombinant DNA technology developed, it has been possible to isolate single genes, analyse them, modify their nucleotide structure, copy them and even of deleting or inserting these copies into the genome of the same or of other species. Currently, analyses of the pig genome (the entire set of genes encoded by its DNA), its transcriptome (the entire complement of mRNA transcripts from the genome) and its proteome (e.g., the entire complement of proteins expressed at a single point in time; Morrison *et al.*, 2002) allows for the simultaneous investigation of the genetic basis (including regulatory steps) of entire biochemical pathways which are reflected in phenotypic traits of interest for pig selection (the hormonal modulation of ovulation, for instance). Compared to QTL-analyses, candidate gene analyses, where only one or few genes are analysed, are less expensive provided these are well characterized and available. As for other livestock, pig genomic research has followed in the footsteps of the Human Genome Project (HUGO; Andersson and Georges, 2004), leading to the European PiGMaP initiative and the USDAS-MARC, developing linkage maps that could be expanded for trait mapping. As well, somatic cell genetics, radiation-hybrid (RH)-panel mapping, and the building up of Expression Sequence Tag (EST)-libraries have been developed (see the review by Rothschild, 2004). Following the construction of a BAC map (http://www.genomic.iastate.edu/newsletter/PigWhitePaper.html, NHGRI), a multi-national whole-genome sequencing is being carried out (Swine Genome Sequencing Consortium, http://www.animalgenome.org/pigs/genomesequence) in parallel with the Sino-Danish Pig Genome Project (Wernersson *et al.*, 2005). Functional genomic analysis research in pigs will lead to a better understanding of the complexity of the pig transcriptome, by expression- and other functional gene analyses, using porcine tissues only (Caetano *et al.*, 2005), efforts that have led to the construction of cDNA microarrays (holding >3,000 elements), for instance towards the porcine intestinal mucosa (Dvorak *et al.*, 2005; Niewold *et al.*, 2005). Such knowledge on functional genomics, transcriptomics and proteomics has not only facilitated the development of gene KO- or transgenic animals (see above) but also enhanced our possibilities to implement MAS by integrating the outputs (e.g. genome markers) arisen from the early linkage maps and/or from functional genomics, with traditional genetic approaches (Rothschild, 2004; Gerrits *et al.*, 2005; Womack, 2005).

The number of pig genetic markers is impressive (Rothschild, 2004) with several hundreds of them associated with performance traits, revealing QTLs for virtually every important performance trait in pigs, from body composition to reproduction (Gerrits *et al.*, 2005). This significant number of QTLs has been reported on nearly all chromosomes for coat colour, growth, carcass and meat quality traits and on several chromosomes for reproduction (Marklund *et al.*, 1999; Bidanel and Rothschild, 2002; Rothschild, 2004; Buske *et al.*, 2006; de Oliveira-Peixoto *et al.*, 2006; Lin *et al.*, 2006; van Wijk *et al.*, 2006; van Wijk *et al.*, 2007). The ultimate marker for MAS is, of course, the mutation underlying the selected phenotype. However, very few mutations underlying quantitative trait variation have been identified, and thus few of these QTLs have been successfully transferred to the commercial pig industry as genetic tests (Evans *et al.*, 2003), particularly those regarding reproduction traits (Rohrer *et al.*, 2006). Fertility in pigs is one of the most difficult and complex traits owing to low heritabilities, long generation intervals, the polygenic nature of reproductive traits and the strong environmental influences on reproduction. In fact, thus far, no gene with a causative mutation has been identified, which underlies a detected QTL effect concerning reproductive traits in pigs. For pig fertility, a polygenic trait with low heritability, many genes account only for a small amount of the phenotype variance, the rest being influenced by the environment.

Once more, the costs of assaying the actual variation in the DNA sequence causing the difference in performance marked by the QTL are still too high for the majority of commercial pig breeders to afford them. Although steadily increasing, the proportion of pig genes that have been mapped is still small and, therefore, the number of positional candidate genes for analyses is currently limited. Moreover, most gene variants have usually been studied in a reference population, under research conditions while few have used commercial farm populations. This has led to the determination that variant alleles were solely present in the studied population under research, but not necessarily present in the general pig population (Gerrits *et al.,* 2005). Despite this discouraging panorama, some fertility-traits have been found as potentially important for introduction in MAS; such as gestation length and the number of stillborn piglets, while other markers have not been found to be valuable, or are questionable (Buske *et al.,* 2006). Litter size, an easy-to-record trait of the highest economic impact for the breeder, is influenced by several traits (as uterine capacity, ovulation and embryo viability) rather than under the control of a single gene. As such, many genes have been linked to this trait, such as the leptin gene (SEPR), the oestrogen receptor gene (ESR), the Follicle stimulating hormone gene beta (FSHb) or the retinol-binding protein-4 gene (RBP4;Buske *et al.,* 2006). In summary, the results obtained thus far indicate that although reproductive performance (fertility) is controlled by the genetic make-up of the boar, dam and offspring, it is mostly affected by environment, a matter very often disregarded when discussing the modulation of reproduction.

In any case, several genetic markers have been used to test the pig susceptibility to malignant hyperthermia (known as the porcine stress syndrome (Fujii *et al.,* 1991); others are currently used for some production traits such as meat quality (Milan *et al.,* 2000), growth and leanness (Jeon *et al.,* 1999) and female reproduction (Rothschild *et al.,* 1996), and many more shall appear. A very different picture is seen when genetic markers are pursued to identify disease resistance genes, mostly due to the lack of evidence that single genes encode resistance to micro-organisms or to effective immunity status. Most likely, complex combinations of genes are involved, something that leads to difficulties in using simple testing (Edfors-Lilja *et al.,* 1998; Wattrang *et al.,* 2005). It is possible that variation in the ability of a pig to respond immunologically to an infection correlates with pig health status and, provided sufficiently reliable pig health markers are identified, this approach could be used for selection in conventional breeding. Most likely, the identification of porcine QTLs for disease resistance that would allow for the selection of animals with innate resistance to pathogens will be important in the near future. Propagation of their genes is, most likely, to be expanded by the use of sound reproductive biotechnologies such as AI initially, but also by nsET.

Reproductive biotechnologies in pigs: which will develop further?

The current economical return on any investment in pig production is, at present, quite low. Therefore, intelligent use of reproductive biotechnologies needs to focus in providing pig producers with long-lasting technologies of low price and sustainable results. Obviously, AI is the major technology and, since it has become common practice in commercial pig production, it makes the maintenance and selection of stud boars profitable. It is already currently possible to genotype stud boars for selected genetic markers and provide a return on investment by selling the semen doses of those boars not having undesirable alleles, common in the breeding population to which the AI-semen is directed. Producers with elite breeding herds would be willing to buy semen from these selected, genotyped boars with two good alleles of a particular gene since they can, using AI, reduce the frequency of an undesirable allele in their population by 50% each generation. Once the elite breeding herd is free from this undesirable allele, such practice of MAS increases the effects (and revenues) when semen from these elite stud boars is used in multiplier and commercial herds. Associated with gender selection, the adoption of MAS at stud sire herds appears as very, very promising, especially if the prediction of spontaneous ovulation could be made reality.

As second choice, the use of nsET should be underlined, a practice that will increase as a possibility, partly to reinforce the above dissemination of genotyped ancestry and – perhaps - of genetically-modified (albeit non-transgenic) individuals, but mostly as an essential tool for the development and use of gene-bank storage; to be used in cases of disease outbreaks, ensuring repopulation of genetics. Obviously, this form of genetic dissemination relies up on the continued success of embryo cryopreservation. Recent advances in cryopreservation and nsET of pig embryos have made ET in pigs a commercially viable technology especially for the international transfer of valuable genetic material. Unfortunately, there is still a long way to go for the commercial application of *in vitro*-produced pig embryos, owing to some yet unsolved problems during IVM and IVC, but mostly for the difficulties in lowering the inherent costs involved. The reason for this moderate need to apply new embryo technologies in pig breeding, compared to other species, seems obvious; pigs have short reproductive cycles and large litters. Therefore, a relatively rapid genetic advancement is still possible using AI combined with traditional breeding.

Acknowledgements

The studies reported have been made possible by grants from FORMAS, formerly the Swedish Council for Research in Forestry and Agriculture (SJFR), and the Swedish Farmer´s Foundation for Agricultural Research (SLF), Stockholm, Sweden.

References

Abeydeera, L.R. (2002) *In vitro* production of embryos in swine. *Theriogenology*, **57**, 257-273.
Abeydeera, L.R., Johnson, L.A., Welch, G.R., Wang, W.H., Boquest, A.C., Cantley, T.C., Rieke, A. and Day, B.N. (1998) Birth of piglets preselected

for gender following *in vitro* fertilization of *in vitro* matured pig oocytes by X and Y chromosome bearing spermatozoa sorted by high speed flow cytometry. *Theriogenology*, **50**, 981-988.

Andersson, L. and Georges, M. (2004) Domestic animal genomics: Deciphering the genetics of complex traits. *National Reviews of Genetics*, **5**, 202-212.

Bathgate, R., Eriksson, B., Maxwell, W.M.C. and Evans, G. (2005) Low dose deep intrauterine insemination of sows with fresh and frozen-thawed spermatozoa. *Theriogenology*, **63**, 553-554.

Bathgate, R., Morton, K.M., Eriksson B.M., Rath, D., Seig, B., Maxwell, W.M. and Evans, G. (2007) Non-surgical deep intra-uterine transfer of *in vitro* produced porcine embryos derived from sex-sorted frozen-thawed boar sperm. *Animal Reproduction Science*, **99**, 82-92.

Berthelot, F., Martinat-Botté, F., Locatelli, A., Perreau, C. and Terqui, M. (2000) Piglets born after vitrification of embryos using the Open Pulled Straw method. *Cryobiology*, **41**,116-124.

Berthelot, F., Martinat-Botté, F., Perreau, C. and Terqui, M. (2001) Birth of piglets after OPS vitrification and transfer of compacted morula stage with intact zona pellucida. *Reproduction, Nutrtion et Development*, **41**, 267-272.

Berthelot, F., Martinat-Botté, F., Vajta, G. and Terqui, M. (2003) Cryopreservation of porcine embryos: state of the art. *Livestock Production Science*, **83**, 73-83.

Betthauser, J., Forsberg, E., Augenstein, M., Childs, L., Eilertsen, K., Enos, J., Forsythe, T., Golueke, P., Jurgella, G., Koppang, R., Lesmeister, T., Mallon, K., Mell, G., Misica, P., Pace, M., Pfister-Genskiw, M., Strelchenko, N., Voelker, G., Watt, S., Thompson, S. and Bishop, M. (2000) Production of cloned pigs from *in vitro* systems. *Nature Biotechnology*, **18**, 1055-1059.

Bidanel, J.P. and Rothschild, M.F. (2002) Current status of quantitative trait locus mapping in pigs. *Pig News and Information*, **23**, 39N-53N.

Bing, Y.Z., Hirao, Y., Iga, N., Che, L.M., Takenouchi, N., Rodriguez-Martinez, H. and Nagai, T. (2002) *In vitro* maturation and glutathione synthesis of porcine oocytes matured in the presence or absence of cysteamine under different oxygen concentrations: role of cumulus cell. *Reproduction, Fertility and Development*, **14**, 125-131.

Bing, Y.Z., Nagai, T. and Rodriguez-Martinez, H. (2001) Effects of cysteamine, FSH and oestradiol-17ß on *in vitro* maturation of porcine oocytes. *Theriogenology*, **55**, 867-876

Bisset, L.R., Boni, J., Lutz, H., Schubach, J. (2007) Lack of evidence for PERV expression after apoptosis-mediated horizontal gene transfer between porcine and human cells. *Xenotransplantation*, **14**, 13-24.

Bolarín, A., Roca, J., Rodriguez-Martinez, H., Hernandez, M., Vazquez, J.M. and Martinez, E.A. (2006) Dissimilarities in sows' ovarian status at the insemination time could explain differences in fertility between farms when frozen-thawed semen is used. *Theriogenology*, **65**, 669-680.

Brackett, B.G., Baranska, W., Sawichi, W. and Korprowski, H. (1971) Uptake of heterologous genome by mammalian spermatozoa and its transfer to ova through fertilization. *Proceedings of the National Academy of Sciences, USA*, **68**, 353-357.

Brevini, T.A.L., Cillo, F. and Gandolfi, F. (2005) Establishment and molecular characterization of pig parthenogenetic embryonic stem cells. *Reproduction, Fertility and Development*, **17**, 235-239.

Brevini, T.A.L., Tosetti, V., Crestan, M., Antonini, S. and Gandolfi, F. (2007) Derivation and characterization of pluripotent cell lines from pig embryos of different origins. *Theriogenology*, **67**, 54-63.

Bucher, P., Morel, P. and Bühler, L.H. (2005) Xenotransplantation: an update on recent progress and future perspectives. *Transplantation International*, **18**, 894-901.

Buske, B., Sternstein, I. and Brockman, G. (2006) QTL and candidate genes for fecundity in sows. *Animal Reproduction Science*, **95**, 167-183.

Bussiere, J.F., Bertaud, G. and Guillouet, P. (2000) Conservation of boar semen by freezing. Evaluation *in vitro* and after insemination. *32emes Journees de la Recherche Porcine en France*, **32**, 429-432.

Caballero, I., Vazquez, J.M., Centurion, F., Rodriguez-Martinez, H., Parrilla, I., Roca, J., Cuello, C. and Martinez, E. (2004) Comparative effects of autologous and homologous seminal plasma on the viability of largely extended boar spermatozoa. *Reproduction in Domestic Animals*, **39**, 370-375.

Caballero, I., Vazquez, J.M., Garcia, E.M., Roca, J., Martinez, E.A., Calvete, J.J., Sanz, L., Ekwall, H. and Rodriguez-Martinez, H. (2006) Immunolocalization and possible functional role of PSP-I/PSP-II heterodimer in highly-extended boar spermatozoa. *Journal of Andrology*, **27**, 766-773.

Caetano, A.R., Edeal, J.B., Burns, K., Johnson, R.K., Tuggle, C.K. and Pomp, D. (2005) Physical mapping of genes in the porcine ovarian transcriptome. *Animal Genetics*, **36**, 322-330.

Cameron, R.D.A., Beebe, L.F. and Blackshaw, A.W. (2006) Cryopreservation and transfer of pig embryos. *Society Reproduction and Fertilility Supplemments*, **62**, 277-291.

Cameron, R.D.A., Beebe, L.F.S., Blackshaw, A.W., Higgins, A. and Nottle, M.B. (2000) Piglets born from vitrified early blastocysts using a simple technique. *Australian Veterinary Journal*, **78**, 195-196.

Cameron, R.D.A., Beebe, L.F.S., Blackshaw, A.W. and Keates, H.L. (2004) Farrowing rates and litter size following transfer of vitrified porcine embryos into a commercial swine herd. *Theriogenology*, **61**, 1533-1543.

Carter, D.B., Lai, L., Park, K.W., Samuel, M., Lattimer, J.C., Jordan, K.R., Estes, D.M., Besch-Williford, C. and Prather, R.S. (2002) Phenotyping of transgenic cloned pigs. *Cloning and Stem Cells*, **4**, 131-145.

Craig, J.A., Zhu, H., Dyce, P.W., Wen, L. and Li, J. (2005) Leptin enhances porcine preimplantation embryo development *in vitro*. *Molecular and Cellular Endocrinology*, **229**, 141-147.

Cuello, C., Berthelot, F., Martinat-Botte, F., Venturi, E., Guillouet, P., Vazquez, J.M., Roca, J. and Martinez, E.A. (2005) Piglets born after non-surgical deep intrauterine transfer of vitrified blastocysts in gilts. *Animal Reproduction Science*, **85**, 275-286.

Cuello, C., Gil, M.A., Parrilla, I., Tornel, J., Vazquez, J.M., Roca, J., Berthelot, F., Martinat-Botte, F. and Martinez, E.A. (2004a) Vitrification of porcine embryos at various developmental stages using different ultra-rapid cooling procedures. *Theriogenology*, **62**, 353-361.

Cuello, C., Gil, M.A., Parrilla, I., Tornel, J., Vazquez, J.M., Roca, J., Berthelot, F., Martinat-Botte, F. and Martinez, E.A. (2004b) *In vitro* development following one-step dilution of OPS-vitrified porcine blastocysts. *Theriogenology*, **62**, 1144-1152.

De Ambrogi, M., Ballester, J., Saravia, F., Caballero, I., Johannisson, A., Wallgren, M., Andersson, M. and Rodriguez-Martinez, H. (2006) Effect of storage in short- and long-term commercial semen extenders on the motility, plasma membrane, and chromatin integrity of boar spermatozoa. *International Journal of Andrology*, 29, 543-552.

De Oliveira Peixoto, J., Facioni Guimaraes, S.E., Savio Lopes, P., Menck Soares, M.A., Vieira Pires, A., Gualberto Barbosa, M.V., de Almeida Torres, R. and de Almeida e Silva, M. (2006) Associations of leptin gene polymorphisms with production traits in pigs. Journal of *Animal Breeding and Genetics*, **123**, 378-283.

Delgado, C., Rosengrant, M., Steinfeld, H., Ehui, S. and Courbois, C. (1999) Livestock to 2020: the next food revolution. Rome Italy: *UN FAO*, 72 pp.

Dobrinski, I. (2006) Germ cell transplantation in pigs-advances and applications. *Society Reproduction and Fertility Supplements*, **62**, 331-339.

Dobrinsky, J.R. (2001) Cryopreservation of pig embryos: adaptation of vitrification technology for embryo transfer. *Reproduction*, **58**, 325-333.

Dobrinsky, J.R., Nagashima, H., Pursel, V.G., Schreier, L.L. and Johnson, L.A. (2001) Cryopreservation of morula and early blastocyst stage swine embryos: Birth of litters after transfers. *Theriogenology*, **55**, 303.

Dobrinsky, J.R., Pursel, V.G., Long, C.R. and Johnson, L.A. (2000) Birth of piglets after transfer of embryos cryopreserved by cytoskeletal stabilization and vitrification. *Biology of Reproduction*, **62**, 564-570.

Dobrinsky, J.R. (1997) Cryopreservation of pig embryos. *Journal of Reproduction and Fertility*, **52**, 301-312.

Ducro-Steverink, D.W.B., Peters, C.G.W., Maters, C.C., Hazeleger, W. and Merks, J.W.M. (2004) Reproduction results and offspring performance after non-surgical embryo transfer in pigs in pigs. *Theriogenology*, **62**, 522-531.

Dvorak, C.M.T., Hyland, K.A., Machado, J.G., Zhang, Y., Fahrenkrug, S.C. and Murtaugh, M.P. (2005) Gene discovery and expression profiling in porcine Peyer´s patch. Veterinary *Immunology and Immunopathology*, **105**, 301-315.

Dziuk, P. (1985) Effect of migration, distribution and spacing of pig embryos on pregnancy and fetal survival. *Journal of Reproduction and Fertility*, **33**, 57-63.

Edfors-Lilja, I., Wattrang, E., Marklund, L., Moller, M., Andersson-Eklund, L., Andersson, L., Andersson, L. and Fossum, C. (1998) Mapping quantitative trait loci for immune capacity in the pig. *Journal of Immunology*, **161**, 829-835.

Ekwall, H., Hernandez, M., Saravia, F. and Rodriguez-Martinez, H. (2007) Cryo-scanning electron microscopy (Cryo-SEM) of boar semen frozen in medium-straws and MiniFlatPacks. *Theriogenology* (in press).

Eriksson, B.M. and Rodriguez-Martinez, H. (2000) Effect of freezing and thawing rates on the post-thaw viability of boar spermatozoa frozen in large 5 ml packages (FlatPack). *Animal Reproduction Science*, **63**, 205–220.

Eriksson, B.M., H. Petersson and Rodriguez-Martinez, H. (2002) Field fertility with exported boar semen frozen in the new FlatPack container. *Theriogenology*, **58**, 1065-1079.

Esaki, R., Ueda, H., Kurome, M., Hirakawa, K., Tomii, R., Yoshioka, H., Ushijima, H., Kurayama, M. and Nagashima, H. (2004) Cryopreservation of porcine embryos derived from *in vitro*-matured oocytes. *Biology of Reproduction*, **71**, 432-437.

Evans G.J., Giuffra, E., Sanchez, A., Kerje, S., Davalos, G., Vidal, O., Illan, S., Noguera, J.L., Varona, L., Velander, I., -Southwood, O.I., de Koning, D.J., Haley, C.s., Plastow, G.S. and Andersson, L. (2003) Identification of quantitative trait loci for production traits in commercial pig populations. *Genetics*, **164**, 621-627.

Fishman, J.A. and Patience, C. (2004) Xenotransplantation: infectious risk revisited. *American Journal of Transplantation*, **4**, 1383-1390.

Fujii, J., Otsu, K., Zorzato, F., de Leon, S., Khanna, V.K., Weiler, J.E., O´Brien, P.J. and MacLennan, D.H. (1991) Identification of a mutation in porcine ryanodine receptor associated with malignant hyperthermia. *Science*, **253**, 448-451.

Fujimura, T., Kurome, M., Murakami, H., Takahagi, Y., Matsunami, K., Shimanuki, S., Suzuki, K., Miyagawa, S., Shirakura, R., Shigehisa, T. and Nagashima, H. (2004) Cloning of the transgenic pigs expressing human decay accelerating factor and N-acetylglucosaminyltransferase III. *Cloning and Stem Cells*, **6**, 294-301.

Garcia, B., Sun, H.T., Yang, H.J., Chen, G. and Zhong, N. (2004) Xenotransplantation of human decay accelerating factor porcine kidney to non-human primates: 4 years experience at a Canadian center.

Transplantation Proceedings, **36**, 1714-1716.

Garner, D.L. (2006) Flow cytometric sexing of mammalian sperm. *Theriogenology*, **65**, 943-957.

Gerrits, R.J., Lunney, J.K., Johnson, L.A., Pursel, V.G., Kraeling, R.R., Rohrer, G.A. and Dobrinsky, J.R. (2005) Perspectives for artificial insemination and genomics to improve global swine populations. *Theriogenology*, **63**, 283-299.

Gil, M.A., Roca, J., Cremades, T., Hernandez, M.,Vazquez, J.M., Rodriguez-Martinez, H. and Martinez, E.A. (2005) Does multivariate analysis of post-thaw sperm characteristics accurately estimate *in vitro* fertility of boar individual ejaculates? *Theriogenology*, **64**, 305-316.

Golovan, S.G., Meidinger, R.G., Ajakaiye, A., Cotrill, M., Wiederkehr, M.Z., Barney, D.J., Plante, C., Pollard, J.W., Fan, M.Z., Hayes, M.A., Laursen, J., Hjorth, J.P., Hacker, R.R., Phillips, J.P. and Forsberg, CW. (2001) Pigs expressing salivary phytase produce low-phosphorous manure. *Nature Biotechnology*, **19**, 741-745.

Grossfeld, R., Klinc, P., Sieg, B. and Rath, D. (2005) Production of piglets with sexed semen employing a non-surgical insemination technique. *Theriogenology*, **63**, 2269-2277.

Guthrie, H.D. and Welch, G.R. (2005) Impact of storage prior to cryopreservation on plasma membrane function and fertility of boar sperm. *Theriogenology*, **63**, 396-410.

Hammer, R.E., Pursel, V.G., RExroad, C.E.jr., Wall, R.J., Bolt, D.J., Ebert, K.M., Palmiter, R.D. and Brinster, R.L. (1985) Production of transgenic rabbits, sheep and pigs by microinjection. *Nature*, **315**, 680-683.

Hazeleger, W., Bouwman, E.G., Noordhuizen, J.P.T.M. and Kemp, B. (2000) Effect of superovulation induction on embryonic development on day 5 and subsequent development and survival after nonsurgical embryo transfer in pigs. *Theriogenology*, **53**, 1063-1070.

Hazeleger, W. and Kemp, B. (1994) Farrowing rate and litter size after transcervical embryo transfer in sows. *Reproduction in Domestic Animals*, **29**, 481-487.

Hazeleger, W. and Kemp, B. (2001) Recent developments in pig embryo transfer *Theriogenology*, **56**, 1321-1333.

Hernandez, M., Ekwall, H., Roca, J., Vazquez, J.M., Martinez, E. & H. Rodriguez-Martinez (2007) Cryo-scanning electron microscopy (Cryo-SEM) of semen frozen in medium-straws from good and sub-standard freezer AI-boars. *Cryobiology*, **54**, 63-70.

Hernandez, M., Roca, J., Ballester, J., Vazquez, J.M., Martinez, E.A., Johannisson, A., Saravia, A. and Rodriguez-Martinez, H. (2005) Differences in SCSA outcome among boars with different sperm freezability. *International Journal of Andrology*, **29**, 583-591.

Holt, W.V, Medrano, A., Thurston, L.M. and Watson, P.F. (2005) The significance of cooling rates and animal variability for boar sperm cryopreservation:

insights from the cryomicroscope. *Theriogenology*, **63**, 370-382.

Holt, W.V. (2000) Fundamental aspects of sperm cryobiology: the importance of species and individual differences. *Theriogenology*, **53**, 47-58.

Isachenko, V., Folch, J., Isachenko, E., Nawroth, F., Krivokharchenko, D., Vajta, G., Dattena, M. and Alabart, J.L. (2003) Double vitrification of rat embryos at different developmental stages using an identical protocol. *Theriogenology*, **60**, 445-452.

Jeon, J.T., Carlborg, O., Tornsten, A., Giuffra, E., Amarger, V., Chardon, P., Andersson-Eklund, L., Andersson, K., Hansson, I., Lundström, K. and Andersson, L. (1999) A paternally expressed QTL affecting skeletal and cardiac muscle mass in pigs maps to the IGF2 locus. *Nature Genetics*, **21**, 157-158.

Johnson, L.A., Rath, D., Vazquez, J.M., Maxwell, W.M.C. and Dobrinsky, J.R. (2005) Preselection of sex of offspring in swine for production: current status of the process and its application. *Theriogenology*, **63**, 615-624.

Johnson, L.A., Weitze, K.F., Fiser, P. and Maxwell, W.M. (2000) Storage of boar semen. *Animal Reproduction Science*, **62**, 143-72.

Karja, N.W., Wongsrikeao, P., Murakami, M., Agung, B., Fahrudin, M., Nagai, T. and Otoi, T. (2004) Effects of oxygen tension on the development and quality of porcine *in vitro* fertilized embryos. *Theriogenology*, **62**, 1585-1595.

Kerr, C.L., Gearhart, J.D., Elliot, A.M. and Donovan, P.J. (2006) Embryonic germ cells: when germ cells become stem cells. *Seminars in Reproductive Medicine*, **24**, 304-313.

Kikuchi, K (2004) Developmental competence of porcine blastocysts produced *in vitro*. *Journal of Reproduction and Development*, **50**, 21-28.

Kikuchi, K., Onishi, A., Kashiwazaki, N., Iwamoto M., Noguchi J., Kaneko H., Akita, T. and Nagai, T. (2002) Successful piglet production after transfer of blastocysrs produced by a modified *in vitro* system. *Biology of Reproduction*, **66**, 1033-1041.

Kitagawa, Y., Suzuki, K., Yoneda, A. and Watanabe, T. (2004) Effects of oxygen concentration and antioxidants on the *in vitro* developmental ability, production of reactive oxygen species (ROS), and DNA fragmentation in porcine embryos. *Theriogenology*, **62**, 1186-1197.

Kobayashi, S., Takei, M., Kano, M., Tomita, M. and Leibo, S.P. (1998) Piglets produced by transfer of vitrified porcine embryos after stepwise dilution of cryoprotectants. *Cryobiology*, **36**, 20-31.

Kumar, S., J.D. Millar and Watson, P.F. (2003) The effect of cooling rate on the survival of cryopreserved bull, ram, and boar spermatozoa: a comparison of two controlled-rate cooling machines, *Cryobiology*, **46**, 246-253.

Kurome, M., Ueda, H., Tommii, R., Naruse, K. and Nagashima, H. (2006) Production of transgenic-clone pigs by the combination of ICSI-mediated gene transfer with somatic cell nuclear transfer. *Transgenic Research*, **15**, 229-240.

Lai, L., Kang, J.X., Li, R., Wang, J., Witt, W.T., Yong, H.Y., Hao, Y., Wax, D.M., Murphy, C.N., Rieke, A., Samuel, M., Linville, M.L., Korte, S.W., Evans, R.W., Starzl, T.E., Prather, R.S. and Dai, Y. (2006) Generation of cloned transgenic pigs rich in omega-3 fatty acids. *Nature Biotechnology*, **24**, 435-436.

Lai, L. and Prather, R.S. (2004) Creating genetically modified pigs by using nuclear transfer. *Reproductive Biology and Endocrinology*, **1**, 82.

Lane, M., Schoolcraft, W.B., Gardner, D.K. (1999) Vitrification of mouse and human blastocysts using a novel cryoloop container-less technique. *Fertility and Sterility*, **72**, 1073-1078.

Langendijk, P., Soede, N.M. and Kemp, B. (2005) Uterine activity, sperm transport, and the role of boar stimuli around insemination in sows. *Theriogenology*, **63**, 500-513.

Lavitrano, M., Bacci, M.L., Forni, M., Lazzereschi, D., DiStefano, C., Fioretti D, Giancotti P, Marfe G, Pucci L, Renzi L, Wang H, Stoppacciaro A, Stassi G, Sargiacomo M, Sinibaldi P, Turchi V, Giovannoni R, Della Casa G, Seren E and Rossi G (2002) Efficient production by sperm-mediated gene transfer of human decay accelerating factor (hDAF) transgenic pigs for xenotransplantation. *Proceedings of the National Academy of Sciences, USA*, **99**, 14230-14235.

Lavitrano, M., Busnelli, M., Cerrito, M.G., Giovannoni, R., Manzini, S. and Vargiolu, A. (2006) Sperm-mediated gene transfer. *Reproduction, Fertility and Development*, **18**, 19-23.

Lavitrano, M., Camaioni, A., Fazio, V.M., Dolci, S., Farace, M.G. and Spadafora, C. (1989) Sperm cells as vectors for introducing foreign DNA into eggs: genetic transformation of mice. *Cell*, **57**, 717-723.

Lavitrano, M., Forni, M., Bacci, M.L., DiStefano, C., Varzi, V., Wang, H. and Seren, E. (2003) Sperm-mediated gene transfer in pig: selection of donor boars and optimization of DNA uptake. *Molecular Reproduction and Development*, **64**, 284-291.

Lavitrano, M., Forni, M., Varzi, V., Pucci, L., Bacci, M.L., DiStefano, C., Fioretti, D., Zoraqi, G., Moioli, B., Rossi, M., Lazzereschi, D., Stoppacciaro, A., Seren, E., Alfani, D., Cortesini, R. and Frati, L. (1997) Sperm-mediated gene transfer: production of pigs transgenic for a human regulator of complement activation. *Transplantation Proceedings*, **29**, 3508-3509.

Lee, G.S., Hyun, S.H., Kim, H.S., Kim, D.Y., Lee, S.H., Lim, J.M., Lee, E.S., Kang, S.K. and Hwang, W.S. (2003) Improvement of a porcine somatic cell nuclear transfer technique by optimizing donor cell and recipient oocyte preparations. *Theriogenology*, **59**, 1949-1957.

Lee, G.S., Kim, H.S., Hyun, S.H., Lee, S.H., Jeon, H.Y., Nam, D.H., Jeon, Y.W., Kim, S., Kim, J.H., Han, J.Y., Ahn, C., Kang, S.K., Lee, B.C. and Hwang, W.S. (2005) Production of transgenic cloned piglets from genetically transformed fetal fibroblasts selected by green fluorescent protein. *Theriogenology*, **63**, 973-991.

Levis, D.G.S., Burroughs. S. and Williams, S. (2002) Use of intra-uterine insemination of pigs: Pros, Cons and Economics. *33rd Annual Meeting American Association of Swine Veterinarians, Kansas City, MO*, 1, 39-62.

Levis, D. (2000) Liquid boar semen production: current extender technology and where do we go from here! In: *Boar semen preservation IV*, pp.121-128. Edited by L.A. Johnson and H.D. Guthrie. Allen Press, Inc, Lawrence, KS USA.

Li, N., Lai, L., Wax, D., Hao, Y., Murphy, C.N., Rieke, A., Samuel, M., Linville, M.L., Korte, S.W., Evans, R.W., Turk, J.R., Kang, J.X., Witt, W.T., Dai, Y. and Prather, R.S. (2006) Cloned transgenic swine via *in vitro* production and cryopreservation. *Biology of Reproduction*, 75, 226-230.

Li, J., Rieke, A., Day, B.N. and Prather, R.S. (1996) Technical note, porcine non-surgical embryo transfer. *Journal of Animal Science*, 74, 2263-2268.

Lin, C.L., Ponsuksili, S., Tholen, E., Jennen, D.G.J., Schellander, K. and Wimmers, K. (2006) Candidate gene markers for sperm quality and fertility of boar. *Animal Reproduction Science*, 92, 349-363.

Liu, D., Kobayashi, T., Onishi, A., Furusawa, T., Iwamoto, M., Suzuki, S., Miwa, Y., Nagasaka, T., Maruyama, S., Kadomatsu, K., Uchida, K. and Nakao, A. (2007) Relation between human decay-accelerating factor (hDAF) expression in pig cells and inhibition of human serum anti-pig cytotoxicity: value of highly expressed hDAF for xenotransplantation. *Xenotransplantation*, 14, 67-73.

Lu, Q., Han, H., Lian, Z., Li, N., Zhang, Q., Zhao, Z., Pei, D., Zhang, X. and Wu, C. (2004) The screening and identification of endogenous retrovirus free CEMPs. *Sci China Council Life Sciences*, 47, 562-566.

Marchal, R., Vigneron, C., Perreau, C., Bali-Papp, A. and Mermillod, P. (2002) Effect of follicular size on meiotic and developmental competence of porcine oocytes. *Theriogenology*, 57, 1523-1532.

Marklund, S., Kijas, J., Rodriguez-Martinez, H., Rönnstrand, L., Funa K., Möller, M., Lange, D., Edfors-Lilja, I. and Andersson, L. (1999) Molecular basis for the dominant white phenotype in the domestic pig. *Genome Research*, 8, 826-833.

Martin, M., Adams, C. and Wiseman, B. (2004) Pre-weaning performance and health of pigs born to cloned (fetal cell derived) swine versus non-cloned swine. *Theriogenology*, 62, 113-122.

Martin, S.I., Wilkinson, R. and Fishman, J.A. (2006) Genomic presence of recombinant porcine endogenous retrovirus in transmitting miniature swine. *Virology Journal*, 3, 91.

Martinat-Botte, F., Berthelot, F., Plat, M. and Madec, F. (2006) Cryopreservation and transfer of pig *in vivo* embryos: state of the art. *Gynecologie Obtétrique et Fertilité*, 34, 754-759.

Martinez, E.A., Caamaño, J.N., Gil, M.A., Rieke, A., Mccauley, T.C., Cantley, T.C., Vazquez, J.M., Roca, J., Vazquez, J.L., Didion, B., Murphy, C.N., Prather, R.S. and Day, D.N. (2004) Successful nonsurgical deep uterine embryo transfer in pigs. *Theriogenology*, **61**, 137-146.

Martinez, E.A., Vazquez, J.M., Roca, J., Lucas, X., Gil, M.A., Parrilla, I., Vazquez, J.L. and Day, B.N. (2001) Successful non-surgical deep intrauterine insemination with low number of spermatozoa in sows by a fiberoptic endoscope technique. *Reproduction*, **122**, 289-296.

Martinez, E.A., Vazquez, J.M., Roca, J., Lucas, X., Gil, M.A., Parrilla, I., Vazquez, J.L. and Day, B.N. (2002) Minimal number of spermatozoa required for normal fertility after deep intrauterine insemination in non-sedated sows. *Reproduction*, **123**, 163-170.

Martinez, E.A., Vazquez, J.M., Roca, J., Cuello, C., Gil, M.A.; Parrilla, I. and Vazquez, J.L. (2005) An update on reproductive technologies with potential short-term application in pig production. *Reproduction in Domestic Animals*, **40**, 300-309.

Maxwell, W.M.C. and Johnson, L.A. (1999) Physiology of spermatozoa at high dilution rates: the influence of seminal plasma. *Theriogenology*, **52**, 1353-1362.

Milan, D., Jeon, J.T., Looft, C., Amarger, V., Robic, A., Thelander, M., Rogel-Gaillard, C., Paul, S., Iannuccelli, N., Rask, L., Ronne, H., Lundström, K., Reinsch, N., Gellin, J., Kalm, E., Roy, P.L., Chardon, P. and Andersson, L. (2000) A mutation in PRKAG3 associated with excess glycogen content in pig skeletal muscle. *Science*, 2000, **288**, 1248-1251.

Milland, J., Christiansen, D. and Sandrin, M.S. (2005) Alpha1,3-galactosyltransferase knockout pigs are available for xenotransplantation: are glycosyltransferases still relevant? *Immunology and Cell Biology*, **83**, 687-693.

Milland, J., Christiansen, D., Lazarus, B.D., Taylor, S.G., Xing, P.X. and Sandrin, M.S. (2006) The molecular basis for galalpha(1,3)gal expression in animals with a deletion of the alpha1,3-galactosyltransferase gene. *Journal of Immunology*, **176**, 2248-2254.

Milovanov, V.K. (1962) Biology of reproduction and artificial insemination of animals. Selhozizdat, Moscow, 696 pp.

Morrison, R.S., Kinishita, Y., Johnson, M.D., Uo, J.T., McBee, J.K., Conrads, T.P. and Venstra, T.D. (2002) Proteomic analysis in the neurosciences. *Molecular Cell Proteomics*, **1**, 553-560.

Müller, M., Brenig, B., Winnacker, E.L. and Brem, G. (1992) Transgenic pigs carrying cDNA copies encoding the murine Mx1 protein which confers resistance to influenza virus infection. *Gene*, **121**, 263-270.

Nagai, T., Funahashi, H., Yoshioka, K. and Kikuchi, K. (2006) Up date of *in vitro* production of porcine embryos. *Frontiers in Bioscience*, **11**, 2565-2573.

Nagashima, H., Hiruma, K., Saito, H., Tomii, R., Ueno, S., Nakayama, N., Matsunari, H. and Kurome, M. (2007) Production of live piglets following cryopreservation of embryos derived from *in vitro*-matured oocytes. *Biology of Reproduction*, Epub ahead of print, PMID: 17267701.

Nagashima, H., Kashiwazaki, N., Ashman, R.J., Gruppen, C.G. and Nottle,

M.B. (1995) Cryopreservation of porcine embryos. *Nature*, **374**, 416.

Niemann, H., Kues, W. and Carnwath, J.W. (2005) Transgenic farm animals: present and future. *Reviews of Science and Technology*, **24**, 285-298.

Niemann, H., Rath, D. and Wrenzycki, C. (2003) Advances in biotechnology: New tools in future pig production for agriculture and biomedicine. *Reproduction in Domestic Animals*, **38**, 82-89.

Niewold, T.A., Kerstens, H.H.D., van der Meulen, J., Smits, M.A. and Hulst, M.M. (2005) Development of a porcine small intestinal cDNA microarray: Characterization and functional analysis of the response to enterotoxigenic E. coli. V*eterinary Innunology and Immunopathology*, **105**, 317-329.

Niimura, S. and Ishida, K. (1980) Histochemical observation of lipid droplets in mammalian eggs during the early development. *Japanese Journal of Animal Reproduction*, **26**, 46-49.

Noble, M.S., Rodriguez-Zas, S., Cook, J.B., Bleck, G.T., Hurley, W.L. and Wheeler, M.B. (2002) Lactational performance of first-partity transgenic gilts expressing bovine alpha-lactalbumin in their milk. *Journal of Animal Science*, **80**, 1090-1096.

Onishi, A., Iwamoto, M., Akita, T., Mikawa, S., Takeda, K., Awata, T., Hanada, H. and Perry, A.C.F. (2000) Pig cloning by microinjection of fetal fibroblast nuclei. *Science*, **289**, 1188-1190.

Paradis, K., Langford, G., Long, Z., Heneine, W., Sandström, P., Switzer, W.M., Chapman, L.E., Lockey, C., Onions, D. and Otto, E. (1999) Search for cross-species transmission of porcine endogenous retrovirus in patients treated with liing pig tissue. *Science*, 285, 1236-1241.

Park, J.I., Hong, J.Y., Yong, H.Y., Hwang, W.S., Lim, J.M. and Lee, E.S. (2005) High oxygen tension during *in vitro* oocyte maturation improves *in vitro* development of porcine oocytes after fertilization. *Animal Reproduction Science*, **87**, 133-141.

Park, K.W., Cheong, H.T., Lai L., Im, G.S., Kuhholzer, B., Bonk, A., Samuel, M., Riehe, A., Day, B.N., Murphy, C.N., Carter, D.B. and Prather, R.S. (2001) Production of nuclear transfer-derived swine that express the enhanced green fluorescent protein. *Animal Biotechnology*, **12**, 173-181.

Paterson, L., DeSouza, P., Ritchie, W., King, T. and Wilmut, I. (2003) Application of reproductive biotechnology in animals: implications and potentials. Application of reproductive cloning. *Animal Reproduction Science*, **79**, 137-143.

Patience, C., Takeuchi, Y. and Weiss, R.A. (1997) Infection of human cells by an endogenous retrovirus of pigs. *Nature Medicine*, **3**, 282-286.

Peña, F.J., Johannisson, A., Wallgren, M. and Rodriguez-Martinez, H. (2004b) Antioxidant supplementation of boar spermatozoa from different fractions of the ejaculate improves cryopreservation: changes in sperm membrane lipid architecture. *Zygote*, **12**, 117-124.

Peña, F.J., Johannisson, A., Wallgren, M. and Rodriguez-Martinez, H. (2004a)

Effect of hyaluronan supplementation on boar sperm motility and membrane lipid architecture status after cryopreservation. *Theriogenology*, **61**, 63-70.

Peña, F.J., Johannisson, A., Wallgren, M. and Rodriguez-Martinez, H. (2003) Antioxidant supplementation *in vitro* improves boar sperm motility and mitochondrial membrane potential after cryopreservation of different fractions of the ejaculate. *Animal Reproduction Science*, **78**, 85-98.

Peña, F.J., Saravia, F., García-Herreros, F., Nuñez, I., Tapia, J.A., Johannisson, A., Wallgren, M. and Rodriguez Martinez, H. (2005) Identification of sperm morphological subpopulations in two different portions of the boar ejaculate and its relation to post thaw quality. *Journal of Andrology*, **26**, 716-723.

Peña, F.J., Saravia, F., Nuñez-Martinez, I., Johannisson, A., Wallgren, M. and Rodriguez Martinez, H. (2006) Do different portions of the boar ejaculate vary in their ability to sustain cryopreservation? *Animal Reproduction Science*, **93**, 101-113.

Phelps, C.J., Koike, C., Vaught, T.D., Boone, J., Wells, K.D., Chen, S.H., Ball, S., Specht, S.M., Polejaeva, I.A., Monahan, J.A., Jobst, P.M., Sharma, S.B., Lamborn, A.E., Garst, A.S., Moore, M., Demetris, A.J., Rudert, W.A., Botttino, R., Bertera, S., Trucco, M., Starzl, T.E., Dai, Y. and Ayares, D.L. (2003) Production of alpha1,3-galactosyltransferase-deficient pigs. *Science*, **299**, 411-414.

Piedrahita, J.A. and Mir, B. (2004) Cloning and transgenesis in mammals: implications for xenotransplantation. *American Journal of Transplantation*, **4**, 43-50.

Polejaeva, I.A., Chen, S.H., Vaught, T.D., Page, R.L., Mullins, J., Ball, S., Dal, Y., Boano, J., Walker, S., Ayares, D., Colman, A. and Campbell, K.H.S. (2000) Cloned pigs produced by nuclear transfer from adult somatic cells. *Nature*, **407**, 505-509.

Polge, C. and Day, B.N. (1968) Pregnancy following non-surgical egg transfer in pigs. *Veterinary Record*, **82**, 712.

Prather, R.S., Sims, M.M. and First, N.L. (1989) Nuclear transplantation in early pig embryos. *Biology of Reproduction*, **41**, 414-418.

Prather, R.S., Sutovsky, P. and Green, J.A. (2004) Nuclear remodelling and reprogramming in transgenic pig production. *Experimental Biology and Medicine*, **229**, 1120-1126.

Pratt, S.L., Sherrer, E.S., Reeves, D.E. and Stice, S.L. (2006) Factors influencing the commercialisation of cloning in the pork industry. *Society Reproduction and fertility Suppl*, **62**, 303-315.

Price, E.M., Prather, R.S. and Foley, C.M. (2006) Multipotent adult progenitor cell lines originating from the peripheral blood of green fluorescent protein transgenic swine. *Stem Cell Development*, **15**, 507-522.

Probst, S. and Rath, D. (2003) Production of piglets using intracytoplasmic sperm injection (ICSI) with flowcytometrically sorted boar semen and artificially activated oocytes. *Theriogenology*, **59**, 961-973.

Pursel, V.G., Mitchell, A.D., Bee, G., Elsasser, T.H., McMurtry, J.P., Wall, R.J., Coleman, M.E. and Schwartz, R.J. (2004) Growth and tissue accretion rates of swine expressing an insulin-like growth factor I transgene. *Animal Biotechnology*, **15**, 33-45.

Rall, W.F. and Fahy, G.M. (1985) Ice-free cryopreservation of mouse embryos at -196°C by vitrification. *Nature*, **313**, 573-575.

Rath, D. (2002) Low dose insemination in the sow – A review. *Reproduction in Domestic Animals,* **37**, 201-205.

Rath, D., Ruiz, S. and Sieg, B. (2003) Birth of female piglets following intrauterine insemination of a sow using flow cytometrically sexed boar semen. *Veterinary Record*, **152**, 400-401.

Robinson, J.A.B. and Buhr, M.M. (2005) Impact of genetic selection on management of boar replacement. *Theriogenology*, **63**, 668-678.

Robl, J.M., Wang, Z., Kasinathan, P. and Kuroiwa, Y. (2007) Transgenic animal production and animal biotechnology. *Theriogenology*, **67**, 127-133.

Roca, J., Carvajal, G., Lucas, X., Vazquez, J.M. and Martinez, E.A. (2003) Fertility of weaned sows after deep intrauterine insemination with a reduced number of frozen-thawed spermatozoa. *Theriogenology*, **60**, 77-87.

Roca, J., H. Rodriguez-Martinez, J.M. Vazquez, A. Bolarin, M. Hernandez, F. Saravia, M. Wallgren and Martinez, E.A. (2006b) Strategies to improve the fertility of frozen-thawed boar semen for artificial insemination. In: *Control of Pig Reproduction VII*. Edited by Ashworth, C.J. and Kraeling, R.R., pp. 261-275. Nottingham Univ. Press, Manor Farm, Thrumpton, UK.

Roca, J., Hernandez, M., Carvajal, G., Vazquez, J. and Martinez, E. (2006a) Factors influencing boar sperm cryosurvival. *Journal of Animal Science*, **84**, 2692-2699.

Roca, J., Gil, M.A., Hernandez, M., Parrilla, I., Vazquez, J.M. and Martinez, E.A. (2004) Survival and fertility of boar spermatozoa after freeze-thawing in extender supplemented with butylated hydroxytoluene. *Journal of Andrology*, **25**, 397-405.

Rodriguez-Martinez, H., Saravia, F., Wallgren, M., Tienthai, P., Johannisson, A., Vazquez, J.M., Martinez, E.A., Roca, J., Sanz, L. and Calvete, J.J. (2005) Boar spermatozoa in the oviduct. *Theriogenology*, **63**, 514-535.

Rodriguez-Martinez, H. (2007) State of the art in farm animal sperm evaluation. *Reproduction, Fertility and Development*, **19**, 91-101.

Rohrer, G.A., Wise, T.H. and Ford, J.J. (2006) Deciphering the pig genome to understand gamete production. *Society of Reproduction and Fertility Supplements*, **62**, 293-301.

Rothschild, M., Jacobson, C., Vaske, D., Tuggle, C., Wang, L., Short, T., Eckardt, G., Sasaki, S., Vincent, A., McLaren, D., Southwood, O., van der Steen, H., Mileham, A., and Plastow, G. (1996) The estrogen receptor locus is associated with a major gene influencing litter size in pigs. *Proceedings of the National Academy of Sciences*, **93**, 201-205.

Rothschild, M.F. (2003) From a sow´s ear to a silk purse: Real progress in porcine genomics. *Cytogenetics Genome Research*, **102**, 95-99.

Rothschild, M.F. (2004) Porcine genomics delivers new tools and results: This little piggy did more that just go to the market. *Gamete Research, Cambridge*, **83**, 1-6.

Rozeboom, K.J., Reicks, D.L. and Wilson, M.E. (2004) The reproductive performance and factors affecting on-farm application of low-dose intrauterine deposit of semen in sows. *Journal of Animal Science*, **82**, 2164-2168.

Santos, F. and Dean, W. (2004) Epigenetic reprogramming during early development in mammals. *Reproduction*, **127**, 643-651.

Saravia, F., Hernandez, M., Wallgren, MK., Johannisson, A. and Rodriguez-Martinez, H. (2007) Cooling during semen cryopreservation does not induce capacitation of boar spermatozoa. *International Journal of Andrology*, (accepted for publication).

Saravia, F., Wallgren, M., Nagy, S., Johannisson, A. and Rodriguez-Martinez, H. (2005) Deep freezing of concentrated boar semen for intra-uterine insemination: effects on sperm viability. *Theriogenology*, **63**, 1320-1333.

Schenk, J.L. and Seidel, G.E. Jr. (2007) Pregnancy rates with cryopreserved sexed spermatozoa: effects of laser intensity, staining conditions and catalase. In: *Reproduction in Domestic Ruminants VI*. pp 165-177. Edited by Juengel, J.I., Murray, J.F. and Smith, M.F. Nottingham University Press, Nottingham, UK.

Seidel, G.E. (2003) Economics of selecting for sex: the most important genetic trait. *Theriogenology*, **59,** 585-598.

Serdiuk, S.I. (1970) Artificial insemination of pigs. Kolos, Moscow, 144 pp.

Soede, N.M., Steverink, D.W.B., Langendijk, P. and Kemp, B., (2000) Optimised insemination strategies in swine AI. In *Boar semen preservation IV*, pp.185-190. Edited by Johnson, L.A. and Guthrie, H.D., Allen Press, Inc, Lawrence, KS USA.

Soede, N.M., Wetzels, C.C.H., Zondag, W., de Koning, M.A.I. and Kemp, B. (1995) Effects of time of insemination relative to ovulation, as determined by ultrasonography, on fertilization rate and accessory sperm count in sows. *Journal of Reproduction and Fertility*, **105**, 135-140.

Steinfeldt, H., Wassenaar, T. and Jutzi, S. (2006) Livestock production systems in developing countries: status, drivers, trends. *Reviews of Science and Technology*, **25**, 505-516.

Steverink, D.W.D., Soede, N.M., Groenland, G.J.R., van Schie, F.W., Noordhuizen, J.P.T.M. and Kemp, B. (1999) Duration of estrus in relation to reproduction results in pigs on commercial farms. *Journal of Animal Science*, **77**, 801-809.

Stringfellow, D.A. (1998) Recommendations for the sanitary holding of *in vivo* derived embryos. In *Manual of the International Embryo Transfer Society*, P 67. Edited by Stringfellow, D.A. and Seidel, S.M., 2nd ed, Savoy USA.

Suh, T.K., Schenk, L.A. and Seidel, J.E. Jr. (2005) High pressure flow cytometric sorting damages sperm. *Theriogenology*, **64**, 1035-1048.

Suzuki, K., Asano, A., Eriksson, B., Niwa, K., Shimizu, H., Nagai, T., and Rodriguez-Martinez, H. (2002) Capacitation status and *in vitro* fertility of boar spermatozoa: Effects of seminal plasma, cumulus-oocytes-complexes-conditioned medium and hyaluronan. *International Journal of Andrology*, **25**, 84-93.

Suzuki, K., Eriksson, B., Shimizu, H., Nagai, T. and Rodriguez-Martinez, H. (2000) Effect of hyaluronan on monospermic penetration of porcine oocytes fertilized *in vitro*. *International Journal of Andrology*, **23**, 13-21.

Tai, H.C., Ezzelarab, M., Hara, H., Ayares, D. and Cooper, D.K. (2007) Progress in xenotransplantation following the gene-knockout technology. *Transplantation International*, **20**, 107-117.

Takahagi, Y., Fujimura, T., Miyagawa, S., Nagashima, H., Shigehisa, T., Shirakura, R. and Murakami, H. (2005) Production of á(1,3)-galactosyltransferase gene knockout pigs expressing both human decay-accelerating factor and N-acetylglucosaminyltransferase III. *Molecular Reproduction and Development*, **71**, 331-338.

Tatemoto, H., Muto, N., Sunagawa, I., Shinjo, A. and Nakada, T. (2004) Protection of porcine oocytes against cell damage caused by oxidative stress during *in vitro* maturation: role of superoxide dismutase activity in porcine follicular fluid. *Biology of Reproduction*, **71**, 1150-1157.

Thibier, M. (2004) Stabilization of numbers of *in vivo* collected embryos in cattle but significant increases of *in vitro* produced bovine embryos in some parts of the world. *Data Retrieval Committee Annual Reports, IETS Newsletter*, **22**, 12-19.

Tseng, Y.L., Kuwaki, K., Dor, F., Shimizu, A., Houser, S., Hisashi, Y., Yamada, K., Robson, S.C., Awwad, M., Schuuman, H.J., Sachs, D.H. and Cooper, D.K. (2005) Alpha-1,3-galactosyltransferase gene-knockout pig heart transplantation in baboons with survival approaching 6 months. *Transplantation*, **80**, 1493-1500.

Vajta, G. and Gjerris, M. (2006) Science and technology of farm animal cloning: state of the art. *Animal Reproduction Science*, **92**, 211-230.

Vajta, G., Holm, P., Greve, T. and Callesen, H. (1997) Vitrification of porcine embryos using the Open Pulled Straw (OPS) method. *Acta Veterinaria Scandinavica*, **38**, 349-352.

Vajta, G., Zhang, Y. and Macháty, Z. (2007) Somatic cell nuclear transfer in pigs: recent achievements and future possibilities. *Reproduction, Fertility and Development*, **19**, 403-423.

Van Wijk, H.J., Buschbell, H., Dibbits, B., Liefers, S.C., Harlizius, B., Heuven, H.C., Knol, E.F., Bovenhuis, H. and Groenen, M.A. (2007) Variance component analysis of quantitative trait loci for pork carcass composition and meat quality on SSC4 and SSC11. *Journal of Animal Science,* **85**, 22-30.

Vazquez, J.M., Martinez, E.A., Parrilla, I., Roca, J., Gil, M.A. and Vazquez, J.L. (2003) Birth of piglets after deep intrauterine insemination with flow cytometrically sorted spermatozoa. *Theriogenology,* **59**, 1605-1614.

Vazquez, J.M., Martinez, E.A., Roca, J., Gil, M.A., Parrilla, I., Cuello, C., Carvajal, G., Lucas, X. and Vazquez, J.L. (2005) Improving the efficiency of sperm technologies in pigs: the value of deep intrauterine insemination. *Theriogenology,* **63**, 536-547.

Waberski, D. (1997) Effects of semen components on ovulation and fertilization *Journal of Reproduction and Fertility,* **52**, 105-109.

Waberski, D., Weitze, K.F., Gleumes, T., Schwartz, M., Willmen, T. and Petzoldt, R. (1994) Effects of time of insemination relative to ovulation on fertility with liquid and frozen semen. *Theriogenology,* **42**, 831-840.

Wagner, H.G. and Thibier, M. (2000) World statistics for artificial insemination in small ruminants and swine. *Proceedings of the 14th International Congress of Animal Reproduction,* Stockholm, Abstracts, **2**, 15:3.

Wall, R.J. (2002) New gene transfer methods. *Theriogenology,* 57, 189-201.

Wallenhorst, S. and Holtz, W. (1999) Transfer of pig embryos to different uterine sites. *Journal of Animal Science,* **77**, 2327-2329.

Waterhouse, K.E., Hofmo, P.O., Tverdal, A. and Miller, R.R. Jr. (2006) Within and between breed differences in freezing tolerance and plasma membrane fatty acid composition of boar sperm. *Reproduction,* **131**, 887-894.

Watson, P.F. (1996) Cooling of spermatozoa and fertilizing capacity. *Reproductionin Domestic Animals,* **31**, 135-140.

Watson, P.F. and Behan, J.R. (2002) Intrauterine insemination of sows with reduced sperm numbers: results of a commercially based field trial. *Theriogenology,* **57**, 1683-1693.

Webster, N.L. Forni, M., Bacci, M.L., Giovannoni, R., Razzini, R., Fantinati, P., Zannoni, A., Fusetti, L., Dalpra, L., Bianco, M.R., Papa, M., Seren, A., Sandrin, M.S., McKenzie, I.F. and Lavitrano, M. (2005) Multi-transgenic pigs expressing three fluorescent proteins produced with high efficiency by sperm mediated gene transfer. *Molecular Reproduction and Development,* **72**, 68-76.

Wernersson, R., Schierup, M.H., Jorgensen, F.G., Gorodkin, J., Panitz, F., Staerfeldt, H-H., Christensen, O.F., Mailund, T., Hornshoj, H., Klein, A., Wang, J., Liu, B., Hu, S., Dong, W., Li, W., Gong, G.K., Yu, J., Wang, J., Bendixen, C., Fredholm, M., Brunak, S., Yang, H. and Bolund, L. (2005) Pigs in sequence space: A 0.66 coverage pig genome survey based on shotgun sequencing. *BMC Genomics,* **6**, 70.

Whitelaw, C.B., Radcliffe, P.A., Ritchie, W.A., Carlisle, A., Ellard, F.M:, Pena, R.N., Rowe, J., Clark, A.J., King, T.J. and Mitrophanous, K.A. (2004) Efficient generation of transgenic pigs using equine infectious anaemia virus (EIAV) derived vector. *FEBS Letters,* **571**, 233-236.

Wilmut, I. (1972) The low temperature preservation of mammalian embryos.

Journal of Reproduction and Fertility, **31**, 513-514.

Woelders, H., Matthijs, A., Zuidberg, C.A. and Chaveiro, A.E. (2005) Cryo-preservation of boar semen: equilibrium freezing in the cryomicroscope and in straws. *Theriogenology*, **63**, 383-395.

Womack, J.E. (2005) Advances in livestock genomics: Opening the barn door. *Genome Research*, **15**, 1699-1705.

Wongtawan, T., Saravia, F., Wallgren, M., Caballero, I. and Rodriguez-Martinez, H. (2006) Fertility after deep intra-uterine artificial insemination of concentrated low-volume boar semen doses. *Theriogenology*, **65**, 773-787.

Wrathall, A.E. and Sutmoller, P. (1998) Potential of embryo transfer to control transmission of Disease. In *Manual of the International Embryo Transfer Society*, p. 17-44. Edited by Stringfellow, D.A. and Seidel, S.M., 2nd ed, Savoy USA.

Wrattrang, E., Almqvist, M., Johansson, A., Fossum, C., Wallgren, P., Pielberg, G., Andersoon, L. and Edfors-Lilja, I. (2005) Confirmation of QTL on porcine chromosomes 1 and 8 influencing leukocyte numbers, haematological parameters and leukocyte function. *Animal Genetics*, **36**, 337-345.

Yonemura, I., Fujino, Y., Irie, S. and Miura, Y. (1996) Transcervical ransfer of porcine embryos under practical conditions. *Journal of Reproduction and Development*, **42**, 89-94.

Yoshioka, K., Suzuki, C. and Rodriguez-Martinez, H. (2005) Replacement of PVA with fetal bovine serum improves formation and hatching of porcine blastocysts produced *in vitro*. *Reproduction, Fertility and Development*, **17**, 280.

Yoshioka, K., Suzuki, C., Itoh, S., Kikuchi, K., Iwamura, K. and Rodriguez-Martinez, H. (2003) Production of piglets derived from *in vitro*-produced blastocysts fertilized and cultured in chemically defined media: effects of theophylline, adenosine, and cysteine during *in vitro* fertilization. *Biology of Reproduction*, **69**, 2092-2099.

Yoshioka, K., Ekwall, H. and Rodriguez-Martinez, H. (2004) Effects of hyaluronan on *in vitro* development of porcine embryos cultured in a chemically defined medium. *Reproduction, Fertility and Development*, **38**, 264-265.

Zhong, R. (2007) Gal knockout and beyond. *American Journal of Transplantation*, **7**, 5-11.

Zhu, A., Dor FJ, and Cooper DK (2007) Pig-to-human primate heart transplantation: immunologic progress over 20 years. *Journal of Heart and Lung Transplantation*, **26**, 210-218.

16

MINORITY BREEDS

MAX WALDO

Waldo Farms Inc., PO Box 8, DeWitt, NE 68341 USA

The two major questions to be addressed in this chapter are:

1. How will Waldo breeds like Duroc and Hampshire improve and be accepted into mainstream pig farming?

2. Are there other breeds that will also move up to major status?

The Waldo family started raising Duroc in 1895. At that time, very little crossing was done. Nearly every farmer raised some pigs, which by custom were one of the several pure US breeds. Those with breeding interests sold or exchanged breeding stock with others. For nearly a century, the US purebred breeders were part of a competitive but cooperative fraternity. In the mid-1900s, the economic value of cross breeding and heterosis was recognized. Some pig farms began to increase in size to become primary enterprises. This change in breeding programs along with the eventual advent of artificial insemination (AI) usage diminished the required numbers of breeding stock. As the commercially-oriented breeding stock business becomes more competitive between private and corporate companies, that traditional attitude of cooperation between breeders has also diminished.

To my knowledge, the Pig Improvement Company (PIC) may have been the first major company to capitalize effectively on the evolving concept of using a terminal program with a specific crossbred maternal female. Their initial Camborough (York x Landrace) that came to the US was lacking in quality, durability and rate of gain. However, the F1 maternal female possessed reproductive advantage over the more dual purpose US breeds that were generally being used in rotational programs by US farmers in the 1950s and 1960s.

Dr. Maurice Bichard (at that time the geneticist for PIC) stated that PIC would not use American breeds because they were inferior. However, they soon recognized the superior growth and durability of the Duroc and Hampshire available in the US and began to utilize both breeds.

For many years, the breeding companies discredited purebreds. They designated their "product" lines by exotic numbers. In reality, the industry uses few synthetic or composite breeds. Nearly all lines are derived from further crossing or back-crossing of originally pure lines. The trend now is back toward a pure or F1 animal.

My assumptions as to why only a limited number of breeds will be utilized for the vast majority of commodity pork world wide:

1.   Considering breed characteristics and merit, the Yorkshire and Landrace will serve as the basis for parent sows. A third maternal White breed would need to be available and widely used in order for any breeder to have a prayer of a chance to produce them profitability. The population size of this third breed would also have to be of sufficient size for effective selection while minimizing inbreeding and associated problems.

2.   Considering heterosis and uniformity, a system must be extremely large to support two pure boar lines to generate F1 AI boars. Assuming one terminal breed becomes the primary sire, the Duroc is likely to be that breed because of its superior performance, ease of carcass processing and muscle quality relative to other terminal sire breeds.

3.   Numerous breeds and populations throughout the world could be utilized, if first subjected to a comprehensive, long term genetic improvement program. Considering the logistical difficulty in adding a good third White breed or two equal terminal breeds to create F1 boars, the world pork industry is likely to focus on the York, Landrace and Duroc. Some niche market situations might utilize other breeds as has been demonstrated by the Berkshire success in some market situations. For other breeds to replace the Yorkshire, Landrace and Duroc, they would first need to be genetically improved to comparable merit.

There is always the danger that, in our collective infinite wisdom, we run the risk of getting on the same bandwagon and driving the selection of a breed too far in the wrong direction. Preserving genetic variability and diversity could prove invaluable, yet minor breeds have little direct application if they lag excessively in their ability to perform. China is an example. With over 50 distinct local breeds developed for local conditions, the breed of choice for their emmerging large-scale commercial industry is basically the Yorkshire X Landrace female crossed to a terminal Duroc boar.

The problems if only three breeds used (2 Maternal 1 Terminal) are:

1.   Parent females produced are from inefficient pure sows.

2.   Rotation of Yorkshire and Landrace has its own disadvantages.

3. Disease resistance and meat quality traits may be lacking in the two primary maternal breeds.

A single terminal sire breed could be selected to lose some important attributes, if an all-inclusive selection program is not designed and used.

Many ill-conceived programs of the past as well as show-pig interests of the present have selected away from many of the most important traits impacting the profitability of commercial pork production.

Refining sub-populations within the existing populations of the major breeds is more likely than breeding a minor breed up to adequate merit to be widely incorporated into the mating systems used by the commodity commercial industry.

A more practical rationale for why minority breeds are likely not to be used can be demonstrated mathematically in the following example.

The UK has a total of about 470,000 sows and it may be assumed that 0.85 or 400,000 might be used in commercial production. If all were in a common breeding system with average efficiency, 2,500 terminal AI boars at a ratio of 160 sows/boar could service the UK population. With a reasonable replacement rate, 1,750 AI boars annually produced from 1,000 sows would satisfy the needs of the UK pork industry. This program with one terminal breed is sustainable.

Theoretically, the UK could also be supplied with F1 boars from two sire line populations of 575 sows each. Seventy-five could be bred pure and 500 in each population crossed to sires of the other breed. This two breed system, if managed efficiently, might operate with two nucleus populations of 75 sows each that is hardly sustainable unless other sources of these breeds existed.

Similar calculations can be made for a three breed maternal program.

400,000	Commercial UK sows
40,000	GP (grandparent) Landrace x Third White breed to produce replacements
	20,000 can be Landrace bred to a third breed.
	20,000 can be that Third breed bred to Landrace
1,500	GGP (great grandparent) Landrace
1,500	GGP Third maternal breed
300	Yorkshire sows to produce AI boars for the 40,000 F1 sows.

These numbers demonstrate how unlikely and impractical it is to expect to incorporate significant number of minor breeds directly into highly productive and profitable breeding programs.

Gene preservation

Identifying variation is the basis for genetic change and improvement. Even the major breeds benefit by genetic exchange. Similarly, a broad genetic base and access to a wide array of native or locally adapted breeds would preserve many inherent gene combinations that, as of today, we may have little understanding or appreciation.

Ken Stalder (Iowa State University)

"There is a world-wide effort in "gene preservation" going on in many species. We should not be so naïve to think that we know what the future will bring and what traits may become important in the future."

One could argue that this was exactly the case just a few years ago with the Berkshire breed. The breed was well on its way to being a zoo exhibit but all of a sudden they were found to have very desirable muscle quality and some have incorporated the breed in a sire line or developed a separate sire line to capture the desirable muscle quality traits that the Berkshire breed possesses.

The muscle quality project at Iowa State University (ISU) is another example where the Duroc breed had the variation for intramuscular fat (IMF) such that a selection project could be started and make good progress using quantitative selection methods at improving IMF in a relatively short period of time. What would have happened if ISU would not have started the project until the year 2050? It is probable that a good deal of the variation for IMF within the Duroc breed could have been lost by then.

These are issues that breeders and the pork industry as a whole need to keep in mind for the long term success of the worldwide pork industry as we compete as a protein source throughout the world."

Observations and personal opinion regarding various breeds

Yorkshire – Large White Probably the most useful and readily improved breed. Disease resistance appears to be average. Meat quality and IMF is average or below. Relatively prolific.

Landrace Milk well and farrow larger pigs with better survival. Structure and durability is often lacking. Meat is generally pale and lacking IMF.

Chester White Historically were prolific, slow growing and fat. Could be a contributor if good breeders would dedicate effort to improvement.

Welsh Interesting, but with several attempts they were not successful in the US. Population is small and may be two similar to the Landrace to make significant contribution.

Chinese White Not the official name, but I'm told there is a White breed in China with potential.

Meishan Prolific, slow growing and devoid of muscle.

Nebraska Index A small population based on Yorkshire and Landrace. Selected for 25+ generations for prolificacy. With this trait "fixed" they can be graded up in performance traits.

Duroc Excel in growth and durability and meat quality. Becoming the terminal breed of choice.

Hampshire Originally considered to be excellent foragers and of high meat content. Those RN negative and selected for performance combine well with Duroc for F1 boars Hamp. Lack appetite and growth in hot climates.

Berkshire Possess unique meat qualities. If used as pure of 0.50 of a market animals, it is difficult for niche market premiums to make up for the lack of performance from today's populations.

Pietrain Extreme high lean. Aggressive breeders. Lack meat eating quality. Have not competed well in the US or Asia if growth rate is a factor. Reportedly may possess some resistance to Circovirus.

Native and Other Minority Breeds – Many may possess certain unique attributes or gene combinations that could benefit future swine breeding.

17

INFLUENCE OF NUTRITIONAL FACTORS ON PLACENTAL GROWTH AND PIGLET IMPRINTING

W. HAZELEGER[1], P. RAMAEKERS[2], C. SMITS[2], AND B. KEMP[1]
[1]*Wageningen University, Department of Animal Sciences, P.O.Box 338, 6700 AH Wageningen, The Netherlands*
[2]*Nutreco Netherlands BV, Swine Research Centre, P.O.Box 220, 5830 AE Boxmeer, The Netherlands*

Introduction

Uterine crowding affects pre- and post-natal development in pigs (for reviews see Kemp *et al.* 2006; Foxcroft *et al.* 2006). Uterine crowding is a consequence of the selection for litter size resulting in a considerable increase in litter size of 0.25 piglets/year in modern commercial dam line sows. This increase in litter size also has consequences for prenatal development. Previously a normal ovulation rate was about 13 to 15-16 for gilts and sows respectively (Van der Lende and Schoenmaker, 1990). Prenatal survival was about 75-80% at Day 28 and 69-74% during the foetal phase. Currently ovulation rates of 25 are considered as normal with embryo survival rates of approximately 60% and foetal survival rates of 50%. It seems therefore that selection for litter size has resulted in high ovulation rates. Since fertilisation rates in pigs are also high (>90% when correctly inseminated) many embryos are available for implantation.

Experiments with limited uterine space in sows show that uterine crowding affects embryonic and foetal survival and growth. Crowding by limiting uterine space by surgical measures leads to reduced embryonic survival, embryonic development at Day 20 and Day 41, reduced placental development at Day 112 and increased coefficients of variation (CV) in placental and foetal weight (Chen and Dziuk, 1993; Père *et al.* 1997; Wu *et al.* 1989). These aspects currently are apparent in modern dam lines, with a reduction in birth weight and limited increase in litter size in comparison with the increase in ovulation rate. It is probable that the increase in "uterine capacity" has not kept up with the increase in ovulation rate, resulting in lower levels of embryonic and foetal survival and negative effects on prenatal development. For example placental development (weight) is decreasing with increasing numbers of viable foetuses in the uterus and, even with moderate uterine crowding, negative effects on foetal development have already been detected (Vonnahme *et al.* 1998; Town *et al.* 2004). Studies of Van der Lende *et al.* (1990) showed that relative weight distribution in a litter and severe retardation ("runting") of foetal development is already established at the end of the embryonic phase (Day 27-35), especially in larger litters.

Postnatal consequences of crowding are reduced birth weight and higher risk of peri-natal mortality and negative effects on post-natal growth. Almost all newborn litters contain mummified foetuses or piglets which are retarded in their development, indicating compromised prenatal development. Piglets with low birth weights are particularly at risk in terms of low post-natal survival, and show reduced and variable weaning weights (Milligan *et al.* 2002). This means that uterine capacity is a limiting factor in determining litter size and post-natal development. Limited uterine capacity implies that some or all foetuses have restricted uterine space available for placental attachment and development. Since placental development is a major factor determining foetal development, the current chapter will focus on nutritional measures to reduce the negative effect of crowding on pre- and post-natal piglet development by enhancing placental development.

Factors affecting placental development

Optimal foetal development can only be achieved if the placenta is well developed. Larger placentas and implantation sites result in heavier foetuses (Van Rens and Van der Lende, 2000). This relationship seems especially to be true for suboptimal situations since the dependency of birth weight on placental weight decreases at higher placental sizes (>300 g) and birth weights (>1500 g) (Van Rens *et al.* 2005).

Placental development around Day 35 is decreased by crowding, often with little or no effect on early foetal development although survival is already reduced in this period (Vallet *et al.* 2003; Town *et al.* 2004). For example at Day 35 average placental weight is decreased from 33±1.8 g to 22±1.9 g by restricting uterine space by unilateral hysterectomy-ovariectomy (UHO), with no effect on average foetal weight (3.7±0.10 and 3.6±0.20 g respectively; Vallet *et al.* 2003). The reduction in foetal development is expressed later in gestation due to the reduced placental development in crowded situations. Père *et al.* (1997) found a 15% reduction in foetal weight in UHO situations at Day 112 of pregnancy, accompanied by an increase in the number of mummies. Town *et al.* (2005) showed that, even in mildly crowded situations, signs of foetal retardation could already be detected based on brain/heart weight ratio in foetuses. This means that placental development is reduced in crowded, space limiting, situations and will have negative consequences for prenatal survival and growth with subsequent consequences for peri- and postnatal survival and growth.

In addition to crowding, other factors are also involved in placental development. The regulation of placental development is complex and aspects such as pre-mating imprinting of important genes such as IGF-2 and H19 appear to be involved (Fowden *et al.* 2006; Vinsky *et al.* 2006, 2007; Wu *et al.* 2006). Therefore, even pre-mating conditions or nutritional status of the sow and boar can have an effect on placental development, with consequences for pre- and post-natal development.

The nutritional status of the mother during (early) pregnancy also has an effect on placental development. Low protein diets have been shown to result in decreased placental and foetal weight, and reduced birth weight and also to result in reduced amounts of important amino acids like arginine, ornithine, glutamine and related substances in placental fluids and foetal plasma (Schoknecht *et al.* 1993, 1994; Wu *et al.* 1998). Placental development therefore appears to be sensitive for nutritional measures.

A critical factor for optimal placental functioning, in addition to the size or exchange surface of the placenta, is the vascularisation of the placenta. A striking example has been shown in the Meishan pig, which produces high numbers of piglets with relative small placentas in a limited uterine space. The placentas in these animals are shorter but much more vascularised than in 'Western' breeds (Ford 1997). Experiments using embryo transfer experiments between highly prolific Meishan gilts and normal Yorkshire gilts showed that size and vascularisation of the placenta are important determinants for placental efficiency, foetal development and survival and appeared to be determined by both maternal and foetal effects (Biensen *et al.* 1998, 1999; Wilson *et al.* 1998; Ford 1997). Furthermore, in numerous other experiments using different species, the importance of vascularisation for placental functioning has been demonstrated (Reynolds and Redmer 2001; Reynolds *et al.* 2005a, 2005b; Wallace *et al.* 2005, 2006; Wu *et al.* 2006). The vascularisation and blood flow in the placenta is important for efficient exchange of nutrients like amino acids, glucose and oxygen and waste products like CO_2 and an efficient placenta can thereby prevent impaired foetal development (Gagnon 2003). Improvement of vascularisation and functioning of the placenta might therefore be of great value in modern pig production.

Regulation of the vascularisation of the placenta

Vascularisation of the placenta is regulated by many factors (like FGF and ANG) with VEGF (vascular endothelial growth factor) as one of the key factors in this process (Charnock-Jones *et al.* 2001; Klagsbrun and D'Amore 1991; Reynolds and Redmer, 2001). In the experiments comparing Yorkshire and Meishan pigs with their highly vascularised placentas, placental VEGF mRNA expression in Meishan pigs was higher compared to Yorkshire pigs (Vonnahme and Ford 2004). Additionally, in Yorkshire pigs, a positive relation was found between the increase in vascularisation and placental expression of the VEGF receptor system between Day 70 and 90 of pregnancy (Vonnahme and Ford, 2003). Besides the role of VEGF in placental angiogenesis, VEGF is also known for its role in foetal angiogenesis (Cheung, 1997).

Hypoxia seems to be one of the major determinants for VEGF expression and angiogenesis (Cheung, 1997; Goldberg and Schneider, 1994; Wheeler *et al.* 1995). Such hypoxic situations are expected in the process of rapid placental development and development of a placental vascular bed. Therefore VEGF is expected to be one of the important factors for vascularisation in placental development.

VEGF activity is stimulated in hypoxic situations, but also stimulated by locally produced nitrogen-oxide (NO). NO is an important factor for many physiological functions including a mediator in immune responses, a neurotransmitter, a cytotoxic free radical and a wide-spread signalling molecule (Ignarro *et al.* 1999; Wu *et al.* 2004). NO also triggers capillary endothelial cell growth and differentiation and stimulates angiogenesis in processes like wound healing, vascular remodelling, ovulation and placental growth and in pathological situations (Battegay, 1995; Milkiewicz *et al.* 2005). NO has proven to stimulate vascular growth in hypoxic situations, but also stimulates embryogenesis and acts as vasodilatator (Murohora *et al.* 1998; Wu *et al.* 2006).

NO is an important component in VEGF activated endothelial cell proliferation (Dulak *et al.* 2000; Zhang *et al.* 2003). Although VEGF is the actual factor that promotes blood vessel growth, VEGF is known by its stimulating effect on NO release in *in vitro* studies (Leung *et al.* 1989; Milkiewickz *et al.* 2006; Murahora *et al.* 1998; Redmer *et al.* 2004). On the other hand, NO induces expression of VEGF (Dulak *et al.* 2000; Zhang *et al.* 2003) and inhibitors of NO reduce the VEGF induced angiogenis. Therefore VEGF and NO seem to act in a close interrelationship in the process of vascularisation/angiogenesis. VEGF and NO, combined with hypoxic situations, seem therefore critical factors for the early development of placental tissue, characterized by rapid growth and vascularisation and formation of the umbilical circulation.

Vascularisation of the placenta and NO

NO is a product of the conversion of L-arginine to citrulline mediated by endothelial NO synthase (eNOS) and tetrahydrobiopterin (BH4) (Wu *et al.* 2006). High levels of arginine can be found in allantoic fluid together with related amino acids like ornithine and glutamine. The amount of these amino acids constitute up to approximately 0.50 of the total N-content in allantoic fluid, and concentrations increase quickly from Day 30 to Day 40 of pregnancy, and decrease sharply thereafter (Wu *et al.* 1995, 1996). This maximum at Day

40 is accompanied by maximal placental NO production (Wu *et al.* 2005). In other studies, using rabbits and rats, arginine also increased vascularisation in ischemic (hypoxic) situations (Muohara *et al.* 1998; Duan *et al.* 2000a, 2000b). Local NO production by arginine and eNOS might therefore play an important role in placental development and vascularisation.

Arginine and related amino acids like proline, glutamine and ornithine are abundant in placental fluids around Day 40 and are important substrates for placental and foetal growth in the pig (Wu *et al.* 2004, 2006). Arginine is, besides being a precursor for nitric oxide (NO), is also a precursor for ornithine which in turn is a precursor for polyamine syntheses. In addition polyamine concentration appears to be maximal around Day 40 of pregnancy, similar to the timing of maximal placental arginine levels and NO production. Polyamines regulate DNA and protein synthesis, cell proliferation and differentiation (Flynn *et al.* 2002) and are therefore important for placental and foetal development.

High levels of arginine and related amino acids in allantoic fluid and the described function of the derivatives of these substances play an important role in the production of NO, activation of VEGF and thereby the vascularisation and development of the placenta. The importance of arginine for foetal development has also been shown in rats. A maternal deficiency in arginine resulted in increased foetal resorption, retardation in intra uterine growth (IUGR), decreased number of vital foetuses and increased perinatal mortality (Greenberg *et al.* 1997; Vosatka *et al.* 1998).

Adding L-arginine (daily 100 g Progenos containing 25 g L-arginine) to the feed of pregnant gilts and sows during the period of placental development (Day 15 – Day 28), increased litter size (total born and born alive) by 0.8 piglet, with no effect on average birth weight or birthweight distribution (Ramaekers *et al.* 2006; see figs 17.1 & 17.2). This period (early pregnancy) of arginine supplementation corresponds to the initial growth phase of the placenta. During this period, formation of blood vessels is one of the most fundamental processes, since blood flow in the foetal placental unit is essential for exchange of nutrients and waste products with the maternal system, as discussed above. Since arginine has a positive effect on litter size, it is interesting to evaluate the effect of arginine in the diet during the period of Day 15-28 of pregnancy on placental and foetal development and survival.

This has been investigated recently in which gilts were stimulated with eCG for higher ovulation rates to increase crowded situations in which the effect of arginine might be more pronounced. The estrus cycle of two groups (total n=50) of cyclic gilts (Hypor) was synchronized with Regumate® during 18 days, the next day followed by 600 or 1000 i.u. eCG (Folligonan®; Intervet; the Netherlands) to stimulate ovulation rates, subsequently (72 h later) followed by 500 i.u. hCG (Chorulon®; Intervet; the Netherlands) to induce ovulation (at 40-42 h after hCG). Gilts were inseminated at 32 and 40h after hCG with a commercial sperm dose. Gilts were treated with Progenos (Trouw Nutrition International; 100 g daily, 25% L-Arginine) or placebo as feed dressing from

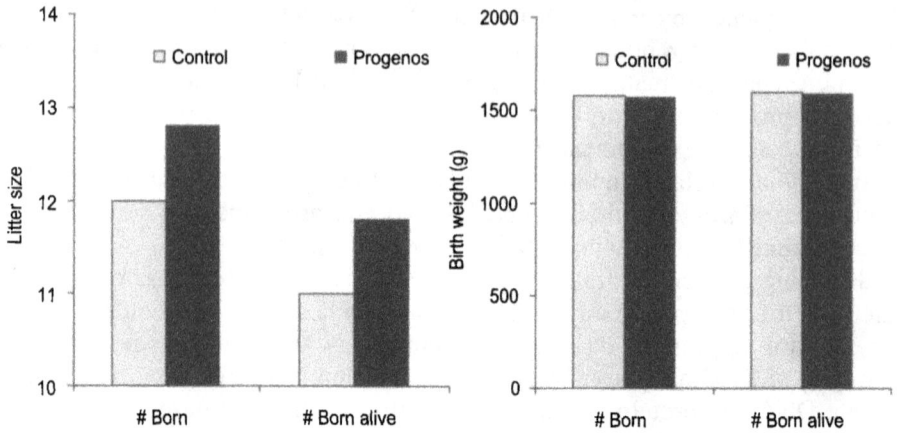

Figure 17.1 Litter size and birth weight of piglets born from Control and Progenos treated multiparous sows (after Ramaekers *et al.* 2006)

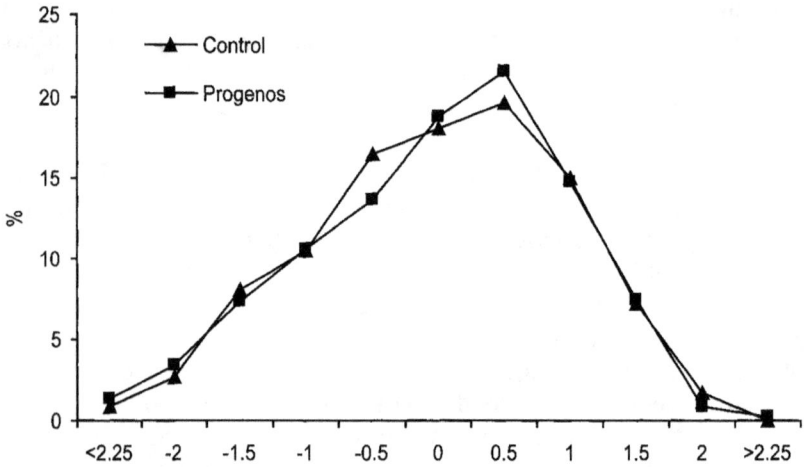

Figure 17.2 Normalised birth weight distribution in Control and Progenos treated sows. (after Ramaekers *et al.* 2006)

Day 15 to Day 28 after 2nd insemination. Pregnant gilts (n=44) were slaughtered at Day 35 of gestation to measure placental and foetal development and foetal survival.

Arginine treatment resulted in 9% higher embryonic survival (p <.05; see table 17.1). This indicates that arginine treatment affects early embryonic survival, that should be accomplished during the period from Day 15 after insemination (start of placental development + arginine treatment) onwards. Analyzing the survival rate in comparison with the amount of crowding, due

Table 17.1 PLACENTAL AND FOETAL DEVELOPMENT AND SURVIVAL AT DAY 35 IN PROGENOS TREATED GILTS

Variable	Control (n=21)	Progenos (n=23)
Foetal survival	0.68 ± 0.05 a	0.77 ± 0.04 b
Foetal weight (g)	4.6 ± 0.2	4.7 ± 0.2
Foetal length (cm)	4.2 ± 0.1	4.2 ± 0.1
Placental weight (g)	30 ± 2	32 ± 2
Placental length (cm)	36 ± 2	37 ± 1
Vascularisation score	2.6 ± .08 [a]	2.9 ± .08 [b]
Placental colour L ("Lightness")[c]	59.3 ± 0.8	58.9 ± 0.5
Placental colour a ("redness")[c]	4.1 ± 0.2 [a]	5.3 ± 0.3 [b]
Placental colour b ("yellowness")[c]	8.1 ± 0.2 [a]	8.8 ± 0.1 [b]

[a,b] different superscripts within a row indicate significance differences (p<0.05)
[c] Control: n=9; Progenos: n=12

to high ovulation rates, it appears that the positive effect on embryonic survival is most obvious in situations with high ovulation rates (see Figure 17.3). Additionally, those animals with relative short implantation sites, indicating crowded intra uterine situations, had a low proportion of viable embryos, and arginine treatment seemed to have a positive effect on post-implantation embryonic survival (expressed as number of vital embryos / total number of implantations; see Figure 17.4). Analyzing placental parameters, placental weight and length were similar in both treatments (see Table 17.1). However, some functional aspects of placental development appeared to be affected. Placenta colour was measured by Minolta L*a*b system on 4 spots approx. 5 cm from the umbilical cord. The redness (a-value) and yellowness (b-value) values of these spots were increased (p<.05) in Progenos treated animals (see Table 17.1).

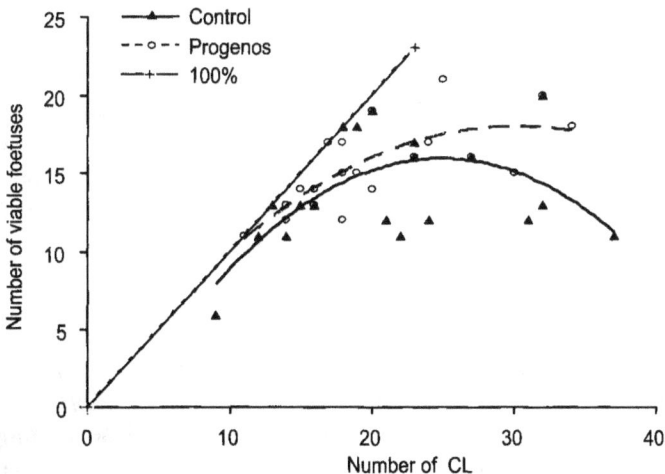

Figure 17.3 Number of vital foetuses related to number of CL in Control and Progenos treated gilts.

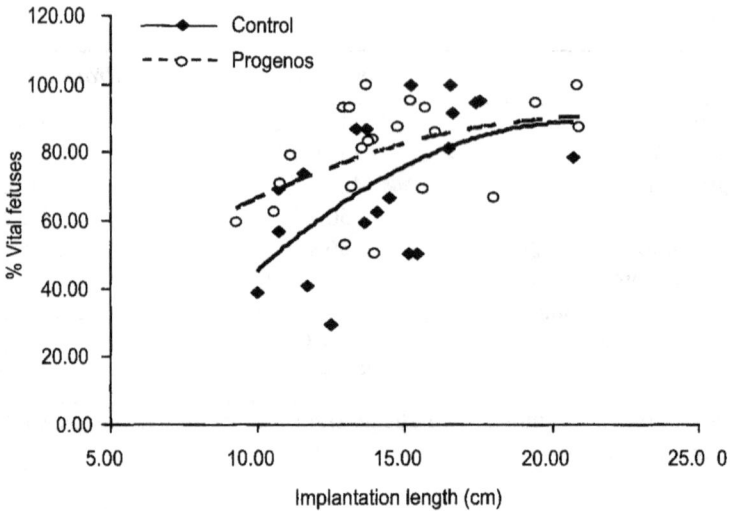

Figure 17.4 Percentage vital fetuses (based on total number of implantations) related to average implantation length in Control and Progenos treated gilts.

A small study on 10 placentas of different sows and experimental groups with extreme "a" and "b" values were analysed by histology for placental vascularisation and placental folding near a spot of colour measuring. No relationship was detected between colour values and vascularisation or placental folding (data not shown). Interpretation of this colour values is therefore difficult and might be related to other aspects of placental morphology. Another aspect studied was the vascularisation of the placenta in the periphery of the necrotic tips. The tips of the placenta are barely reached by branching blood vessels coming from the central blood vessels of the placenta. However, numerous small blood vessels are frequently detected near the necrotic tip. Since the presence of necrotic tips indicates local hypoxic situations, it might be supposed that NO production by arginine might stimulate vascularisation in these locations. The amount of vascularisation of the periphery of the necrotic tips was therefore scored in classes from 1 to 5 on pictures of these placentas. Arginine treatment indeed increased the average vascularisation score from 2.6 ± 0.08 to 2.9 ± 0.08 ($p<0.05$). The effects of arginine on placental colour and vascularisation of the necrotic tips indicate an increased functionality of the placenta due to this treatment.

Analysing data on placental and foetal development (weight and length) on Day 35, no differences could be detected (see Table 17.1) for control and Progenos-treated gilts, indicating that the "rescued" embryos have normal functioning and developing placentas, resulting in normal developing foetuses. Data of Ramaekers *et al.* (2006) also show no differences in average birth weight or distribution in birth weight variance in sows with control and Progenos treatment. (see figures 17.1 & 17.2).

These observations lead to the conclusion that arginine treatment improves functioning of the placenta at least until Day 35 of pregnancy, resulting in better embryonic survival due to "rescue" of retarded embryos, without a negative effect on average birth weight. Probably the "rescued" embryos develop in a normal way, comparable with the other embryos. Since no negative effect on average birth weight due to an increased litter size could be detected, it also means that all developing embryos/foetuses have profit from arginine treatment. The mechanism of better embryo survival after arginine treatment is therefore probably that the most retarded embryos are in the most hypoxic conditions. In these hypoxic situations L-Arginine can have a positive effect on vascularisation, as indicated by increased vascularisation near the hypoxic necrotic tip of the placenta. The positive effects of arginine treatment on prenatal survival and development are therefore probably related to the increased functionality of the placenta.

Interestingly, Mateo *et al.* (2007) found that adding arginine to the feed of pregnant gilts later in gestation between d 30 and 114, caused an increase in number of live-born piglets from 9.4 to 11.4 piglets per litter and an increase in total live litter weight from 13.2 to 16.4 kg. Thus arginine treatment improves the survival of porcine embryos/foetuses and enhances the provision of nutrients to the foetuses for supporting their *in utero* growth also later in gestation. This might be explained by the role of NO as vasodilatating factor, resulting in increasing placental blood flow and thereby exchange of nutrients and waste products. A similar effect has been found by providing underfed ewes sildenafil citrate (Viagra, causing increased NO levels) during Day 28 to day 112 of pregnancy, resulting in prevention of IUGR, probably by improved uteroplacental blood flow (Wu *et al.* 2006). Arginine might also increase the production of polyamines resulting in better functioning of the placenta and foetus as mentioned before, but arginine treatment might also affect growth hormone levels and metabolic processes like insulin resistance, as recently has been shown in man (de Castro Barbosa *et al.* 2006).

In general it can be concluded that vascularisation and functioning of the placenta seems to be enhanced by arginine addition to the diet during the period of early placental development.

Application during pregnancy results in improved embryonic and foetal survival, probably due to increased placental functioning. No negative effects on birth weight due to surviving small embryos are found. The mechanism is expected to be NO production by conversion of arginine affecting vascularisation and blood supply, although arginine metabolism might also affect placental development by other mechanisms like effects on metabolic hormones and the production of polyamines, important for differentiation and placental growth during pregnancy.

Other factors affecting prenatal development

In addition to the effect of arginine on NO and polyamine production, other (nutritional) factors might favour placental and foetal development. Polyamine production is also depending on the decarboxylation of S-adenosylmethionine (SAM) for the conversion of ornithine to putrescine, a precursor for the polyamines spermine and spermidine (Wu *et al.* 2006). SAM is also the critical component as methyl donor in DNA and protein methylation reactions, essential for proper regulation of gene function (DNA conformation, silencing etc.) and development (Jaenisch and Bird 2003) and responsible for epigenetic alterations (genomic imprinting; covalent modifications of DNA and histones). Next to this, the methylating property of SAM is also important in the production of creatine, an important component in energy metabolism of muscle and neuronal cells, and in the production of carnitine, an important factor for fatty acid metabolism. In these methylation processes SAM is converted to homocysteine, which in turn has to be converted back to methionine or to be converted to cysteine and taurine. This process is sometimes a rate limiting step for proper methylation processes.

Elevated levels of homocysteine have been related to several clinical conditions like neural tube defects, spontaneous abortion, placental abruption, low birth weight, renal failure, rheumatic arthritis and complications of diabetes for humans (Mills *et al.* 1995; Nelen *et al.* 2000), but also to a reduction in NO-induced vasodilatation and angiogenesis that could be counteracted by arginine administration (Duan *et al.* 2000b). Homocysteine can be converted by re-methylation to methionine or can be degraded to cysteine and subsequently taurine by trans-sulphuration. Experiments have demonstrated that accumulation of high levels of homocysteine or S-adenosyl homocysteine in cells led to changes in fetal methylation processes (MacLennan *et al.* 2004). An effect of homocysteine on embryo-foetal development might therefore be expected, but as far as is known, no experiments have been published which described the effect of homocysteine level in the sow in relation to litter size.

Many factors are involved in the re-methylation of homocysteine to methionine. The most important factors are Folic acid, Vitamine B6 and B12, serine, choline and betaine. (Wu *et al.* 2006). Supplemental Folic acid has been associated with increased litter sizes of approximately 10% (Fuchs *et al.* 1996; Lindemann, 1993; Matte *et al.* 1984). This effect seems to be related to the first trimester of pregnancy (Tremblay *et al.* 1989) and might be related to early embryonic development and oestrogen production and uterine PGE2 production around the period of implantation (Duquette *et al.* 1997; Matte *et al.* 1996), although PGE2 is not always affected by Folic acid (Guay *et al.* 2004). Recent observations demonstrated the occurrence of secreted and membrane bound folate binding proteins, sequential occurring in endometrial flushings, allantoic fluid and placental tissue of pregnant pigs (Kim and Vallet, 2004, 2007). Vitamin B12 is an important factor for proper folic acid functioning since it stimulates the conversion of 5-methyl TH4-folate to TH4-folate and methylation of homocysteine to methionine (Selhub, 1999). High levels of Vitamine B12 (>100 ug/kg) resulted in an increase of 1 piglet at birth (Matte *et al.* 2006).

The role of Vitamin B6 in affecting litter size has not been reported, but is an essential component is the metabolism of homocysteine by methylation to methionine and trans-sulphuration to cysteine. Also amino acids like serine, choline and betaine play an important role in these methylation processes and might therefore be a target for nutritional approaches to affect placental and/or foetal development (Selhub, 1999; Wu *et al.* 2006). Nutritional intervention with the co-factors required for optimal metabolism of the methionine-homocysteine pathways offers a new possibility for optimizing methyl- and sulphur-group metabolism, and might play a significant role in improving the number and quality of piglets born alive of high prolific sow herds.

Carnitine, also a product of the methylation process of SAM, has shown to increase birth weight and placental development without effect on litter size or prenatal survival (Doberenz *et al.* 2006; Musser *et al.* 1999), although increased number of piglets born alive have been reported (Musser *et al.* 1999). The role of carnitine is expected to increase fatty acid oxidation, and thereby sparing protein degradation. Carnitine stimulates fatty acid degradation and transport of C2 components to mitochondrion. The study of Doberenz *et al.* (2006) did not show a consistent effect of carnitine on these aspects, however it appeared that IGF-1 and IGF-2 levels were increased, at least from Day 60 of gestation onwards with declining, hardly detectable, levels around Day 100 of pregnancy (Musser *et al.* 1999, Doberenz *et al.* 2006). Before Day 60 no effects on IGF-1 levels were found (Waylan *et al.* 2005). In the study of Doberenz *et al.* (2006) not only placental weight at birth was increased (+22%) but also DNA (+38%) and protein (+45%) content and relative glucose transporter-1 concentration (+62%). Since IGF-1 and IGF-2 are key regulators of placenta development and also affect foetal development, it might be concluded that the carnitine treatment stimulates the IGF-complex, resulting in increased placental development and increased foetal growth. Based on this knowledge, it might be expected that carnitine affects placental development only at later stages of gestation, corresponding with the lack of effect of carnitine supplementation on IGF-1 levels before Day 60.

Foetal imprinting

Post-natal development is affected by prenatal growth retardation. In human and several other species strong evidence exists that pre-natal growth retardation has consequences for post-natal performance and health. Well-known effects of IUGR are the risks of development of metabolic (like diabetes type-II) and vascular diseases (like coronary heart disease) in human, poor athletic performance in horses, leading to the theory of "fetal origins of adult disease" or the "Barker theory" (Barker and Clark; 1997). Placental insufficiency in rats, causing similar postnatal effects as in other species, results in postnatal genome-wide DNA hypo-methylation in the liver (MacLennan *et al.* 2004). This is accompanied by an increase in S-adenosyl-homocysteine and homocysteine and a decreased mRNA level of methionine adenosyl transferase, indicating the deviations in the methylation processes. Alterations in methylation processes are expected to be the cause for permanent modifications in the chromatine structure (imprinting), with consequences for post-natal performance.

In pigs it has been shown that growth retardation during the foetal phase due to crowding affects post-natal growth performance and survival, muscle development and deviating allometric organ development (Dwyer *et al.* 1994; Foxcroft *et al.* 2006; Milligan *et al.* 2002; Schoknecht *et al.* 1993; Town *et al.* 2005; Vallet and Freking, 2006). A well-known example is the effect of IUGR on muscle development and differentiation (primary and secondary fibres: number and ratio) during the foetal period. The number of primary fibres is established rather early during gestation and relative insensitive to environmental influences but the number of secondary fibres, developing between D50 and 85 of pregnancy, is sensitive for many factors and responsible for most of the variation in number of muscle fibres established during pregnancy (Dwyer *et al.* 1993, 1994). An increase in maternal feed intake during early pregnancy increases the ratio of secondary to primary fibres (Dwyer *et al.* 1994) and according to Town *et al.* (2004) the number of secondary fibres is also depending on the amount of crowding in the uterus. After parturition, daily growth up to 25 kg is depending on birth weight and not on number of muscle fibres, but the growth from 25 kg to slaughter weight is depending on the number of muscle fibres (Dwyer *et al.* 1993). Littermates with a high number of fibres had a higher weight gain then their littermates with a low number of fibres. Another example of the consequences of foetal growth retardation in pigs is that the ovarian development (weight and number of primary and secondary follicles is decreased in growth retarded piglets at birth (Da Silva-Buttkus *et al.* 2003), although the consequences for later live are yet unknown.

Based on these examples it is obvious that IUGR in pigs is an important obstacle for optimal postnatal development and can affect porcine productivity in many ways. Measures to reduce IUGR are therefore important for optimal pig production.

Although many more factors are involved in optimal placental and and foetal development, the take-home message is that it is possible to implement nutritional measures (although not fully defined yet) during pregnancy to optimise placental development, counteracting the negative effect of crowding as found in modern pig production. Further development of such measures will help in understanding the mechanisms involved in IUGR related complications and offer opportunities for optimising piglet survival and development in the near future.

References

Barker, D. J. and Clark, P.M. (1997). Fetal undernutrition and disease in later life. *Reviews of Reproduction,* **2**(2), 105-112.

Battegay, E. J. (1995). Angiogenesis: mechanistic insights, neovascular diseases, and therapeutic prospects. *Journal of Molecular Medicine,* **73**(7), 333.

Biensen, N. J., Wilson M.E., and Ford S.P. (1998). The impact of either a Meishan or Yorkshire uterus on Meishan or Yorkshire fetal and placental development to days 70, 90, and 110 of gestation. *Journal of Animal Science,* **76**(8), 2169-2176.

Biensen, N. J., Wilson M.E., and Ford S.P. (1999). The impacts of uterine environment and fetal genotype on conceptus size and placental vascularity during late gestation in pigs. *Journal of Animal Science,* **77**(4), 954-959.

Bremer, J. (1961). Carnitine in intermediary metabolism - The biosynthesis of palmitoylcarnitine by cell subfractions. *Journal of Biological Chemistry,* **238**, 2774-2779.

Charnock-Jones, D. S., Clark D.E., Licence D., Day K., Wooding F.B.and Smith S.K.(2001). Distribution of vascular endothelial growth factor (VEGF) and its binding sites at the maternal-fetal interface during gestation in pigs. *Reproduction,* **122**(5), 753-760.

Cheung, C. Y. (1997). Vascular Endothelial Growth Factor: Possible Role in Fetal Development and Placental Function. *Reproductive Sciences,* **4,** 169-177.

Da Silva-Buttkus, P., van den Hurk R., te Velde E.R. and Taverne M.A.M. (2003). Ovarian development in intrauterine growth-retarded and normally developed piglets originating from the same litter. *Reproduction* **126**(2), 249-258.

de Castro Barbosa, T., Lourenco Poyares L., Fabres Machado U. and Nunes M.Y. (2006). Chronic oral administration of arginine induces GH gene expression and insulin resistance. *Life Sciences* **79**(15), 1444.

Doberenz, J., Birkenfeld C., Kluge H. and Eder K. (2006). Effects of L-carnitine supplementation in pregnant sows on plasma concentrations of insulin-like growth factors, various hormones and metabolites and chorion characteristics. *Journal of Animal Physiology and Animal Nutrition,* **90**(11-12), 487-499.

Duan, J., Murohara T., Ikeda H. , Katoh A., Shintani S., Sasaki K., Kawata H., Yamamoto N. and Imaizumi T. (2000a). Hypercholesterolemia Inhibits Angiogenesis in Response to Hindlimb Ischemia: Nitric Oxide-Dependent Mechanism. *Circulation* **102**(90003), 370III—376.

Duan, J., Murohara, Ikeda H., Sasaki K., Shintani S., Akita T., Shimada T. and Imaizumi T. (2000b). Hyperhomocysteinemia Impairs Angiogenesis in Response to Hindlimb Ischemia. *Arteriosclerosis, Thrombosis, and Vascular Biology.* **20**, 2579.

Dulak, J., Jozkowicz A., Dembinska-Kiec A., Guevara I., Zdzienicka A., Zmudzinska-Grochot D.. Florek I., Wojtowicz A., Szuba A. and Cooke J.P. (2000). Nitric Oxide Induces the Synthesis of Vascular Endothelial Growth Factor by Rat Vascular Smooth Muscle Cells. *Arteriosclerosis, Thrombosis, and Vascular Biology*, **20**(3), 659-666.

Duquette, J., Matte J.J., Farmer C., Girard C.L. and Laforest J.P.(1997). Pre- and post-mating dietary supplements of folic acid and uterine secretory activity in gilts. *Canadian Journal of Animal Science*, **77**, 415–420.

Dwyer, C. M., Fletcher J.M. and Stickland N.C. (1993). Muscle cellularity and postnatal growth in the pig. *Journal of Animal Science*, **71**(12), 3339-3343.

Dwyer, C. M., Stickland N.C. and Fletcher J.M. (1994). The influence of maternal nutrition on muscle fiber number development in the porcine fetus and on subsequent postnatal growth. *Journal of Animal Science*, **72**(4), 911-917.

Flynn, N. E., Meininger C.J. ,Hayes T.E. and Wu G. (2002). The metabolic basis of arginine nutrition and pharmacotherapy. *Biomedecine & Pharmacotherapy*, **56**(9), 427.

Ford, S.P. (1997). Embryonic and fetal development in different genotypes in pigs. *Journal of reproduction and fertility. Supplement*, **52**, 165-176.

Fowden, A. L., Ward J.W. ,Wooding F.P.B., Forhead A.J. and Constancia M.(2006). Programming placental nutrient transport capacity. *The Journal of Physiology (London)*, **572**(1), 5-15.

Foxcroft, G., Dixon, W.T. Novak, S., Putman, C.T., Town, S.C. and Vinsky, M.D.A. (2006). Prenatal programming of postnatal growth performance. *Proceedings University of Alberta – University of Minnesota Leman Pre-Conference Reproduction Workshop; September 2006; Minnesota 57-68.*

Fuchs, B., Orda, J. and Wiliczkiewicz, A. (1996). [Effects of folic acid upplementation in pregnant sows on the fetal mortality. *Medycyna Veterynaryjna* **52**, 51–53.

Gagnon, R. (2003). Placental insufficiency and its consequences. *European Journal of Obstetrics & Gynecology and Reproductive Biology* **110**(Supplement 1), S99.

Goldberg, M. A. and Schneider T.J. (1994). Similarities between the oxygen-sensing mechanisms regulating the expression of vascular endothelial growth factor and erythropoietin. *The Journal of Biological Chemistry*, **269**(6), 4355-4359.

Greenberg, S. S., Lancaster J.R., Xie J., Hua L., Freeman T., Kapusta D.R., Giles T.D.and Powers D.R. (1997). Effects of NO synthase inhibitors, arginine-deficient diet, and amiloride in pregnant rats. *The American Journal of Physiology – Regulatory, Intergrative and Comparative Physiology*, **273**(3), R1031-1045.

Guay, F., Matte J.J., Girard C.L., PalinM.-F., Giguere A. and Laforest J.P. (2004).

Effect of folic acid plus glycine supplement on uterine prostaglandin and endometrial granulocyte-macrophage colony-stimulating factor expression during early pregnancy in pigs. *Theriogenology*, **61**(2-3), 485.

Ignarro, L.J., Cirino G., Casini and C. Napoli C. (1999). Nitric Oxide as a Signaling Molecule in the Vascular System: An Overview. *Journal of Cardiovascular Pharmacology*. 34(6),879-886

Jaenisch, R. and Bird A. (2003). Epigenetic regulation of gene expression: how the genome integrates intrinsic and environmental signals. *Nature Genetics*, **33**, 245 – 254.

Kemp, B., Foxcroft G. and Soede N.(2006). Physiological determinants of litter size in contemporary dam-line sows. *Proceedings University of Alberta – University of Minnesota Leman Pre-Conference Reproduction Workshop; September 2006; Minnesota*, 33-44.

Kim J. and Vallet J. (2004). Secreted and Placental Membrane Forms of Folate-Binding Protein Occur Sequentially During Pregnancy in Swine. *Biology of Reproduction*, **71**(4),1214–1219.

Kim, J. G. and Vallet J.L. (2007). Placental expression of the membrane form of folate binding protein during pregnancy in swine. *Theriogenology*, **67**(7), 1279.

Klagsbrun, M. and D'Amore P.A. (1991). Regulators of Angiogenesis. *Annual Review of Physiology*, **53**(1), 217-239.

Leung, D.W., Cachianes G., Kuang W.-J., Goeddel D.V. and Ferrara N. (1989). Vascular endothelial growth factor is a secreted angiogenic mitogen. *Science* **246** (4935), 1306-1309.

Lindemann, M. D. (1993). Supplemental folic acid: A requirement for optimizing swine reproduction. *Journal of Animal Science*, **71**, 239–246.

MacLennan, N. K., James S.J., Melnyk S., Piroozi A., Jernigan S., Hsu J.L., Pham T.D. and Lane R.H. (2004). Uteroplacental insufficiency alters DNA methylation, one-carbon metabolism, and histone acetylation in IUGR rats. *Physiological Genomics* **18**(1), 43-50.

Mateo, R. D., Wu G., Bazer F.W., Park J.C., Shinzato I. and Kim S.W. (2007). Dietary L-Arginine Supplementation Enhances the Reproductive Performance of Gilts. *Journal of Nutrition*, **137**(3), 652-656.

Matte, J.J. ,Farmer C., Girard C.L. and Laforest J.P. (1996). Dietary folic acid, uterine function and early embryonic development in sows. *Canadian Journal of Animal Science*, **76**, 427–433

Matte, J.J., Girard C.L. and Brisson G.J. (1984). Folic acid and reproductive performances of sows. *Journal of Animal Science*, **59**, 1020–1025.

Milkiewicz, M., Ispanovic M.E., Doyle J.L. and Haas T.L. (2006). Regulators of angiogenesis and strategies for their therapeutic manipulation. *The International Journal of Biochemistry & Cell Biology*, **38**(3), 333.

Milkiewicz, M., Hudlicka O., Brown M.D. and Siligram H. (2005). Nitric oxide, VEGF, and VEGFR-2: interactions in activity-induced angiogenesis in rat skeletal muscle. *The American Journal of Physiology – Heart and*

Circulatory Physiology , (1), H336-343.

Milligan, B. N., Dewey C.E. and de GRau A.F. (2002). Neonatal-piglet weight variation and its relation to pre-weaning mortality and weight gain on commercial farms. *Preventive Veterinary Medicine*, **56**(2), 119.

Mills, J. L., Lee Y.J. , Conley M.R., Kirke P.N. . McPartin J.M., Weir D.G. and Scott J.M. (1995). Homocysteine metabolism in pregnancies complicated by neural-tube defects. *The Lancet*, **345**(8943), 149.

Murohara, T., Asahara T., Silver M., Bauters C., Masuda H., Kalka C., Kearny M., Chen D., Chen D., Symes J.F., Fishman M.C., Huang P.L.and Isner J.M. (1998). Nitric Oxide Synthase Modulates Angiogenesis in Response to Tissue Ischemia. *The Journal of Clinical Investigation*, **101**(11), 2567-2578.

Musser, R. E., Goodband R.D., Tokach M.D., Owen K.Q., Nelssen J.L., Blum S.A., Dritz S.S. and Civiz C.A.(1999). Effects of L-carnitine fed during gestation and lactation on sow and litter performance. *Journal of Animal Science*, **77**(12), 3289-3295.

Nelen, W. L. D. M., Blom H.J., SteegersE.A.P., den Heijer M., Thomas C.M.G. and Eskes T.K.A.B. (2000). Homocysteine and folate levels as risk factors for recurrent early pregnancy loss. *Obstetrics & Gynecology*, **95**(4), 519.

Père, M. C., Dourmad J.Y. and Etienne M. (1997). Effect of number of pig embryos in the uterus on their survival and development and on maternal metabolism. *Journal of Animal Science*, **75**(5), 1337-1342.

Ramaekers, P., Kemp B. and van der Lende T.(2006). Progenos in sows increases number of piglets born. *Journal of Animal Science*, **84** (Suppl. 1), 394.

Redmer, D. A., Wallace J.M. and Reynolds L.P. (2004). Effect of nutrient intake during pregnancy on fetal and placental growth and vascular development. *Domestic Animal Endocrinology*, **27**(3), 199-217.

Reynolds, L. P., Borowicz P.P., Vonnahme K.A., Johnson M.L., Grazul-Bilska A.T., Wallace J.M., Caton J.S. and Redmer D.A. (2005). Animal models of placental angiogenesis. *Placenta,* **26**(10), 689.

Reynolds, L. P., P. P. Borowicz P.P., Vonnahme K.A., Johnson M.L., Grazul-Bilska A.T., Redmer D.A. and Caton J.S. (2005). Placental angiogenesis in sheep models of compromised pregnancy. *The Journal of Physiology (London)*, **565**(1), 43-58.

Reynolds, L. P., Borowicz P.P., Vonnahme K.A., Johnson M.L., Grazul-Bilska A.T., Wallace J.M., Caton J.S. and Redmer D.A. (2005). Animal models of placental angiogenesis. *Placenta*, **26**(10), 689.

Reynolds, L. P., J. S. Caton, D. A. Redmer, A.T. Grazul-Bilska, K.A. Vonnahme, P.P. Borowicz, J.S. Luther, J.M. Wallace, G. Wu and T.E. Spencer (2006). Evidence for altered placental blood flow and vascularity in compromised pregnancies. *The Journal of Physiology (London)*, **572**(1), 51-58.

Reynolds, L. P. and Redmer D.A. (2001). Angiogenesis in the Placenta. *Biology of Reproduction*, **64**(4), 1033-1040.

Schoknecht, P. A., Newton G.R., Wise D.E. and Pond W.G. (1994). Protein restriction in early pregnancy alters fetal and placental growth and allantoic fluid proteins in swine. *Theriogenology,* **42**(2), 217.

Schoknecht, P. A., Pond W.G., Mersmana H.J. and Maurer R.R. (1993). Protein Restriction during Pregnancy Affects Postnatal Growth in Swine Progeny. *Journal of Nutrition,* **123**(11), 1818-1825.

Selhub, J. (1999). HOMOCYSTEINE METABOLISM. *Annual Review of Nutrition,* **19**(1), 217-246.

Town, S. C., Putman C.T., Turchinsky N.J., Dixon W.J. and Foxcroft G.R. (2004). Number of conceptuses in utero affects porcine foetal muscle development. *Reproduction,* **128**(4), 443-454.

Town, S. C., J. L. Patterson J.L., C.Z. Pereira C.Z., G. Gourley G.and G.R. Foxcroft G.R. (2005). Embryonic and foetal development in a commercial dam-line genotype. *Animal Reproduction Science,* **85**(3-4), 301.

Tremblay, G.F., Matte, J.J. and Brisson G.J. (1989). Survival rate and development of fetuses during the first 30 days of gestation after folic acid addition to a swine diet. *Journal of Animal Science,* **67** (3), 724-732

Vallet, J. L. and Freking B.A. (2006). Changes in fetal organ weights during gestation after selection for ovulation rate and uterine capacity in swine. *Journal of Animal Science,* **84**(9), 2338-2345.

Vallet, J. L., Freking B.A., Leymaster K.A. and Christenson R.K. (2005). Allelic variation in the secreted folate binding protein gene is associated with uterine capacity in swine. *Journal of Animal Science,* **83**(8), 1860-1867.

Vallet, J. L., Klemcke H.G., Christenson R.K. and Pearson P.L. (2003). The effect of breed and intrauterine crowding on foetal erythropoiesis on day 35 of gestation in swine. *Journal of Animal Science,* **81**(9), 2352-2356.

Van der Lende, T. and Schoenmaker G.J. (1990) . Weight distribution within litters at the early foetal stage and at birth in relation to embryonic mortality in the pig. *Livestock Production Science,* **26**, 53-65

van Rens, B. T. T. M., de Koning G., Bergsma R. and van der Lende T. (2005). Preweaning piglet mortality in relation to placental efficiency. *Journal of Animal Science,* **83**(1), 144-151.

van Rens, B. T. T. M. and van der Lende T. (2000). Foetal and placental traits at day 35 of pregnancy in relation to the estrogen receptor genotype in pigs. *Theriogenology,* **54**(6), 843.

Vinsky, M. D., Murdoch G.K., Dixon W.T. , Dyck M.K. and Foxcroft G.R. (2007). Altered epigenetic variance in surviving litters from nutritionally restricted lactating primiparous sows. *Reproduction, Fertility and Development,* **19**(3), 430-435.

Vinsky, M. D., Novak S., Dixon W.T., Dyck M.K. and Foxcroft G.R. (2006). Nutritional restriction in lactating primiparous sows selectively affects female embryo survival and overall litter development. *Reproduction,*

Fertility and Development, **18**(3), 347-355.

Vonnahme, K. A. and Ford S.P. (2004). Differential Expression of the Vascular Endothelial Growth Factor-Receptor System in the Gravid Uterus of Yorkshire and Meishan Pigs. *Biology of Reproduction*, **71**(1), 163-169.

Vonnahme, K. A., Hess B.W., Hansen T.R., McCormick R.J., Rule D.C., Moss G.E., Murdoch W.J., Nijland M.J., Skinner D.C., Nathanielz P.W. and Ford S.P. (2003). Maternal Undernutrition from Early- to Mid-Gestation Leads to Growth Retardation, Cardiac Ventricular Hypertrophy, and Increased Liver Weight in the Fetal Sheep. *Biology of Reproduction*, **69**(1), 133-140.

Vonnahme, K. A., Wilson M.E., Foxcroft G.R. and Ford S.P. (2002). Impacts on conceptus survival in a commercial swine herd. *Journal of Animal Science*, **80**(3), 553-559.

Vosatka, R. J., Hassoun P.M. and Harvey-Wilkes K.B. (1998). Dietary l-arginine prevents fetal growth restriction in rats. *American Journal of Obstetrics and Gynecology*, **178**(2), 242.

Waylan, A. T., Kayser J.P.,Gnad J.J., Higgins J.D., Starkey E.K., Sissom E.K., Woodrow J.C. and Johnson B.J. (2005). Effects of L-carnitine on fetal growth and the IGF system in pigs. *Journal of Animal Science*, **83**(8), 1824-1831.

Wallace, J. M., Luther J.S., Milne J.S., Aitken R.P., Redmer D.A., Reynolds L.P. and Hay W.W. Jr (2006). Nutritional Modulation of Adolescent Pregnancy Outcome - A Review. *Placenta*, **27**(Supplement 1), 61.

Wallace, J. M., Regnault T.R.H., Limesand S.W., Hay W.W. Jr. and Anthony R.V. (2005). Investigating the causes of low birth weight in contrasting ovine paradigms. *The Journal of Physiology*, **565**(1), 19-26.

Wheeler, T., Elcock C.L. and Anthony R.V. (1995). Angiogenesis and the placental environment. *Placenta*, **16**(3), 289.

Wilson, M. E., Biensen N.J., Youngs C.R. and Ford S.P. (1998). Development of Meishan and Yorkshire littermate conceptuses in either a Meishan or Yorkshire uterine environment to day 90 of gestation and to term. *Biology of Reproduction*, **58**(4), 905-910.

Wu, G., Bazer F.W., Cudd T.A., Meininger C.J. and Spencer T.E. (2004). Maternal Nutrition and Fetal Development. *Journal of Nutrition*, **134**(9), 2169-2172.

Wu G, Bazer F.W., Hu J., Johnson G.A. and Spencer T.E. (2005). Polyamine Synthesis from Proline in the Developing Porcine Placenta. *Biology of Reproduction*, **72**(4): 842–850.

Wu, G., Bazer F.W. and W. Tou W. (1995). Developmental Changes of Free Amino Acid Concentrations in Fetal Fluids of Pigs. *Journal of Nutrition*, **125**(11), 2859-2868.

Wu, G., Bazer F.W., Tou W. and Flynn S.P. (1996). Unusual abundance of arginine and ornithine in porcine allantoic fluid. *Biology of Reproduction*, **54**(6), 1261-1265.

Wu, G., Bazer F.W., Wallace J.W. and Spencer T.E. (2006). BOARD-INVITED REVIEW: Intrauterine growth retardation: Implications for the animal sciences. *Journal of Aimial Science,* **84**(9), 2316-2337.

Wu, G., Pond W.G., Ott T.and Bazer F.W. (1998). Maternal Dietary Protein Deficiency Decreases Amino Acid Concentrations in Fetal Plasma and Allantoic Fluid of Pigs. *Journal of Nutrition,* **128**(5), 894-902.

Zhang, R., Wang L., Zhang L., Chen J., Zhu Z., Zhang Z. and Chopp M. (2003). Nitric Oxide Enhances Angiogenesis via the Synthesis of Vascular Endothelial Growth Factor and cGMP After Stroke in the Rat. *Circulation Research,* **92**(3), 308-313.

18

GILT MANAGEMENT, OOCYTE QUALITY AND EMBRYO SURVIVAL

W. H. E. J. VAN WETTERE[a] AND P. E. HUGHES[b]

[a] Discipline of Agricultural and Animal Science, School of Agriculture, Food and Wine, The University of Adelaide, Roseworthy, South Australia 5371, Australia; [b] Pig and Poultry Production Institute, Roseworthy, South Australia 5371, Australia

Introduction

The reproductive performance of replacement gilts exerts a major impact on overall breeding herd productivity. Facilitating gilt entry into the breeding herd and maximising subsequent reproductive performance depends on the combination of appropriate nutritional management and effective puberty stimulation and mating strategies (Evans and O'Doherty, 2001). However, incidences of reproductive disorders and failure, including delayed or asynchronous puberty attainment and low first litter sizes, are common within cohorts of replacement gilts (Whittemore, 1996). As a result, approximately 0.20 of premature culling of sows from the breeding herd occurs at parity 0, with 0.65 of these culls attributed to reproductive disorders or failure (Lucia, Dial and Marsh, 2000). In order to address this problem and improve gilt reproductive performance, it is vital to understand the nutritional and socio-sexual signals that control and coordinate sexual activity and reproductive processes in the female pig (Cosgrove and Foxcroft, 1996; Prunier and Quesnel, 2000; Martin, Rodger and Blache, 2004). This chapter will focus on some of the strategies used to control the onset of puberty and maintain cyclicity in the gilt, as well as discussing current knowledge concerning the effects of nutrition and environment on puberty attainment and first litter size.

Sexual maturation and puberty attainment

The onset of puberty in the gilt depends on the progressive maturation of the hypothalamic-pituitary-ovarian axis, with the period from birth to puberty characterised by specific patterns of gonadotrophin and steroid release, as well as gradual development of the ovarian follicle population (Camous, Prunier and Pelletier, 1985; Christenson, Ford and Redmer, 1985). Integration of the pituitary-ovarian axis occurs during the first three to four months of post-natal life, and is followed by a 'waiting phase', which lasts from approximately 120 days of age until the onset of puberty (Camous *et al.*, 1985; Pressing, Dial, Esbenshade and Stoud, 1992). During this 'waiting phase', circulating oestrogen concentrations rise (Pressing *et al.*, 1992) and ovarian follicle growth appears to occur in successive waves (Bolamba, Matton, Estrada and Dufour, 1994). However, suppressed activity of the gonadotrophin releasing hormone (GnRH) pulse generator and thus inadequate episodic luteinising hormone (LH) secretion prevent the final stages of follicle growth and ovulation (Camous *et al.*, 1985; Plant, 2002). Similar to studies involving sheep (Ryan, Goodman, Karsch, Legan and Foster, 1991), the frequency of episodic LH release increases, while the amplitude of LH pulsing decreases, approximately 30 days prior to the onset of puberty in gilts (Lutz, Rampacek, Kraeling, Pinkert, 1984). This increase in LH pulse frequency stimulates the initiation of the first follicular phase, during which growth of ovarian follicles to the preovulatory stage results in increased oestrogen secretion and activation of positive oestrogen feedback, thus stimulating the specific neural pathways that elicit the continuous surge of GnRH discharge responsible for the first pre-ovulatory LH surge and ovulation.

Although a degree of controversy exists, it is probable that a decline in the sensitivity of the GnRH pulse generator to negative oestrogen feedback, combined with a gonad independent reduction in inhibitory inputs and activation of stimulatory inputs to the GnRH neuronal network, is responsible for the increase in the frequency of LH pulse release that precedes puberty in ewe lambs, heifers and gilts (Pelletier, Carrez-Camous and Thiery, 1981; Lutz *et al.*, 1984; Day, Imakawa, Wolfe, Kittok, Kinder, 1987; Huffman, Inskeep and Goodman, 1987; Dyer, Bishop and Day, 1990; Elsaesser, Parvizi, and Foxcroft, 1998; Plant, 2002; Ojeda, Lomniczi, Mastronardi, Heger, Roth, Parent, Matagne, and Mungenast, 2006). Current understanding of the developmental cues responsible for initiating this sequence of neuroendocrine events is limited. However, the available literature strongly supports the existence of an inherent, genetically programmed pubertal clock (Plant, 2002), and Ojeda *et al.* (2006) suggested that the neuroendocrine control of puberty is in fact regulated by a hierarchically arranged and complex network of genes controlled by a *"few highly connected upper-echelon gene hubs"*, the activation of which is genetically determined. Therefore, whilst socio-sexual, seasonal and nutritional factors modulate the timing of puberty, it is likely that they do so by influencing the activity of the GnRH neuronal system and development of the ovary, as opposed to acting as the ultimate cue for pubertal development to occur (Plant, 2002).

Control of puberty attainment and subsequent cyclicity

Although gilts normally attain puberty when they are approximately 200 to 220 days old, age at first oestrus varies enormously, ranging from 105 to 350 days (Hughes, 1982). Reducing this variation is extremely beneficial to the productivity of a breeding herd, facilitating gilt entry into the breeding herd and increasing the efficiency of cull sow replacement. Equally important, early puberty attainment is associated with younger mating ages, a shorter non-productive period prior to breeding herd entry, and a reduction in rearing costs (Brooks and Smith, 1980; Hughes, 1982; Aherne and Kirkwood, 1985; Koketsu, Takahashi and Akachi, 1999). Boar contact, and to a lesser extent exogenous hormone administration, are the most common methods used to stimulate and control gilt puberty.

THE BOAR EFFECT AND PUBERTY ATTAINMENT

The efficacy of boar exposure as a stimulus for early puberty attainment, often referred to as the 'boar effect', has been clearly demonstrated (Hughes and Cole, 1976; Hughes, 1994) and extensively reviewed (see Hughes, Pearce and Paterson, 1990). The precise mechanisms responsible for early puberty attainment in response to boar stimulation are incompletely understood. However, it is generally accepted that olfactory cues, namely priming pheromones (e.g. 3α-androstenol), present in saliva secreted by the boar's submaxillary salivary glands, act synergistically with tactile and possibly auditory and visual stimuli to accelerate and synchronise puberty in the gilt (Hughes, Pearce and Paterson, 1990). Although data is limited in the gilt, there is ample evidence from other mammalian species that priming pheromones released by mature males affect female reproductive status via changes in the activity of the GnRH neurons (Rissman, 1996; Rissman, Li, King and Millar, 1997; Bakker, Kelliher and Baum, 2001), and the initial response of pre-pubertal female mice (Bronson and Desjardins, 1974) and ewe lambs (Knights, Baptiste, Lewis, 2002) to the introduction of a reproductively mature male is an alteration in the pattern of LH release. Consequently, it is generally believed that boar-originating stimuli alter the pattern of LH secretion in the gilt (Hughes, Pearce and Paterson, 1990; Kingsbury and Rawlings, 1993), and that this actuates an increase in ovarian follicle growth, causing oestrogen concentrations to rise, thus triggering the cascade of endocrine events that culminate with the onset of oestrus and the first ovulation (Paterson, 1982; Deligeorgis, English, Lodge and Foxcroft, 1984; Esbenshade Paterson, Cantley and Day, 1992).

The exhibition of a pubertal response depends on the gilt receiving sufficient boar stimulation, and variations in gilt response are attributed primarily to the amount of physical gilt-boar interaction that occurs and the frequency at which boar exposure is applied. The transfer of non-volatile, peptide bound priming pheromones from the boar's submaxillary gland to the sensory neurones of the gilt's vomeronasal organ (VNO) requires the gilt to make physical contact with boar saliva (Pearce and Paterson, 1992; Austin, 2004; Brennan and Keverne, 2004). Allowing gilts and boars to interact in a reciprocal manner may also promote the exhibition of boar courtship behaviours as well as saliva production by the submaxillary glands (Booth, 1980). Further, although data is absent in the pig, studies involving rodent species indicate that priming pheromones in female urine and/or faeces increase testosterone secretion and stimulate LH release in the male, potentially increasing the production of male priming pheromones (Bronson and Maruniak, 1975; Lombardi, Vandenbergh and Whitsett, 1976; Maruniak, Coquelin and Bronson, 1978; Bakker, 2003; Anand, Turek and Horton, 2004). Therefore, it is not surprising that maximising physical gilt to boar contact increases the efficacy of boar exposure. Compared to fenceline boar contact, the provision of full, physical boar contact can double the proportion of gilts attaining puberty (Pearce and Paterson, 1992), as well as reduce age at puberty by approximately 10 days (Patterson, Willis, Kirkwood and Foxcroft, 2002a; Table 18.1).

Table 18.1 TIMING OF THE PUBERTAL RESPONSE OF 160 DAY OLD GILTS TO EITHER FULL OR FENCELINE DAILY BOAR CONTACT (AFTER PATTERSON *et al.*, 2002)

	Days to puberty	*Age at puberty (days)*
Full boar contact	21.8[a] ± 2.7	180.9[a] ± 2.5
Fence - line boar contact	32.0[b] ± 3.2	191.1[b] ± 2.8

[ab] within columns indicates significant difference; $P < 0.05$

Gilt perception of boar component stimuli also requires sufficient contact with the boar, as well as regular reinforcement of the stimulus (Paterson, Hughes and Pearce, 1989; Philip and Hughes 1995). Although a number of studies (e.g. Paterson *et al.*, 1989; Hughes, 1994) indicate that conducting boar stimulation daily for at least 15 – 20 minutes until the onset of first oestrus optimises the effectiveness of boar stimuli, more recent studies demonstrate an enhanced gilt response to boar stimulation when it occurs twice daily (Hughes, 1994; Hughes and Thorogood, 1999; Table 18.2).

The efficacy with which boar contact accelerates puberty attainment ultimately depends on the sexual maturity, or physiological age, of the gilt (Eastham, Dyck and Cole, 1986). Chronological age, as opposed to gilt liveweight and body composition, is a more favourable predictor of the timing of the pubertal response to boar stimulation (Hughes and Varley, 1980; Hughes and Varley, 2003). Early studies indicated that the hypothalamic-pituitary-ovarian axis was sufficiently integrated by 23 weeks of age to allow a rapid and synchronous pubertal response, with optimal response to boar stimulation occurring at this age (Hughes and Cole, 1976; Kirkwood and Hughes, 1979; Eastham *et al.*, 1986; Paterson *et al.*, 1989). However, there is general agreement that current selection strategies based on increasing lean growth also result in greater mature weight (O'Dowd, Hoste, Mercer, Flower, and

Table 18.2 TIMING OF THE PUBERTAL RESPONSE OF 160 DAY OLD GILTS TO DIFFERENT FREQUENCIES OF FULL BOAR CONTACT (AFTER HUGHES AND THOROGOOD, 1999)

Proportion pubertal by:	*Frequency of daily boar contact*		
	0	*1*	*2*
Day 15	0.00[a]	0.12[b]	0.53[c]
Day 30	0.00[a]	0.25[b]	0.73[c]
Day 45	0.07[a]	0.69[b]	0.87[c]
Mean days to puberty	45.0	32.4[b]	16.0[c]

[abc] within rows indicates significant difference; $P < 0.05$

Edwards, 1997; Evans and O'Doherty, 2001). Therefore, although modern gilts are now heavier at 23 weeks of age (van Wettere, Revell, Mitchell and Hughes, 2006), they are also leaner, at a lower proportion of their mature weight and therefore physiologically less mature than their smaller, fatter predecessors of 3 decades ago (reviewed by Whittemore, 1996; Edwards, 1998; Evans and O'Doherty, 2001; Slevin and Wiseman, 2003). Although the consequences of current selection on sexual maturation and reproductive performance have yet to be fully explored (Gaughan, Cameron, Dryden and Josey, 1995), it has recently been demonstrated (van Wettere *et al.*, 2006) that maximal response to puberty stimulation occurs at 26 weeks of age or older, approximately 3 weeks later than previously reported in the literature (Hughes and Cole, 1976; Kirkwood and Hughes, 1979; Eastham *et al*, 1986). In the study of van Wettere *et al.* (2006), puberty occurred more rapidly, and with greater synchrony between contemporary females, when boar stimulation commenced at 26 or 29 weeks of age as opposed to 23 weeks of age (Table18. 3). Faster gilt response to boar stimulation is indicative of a more developed hypothalamic-pituitary-ovarian axis (Kirkwood and Hughes, 1979; Deligeorgis, Lunney and English, 1984), suggesting this hormonal axis is not fully developed at 23 weeks in current genotypes. This supports the notion that today's genotypes are later maturing, emphasising the need to adapt current recommendations if the effectiveness of boar stimulation is to be improved, or at least maintained.

EXOGENOUS HORMONE TREATMENTS AND PUBERTY ATTAINMENT

Between 100 and 135 days of age gilts acquire the ability to respond to appropriate doses of exogenous GnRH or gonadotrophins with follicular growth and ovulation (Paterson, 1982; Dial, Dial, Wilkinson and Dziuk, 1984; Esbenshade, Ziecik and Britt, 1990; Pressing *et al.*, 1992). The success with which human chorionic gonadotrophin (hCG) stimulates ovulation depends

Table 18.3 ATTAINMENT OF PUBERTY IN GILTS IN RESPONSE TO 20 MINUTES OF FULL, DAILY CONTACT WITH A VASECTOMIZED BOAR IN A DETECTION-MATING AREA WHEN BOAR EXPOSURE COMMENCED AT 23, 26 OR 29 WEEKS OF AGE (AFTER VAN WETTERE *et al.*, 2006)

	Age at start of boar exposure		
	23 weeks	*26 weeks*	*29 weeks*
Proportion of gilts pubertal by:			
Day 10	0.24[a]	0.67[b]	0.70[b]
Day 20	0.70[a]	0.81[b]	0.93[c]
Day 35	0.82	0.98	1.00
Mean days to puberty	18.9 ±1.5 [b]	10.6 ±1.2 [a]	8.3 ± 0.9 [a]

[abc] means, in the same row, with different superscripts are significantly different (P < 0.05)

on the presence of sufficient number of medium (4 – 5 mm) and large (≥ 6mm) follicles to secrete, and maintain, oestradiol concentration above the threshold required to stimulate GnRH surge release (Driancourt. Prunier, Bidanel and Martinat-Botte, 1992). Additionally, ovulation rate at the hormonally induced oestrus reflects the number of large follicles present at the time of injection (Table 18.4).

Table 18.4 THE NUMBER OF SURFACE FOLLICLES PER OVARY OF 170 DAY OLD, PREPUBERTAL GILTS COINCIDENT WITH hCG INJECTION, AND THE NUMBER OF CORPORA LUTEA (CL'S) PER OVARY 8 DAYS POST HCG (AFTER BOLAMBA *et al.*, 1991)

Ovarian type coincident with hCG injection	*Follicles per ovary coincident with hCG injection*				*CL's per ovary*
	(Total)	*(1-3mm)*	*(4-5 mm)*	*(≥6mm)*	*(day 8 post hCG)*
'honeycomb'	59.4 ± 2.3[b]	42.3 ± 0.8[b]	14.6 ± 1.7	2.6 ± 0.3[a]	5.2 ± 0.7[a]
'grape-like'	52.2 ± 1.5[a]	26.7 ± 0.9[a]	15.6 ± 0.8	10.0 ± 0.5[b]	10.6 ± 2.3[b]

[ab] within same column indicates significant difference (P < 0.05)

After 100 days, gilt age has little effect on the proportion of animals ovulating in response to exogenous gonadotrophins (Paterson, 1982). However, although 0.90 of animals exhibit an ovulatory response, roughly 0.30 fail to exhibit behavioural oestrus, with an equal number failing to exhibit normal subsequent oestrous cyclicity (Paterson, 1982; Tilton, Bates and Prather, 1995). Equally problematic is the enormous variation in both farrowing rate and litter size of gilts bred at the hormonally induced oestrus. These high rates of pregnancy failure are attributed to premature regression of corpora lutea (Paterson, 1982), with exogenous hormone treatments also associated with increased incidences of oocyte and embryo abnormalities (Guthrie, Pursel and Wall, 1997) and an increased proportion of degenerating embryos (Ziecik, Biallowicz, Kaczmarek, Demianowicz, Rioperez, Wasielak, and Bogacki, 2005).

The provision of boar contact can considerably improve the efficacy of exogenous gonadotrophin treatments. The provision of daily fence-line boar contact prior to PG600 injection increases the proportion of gilts exhibiting oestrus within 7 days of treatment, as well as eliciting a 13% increase in the proportion of gilts that ovulated (Breen, Farris, Rodriguez-Zas and Knox, 2005). Daily boar contact significantly increases the proportion of gilts that maintain ovarian cyclicity following hormonally induced puberty (Paterson, 1982), while the provision of boar contact for 3 days following PMSG injection, as well as during fixed time mating, significantly increases pregnancy rates of 18, 21 and 24 week old gilts (Smits, Luxford, Morley, Hughes and Kirkwood, 2001; Table 18.5). However, it is unclear whether the improved pregnancy rates of boar exposed gilts observed by Smits *et al.* (2001) reflect improved semen transport in response to boar stimuli effects on oxytocin release and uterine contractions (Langendijk, Soede and Kemp, 2005) or an additive effect of boar component stimuli on LH release and ovarian follicle growth following hormone treatment.

Table 18.5 EFFECTS OF EXOGENOUS HORMONE TREATMENT AND BOAR EXPOSURE ON THE TIMING OF PUBERTY ATTAINMENT IN VERY YOUNG GILTS (AFTER SMITS *et al.*, 2001)

Age at hormone treatment*	18 weeks		21 weeks		24 weeks	
Boar contact	No	Yes	No	Yes	No	Yes
No. gilts mated	88	96	106	107	95	98
Proportion gilts pregnant	0.12	0.26	0.18	0.35	0.35	0.55

*1,000 IU PMSG at 0 hrs followed by 500 IU hCG at +72 hrs and single insemination at +102-104 hrs

GILT GROWTH, NUTRITION AND PUBERTY ATTAINMENT

It is logical to assume that puberty will only occur once the gilt has developed sufficiently to support the extremely high energetic costs of gestation and lactation (Messer and I'Anson, 2000; Schneider, 2004; Gamba and Pralong, 2006). As the animal grows, alterations in the partitioning of ingested nutrients and a reduction in basal metabolic rate increase the availability of energy for use in the creation of energy stores and / or reproductive activity (Foster and Nagatini, 1999). This developmental change in metabolism, mediated by altered levels of metabolites and metabolic hormones, may be responsible, at least in part, for permitting the pre-pubertal increase in the frequency of GnRH release (Hall, Staigmiller, Bellows, Short, Moseley, and Bellows 1995; Foster and Nagatini, 1999; Messer and I'Anson, 2000). However, a definitive relationship between reproductive development and gilt growth characteristics remains to be established. Under ad-libitum feeding conditions, gilt age at puberty appears independent of prepubertal growth rate (Beltranena, Aherne, Foxcroft and Kirkwood, 1991), and a number of authors (e.g. Rozeboom, Pettigrew, Moser, Cornelius and Kandelgy, 1995; Paterson, Ball, Willis, Aherne and Foxcroft, 2002b) report enormous variation in gilt liveweight, body composition and growth rate at puberty. The data of Burnett, Walker and Kilpatrick (1988) also demonstrate a decreasing influence of gilt liveweight and body tissue reserves on sexual maturity with increasing age, supporting the view that minimum levels of weight or body tissue reserves play a permissive rather than a stimulatory role in the progression of sexual maturation (Kirkwood and Aherne, 1985).

However, activity of the reproductive neuroendocrine axis is permitted by the availability of sufficient metabolic fuels (Schneider, 2004). Consequently it is not surprising that reproductive processes are reduced, or postponed, during periods of extreme growth or when nutrient availability is limited (Cosgrove and Foxcroft, 1996; Wade and Jones, 2004). Within genotypes, the rate of sexual maturation, and hence puberty attainment, is sensitive to altered nutritional status during the prepubertal period. Nutritional restriction of pre-pubertal growth rate delays, but does not prevent, the attainment of puberty (Beltranena, Aherne and Foxcroft, 1993; Prunier, Martin, Mounier and Bonneau, 1993). Similarly, a delayed pubertal response and a reduction in circulating concentrations of oestradiol and metabolic hormones are observed in gilts that are restrictively fed between 80 to 220 days of age (Prunier *et al.*, 1993) and 70 to 175 days of age (Figure 18.1).

Nutritionally induced alterations in metabolic status can also affect reproductive function without changing either gilt weight or P2 backfat (Booth, Craigon and Foxcroft, 1994), suggesting that the availability of metabolic fuels rather than any aspect of body size controls activity of the neuroendocrine reproductive axis (Schneider and Wade, 1987). However, body reserves can act as a buffer against the suppressive effects of feed deprivation on reproduction (Schneider and Wade, 1987). This is supported by a recent study (van Wettere, Revell, Mitchell and Hughes, 2005a), in which gilts were fed to reach a target liveweight of either 70 kg (Light) or 100 kg (Heavy) at 161 days of age, with half the gilts in each group then fed to gain liveweight at either 1.0 (high) or 0.5 (low) kg / day from 161 days of age until they attained puberty. In this study, puberty attainment in response to boar contact was delayed by long-term feed restriction, with Light gilts reaching puberty approximately 7.4 days later than Heavy gilts ($P < 0.05$), whereas feed intake during the 14 days prior to, and coincident with, boar exposure had no effect on the timing of puberty.

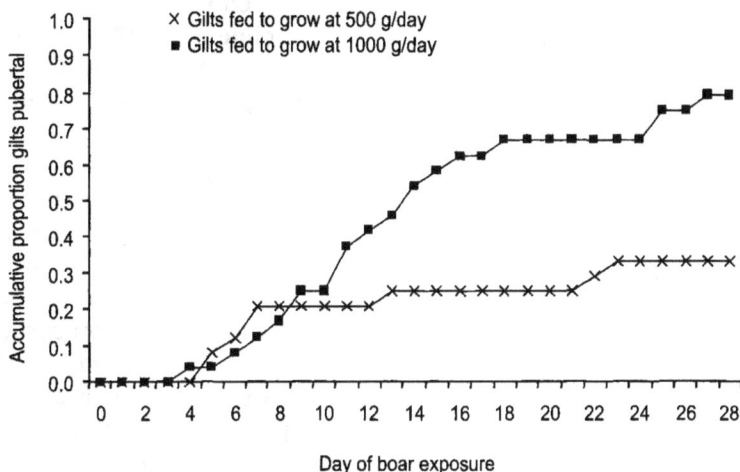

Figure 18.1 Effect of restrictive feeding between 70 and 175 days of age on the accumulative proportion of gilts attaining puberty in response to daily contact with a mature boar commencing at 175 days of age (van Wettere, Revell, Mitchell and Hughes, 2007, unpublished data).

The data presented in Table 18.6 indicate that feeding Heavy gilts on either a high or a low feeding level has little effect on the timing of first oestrus, supporting the buffering effect proposed by Schnieder and Wade (1987). However, based on this study, it is also evident that dietary repletion of long-term restrictively fed (Light) gilts fails to promote puberty attainment, which is in agreement with similar work in sheep demonstrating no effect of realimentation on age at puberty in long-term restrictively fed ewe lambs (Boulanouar, Ahmed, Kloptenstein, Brink and Kinder, 1995).

Table 18.6 EFFECTS OF PRE-PUBERTAL GROWTH CURVES ON PUBERTY ATTAINMENT IN RESPONSE TO BOAR STIMULATION COMMENCING AT 175 DAYS OF AGE (VAN WETTERE *et al.*, UNPUBLISHED DATA 2007)

Treatment	Accumulative proportion of gilts pubertal by:				Mean days to puberty
	Day 7	*Day 14*	*Day 21*	*Day 28*	
$Heavy_1 - High_2$	0.08	0.67	0.83	0.83	15.0
$Heavy_1 - Low_2$	0.25	0.58	0.67	0.67	17.8
$Light_1 - Low_2$	0.25	0.25	0.25	0.42	23.8
$Light_1 - High_2$	0.17	0.33	0.42	0.42	24.5

[1] Target liveweight at 161 days of age: Heavy (100kg) versus Light (70 kg)
[2] Target liveweight gain from 161 days of age onwards: High (1.0 kg/day) versus Low (0.5 kg/day)

The suppressive effect of restrictive feed intake on the release of both GnRH and LH is well established (Schillo, 1992; Schneider, 2004), and studies involving pre-pubertal rats, ewe lambs and gilts indicate that a suppressed pattern of tonic LH release is responsible, at least partially, for delayed puberty attainment in restrictively fed animals (Bronson, 1986; Schillo, 1992; Prunier *et al.*, 1993). Ovarian follicle growth is also significantly reduced by acute (Booth, Cosgrove and Foxcroft, 1996) and moderate feed restriction (van Wettere, Mitchell, Revell and Hughes, 2005b and 2005c) of pre-pubertal gilts. More specifically, compared to gilts fed to grow at 800 g / day (Heavy), feeding gilts to grow at a rate of 500 g / day (Light) from 70 days of age onwards reduced the proportion of 3 – 6 mm follicles and increased the proportion of 1-2.9 mm follicles present on the ovary at 161 (van Wettere *et al.*, 2005b) and 175 days of age (Figure 18.2). This reduction in follicle growth could impair the ability of Light gilts to initiate an ovarian response to the rise in LH secretion associated with boar stimulation (Kingsbury and Rawlings, 1993), and may partially explain the delayed attainment in puberty observed in both current (van Wettere *et al.*, 2005a) and previous studies (e.g. Beltranena *et al.*, 1993; Prunier *et al.*, 1993).

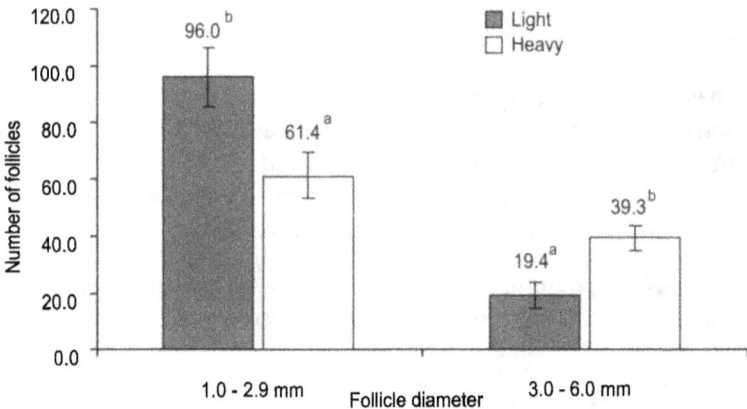

Figure 18.2 Mean number of surface antral follicles with a diameter of 1.0 – 2.9 mm or 3.0 – 6.0 mm on the ovaries of 175 day-old, non-cycling gilts fed to gain liveweight at a rate of 500 g / day (Light) or 800 g / day (Heavy) between 70 and 175 days of age (van Wettere, Mitchell, Revell and Hughes, 2007, unpublished data)

MAINTAINING GILT CYCLICITY

Currently it is common for gilts to receive their first mating at their second or third oestrus, making it necessary for gilts to continue to cycle normally after puberty stimulation. There is ample evidence to indicate that the provision of boar contact to gilts after hormonal induction of puberty will increase the proportion of gilts that continue to exhibit normal (18-25 day) oestrous cycles (Hughes and Cole, 1978; Paterson and Lindsay, 1981). Conversely, it has generally been accepted that ongoing boar contact is not necessary to maintain cyclicity in gilts reaching puberty in response to the boar effect. However, a recent study in Australia strongly suggests that, regardless of the means used to induce early puberty attainment, it is necessary to continue boar contact after puberty in order to ensure 'normal' oestrous cycles (Table 18.7).

Table 18.7 THE EFFECTS OF BOAR EXPOSURE ON OESTROUS CYCLICITY IN POSTPUBERTAL GILTS (SISWADI, 1996)

Boar contact	Mean number of cycles / 100 days	Proportion of abnormal cycles		
		Short (< 18 days)	Long (> 25 days)	Total
No	3.0	0.04	0.29[b]	0.33[b]
Yes	4.9	0.00	0.03[a]	0.03[a]

[a,b] within same column indicates significant difference; $P < 0.01$

Factors affecting first litter size

Early parity sows constitute a large proportion of modern breeding herds (Hughes and Varley, 2003). Therefore breeding herd profitability has become increasingly dependent on litter size at first farrowing. The number of piglets born is a function of ovulation rate, fertilisation rate and the number of embryos surviving to term (prenatal survival rate). Under normal conditions, losses of potential piglets due to fertilisation failure are small, with fertilisation rate generally accepted to be between 95% and 100% (Lambert, Williams, Lynch, Hanrahan, McGeady, Austin, Boland, and Roche *et al.*, 1991; van der Lende, Soede and Kemp, 1994). Therefore, the number of ova shed at the oestrus of mating and the extent of pre-natal mortality are the primary determinants of gilt prolificacy. Most pre-natal mortalities occur prior to day 35 of pregnancy, with approximately 20 to 30% of embryos lost during this period (Pope and First, 1985; van der Lende and Schoenmaker, 1990, Lambert *et al.*, 1991). A number of factors are believed to affect embryo mortality rates, and these have been recently reviewed by Ashworth and Pickard (1998) and van Wettere *et al.* (2005b). Here we present the most recent information relating to the effect of oestrus number and age at first mating on first litter size, and the role of pre- and post-mating feed intake in altering the number and quality of the oocytes shed at ovulation, as well as embryo survival. Factors involved in foetal loss in the gilt (between days 30-35 and term) have been recently reviewed elsewhere (Vallet, Leymaster and Christenson, 2002; Ford, 2003).

OESTRUS NUMBER, AGE AT FIRST MATING AND FIRST LITTER SIZE

It is common practice within the pork industry for gilts to be mated at their second or third oestrus (Martinat-Botte, Bariteau, Badouard and Terqui, 1985; Whittemore, 1996), a strategy that has been adopted despite the equivocal nature of the available literature. Delayed mating beyond the first (pubertal) oestrus has been associated with higher ovulation rates (Warnick Wiggins, Casida, Grummer, and Chapman, 1951; Archibong, England and Stormshak, 1987). Although Kirkwood and Aherne (1985) concluded that ovulation rates will increase with each oestrous period when gilts reach puberty at a young age (150 – 210 days), it is unclear whether advancing sexual age (i.e. number of oestrous cycles experienced) or chronological age (i.e. days) has the greater effect on first litter size (Brooks and Smith, 1980; Archibong, Maurer, England and Stormshak, 1992). However in a recent study (van Wettere *et al.*, 2006) in which boar exposure started at 23, 26, or 29 weeks of age and gilts were mated either at their pubertal or second oestrus, chronological age at mating had no significant effect on ovulation rate (Table 18.8). In contrast, a numerical, but not significant, increase in ovulation rate (0.6 ova) was associated with delaying mating gilts until the second oestrus (van Wettere *et al.*, 2006). The results of this study also suggest that modern gilts have higher ovulation rates than their counterparts of 20-30 years ago, shedding approximately 3 more ova at the pubertal oestrus, suggesting that even if gilts are mated at their pubertal oestrus ovulation rate is not likely to limit first litter size, (ovulation rates of older genotypes reported by Paterson and Lindsay, 1980; Brooks and Smith, 1980; Archibong *et al.*, 1992).

Table 18.8 OVULATION RATE AND EMBRYO NUMBER ON DAY 22 POST-MATING FOR GILTS STARTING BOAR CONTACT AT EITHER 161, 182 OR 203 DAYS OF AGE AND FIRST MATED AT EITHER THE FIRST (PUBERTAL) OR SECOND OESTRUS (AFTER VAN WETTERE *et al.*, 2006)

| Age at start of boar exposure | Ovulation rate | | | Number of embryos | | |
| | Mating oestrus | | | Mating oestrus | | |
	1st	*2nd*	*Pooled*	*1st*	*2nd*	*Pooled*
161 days	14.8	15.5	15.1	11.8	11.8	11.8
182 days	14.6	15.7	15.1	10.7	12.0	11.3
203 days	15.3	15.5	15.4	11.5	13.2	12.4
Pooled	14.9	15.5		11.3	12.3	
Pooled S.E.M		0.5			0.7	

However, there is evidence to support a decrease in embryo mortality when mating is delayed (Menino, Archibong, Li, Stormshak and England, 1989), and although our own studies suggest that embryo number and survival are unaffected by age at mating, embryo number tends to increase when gilts are mated at second compared to first oestrus (Table 18.8). Recently, Bagg, Vassena, Papasso-Brambilla, Grupen, Armstrong and Gandolfi (2004) concluded that improved oocyte competence may depend on the completion of a number of oestrous cycles, and compared to reproductive performance at the third oestrus, the pubertal oestrus was associated with inferior oocyte quality (Herrick, Brad, Krisher and Pope, 2003) and higher incidences of abnormal embryo development (Menino *et al.*, 1989; Archibong *et al.*, 1992). In spite of this, it remains controversial whether delaying mating does increase first litter size. Mating gilts at their third compared to their first oestrus has been found to increase first litter size by 2.6 (MacPherson, Deb Hovell and Jones, 1977) or 1.0 (Young and King, 1981) piglets. However, other studies demonstrate no effect of oestrus number or age at mating on first litter size or productivity over multiple parities (Brooks and Smith, 1980; Young, King, Walton, McMillan and Klevorick, 1990a and 1990b), and Aumaitre, Dourman and Dagorn (2000) concluded that annual breeding herd productivity does in fact decline when gilts are mated at their second or third oestrus.

NUTRITION AND OVULATION RATE

A number of extensive reviews have discussed the endocrine, intra-follicular and cellular processes involved in the growth of ovarian follicles through to the ovulatory stage (e.g. Foxcroft and Hunter, 1985; Cardenas and Pope, 2002; Knox, 2005). In brief, prior to the onset of puberty and during the luteal phase of the oestrous cycle, a morphologically heterogeneous pool of approximately 50 – 100 follicles, with a diameter of 1 to 6 mm, is present on the surface of the ovary. Continued follicle growth and development (selection) is stimulated by the increased frequency of LH pulsing that occurs on days 13 – 14 of the oestrous cycle (Knox, 2005), and on day 18 of the oestrous cycle the oestrogenic activity of the most mature follicle(s) stimulates the commencement of the pre-ovulatory LH surge (Foxcroft and Hunter, 1985, Hunter and Weisak, 1990). Follicle selection occurs continuously during the 5 – 7 day follicular phase that precedes ovulation, resulting in enormous variation in follicle maturity coincident with the initiation of the LH surge, and the formation of a morphologically and biochemically heterogeneous pool of ovulatory follicles (Grant, Hunter and Foxcroft, 1989; Hunter and Weisak, 1990; Cardenas and Pope; 2002; Hunter, Robinson, Mann and Webb, 2004; Knox, 2005).

The size of the proliferating follicle pool coincident with follicle selection has been identified as one determinant of ovulation rate in the pig (Hunter and

Weisak, 1990). Studies involving cycling gilts indicate that altered feed intake affects the growth of 1 – 6 mm follicles and influences the dynamics of the proliferating pool (Quesnel, Pasquier, Mounier, and Prunier, 2000). Equally, feeding pre-pubertal gilts to gain liveweight at a rate of 1000 g / day (High) as opposed to 500 g / day (Low) for at least two weeks prior to the start of boar stimulation significantly increases the number of 3 – 6 mm follicles present on the ovary of 175 day old gilts (Table 18.9; van Wettere *et al.*, 2005c). When boar exposure was commenced at 175 days of age, the same feeding treatments also increased pubertal ovulation rate (Table 18.9; van Wettere *et al.*, 2005a). Interestingly, studies involving post-pubertal gilts demonstrate that although ovulation rates decrease by an average of 4.5 ova in response to feed restriction during both the luteal and follicular phases (Prunier and Quesnel, 2000), they are unaffected by reduced feed intake during the luteal phase alone (Table 18.10; Almeida, Kirkwood, Aherne and Foxcroft, 2000; Novak, Almeida, Cosgrove, Dixon and Foxcroft, 2003). Nutritional status during follicle selection therefore appears to be a critical determinant of ovulation rate (Prunier and Quesnel, 2000), indicating that nutritional effects on ovulation rate reflect alterations in the follicle's final stages of growth. Consequently, as long as nutrient supply is high during the 5-7 day period prior to ovulation, ovulation rate is unlikely to limit first litter size in modern gilts, even if they are mated at the pubertal oestrus.

Table 18.9 EFFECTS OF HIGH VERSUS LOW PERI - PUBERTAL GROWTH RATES ON THE DYNAMICS OF THE ANTRAL FOLLICLE POOL COINCIDENT WITH BOAR STIMULATION COMMENCING AT 175 DAYS OF AGE AND OVULATION RATE, EMBRYO NUMBER AND EMBRYO SURVIVAL FOR GILTS MATED AT THE PUBERTAL OESTRUS

Source*		Liveweight gain from 161 days of age to puberty	
		Low (500 g / day)	*High (1000 g / day)*
Study 1	Number of 1 – 2.9 mm follicles	92.8[b]	59.5[a]
	Number of 3 – 6 mm follicles	22.7[a]	34.3[b]
Study 2	Ovulation rate	13.1[a]	15.3[b]
	Number of embryos	10.1	11.4
	Embryo survival (%)	78.0	76.1

*Study 1: van Wettere et al., 2005c; Study 2: van Wettere et al., 2005a
[ab] within same row indicates significant difference; P < 0.05

Nutritional effects on follicle growth and ovulation rate appear to be mediated by alterations in LH pulsing, with changes in the general availability of metabolic fuels, rather than one specific fuel, eliciting alterations in reproductive activity (Schneider and Wade, 1987; Foster and Nagatini, 1999).

Table 18.10 EFFECT OF RESTRICTED FEED INTAKE DURING TWO PERIODS OF THE LUTEAL PHASE (DAY 1 –7 VERSUS DAY 8 – 15) ON SUBSEQUENT OVULATION AND EMBRYO SURVIVAL IN POST-PUBERTAL GILTS (AFTER ALMEIDA *et al.*, 2000)

Feeding level during oestrous cycle[1]			Ovulation rate	Embryo survival (%)
Day 1 - 7	*Day 8 - 15*	*Day 16 to oestrus*		
High	High	High	17.1	83.6[b]
Restricted	High	High	17.7	81.7[b]
High	Restricted	High	18.5	68.3[a]

[ab] superscripts within column indicate significant difference; $P < 0.05$
[1] Restricted = 2.1 x Maintenance; High = 2.8 x Maintenance (95% of ad-lib intake)

Alterations in the availability of metabolic fuels directly affect activity of the neuronal subpopulations responsible for the regulation of GnRH release (Gamba and Pralong, 2006). However, circulating oestradiol concentrations are also lower in gilts on a high level of feeding, suggesting the stimulatory effects of high feed intake on LH pulse frequency are also due to a reduction in oestradiol negative feedback at the hypothalamus (Ferguson *et al.*, 2003). A direct effect of metabolic hormones and metabolites, in particular insulin and IGF-I, on ovarian follicle growth is also well established (Cosgrove and Foxcroft, 1996; Cardenas and Pope, 2002), and in the absence of any changes in gonadotrophins, short-term changes in nutrient intake can influence folliculogenesis (Messer and I'Anson, 2000; Webb *et al.*, 2004). The available literature supports a synergistic relationship between the insulin-related regulatory system and gonadotrophins during follicle growth and development (reviewed by Giudice, 1992; Poretsky, Cataldo, Rosenwaks and Giudice, 1999). Changes in the bioavailability of intrafollicular IGF-I, and associated binding proteins, are involved in the regulation of follicle responsiveness to gonadotrophins (Prunier and Quesnel, 2000; Webb *et al.*, 2004). High energy diets appear to increase the bioavailability of IGF-1 in small (1 – 4 mm) bovine follicles due to a reduction in intrafollicular expression of IGF binding proteins (IGFBP) – 2 and – 4 (Armstrong, McEvoy, Baxter, Robinson, Hogg, Woad, Webb, and Sinclair, 2001). Equally, systemic levels of IGFBP-3, a binding protein known to promote IGF-I bioavailability, are positively correlated with dietary intake in cattle (Armstrong, Gong, Gardner, Baxter, Hogg and Webb, 2002), and the IGFBP-3 present in follicular fluid is predominantly of extra-follicular origin (Wandji, Gadsby, Simmen, Barber and Hammond, 2000).

PRE-MATING FEED INTAKE, OOCYTE QUALITY AND EMBRYO SURVIVAL

Recent evidence indicates that nutrition prior to mating has a greater effect on embryo survival than feed intake after mating (Ashworth, Antipatis and Beattie,

1999a and Ashworth, Beattie, Antipatis and Vallet, 1999b; Table 18.11), and in the absence of any change in ovulation rate, moderate feed restriction between days 8 and 15 of the oestrous cycle reduces embryo survival (Table 18.10; Almeida *et al.*, 2000). The effect of pre-mating nutrition on embryo survival appears to reflect alterations in the developmental competence of the oocytes released at ovulation (Zak, Cosgrove, Aherne and Foxcroft, 1997; Ferguson, Ashworth, Edwards, Hawkins, Hepburn and Hunter., 2003). Using similar dietary regimes to Ashworth *et al.* (1999a and b), Ferguson *et al* (2003) demonstrated that maintenance feeding (1.35 kg /day) during the oestrous cycle before ovulation significantly reduced the proportion of oocytes in the presumptive ovulatory follicle pool that were able to reach metaphase II (MII) in vitro compared to oocytes obtained from gilts on a high feeding level (3.5 kg / day) (Table 18.12). The negative impacts of maintenance feeding on oocyte quality are convincing; however, the effects of moderate feeding levels on oocyte developmental competence and subsequent embryo survival are less clear. Reducing growth rate from 1.0 kg / day (High) to 0.5 kg / day (Low) for 14 days resulted in a 14% reduction in the proportion of oocytes derived from 175 day old, pre-pubertal gilts that were able to reach MII in vitro (P < 0.05) (van Wettere *et al.*, 2007, unpublished data). However, despite a significant reduction in pubertal ovulation rate, low growth rates did not reduce either embryo number or survival rate compared to high growth rates (van Wettere *et al.*, 2007; unpublished data).

Table 18.11 EFFECT OF FEED INTAKE, BEFORE AND AFTER MATING, ON OVULATION RATE, EMBRYO NUMBER AND EMBRYO SURVIVAL ON DAY 12 OF GESTATION IN MEISHAN GILTS (ADAPTED FROM ASHWORTH *et al.*, 1999a)

Feed intake before mating*	Feed intake after mating*	Ovulation rate	Number of embryos	Embryo survival (%)
Maintenance	Maintenance	19.0	13.8	73.4
	High	19.3	14.5	75.9
High	Maintenance	22.4	22.0	99.0
	High	22.3	20.0	89.6

*Maintenance: 1.15 kg day $^{-1}$, High: 3.5 kg day $^{-1}$; Gilts fed a complete diet containing 20.2% crude protein and 1.2% lysine and supplying 14 MJ DE kg^{-1}; diet before mating fed for entire oestrous cycle before mating; diet after mating commenced on the first day after detection of oestrus.

The effect of pre-mating nutrition on oocyte quality appears to be mediated, at least in part, by alterations in the intra-follicular environment. Concentrations of oestradiol are lower in the follicular fluid of maintenance fed gilts (Table 18.12), and using a lactation feeding regime that reduced embryo survival in the subsequent gestation, Zak *et al.* (1997) demonstrated that restrictive feeding

Table 18.12 EFFECT OF FEED INTAKE DURING THE LUTEAL AND FOLLICULAR PHASE ON OOCYTE NUCLEAR MATURATION AND FOLLICULAR FLUID CONCENTRATION OF OESTRADIOL ON DAY 19 OF THE OESTROUS CYCLE (ADAPTED FROM FERGUSON *et al.*, 2003)

Dietary treatment[1]	*Proportion of matured oocytes at:*			*Follicular fluid concentrations of oestradiol (ng/ml)*
	GV_2	MI_2	MII_2	
Maintenance	10.3	21.7[b]	68.2[a]	173[a]
High	7.0	1.5[a]	88.3[b]	332[b]

[ab] superscripts within column indicate significant difference; $P < 0.05$

[1] Maintenance = 1.35 kg day^{-1}; High = 3.5 kg day^{-1}

[2] GV = germinal vesicle; MI = metaphase I; MII = metaphase II

reduces the ability of follicular fluid to support meiotic progression to MII stage in pre-pubertal oocytes. An association between oocyte quality and intra-follicular steroid concentrations has been proposed (Ding and Foxcroft, 1992; Grupen, McIlfatrick, Ashman, Boquest, Armstrong and Nottle, 2003), with alterations in circulating gonadotrophins eliciting differences in oocyte developmental competence (Gandolfi, Luciano, Modina, Ponzini, Pocar, Armstrong, and Lauria, 1997; Ferguson *et al.*, 2003; Algriany, Bevers, Schoevers, Colenbrander and Dieleman 2004). As previously discussed, there is now substantial evidence to support a direct effect of nutritionally induced alterations in metabolic hormones on ovarian function (Hunter, 2004), and a central role for the insulin-related regulatory system in the regulation of ovarian steroidogenesis has been established. Almeida *et al.* (2000) also reported a reduction in peripheral progesterone concentrations at 48 and 72 hours after first detection of oestrus, suggesting that nutritional effects on the physiological development of recruited and pre-ovulatory follicles also alter progesterone production and/or secretion by the corpora lutea (Mao and Foxcroft, 1998).

PERI-AND POST-MATING FEED INTAKE AND EMBRYO SURVIVAL

Variations in nutritional and endocrine signals also elicit alterations in the pattern of uterine secretion (Vallet, Leymaster and Christenson., 2002). A link between feed intake during early pregnancy and both uterine protein secretion and embryo development has been established (Soede *et al.*, 1999), with nutritional effects on embryo survival appearing to be mediated by alterations in peri-ovulatory progesterone concentrations (Jindal, Cosgrove, Aherne and Foxcroft, 1996 and Jindal, Cosgrove and Foxcroft, 1997; Almeida *et al.*, 2000). Peripheral progesterone concentrations reflect the balance between synthesis and secretion of progesterone by the luteal cells and metabolic clearance rate (Jindal *et al.*, 1996; Mburu, Einarsson, Kindahl, Madej and Rodgriguez-Martinez, 1998). The increase in progesterone concentration is delayed, and embryonic mortality is increased, when gilts receive a high (2 times maintenance) compared to a low (1.5 times maintenance) plane of feeding from day 1 of gestation (Table 18.13; Jindal *et al.*, 1996), and the detrimental effects of higher feed intakes on embryo survival are reversed by supplemental progesterone administration on days 1 to 4 of gestation (Jindal *et al.*, 1997). However, reducing feed intake on day 1, as opposed to day 3, of gestation appears to have a greater effect in terms of decreasing embryo mortality (Table 18.13; Jindal *et al.*, 1996), and it was suggested that low feed intake during the peri-ovulatory period results in an earlier rise in progesterone secretion relative to ovulation, potentially reducing asynchrony between the conceptus and the uterine environment. Similarly, exogenous progesterone administration on day 2 and 3 of pregnancy elicited an earlier rise in the secretion of total protein and retinol binding protein (Vallet, Christenson, Trout and Klemcke, 1998). Together, these studies suggest that changes in progesterone concentrations, induced by feeding level or exogenous supplementation, affect embryo survival via alterations in uterine development that affect the survival of the less developed embryos (Vallet and Christenson, 2004).

Table 18.13 EFFECT OF LOW OR HIGH POST-MATING FEEDING TREATMENT ON EMBRYO SURVIVAL IN GILTS (ADAPTED FROM JINDAL *et al.*, 1996)

Treatment[1]	Ovulation rate	Embryo survival
Low feed intake from day 1 of pregnancy	14.50	85.93[b]
Low feed intake from day 3 of pregnancy	14.95	77.35[ab]
High feed intake from day 1 of pregnancy	14.95	66.96[a]

[ab] within column indicates significant difference; $P < 0.05$
[1] Low = 1.5 times maintenance; High = 2 times maintenanc

Although, low level feeding during early pregnancy reduces embryo mortality by approximately 10% (van der Lende *et al.*, 1994), maintaining a low level of feeding throughout the first 30 days of gestation significantly reduces pregnancy rates (Dyck and Strain, 1983; Virolainen, Tast, Sorsa, Love and Peltoniemi, 2004). Recent studies also demonstrate a negative effect of severe feed restriction on reproductive performance. Fasting sows for 2 days after ovulation delays ova transport within the oviduct, reduces blastocyst cleavage rates and reduces the viability of the sperm reservoir (Mburu *et al.*, 1998; Mwanza, Englund, Kindahl, Lundeheim, Einarsson, 2000), while studies involving non-pregnant gilts demonstrate that fasting during day 2 of the oestrous cycle increases the sensitivity of corpora lutea to luteolytic factors, such as prostaglandin $F_{2\alpha}$

(Galeati, Forni, Govoni, Spinaci, Zannoni, De Ambrogi, Volpe, Seren, Tamanini, 2006).

Conclusions

It is now clear that modern lean genotypes reach sexual maturity later than did their counterparts of 20 or 30 years ago. Data presented here identify that current gilt management recommendations need to be modified accordingly, particularly as they relate to the timing of initiation of puberty stimulation using boar exposure. Specifically, it is suggested that the pubertal response of modern gilts is optimised when first boar exposure commences at 26 – 29 weeks of age. In addition, while the limited information currently available indicates that, with the exception of gilt age, the optimum conditions under which to apply boar exposure are unchanged, there is new data indicating advantages for increasing the frequency of boar exposure beyond once daily. Furthermore, our own results clearly demonstrate that reduced growth rate during the grower/ finisher phase will significantly reduce the gilt's pubertal response.

Ovulation rates in modern genotypes appear to be 2 – 3 ova higher than those reported in previous decades. Indeed current results suggest that ovulation rates will now be sufficient at the pubertal oestrus to maximise potential litter size. However, embryo loss remains a major driver of first litter size and appears to be largely determined by gilt nutrition. New data suggest that pre-mating nutrition, acting via changes in oocyte quality, is the major factor here, while early gestation feeding effects are equivocal and need to be re-evaluated.

References

Aherne, F. X., and R. N. Kirkwood. 1985. Nutrition and sow prolificacy. *Journal of Reproduction and Fertility*, **Suppl 33**, 169-183.

Algriany, O., M. Bevers, E. Schoevers, B. Colenbrander, and S. Dieleman. 2004. Follicle size-dependent effects of sow follicular fluid on in vitro cumulus expansion, nuclear maturation and blastocyst formation of sow cumulus oocytes complexes. *Theriogenology*, **62**, 1483-1497.

Almeida, F. R., R. N. Kirkwood, F. X. Aherne, and G. R. Foxcroft. 2000. Consequences of different patterns of feed intake during the estrous cycle in gilts on subsequent fertility. *Journal of Animal Science*, **78**, 1556-1563.

Anand, S., F. W. Turek, and T. H. Horton. 2004. Chemosensory stimulation of luteinizing hormone secretion in male siberian hamsters (phodopus sungorus). *Biology of Reproduction*, **70**, 1033-1040.

Archibong, A. E., D. C. England, and F. Stormshak. 1987. Factors contributing to early embryonic mortality in gilts bred at first estrus. *Journal of Animal Science,* **64**, 474-478.

Archibong, A. E., R. R. Maurer, D. C. England, and F. Stormshak. 1992. Influence of sexual maturity of donor on in vivo survival of transferred porcine embryos. *Biology of Reproduction,* **47**, 1026-1030.

Armstrong, D. G., McEvoy, T.G., Baxter, G., Robinson, J.J., Hogg, C.O. Woad, K.J., Webb, R., Sinclair, K.D (2001) Effect of dietary energy and protein on bovine follicular dynamics and embryo production in vitro: Associations with the ovarian insulin-like growth factor system. *Biology of Reproduction,* **64**, 1624-1632.

Armstrong, D. G., Gong, J. G., Gardner, J. O., Baxter, G., Hogg, C. O. and Webb, R. (2002) Steroidogenesis in bovine granulosa cells: The effect of short-term changes in dietary intake. *Reproduction,* **123**, 371-378.

Ashworth, C.J. and Pickard, C.J. (1998) Embryo survival and prolificacy. In *Progress in Pig Science,* pp.303-306. Edited by J. Wiseman, M.A. Varley and Chadwick, J. P. Nottingham University Press, Nottingham.

Ashworth, C. J., C. Antipatis, and L. Beattie. 1999. Effects of pre- and post-mating nutritional status on hepatic function, progesterone concentration, uterine protein secretion and embryo survival in meishan pigs. *Reproduction, Fertility and Development,* **11**, 67-73.

Ashworth, C. J., L. Beattie, C. Antipatis, and J. L. Vallet. 1999. Effects of pre- and post-mating feed intake on blastocyst size, secretory function and glucose metabolism in meishan gilts. *Reproduction, Fertility and Development,* **11**, 323-327.

Aumaitre, A.L., Dourmad, J.Y. and Dagorn, J. (2000). Management systems for high productivity of sows in Europe, *Pig News and Information,* **21,** 89-98.

Austin, C. J., L. Emberson, and P. Nicholls. 2004. Purification and characterization of pheromaxein, the porcine steroid-binding protein. A member of the secretoglobin superfamily. *European Journal of Biochemistry,* **271**, 2593-2606.

Bagg, M. A. Vassena, R., Papasso-Brambilla, E., Grupen, C. G., Armstrong, D. T. and Gandolfi, F. (2004) Changes in ovarian, follicular, and oocyte morphology immediately after the onset of puberty are not accompanied by an increase in oocyte developmental competence in the pig. *Theriogenology,* **62**, 1003-1011.

Bakker, J., K. R. Kelliher, and M. J. Baum. 2001. Mating induces gonadotropin-releasing hormone neuronal activation in anosmic female ferrets. *Biology of Reproduction,* **64**, 1100-1105.

Bakker, J. 2003. Sexual differentiation of the neuroendocrine mechanisms regulating mate recognition in mammals. *Journal of Neuroendocrinology,* **15**, 615-621.

Beltranena, E., F. X. Aherne, and G. R. Foxcroft. 1993. Innate variability in sexual development irrespective of body fatness in gilts. *Journal of Animal Science*, **71**, 471-480.

Beltranena, E., F. X. Aherne, G. R. Foxcroft, and R. N. Kirkwood. 1991. Effects of pre- and postpubertal feeding on production traits at first and second estrus in gilts. *Journal of Animal Science,* **69**, 886-893.

Bolamba, D., P. Matton, R. Estrada, and J. J. Dufour. 1994. Ovarian follicular dynamics and relationship between ovarian types and serum concentrations of sex steroids and gonadotrophin in prepubertal gilts. *Animal Reproduction Science*, **36**, 291-304.

Bolamba, D., P. Matton, M. A. Sirard, R. Estrada, and J. J. Dufour. 1991. Ovarian morphological conditions and the effect of injection of human chorionic gonadotropin on ovulation rates in prepuberal gilts with two morphologically different ovarian types. *Journal of Animal Science,* **69**, 3774-3779.

Booth, P. J., J. R. Cosgrove, and G. R. Foxcroft. 1996. Endocrine and metabolic responses to realimentation in feed-restricted prepubertal gilts: Associations among gonadotropins, metabolic hormones, glucose, and uteroovarian development. *Journal of Animal Science,* **74**, 840-848.

Booth, P. J., J. Craigon, and G. R. Foxcroft. 1994. Nutritional manipulation of growth and metabolic and reproductive status in prepubertal gilts. *Journal of Animal Science*, **72**, 2415-2424.

Boulanouar, B., M. Ahmed, T. Klopfenstein, D. Brink, and J. Kinder. 1995. Dietary protein or energy restriction influences age and weight at puberty in ewe lambs. *Animal Reproduction Science*, **40**, 229-238.

Breen, S. M., K. L. Farris, S. L. Rodriguez-Zas, and R. V. Knox. 2005. Effect of age and physical or fence-line boar exposure on estrus and ovulation response in prepubertal gilts administered pg600. *Journal of Animal Science,* **83**, 460-465.

Brennan, P. A., and E. B. Keverne. 2004. Something in the air? New insights into mammalian pheromones. *Current Biology*, **14**, R81-89.

Bronson, F. H., and C. Desjardins. 1974. Circulating concentrations of fsh, lh, estradiol, and progesterone associated with acute, male-induced puberty in female mice. *Endocrinology*, **94**, 1658-1668.

Bronson, F. H., and J. A. Maruniak. 1975. Male-induced puberty in female mice: Evidence for a synergistic action of social cues. *Biology of Reproduction*, **13**, 94-98.

Bronson, F. H., and E. F. Rissman. 1986. The biology of puberty. *Biological Reviews of the Cambridge Philosophical Society*, **61**, 157-195.

Brooks, P. H., and D. A. Smith. 1980. The effect of mating age on the reproductive performance, food utilisation and liveweight change of the female pig. *Livestock Production Science*, **7**, 67-78.

Burnett, P. J., Walker, N. and Kilpatrick, D. J. (1988) The effect of age and growth traits on puberty and reproductive performance in the gilt. *Animal*

Production, **46**, 427 – 436.

Camous, S., A. Prunier, and J. Pelletier. 1985. Plasma prolactin, lh, fsh and estrogen excretion patterns in gilts during sexual development. *Journal of Animal Science*, **60**, 1308-1317.

Cardenas, H. & Pope, W.F. (2002) Control of ovulation rate in swine. *Journal of Animal Science*, **80 (E. Suppl. 1)**, E36-E46.

Christenson, R. K., J. J. Ford, and D. A. Redmer. 1985. Maturation of ovarian follicles in the prepubertal gilt. *Journal of Reproduction and Fertility*, **Suppl 33**, 21-36.

Cosgrove, J. R., and G. R. Foxcroft. 1996. Nutrition and reproduction in the pig: Ovarian aetiology. *Animal Reproduction Science*, **42**, 131-141.

Day, M. L., K. Imakawa, P. L. Wolfe, R. J. Kittok, and J. E. Kinder. 1987. Endocrine mechanisms of puberty in heifers. Role of hypothalamo-pituitary estradiol receptors in the negative feedback of estradiol on luteinizing hormone secretion. *Biology of Reproduction*, **37**, 1054-1065.

Deligeorgis, S. G., Lunney, D. C. and English, P. R. (1984) A note on the efficacy of complete v. partial boar exposure on puberty attainment in the gilt. *Animal Production*, **39**, 145-147.

Dial, G. D., O. K. Dial, R. S. Wilkinson, and P. J. Dziuk. 1984. Endocrine and ovulatory responses of the gilt to exogenous gonadotropins and estradiol during sexual maturation. *Biology of Reproduction*, **30**, 289-299.

Ding, J., and G. Foxcroft. 1992. Follicular heterogeneity and oocyte maturation in vitro in pigs. *Biology of Reproduction*, **47**, 648-655.

Driancourt, M. A., A. Prunier, J. P. Bidanel, and F. Martinat-Botte. 1992. Hcg induced oestrus and ovulaiton rate and fsh concentrations in prepubertal gilts from lines differing by their adult ovulation rates. *Animal Reproduction Science*, **29**, 297-305.

Dyck, G. W. and Strain, J. H. (1983) Post-mating feeding level effects on conception rate and embryonic survival in gilts. *Canadian Journal of Animal Science*, **63**. 579.

Dyer, R. M., M. D. Bishop, and M. L. Day. 1990. Exogenous estradiol reduces inhibition of luteinizing hormone by estradiol in prepubertal heifers. *Biology of Reproduction*, 42: 755-761.

Eastham, P. R., G. W. Dyck, and D. J. A. Cole. 1986. The effect of age at stimulation by relocation and first mature boar contact on the attainment of puberty in the gilt. *Animal Reproduction Science*, **12**, 31-38.

Elsaesser, F., N. Parvizi, and G. Foxcroft. 1998. Ovarian modulation of the oestradiol-induced lh surge in prepubertal and sexually mature gilts. *Journal of Reproduction and Fertility*, **113**, 1-8.

Eppig, J. J. 2001. Oocyte control of ovarian follicular development and function in mammals. *Reproduction*, **122**, 829-838.

Esbenshade, K. L., A. M. Paterson, T. C. Cantley, and B. N. Day. 1982. Changes in plasma hormone concentrations associated with the onset of puberty in the gilt. *Journal of Animal Science*, **54**, 320-324.

Esbenshade, K. L., A. J. Ziecik, and J. H. Britt. 1990. Regulation and action of gonadotrophins in pigs. *Journal of Reproduction and Fertility*, **Suppl 40**, 19-32.

Evans, A. C. O., and J. V. O'Doherty. 2001. Endocrine changes and management factors affecting puberty in gilts. *Livestock Production Science*, **68**, 1-12.

Ferguson, E. M. *et al.* 2003. Effect of different nutritional regimens before ovulation on plasma concentrations of metabolic and reproductive hormones and oocyte maturation in gilts. *Reproduction*, **126**, 61-71.

Ford, S.P. (2003) Placental development, efficiency and embryonic mortality. In *Perspectives in Pig Science* pp 279 – 292. Edited by J. Wiseman, M.A. Varley and B. Kemp. Nottingham University Press, Nottingham.

Foster, D. L., and S. Nagatani. 1999. Physiological perspectives on leptin as a regulator of reproduction: Role in timing puberty. *Biology of Reproduction*, **60**, 205-215.

Galeati, G. *et al.* Food deprivation stimulates the luteolytic capacity in the gilt. *Domestic Animal Endocrinology*, **In Press**, Corrected Proof.

Gamba, M., and F. P. Pralong. 2006. Control of gnrh neuronal activity by metabolic factors: The role of leptin and insulin. *Molecular and Cellular Endocrinology*, **254-255**, 133-139.

Gandolfi, F., Luciano, A. M., Modina, S., Ponzini, A., Pocar, P., Armstrong, D. T. and Lauria, A. (1997) The in vitro developmental competence of bovine oocytes can be related to the morphology of the ovary. *Theriogenology*, **48**, 1153-1160.

Gaughan. J.B., Cameron, R.D.A., Dryden, G.M. and Josey, M.L. (1995) Effect of selection for leanness on overall reproductive performance in Large White sows. *Animal Science*, **61**, 561–564.

Gaughan, J. B., R. D. Cameron, G. M. Dryden, and B. A. Young. 1997. Effect of body composition at selection on reproductive development in large white gilts. *Journal of Animal Science,* **75**, 1764-1772.

Giudice, L. C. 1992. Insulin-like growth factors and ovarian follicular development. *Endocrine Reviews*, **13**, 641-669.

Grant, S.A., Hunter, M. G. and Foxcroft, G. R. (1989) Morphological and biochemical characteristics during ovarian follicular development in the pig. *Journal of Reproduction and Fertility*, **117**, 171 – 183.

Grupen, C. G. *et al.* 2003. Relationship between donor animal age, follicular fluid steroid content and oocyte developmental competence in the pig. *Reproduction, Fertility, and Development*, **15**, 81-87.

Guthrie, H. D., V. G. Pursel, and R. J. Wall. 1997. Porcine follicle-stimulating hormone treatment of gilts during an altrenogest-synchronized follicular phase: Effects on follicle growth, hormone secretion, ovulation, and fertilization. *Journal of Animal Science,* **75**, 3246-3254.

Hall, J. B., Staigmiller, R. B., Bellows, R. A., Short, R. E., Moseley, W. M. and Bellows, S. E. (1995). Body composition and metabolic profiles associated with puberty in beef heifers. *Journal of Animal Science*, **73**, 3409-3420.

Herrick, J. R., A. M. Brad, R. L. Krisher, and W. F. Pope. 2003. Intracellular adenosine triphosphate and glutathione concentrations in oocytes from first estrous, multi-estrous, and testosterone-treated gilts. *Animal Reproduction Science*, **78**, 123-131.

Huffman, L. J., E. K. Inskeep, and R. L. Goodman. 1987. Changes in episodic luteinizing hormone secretion leading to puberty in the lamb. *Biology of Reproduction*, **37**, 755-761.

Hughes, P.E. and Cole, D.J.A. (1976) Reproduction in the gilt. 2. The influence of gilt age at boar introduction on the attainment of puberty. *Animal Production*, **23**, 89-94.

Hughes, P.E. & Cole, D.J.A. (1978) Reproduction in the gilt. 3. The effect of exogenous oestrogen on the attainment of puberty & subsequent reproductive performance. *Animal Production*, **27**, 11-20.

Hughes, P.E. (1982) Factors affecting the natural attainment of puberty in the gilts. In: *Control of Pig Reproduction*, pp 117 – 138. Edited by D.J.A. Cole and G.R. Foxcroft. Butterworth Scientific.

Hughes, P. E. 1994. The role of contact frequency in modifying the efficacy of the boar effect. *Animal Reproduction Science*, **35**, 273-280.

Hughes, P. E., G. P. Pearce, and A. M. Paterson. 1990. Mechanisms mediating the stimulatory effects of the boar on gilt reproduction. *Journal of Reproduction and Fertility*, **Suppl 40**, 323-341.

Hughes, P. E., and K. L. Thorogood. 1999. A note on the effects of contact frequency and time of day of boar exposure on the efficacy of the boar effect. *Animal Reproduction Science*, **57**, 121-124.

Hughes, P.E. and Varley, M. A. (2003) Lifetime performance of the sow. In *Perspectives in Pig Science*, pp. 333–355 Edited by J. Wiseman, M.A. Varley and B. Kemp, Nottingham University Press, Nottingham.

Hunter, M. G. and Weisak, T. (1990) Evidence for and implications of follicular heterogeneity in pigs. *Journal of Reproduction and Fertility*, **Suppl 40**, 163 – 177.

Hunter, M. G. 2000. Oocyte maturation and ovum quality in pigs. *Reviews of Reproduction*, **5**, 122-130.

Hunter, M. G., R. S. Robinson, G. E. Mann, and R. Webb. 2004. Endocrine and paracrine control of follicular development and ovulation rate in farm species. *Animal Reproduction Science*, **82-83**, 461-477.

Hunter, M. G., and T. Wiesak. 1990. Evidence for and implications of follicular heterogeneity in pigs. *Journal of Reproduction and Fertility*, **Suppl 40**, 163-177.

Jindal, R., J. R. Cosgrove, F. X. Aherne, and G. R. Foxcroft. 1996. Effect of nutrition on embryonal mortality in gilts: Association with progesterone. *Journal of Animal Science*, **74**, 620-624.

Jindal, R., J. R. Cosgrove, and G. R. Foxcroft. 1997. Progesterone mediates nutritionally induced effects on embryonic survival in gilts. *Journal of Animal Science*, **75**, 1063-1070.

Kingsbury, D. L., and N. C. Rawlings. 1993. Effect of exposure to a boar on circulating concentrations of lh, fsh, cortisol and oestradiol in prepubertal gilts. *Journal of Reproduction and Fertility*, **98**, 245-250.

Kirkwood, R.N. and Hughes, P.E. (1979) The influence of age at first boar contact on puberty attainment in the gilt, *Animal Production*, **29**, 231–239.

Kirkwood, R. N., and F. X. Aherne. 1985. Energy intake, body composition and reproductive performance of the gilt. *Journal of Animal Science*, **60**, 1518-1529.

Knights, M., Q. S. Baptiste, and P. E. Lewis. 2002. Ability of ram introduction to induce lh secretion, estrus and ovulation in fall-born ewe lambs during anestrus. *Animal Reproduction Science*, **69**, 199-209.

Knox, R. V. 2005. Recruitment and selection of ovarian follicles for determination of ovulation rate in the pig. *Domestic Animal Endocrinology*, **29**, 385-397.

Koketsu, Y., H. Takahashi, and K. Akachi. 1999. Longevity, lifetime pig production and productivity, and age at first conception in a cohort of gilts observed over six years on commercial farms. *Journal of Veterinary Medical Science*, **61**, 1001-1005.

Krisher, R. L., A. M. Brad, J. R. Herrick, M. L. Sparman, and J. E. Swain. 2007. A comparative analysis of metabolism and viability in porcine oocytes during in vitro maturation. *Animal Reproduction Science*, **98**, 72-96.

Lambert, E., Williams, D. H., Lynch, P. B., Hanrahan, T. J., McGeady, T. A., Austin, F. H., Boland, M. P. and Roche, J. F. (1991) The extent and timing of prenatal loss in gilts. *Theriogenology*, **36**, 655-665.

Langendijk, P., N. M. Soede, and B. Kemp. 2005. Uterine activity, sperm transport, and the role of boar stimuli around insemination in sows. *Theriogenology*, **63**, 500-513.

Lombardi, J. R., J. G. Vandenbergh, and J. M. Whitsett. 1976. Androgen control of the sexual maturation pheromone in house mouse urine. *Biology of Reproduction*, **15**, 179-186.

Lucia, T., G. D. Dial, and W. E. Marsh. 2000. Lifetime reproductive performance in female pigs having distinct reasons for removal. *Livestock Production Science*, **63**, 213-222.

Lutz, J. B., G. B. Rampacek, R. R. Kraeling, and C. A. Pinkert. 1984. Serum luteinizing hormone and estrogen profiles before puberty in the gilt. *Journal of Animal Science*, **58**, 686-691.

MacPherson, R.M., Deb Hovell, I.F.D. and Jones, A.S. (1977) Performance of sows first mated at puberty or second or third oestrus, and carcass assessment of once-bred gilts. *Animal Production*, **24**, 333-342.

Mao, J., and G. R. Foxcroft. 1998. Progesterone therapy during early pregnancy and embryonal survival in primiparous weaned sows. *Journal of Animal Science,* **76**, 1922-1928.

Martin, G. B., J. Rodger, and D. Blache. 2004. Nutritional and environmental effects on reproduction in small ruminants. *Reproduction, Fertility, and Development,* **16**, 491-501.

Martinat-Botte, R., Bariteau, F., Badouard, B. and Terqui, M. (1985). Control of pig reproduction in a breeding programme. *Journal of Reproduction and Fertility,* **Suppl 33**, 211 – 228.

Maruniak, J. A., A. Coquelin, and F. H. Bronson. 1978. The release of lh in male mice in response to female urinary odors: Characteristics of the response in young males. *Biology of Reproduction,* **18**, 251-255.

Mburu, J. N., S. Einarsson, H. Kindahl, A. Madej, and H. Rodriguez-Martinez. 1998. Effects of post-ovulatory food deprivation on oviductal sperm concentration, embryo development and hormonal profiles in the pig. *Animal Reproduction Science,* **52**, 221-234.

Menino, A. R., Jr., A. E. Archibong, J. R. Li, F. Stormshak, and D. C. England. 1989. Comparison of in vitro development of embryos collected from the same gilts at first and third estrus. *Journal of Animal Science,* **67**, 1387-1393.

Messer, N. A., and H. I'Anson. 2000. The nature of the metabolic signal that triggers onset of puberty in female rats. *Physiology and Behaviour,* **68**, 377-382.

Mwanza, A. M., P. Englund, H. Kindahl, N. Lundeheim, and S. Einarsson. 2000. Effects of post-ovulatory food deprivation on the hormonal profiles, activity of the oviduct and ova transport in sows. *Animal Reproduction Science,* **59**, 185-199.

Novak, S., F. R. Almeida, J. R. Cosgrove, W. T. Dixon, and G. R. Foxcroft. 2003. Effect of pre- and postmating nutritional manipulation on plasma progesterone, blastocyst development, and the oviductal environment during early pregnancy in gilts. *Journal of Animal Science,* **81**, 772-783.

O'Dowd. S., Hoste, S., Mercer, J.T., Flower, V.R. and Edwards, S.A., (1997) Nutrition modification of body composition and the consequences for reproductive performance and longevity in genetically lean sows. *Livestock Production Science,* **52**, 155–165.

Ojeda, S. R., Lomniczi, A., Mastronardi, C., Heger, S., Roth, C., Parent, A. S., Matagne, V. and Mungenast, A. E. (2006). Minireview: The neuroendocrine regulation of puberty: Is the time ripe for a systems biology approach? *Endocrinology,* **147**, 1166-1174.

Paterson, A.M. and Lindsay, D.R. (1980). Induction of puberty in gilts. 1. The effectsof rearing conditions on reproductive performance and response to mature boars after early puberty. *Animal Production,* **31**, 291-297.

Paterson, A.M. & Lindsay, D.R. (1981). Induction of puberty in gilts. 2. The effect of boars on maintenance of cyclic activity in gilts induced to ovulate with pregnant mare's serum gonadotrophin and human chorionic gonadotrophin. *Animal Production*, **32**, 51-54

Paterson, A. M. 1982. The controlled induction of puberty. In *Control of Pig Reproduction*, pp 139-160. Edited by D. J. A. Cole and G. R. Foxcroft. Butterworths, London.

Paterson, A. M., P. E. Hughes, and G. P. Pearce. 1989. The effect of season, frequency and duration of contact with boars on the attainment of puberty in gilts. *Animal Reproduction Science*, **21**, 115-124.

Patterson, J. L., R. O. Ball, H. J. Willis, F. X. Aherne, and G. R. Foxcroft. 2002. The effect of lean growth rate on puberty attainment in gilts. *Journal of Animal Science*, **80**, 1299-1310.

Patterson, J. L., H. J. Willis, R. N. Kirkwood, and G. R. Foxcroft. 2002. Impact of boar exposure on puberty attainment and breeding outcomes in gilts. *Theriogenology*, **57**, 2015-2025.

Pearce, G. P., and A. M. Paterson. 1992. Physical contact with the boar is required for maximum stimulation of puberty in the gilt because it allows transfer of boar pheromones and not because it induces cortisol release. *Animal Reproduction Science*, **27**, 209-224.

Pelletier, J., S. Carrez-Camous, and J. C. Thiery. 1981. Basic neuroendocrine events before puberty in cattle, sheep and pigs. *Journal of Reproduction and Fertility*, **Suppl 30**, 91-102.

Philip, G., and P. E. Hughes. 1995. The effects of contact frequency and season on the efficacy of the boar effect. *Animal Reproduction Science*, **40**, 143-150.

Plant, T. M. 2002. Neurophysiology of puberty. *Journal of Adolescent Health*, **31**, 185-191.

Pope, W. F., and N. L. First. 1985. Factors affecting the survival of pig embryos. *Theriogenology*, **23**, 91-105.

Poretsky, L., N. A. Cataldo, Z. Rosenwaks, and L. C. Giudice. 1999. The insulin-related ovarian regulatory system in health and disease. *Endocrine Reviews*, **20**, 535-582.

Pressing, A., G. D. Dial, K. L. Esbenshade, and C. M. Stroud. 1992. Hourly administration of gnrh to prepubertal gilts: Endocrine and ovulatory responses from 70 to 190 days of age. *Journal of Animal Science*, **70**, 232-242.

Prunier, A., C. Martin, A. M. Mounier, and M. Bonneau. 1993. Metabolic and endocrine changes associated with undernutrition in the peripubertal gilt. *Journal of Animal Science*, **71**, 1887-1894.

Prunier, A., and H. Quesnel. 2000. Influence of the nutritional status on ovarian development in female pigs. *Animal Reproduction Science*, **60-61**, 185-197.

Quesnel, H., A. Pasquier, A. M. Mounier, and A. Prunier. 2000. Feed restriction in cyclic gilts: Gonadotrophin-independent effects on follicular growth. *Reproduction, Nutrition, Development*, **40**, 405-414.

Rissman, E. F. 1996. Behavioral regulation of gonadotropin-releasing hormone. *Biology of Reproduction,* **54,** 413-419.

Rissman, E. F., X. Li, J. A. King, and R. P. Millar. 1997. Behavioral regulation of gonadotropin-releasing hormone production. *Brain Research Bulletin,* **44,** 459-464.

Rozeboom, D. W., J. E. Pettigrew, R. L. Moser, S. G. Cornelius, and S. M. el Kandelgy. 1995. Body composition of gilts at puberty. *Journal of Animal Science,* **73,** 2524-2531.

Ryan, K. D., R. L. Goodman, F. J. Karsch, S. J. Legan, and D. L. Foster. 1991. Patterns of circulating gonadotropins and ovarian steroids during the first periovulatory period in the developing sheep. *Biology of Reproduction,* **45,** 471-477.

Schillo, K. K. 1992. Effects of dietary energy on control of luteinizing hormone secretion in cattle and sheep. *Journal of Animal Science,* **70,** 1271-1282.

Schneider. J.E. and Wade, G. N. (1987) Body composition, food intake, and brown fat thremogenesis in pregnant Djungarian hamsters. *American Journal of Physiology,* **253,** R314–R320.

Schneider, J. E. 2004. Energy balance and reproduction. *Physiology and Behaviour,* **81,** 289-317.

Slevin, J. and Wiseman, J. (2003) Physiological development in the gilt. In *Perspectives in Pig Science,* pp 293 – 332. Edited by J. Wiseman, M.A. Varley and B. Kemp, Nottingham University Press, Nottingham.

Smits, R.J., Luxford, B.G., Morley, W.C., Hughes, P.E. and Kirkwood, R.N. (2001). Boar exposure increases conception rate in induced 'once-bred' pre-pubertal gilts. In *Manipulating Pig Production VIII* pp 194, Edited by P.D. Cranwell. Australasian Pig Science Association, Werribee.

Soede, N. M., T. van der Lende, and W. Hazeleger. 1999. Uterine luminal proteins and estrogens in gilts on a normal nutritional plane during the estrous cycle and on a normal or high nutritional plane during early pregnancy. *Theriogenology,* **52,** 139-152.

Tilton, S. L., R. O. Bates, and R. S. Prather. 1995. Evaluation of response to hormonal therapy in prepubertal gilts of different genetic lines. *Journal of Animal Science,* **73,** 3062-3068.

Vallet, J. L., R. K. Christenson, W. E. Trout, and H. G. Klemcke. 1998. Conceptus, progesterone, and breed effects on uterine protein secretion in swine. *Journal of Animal Science,* **76,** 2657-2670.

Vallet, J.L., Leymaster, K.A. and Christenson, R.K. (2002). The influence of uterine function on embryonic and fetal survival. *Journal of Animal Science,* **80 (E. Suppl. 2),** E115-E125.

Vallet, J. L., and R. K. Christenson. 2004. Effect of progesterone, mifepristone, and estrogen treatment during early pregnancy on conceptus development and uterine capacity in swine. *Biology of Reproduction,* **70,** 92-98.

van der Lende, T., and G. J. W. Schoenmaker. 1990. The relationship between ovulation rate and litter size before and after day 35 of pregnancy in gilts and sows: An analysis of published data. *Livestock Production Science*, **26**, 217-229.

van der Lende, T., Soede, N. M. and Kemp, B. (1994) Embryo mortality and prolificacy in the pig. In Principles of Pig Science pp297 – 317. Edited by D. J. A. Cole, J. Wiseman, M.A. Varley. Nottingham University Press, Nottingham.

van Wettere, W. H. E. J., M. Mitchell., D. K. Revell, and P. E. Hughes (2005a). Liveweight gain affects gilt reproductive performance. In *Manipulating Pig Production X,* pp 200. Edited by J. E. Paterson. Australasian Pig Science Association, Werribee

van Wettere, W. H. E. J., M. Mitchell., D. K. Revell, and P. E. Hughes (2005b). Management and nutritional factors affecting puberty attainment and first litter size in replacement gilts. In *Manipulating Pig Production X,* pp 180 – 192. Edited by J. E. Paterson. Australasian Pig Science Association, Werribee.

van Wettere, W. H. E. J., M. Mitchell, S. Schulz., D. K. Revell, and P. E. Hughes (2005c). The effect of growth characteristics on the ovarian follicle population of pre-pubertal gilts. *Manipulating Pig Production X,* pp 201. Edited by J. E. Paterson. Australasian Pig Science Association, Werribee.

van Wettere, W. H. E. J., M. Mitchell., D. K. Revell, and P. E. Hughes (2005b). Management and nutritional factors affecting puberty attainment and first litter size in replacement gilts. In *Manipulating Pig Production X,* pp 180 – 192. Edited by J. E. Paterson. Australasian Pig Science Association, Werribee.

van Wettere, W. H. E. J., D. K. Revell, M. Mitchell, and P. E. Hughes (2006). Increasing the age of gilts at first boar contact improves the timing and synchrony of the pubertal response but does not affect potential litter size. *Animal Reproduction Science*, **95**, 97-106.

Virolainen, J. V., A. Tast, A. Sorsa, R. J. Love, and O. A. Peltoniemi. 2004. Changes in feeding level during early pregnancy affect fertility in gilts. *Animal Reproduction Science*, **80**, 341-352.

Wade, G. N., and J. E. Jones. 2004. Neuroendocrinology of nutritional infertility. American *Journal of Physiology. Regulatory, Integrative and Comparative Physiology*, **287**, R1277-1296.

Wandji, S. A., J. E. Gadsby, F. A. Simmen, J. A. Barber, and J. M. Hammond. 2000. Porcine ovarian cells express messenger ribonucleic acids for the acid-labile subunit and insulin-like growth factor binding protein-3 during follicular and luteal phases of the estrous cycle. *Endocrinology*, **141**, 2638-2647.

Warnick, A. C., Wiggins, E. L., Casida, L. E., Grummer, R. H. and Chapman, A. B. (1951). Variation in puberty phenomena in inbred gilts. *Journal of Animal Science,* **10**, 479 – 493.

Webb, R., P. C. Garnsworthy, J. G. Gong, and D. G. Armstrong. 2004. Control of follicular growth: Local interactions and nutritional influences. *Journal of Animal Science*, **82 E-Suppl**: E63-74.

Whittemore, C. T. 1996. Nutrition reproduction interactions in primiparous sows. *Livestock Production Science*, **46**, 65-83.

Wiesak, T., M. G. Hunter, and G. R. Foxcroft. 1990. Differences in follicular morphology, steroidogenesis and oocyte maturation in naturally cyclic and pmsg/hcg-treated prepubertal gilts. *Journal of Reproduction and Fertility*, **89**, 633-641.

Young, I. G. and King, G. J. (1981) Reproductive performance of gilts bred on first versus third estrus. *Journal of Animal Science,* **53**, 19–25.

Young, B.A., King, G.J., Walton, J.S., McMillan, I. and M. Klevorick, (1990a) Age, weight, backfat and time of mating effects on performance of gilts, *Canadian Journal of Anima. Science,* **70**, 469–481.

Young, I.G., King, G.J., Walton, J.S., McMillan, I. and M. Klevorick, (1990b), Reproductive performance over four parities of gilts stimulated to early estrus and mated at first, second or third observed estrus. *Canadian Journal of Anima. Science,* **70**, 483–492.

Zak, L. J., X. Xu, R. T. Hardin, and G. R. Foxcroft. 1997. Impact of different patterns of feed intake during lactation in the primiparous sow on follicular development and oocyte maturation. *Journal of Reproduction and Fertility*, **110**, 99-106.

Ziecik, A. J., Biallowicz, M., Kaczmarek, M., Demianowicz, W., Rioperez, J., Wasielak, M. and Bogacki, M. Influence of estrus synchronization of prepubertal gilts on embryo quality. *Journal of Reproduction and Development*, **51**, 379-384.

19

INTERMITTENT SUCKLING: TACKLING LACTATIONAL ANOESTRUS AND ALLEVIATING WEANING RISKS FOR PIGLETS

LANGENDIJK P[1], BERKEVELD M[2], GERRITSEN R[1], SOEDE NM[1], KEMP B[1]

[1]Wageningen university, The Netherlands; [2]Utrecht university, The Netherlands

Background

In most conventional pig management systems piglets are weaned around 3 to 4 wk of age. At this young age, piglets have great difficulties adapting to a different physical and social environment and to a different source of nutrition. In addition, the level of dry feed intake at weaning is generally inadequate to enable the piglet to cope with the challenge of weaning. As a result behavioural problems, but also maladaptation of the gastro-intestinal tract, can occur, increasing the risk of gastro-intestinal problems after weaning. These problems can vary from mild to serious diarrhoea and even death of the piglet. To prevent or reduce the risk of developing postweaning problems, piglets ideally would have to be weaned at an age or at the time when they are able to face the challenge of weaning, i.e. when they can cope with weaning in a behavioural sense and when their level of voluntary feed intake is sufficient to rely on the postweaning diet as their sole nutritional source. Postponing the time of weaning, however, would compromise the number of litters produced per year, and therefore be economically unacceptable. Therefore, it would only be profitable to postpone the time of weaning if the timing of the next conception is independent of the time of weaning, which means sows will have to conceive during their ongoing lactation. This chapter reviews studies that have attempted to induce oestrus during lactation, by imposing various types of limited nursing regimes. Most of these regimes separate sows and piglets for a number of hours per day. Effects of such nursing regimes on how piglets adapt to weaning are discussed, as well as effects on follicle development, ovulation and conception by the sow.

Piglet adaptation to weaning in a limited nursing model

Weaning is associated with an abrupt dietary change, transport, altered housing, and mixing with unfamiliar penmates. This results in reduced nutrient intake, reduced growth and a greater susceptibility to diarrhoea (van Beers-Schreurs *et al.*, 1992; Nabuurs, 1998). To familiarise piglets with creep feed before weaning, it is often provided during lactation. In a study by English *et al.* (1980), litters that consumed an average of 610 g of creep feed per piglet during lactation had an improved performance after weaning (d 28) compared to control litters that were not supplemented with creep feed during lactation. The level of 610 g is of course arbitrary if used as a criteria to assess whether feed intake is adequate to face postweaning challenges, but it does illustrate the importance of experience with a certain level of feed intake. A number of studies have shown that feed intake early postweaning is strongly related to (cumulative) feed intake during lactation (Kuller *et al.*, 2004; Bruininx *et al.*, 2002; Berkeveld *et al.*, 2007; also see Figure 19.1). Postweaning feed intake, or more precisely the level of supply of nutrients to the small intestine, is important in preventing the reduced gut function and morphological changes that are normally associated with weaning. Van Beers-schreurs *et al.* (1998) and Marion *et al.* (2002) have elegantly shown that small intestine morphology characteristics such as villus height are strongly related to energy intake postweaning. For example, piglets that are fed to requirement with sow's milk after weaning, do not show changes in gut morphology postweaning. However, in piglets that are fed sow's milk at a low energy level equal to the voluntary energy intake of litter mates that are weaned on a dry feed diet, the same changes in gut morphology as those litter mates are observed (Van Beers-schreurs *et al.*, 1998). Only at a high energy intake level (2.5M), do piglets fed a pelleted diet show more villus atrophy than piglets fed cow's milk at the same energy level, although absorption does not seem to be affected (Pluske *et al.*, 1996). Indeed, Pluske *et al.* (1997) reviewed a number of factors that influence small intestine function postweaning and concluded that the level of nutrient supply to the gut, rather than the nature of the nutrients (milk or dry feed) is important in maintaining gut morphology and function postweaning. Because postweaning feed intake is strongly related to intake of creep feed pre-weaning, it is important to ensure a sufficient level of feed intake during lactation.

However, creep feed intake during a conventional lactation is usually low (Pajor *et al.*, 1991). For example, cumulative feed intake during a 21 d lactation was 18 g per piglet (Berkeveld *et al.*, 2007) and 314 g per piglet during a 27 d lactation (Kuller *et al.*, 2004). Moreover, variation in feed intake is considerable: between litters cumulative feed intake varied from 0 to 140 g per piglet at d 21 (Berkeveld *et al.*, 2007) and from 20 g to 1440 g per piglet at d 27 of lactation (Kuller *et al.*, 2004). Under natural conditions, the weaning process may take 8 to 12 wk, starting early in lactation. During this period, young piglets make a gradual transition from a diet based on sow's milk to a non milk diet, ultimately achieving nutritional independence from the sow (for review, see Miller and Slade, 2003). For example, if the contact between a

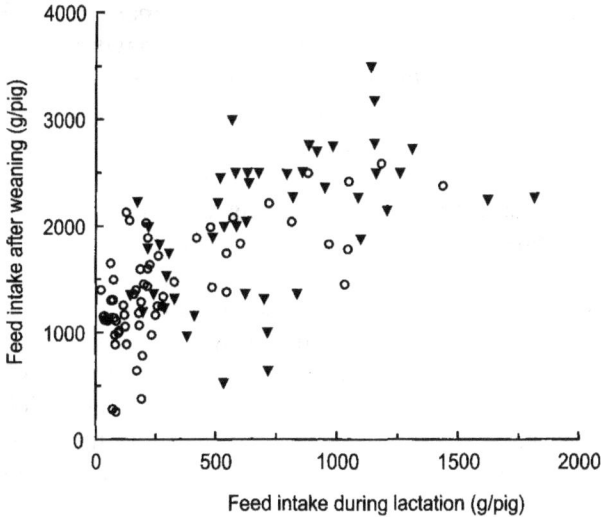

Figure 19.1 Relation between cumulative feed intake before weaning and feed intake in the first seven days after weaning for control litters (continuous lactation until weaning at 27 d, o) and for intermittent suckling litters (continuous lactation until 14 d, then 12 h per day separation from the sow until weaning at 27 d, ▼). (After Kuller *et al.* (2004); courtesy of Journal of Animal Science)

sow and her litter is sow-controlled, the time spent away from the litter as well as the frequency of sucklings starts to decrease considerably during the second week of lactation. By four weeks of lactation, the average sow in a sow-controlled suckling system may spend about 0.75 of the day away from the piglets and suckling frequency may decrease to 13 per day (Rantzer *et al.*, 1995). Variation between sows, nevertheless, is high. The more sows avoid their litters during lactation, the lower the number of sucklings and the higher the dry feed intake by the piglets during lactation (Pajor *et al.*, 2002). To achieve an adequate level of feed intake at the time of weaning, piglets need to be stimulated to ingest dry feed during lactation, which can be done by controlling the level of contact between the sow and her litter.

Over recent years, joint research teams from Wageningen University and Utrecht University have worked on the intermittent suckling (IS) model that is meant to overcome or reduce the problems normally associated with weaning. In the IS model, sows and piglets are separated for a certain period each day, from two weeks of lactation until weaning, with the aim of increasing feed intake by the piglets. Intermittent suckling does indeed stimulate creep feed intake compared to piglets that are weaned after a conventional lactation: separation of piglets from the sow for 12 h per day increased cumulative feed intake from 18 to 96 g per piglet at d21 of lactation (Berkeveld *et al.*, 2007), and from 314 to 686 g at d27 of lactation (Kuller *et al.*, 2004). Although creep feed intake during lactation is increased by IS at weaning (d 27) a total feed intake of 600 g/piglet is accomplished in only a proportion (60%) of the litters (Kuller *et al.*, 2004). Ideally, feed intake should have reached an adequate level, whatever that may be, before piglets are weaned. Alternatively, the time of weaning should be postponed until all litters have achieved such a level of feed intake. In recent work, Berkeveld *et al.* (2007) have implemented the intermittent suckling regime within a lactation period of six weeks. Obviously, cumulative feed intake at the time of weaning was much higher (3808 g vs 18 g per piglet) compared to piglets that were weaned after a 21 d continuous lactation. In the litters that were weaned at six weeks, the relative contribution of dry feed intake to energy intake, increased progressively throughout lactation (Figure 19.2), and was around 0.50 at the time the piglets on the IS regime were weaned. At the time of weaning, the piglets on the IS regime were acquainted sufficiently with dry feed to ensure an adequate level of feed intake once they were weaned definitively, because they had the same level of feed intake shortly after weaning as the control piglets that had been weaned three weeks earlier (Figure 19.3). As a consequence, the piglets on the IS regime showed only a slight growth check at weaning. The piglets that were weaned at 21 d of a continuous lactation experienced a reduction in growth of 98 % at 2 d after weaning, whereas the piglets that were weaned later and after a limited nursing regime experienced a reduction in growth of only 14%. It has to be noted that after initiating the IS regime piglets do experience a temporary reduction in growth of around 25 to 30 % due to a lower milk intake (Kuller *et al.*, 2004; Berkeveld *et al.*, 2007), but this is much less than the growth check observed in conventionally weaned piglets. The resulting weight gain over the first eight weeks after farrowing was similar for conventionally weaned pigs and for pigs on the intermittent suckling regime weaned at six weeks. Postponing the time of weaning and subjecting piglets to a limited nursing regime, therefore, does not improve overall growth rate, but ensures a gradual adaptation to the postweaning situation, reducing the potential risk of weaning associated changes in gut morphology and function and the risk of developing post-weaning diarrhoea.

In the work of Berkeveld *et al.* (2007), sows and piglets were separated either for a continuous period of 12 h per day, or for two periods of six hours per day. In the latter model, feed intake was lower, but weight gain was higher during the limited nursing regime. Eating behaviour increases in the course of a limited nursing regime (Berkeveld *et al.*, unpublished results), which correlates with the increase in feed intake. In contrast to the limited nursing model with one 12-h block of separation, explorative behaviour was not seen to increase with two 6-h blocks of separation. This indicates that, when a similar total separation period is distributed over shorter blocks of time, piglets tend to rely more on milk intake and also have a higher total milk intake. Extrapolating this to a sow-controlled suckling system, feed intake is probably less stimulated in such a system because the suckling bouts are more evenly distributed

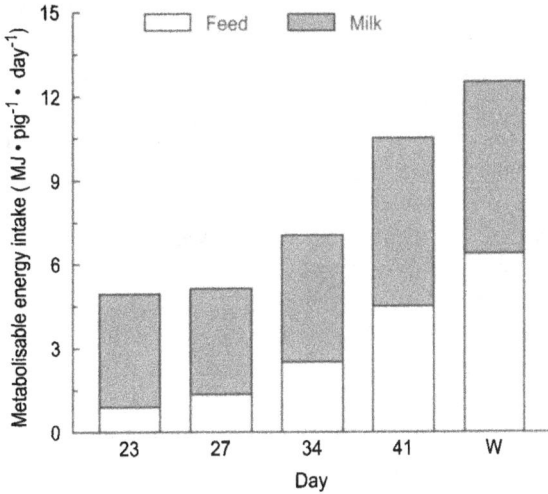

Figure 19.2 Relative contribution of dry feed and milk to metabolisable energy intake in piglets that were subjected to an intermittent suckling regime (separation from the sow for 12 h each day from d14 of lactation), from d23 of lactation until weaning around d41. Because not all litters were weaned at d41, data are also presented for the day of weaning (W). Energy requirements for maintenance and growth, as well as metabolisable energy derived from feed were estimated based on the Dutch CVB Report (1995). Feed intake was recorded at the litter level. Energy intake from milk was calculated as the difference between energy requirement and estimated energy intake from feed.

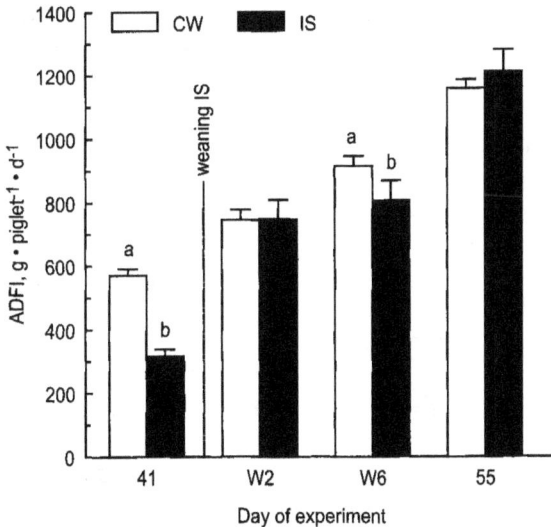

Figure 19.3 Average daily feed intake around the time of weaning (d41) of litters that were subjected to a limited nursing regime (IS), compared to conventionally weaned litters (CW). The CW litters were weaned on d 21 after a continuous lactation and moved to fattening pens at d 41. IS litters were separated from the sow for 12 h each day from d14 of lactation until weaning (d41) and then moved to fattening pens. W2 = 2 d after moving to fattening pens (CW, d 41) or weaning (IS). W6 = 6 d after moving to fattening pens (CW, d 41) or weaning (IS). [a,b] Differences between treatments (P<0.05) are indicated with different superscripts. (After Berkeveld et al. (2007); courtesy of Journal of animal Science)

throughout the day, as compared to a limited nursing regime where separation from the sow occurs during one block. Besides the type of limited nursing regime that is imposed, factors such as age of the piglets and the duration of the regime probably influence the development in feed intake pattern of the piglets. It is interesting to question both from a biological and from a practical perspective at what age an intermittent suckling regime is most effective, and whether the period needed to achieve a sufficient level of feed intake depends on the age of the piglets at the onset of the regime. To our knowledge, the effect of age at the onset of limited nursing on the development of feed intake patterns has not yet been investigated. Preliminary data from our group indicates that piglets subjected to a limited nursing regime at 21d of age have a higher feed intake in the first week of the regime than piglets subjected to the regime from 14 d of age (740 ± 118 g vs 287 ± 58 g per piglet). In the study by Berkeveld *et al.* (2007), piglets were weaned at six weeks of age after three weeks of limited nursing. Judging by the level of feed intake at and shortly after weaning, one can hypothesise that a shorter period of limited nursing would be as effective in terms of feed intake. Experimental work is currently being undertaken to establish whether there is an interaction between the age at which a limited nursing regime commences and the required duration of the regime before piglets are weaned.

As mentioned earlier, weaning induces a disruption of nutrient intake and thus supply of nutrients to the gut, causing morphological changes in the gut epithelium. A continuous supply of nutrients after weaning has been shown to prevent crypt hypertrophy and the decrease of villus height associated with weaning (Kelly *et al.*, 1991a,b; van Beers-Schreurs *et al.*, 1998). Because IS disrupts the continuous supply of nutrients to the small intestine as with weaning, one might suggest that IS also negatively affects intestinal structure or function. Hartke *et al.* (2005) observed no detrimental effects on intestinal morphology, after a single short-term (6 or 12 h) fasting, but piglets in the IS model are subjected to repeated nonsuckling periods. Nabuurs *et al.* (1996), however, demonstrated that 2 wk of IS (8 h/d) without supplementary feeding did not induce changes of intestinal structure or net fluid absorption of the small intestine in a small intestinal segment perfusion test at weaning (around d 30). Interestingly, 2 wk of IS (8 h/d) with supplementary feeding (Nabuurs *et al.*, 1996) resulted in a reduced villus height and net fluid absorption at weaning (around d 30) but partially protected piglets from the decrease in villus height and intestinal net absorption 4 d after weaning compared with piglets from continuously suckled or IS litters without supplementary feeding. Altogether, it appears that IS does not per se negatively affect intestinal structure or function. Moreover, IS with supplementary feeding partially prevents negative effects of weaning on intestinal structure and function, due to the faster and higher intake of feed after weaning.

Because the IS model involves separation of piglets from their dam at a rather early age, one can question whether this involves separation distress. From behavioural studies, in which scan sampling as well as video recording were used to monitor behavioural changes as affected by the intermittent suckling regime, it appears that the activity of the litter largely coincides with the presence of the dam (Berkeveld *et al.*, unpublished results). When subjected to a limited nursing regime, about 70 % of all activity was recorded in the presence of the sow, a pattern which was established within two days after the onset of the limited nursing regime. Around 0.50 of the litter activity during presence of the sow consisted of nursing behaviour. Roughly, this means an equal percentage of the non-nursing activity was spent during the presence and the absence of the sow. In other words, the piglets were much more active during the presence of the sow, because of suckling related activity. As a consequence, it seemed that the litters were less active during absence of the sow, and this could indicate distress. Colson *et al.* (2006) observed an increase of lying in litter cohesion when piglets were weaned at 21 d instead of 28 d of age, and suggested this being an indicator of distress. In the experiments of Berkeveld *et al.* (2007), piglets were also seen to lie in litter cohesion, although this was not recorded as such. The litters being less active in the absence of the sow, therefore, could indicate distress, but might also be a consequence of the fact that most activity at this age is suckling related. It is therefore difficult to conclude whether the limited nursing regime did cause distress. Nevertheless, aggressive behaviour, belly-nosing and other types of behaviour that could indicate distress (Dybkjaer *et al.*, 1992) were not observed during the limited nursing regime, whereas conventionally weaned piglets did show an increase in belly-nosing and aggressive behaviour after weaning (Berkeveld *et al.*, unpublished).

In conclusion, limited nursing regimes stimulate feed intake during lactation and thus allow piglets to adapt to a non-milk diet, especially in terms of level of feed intake. This alleviates the drop in nutrient intake normally observed around weaning and thus potentially reduces the risk of developing diarrhoea and other weaning related problems. If, in addition to a limited nursing regime, piglets are weaned at a later age there is hardly any change in growth rate. A limited nursing regime does have a profound effect on piglet behaviour, in the sense that piglets synchronise most of their activity with the presence of the sow. Piglets adapt their behavioural pattern very soon after the start of a limited nursing regime, and so far no obvious behavioural signs of a negative effect of a limited nursing regime have been observed.

The occurrence of oestrus during lactation

A sow is classically considered to have 'lactational anoestrus'; lactation blocks outgrowth of follicles and, consequently, prevents occurrence of oestrus and ovulation. Immediately after weaning, follicles will typically grow from 2-3 mm to 7-8 mm in a period of 4 to 7 d (Soede *et al.*, 2000).

Follicle development during the follicular phase is FSH- and LH-dependent. FSH appears to be more important for early to late antral follicular growth, and LH more important for final follicle growth, such as during the follicular phase of the oestrous cycle (Guthrie *et al.*, 1990; Duggan *et al.*, 1982). LH, which is secreted by the pituitary in a pulsatile manner, shows a distinct change at the time of weaning. During lactation, the rate of secretion is very low, with zero to a few pulses of LH per day. At weaning, this quiescent secretion pattern changes abruptly into a secretion pattern characterised by a high frequency and low amplitude of LH pulses (Figure 19.4). The pattern of LH secretion determines the rate at which follicular development proceeds. In primiparous sows, van den Brand *et al.* (2000) showed that the higher the frequency of LH pulses on the day of weaning, the shorter the interval from weaning to oestrus. A few days after weaning, pulsatile LH secretion decreases again due to the negative feedback by steroids produced by the growing follicles. Around four days after weaning, the pre-ovulatory LH surge takes place and induces the ovulation process and luteinisation of the follicles.

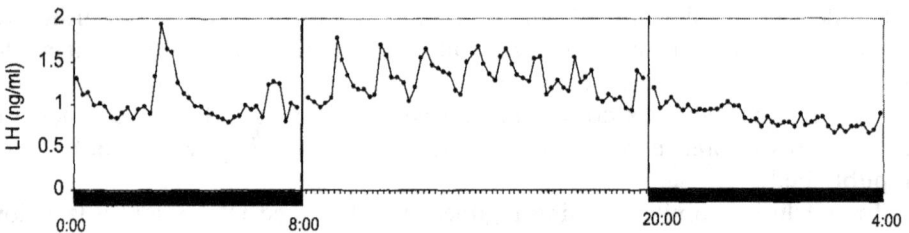

Figure 19.4 LH concentration in peripheral plasma samples taken at 12-min intervals during a period of 28 hours from one sow. The first period of 8 hrs (0.00 to 8.00), the sow was still nursing. At 8.00 (d14 of lactation), the sow was separated from the piglets to prevent suckling during the 12 following hours. At 20.00, piglets were allowed to suckle again until the end of the sampling period.

The reason for the quiescent secretion pattern of LH during lactation has been studied extensively during the 1980s and the 1990s. Inhibition of LH secretion by elevated levels of prolactin during lactation has been proposed to be one of the mechanisms responsible for lactational anoestrus in the sow. However, studies using bromocriptine (an inhibitor of prolactin secretion) in lactating sows, have failed to show an increase in LH secretion similar to that observed after weaning (Bevers *et al.*, 1985). Moreover, administration of prolactin after weaning failed to suppress the recovery of LH secretion (Booman *et al.*, 1982). Inhibition of LH secretion appears to be caused by neurotransmitters that are released as a result of the suckling activity of the piglets (Armstrong *et al.*, 1988). Suckling causes an acute release of endogenous opiates (EOPs) that suppress the release of GnRH by the hypothalamus and thus inhibit the secretion of LH. This is an immediate effect; both the removal of piglets from the sow and the blocking of EOP-activity with naloxone infusion in suckled sows cause an acute increase in the secretion of LH. When naloxone infusion stops or piglets are joined with the sow again LH secretion drops (Armstrong *et al.*, 1988). This explains why during lactation LH secretion is at a very low level, insufficient to support follicle growth up to ovulatory size.

The inhibition of LH secretion is not immediately established after parturition. It takes a few days before the inhibitory effects of EOPs become apparent (De Rensis *et al.*, 1993). From then on, LH secretion is low, but gradually increases throughout lactation (Quesnel and Prunier, 1995). The gradual restoration of LH secretion is due to the gradual decrease in suckling frequency, and possibly also due to the gradual escape from the inhibitory effects of suckling induced release of EOPs. Furthermore, the capacity of the hypothalamus-pituitary axis to secrete LH increases throughout lactation (Sesti and Britt, 1993[a]), which explains why sows weaned at 12 d of lactation have a lower LH pulse frequency before and after weaning than sows weaned at 21 d of lactation (Koketsu *et al.*, 1998). Besides the suckling-induced inhibition of LH secretion, metabolic factors that are related to the negative energy balance of the sow, inhibit the onset of cyclicity by influencing gonadotrophic secretion, and by influencing follicle development directly at the ovarian level. As outlined in a review by Quesnel and Prunier (1995) the metabolic factor becomes more important in the third and fourth week of lactation. As a result of changes in the hypothalamus-pituitary axis, as well as recovery from negative energy balance, especially in longer lactations (e.g. 5 to 7 weeks) sows can show some follicle growth towards the end of lactation (Kunavongkrit *et al.*, 1982) and even, occasionally, lactational oestrus.

Even if pulsatile LH secretion could be restored during lactation to a level sufficient for follicle growth, ovulation does not necessarily occur. For ovulation to occur, elevated oestradiol levels have to induce the pre-ovulatory LH surge from the hypothalamus-pituitary axis, by means of positive feedback. The capacity of the hypothalamus-pituitary unit to produce an LH surge in reaction to positive feedback, changes throughout lactation. Sesti and Britt (1993[b]), showed that early in lactation (d 14), 1 out of 8 sows showed an LH surge after oestradiol injection. Later in lactation (d 28) oestradiol injection induced an LH surge in 7 out of 8 sows. This indicates that with progressing lactation, the positive feedback communication between ovaries and the hypothalamus matures. Besides the maturation of the positive feedback system, the capacity of the hypothalamus-pituitary axis to secrete LH increases in the course of lactation (Sesti and britt, 1993[a]). Kirkwood *et al.* (1984) showed that sows weaned at 10 d of lactation had LH surges with a lower maximum than sows that were weaned at 35 d of lactation. Even if sows were able to show follicle growth during lactation, both the induction of the LH surge through positive feedback, as well as the magnitude of the resulting LH surge might be limited, depending on the stage of lactation.

As discussed previously, removal of the suckling stimulus induces increased LH secretion. There are several studies in which limited nursing, by separation of sow and litter during a fixed period of the day, has been used with the aim of inducing lactational oestrus. As apparent from Table 19.1, the number of sows in which estrus or ovulation could be established, varies considerably between studies. As such, the effect of limited nursing regimes is unpredictable. Moreover, the time between the start of an IS treatment and the manifestation of estrus varies between and within studies. In other words, the timing of oestrus is not as predictable as in weaned sows.

Table 19.1 NUMBER OF SOWS THAT SHOWED OESTRUS IN RESPONSE TO A LIMITED NURSING REGIME OR GROUPING DURING LACTATION, IN STUDIES FROM THE 1970S AND 1980S

Source	Start of limited nursing regime (d of lactation)		Parity	Duration of separation	Sows in oestrus during lactation
Crighton *et al.* (1970)	d21	10 d	Multi	12 h	1/5
Smith (1961)	d21				10/10
	d31-35				4/5
Thompson *et al.* (1981)	d21	12 d	primi	3-4	8/11
			multi	nursings/d	45/52
Stevenson & Davis (1984)	d14	14 d	Mixed	6 h	13/20
				12 h	5/10
Henderson & Hughes (1984)	d10	25 d	Multi	12 h	1/32
Grinwich and McKay (1985)	d21	14 d	2	3 h	1/10
			2	22 h	8/11
			1	22 h	12/20
Newton *et al.* (1987[a])	d13-21	8 d	mixed	3 h + boar	13/20
			mixed	6 h + boar	15/19
			mixed	6 h	4/28
			mixed	6 h + boar	5/28
Newton *et al.* (1987[b])	d20	8 d	multi	6 h	10/10
			primi	6 h	0/4
Kuller *et al.* (2004)	d14	14 d	multi	12 h	11/49
Petchey and Jolley (1979)[1]	d14	39 d		Grouping	63/129
Rowlinson & Bryant (1982)[1]	d20	22 d		Grouping + boar contact	29/46

[1]In these studies sows were grouped at d14 of lactation, rather than imposing a limited nursing regime.

There are several factors that may affect the number of sows that show oestrus within an IS regime. First, similar to the effect of lactation length on weaning to oestrus interval, the stage of lactation at which IS commences affects the rate at which sows come into oestrus (Smith, 1961; Newton *et al.*, 1987[a]). As explained earlier, this is related to the recovery of pulsatile LH secretion with progressing lactation and the maturation of the positive feedback of oestradiol on the LH surge system. Second, the time during which sows and piglets are separated might be a determinant of the proportion of sows that ovulate. Although Table 1 does not show a clear effect of the duration of separation, Matte *et al.* (1992) reviewed a number of studies that used intermittent suckling and concluded that the duration of separation did affect the interval to oestrus once the sows were weaned. A third important factor is social contact. Both contact with a boar and contact with other sows in a group appear to influence the onset and expression of oestrus (Petchey and Jolly, 1979; Rowlinson and Bryant, 1982). Possible mechanisms that explain these social effects are reviewed by Kemp *et al.* (2005). Fourth, litter size is also a factor that probably influences onset of follicle growth in an IS model. Varley and Foxcroft (1990) showed that reducing the number of suckling piglets from 10 to 5 increased pulsatile LH secretion. The increase was marked even more when the 'spare' teats were covered. Litter size probably affects suckling intensity, but also energy balance of a sow. The fifth factor that could influence LH secretion and resumption of follicle growth once a limited nursing regime is imposed, is the way sows and piglets are separated. Although based on a limited number of observations, Langendijk *et al.* (2007[a]) showed that sows that remained in the farrowing crate but were separated from their piglets using a partition had a different LH release pattern than sows that were moved to a different unit. Mattioli *et al.* (1988) made the same observation in a similar experiment. Sixth, parity also affects the number of sows that show oestrus. Primiparous sows are more sensitive to the suckling-inhibition of LH secretion (Britt, 1986), and are therefore less likely to show follicle growth and oestrus under an IS regime (Newton *et al.*, 1987[b]). Furthermore, primiparous sows suffer more from a negative energy balance (Van den Brand *et al.*, 2000), which can affect LH pulsatile secretion. Intermittent suckling can reduce the metabolic burden in primiparous sows, potentially improving LH secretion and follicle development, be it after weaning. The seventh factor is genotype. There are no studies comparing genotypes, but a varied source of genotypes have been used in the documented studies and all show a varied percentage of sows ovulating during lactation. In our work we have used two different genotypes, using exactly the same limited nursing model: starting at 14 d of lactation and separating sows and litters for 12 h per day, although not in the same experiment. Of the first genotype, only one-third of the sows ovulated during lactation whereas, as a proportion, up to 1.00 of the other genotype ovulated during lactation (Table 19.2). Altogether, these factors explain the variation in number of sows that show oestrus in different studies.

Table 19.2 OESTRUS AND OVULATION IN SOWS THAT WERE SUBJECTED TO A
LIMITED NURSING REGIME DURING LACTATION, IN RECENT STUDIES

Source	Line of sows	Stage of lactation[1]	Sows in oestrus	Sows that ovulated	Timing of oestrus[7]
Langendijk et al. (2007)	TOPIGS40	D13-18	10/12	10/12	wk3
Gerritsen et al. (2007)	TOPIGS40	D14 12h	14/14	13/14[2]	wk3
		2×6h	12/13	10/13[3]	wk3
Gerritsen et al. (unpublished)	TOPIGS40	D14	23/27	17/27[4]	wk3
		D21	25/26	22/26[5]	wk4
Langendijk et al. (unpublished)	TOPIGS20 Landrace × York	D14	11/32	9/32[6]	wk3

[1]The limited nursing regimes commenced at the stage of lactation as indicated and sows were separated from the piglets for 12 continuous hours per day unless stated differently. The limited nursing regime was imposed during at least two weeks, depending on the study, and sows that had not ovulated in that period were considered as anovulatory. [2]One sow did not ovulate and developed cystic follicles. [3]Two sows did not ovulate and developed cystic follicles and two sows did ovulate a number of follicles but also developed cystic follicles. [4]Three sows developed cystic ovaries and 7 sows did not grow pre-ovulatory follicles. [5]One sow developed cystic follicles and three sows did not grow pre-ovulatory follicles. Of the sows that ovulated, two did so before D21 and three already had large follicles at D21. [6]Two sows developed cystic follicles. [7]Timing of oestrus indicates in which week (wk) of lactation oestrus occurred.

In order for the IS model to have practical applications, oestrus should occur in an acceptable number of sows and in a predictable manner. Recently, we have started studies to investigate the regulation of the hypothalamus-pituitary-ovarian axis, follicular development, ovulation and the establishment of pregnancy during lactation. For this purpose, the IS model with 12 h separation per day from d14 of lactation onwards, was used in a Dutch commercial line (TOPIGS40 line, Topigs, The Netherlands). This line was chosen for the studies because it was expected to be highly responsive to limited nursing in terms of follicle development and ovulation, since this line has been intensively selected for a shorter weaning-to-oestrus interval, and this line appears to manifest lactational oestrus more frequently in practice than other lines.

With the TOPIGS40 line, using a limited nursing regime from d 14 of lactation resulted in 0.85 to 1.00 of the sows showing oestrus, and 0.63 to 0.93 of all the sows actually ovulating in the different studies. Postponing the start of intermittent suckling from d 14 to d 21 improved the percentage of sows that ovulated from 0.63 to 0.85 (Gerritsen et al., unpublished). Over the three studies with TOPIGS40 sows that are listed in Table 19.3, out of the 92 sows that were enrolled in the different limited nursing regimes, 0.91 showed oestrus and 0.78 ovulated. In none of these studies was a boar used for oestrus detection or to stimulate oestrus. Therefore, these results are promising and indicate that, if intermittent suckling does not start too early after farrowing (beyond three weeks) and is imposed for 12 consecutive hours per day, the potential percentage of sows that show oestrus and ovulate lies between 0.90 and 1.00, which is similar to results obtained after conventional weaning. This might indicate that modern sow types are less sensitive to inhibitory factors that are responsible for the classic lactational anoestrus than two decades ago. On the other hand, genotype might still be an important factor that determines responsiveness of sows to limited nursing, considering the difference in results between the TOPIGS20 and the TOPIGS40 sows (Table 19.2).

Table 19.3 PREGNANCY RATE AND EMBRYO SURVIVAL FOR SOWS THAT OVULATED AND WERE MATED DURING A LIMITED NURSING REGIME AND FOR SOWS THAT OVULATED AFTER A CONVENTIONAL LACTATION (CONTROL), IN RECENT STUDIES.

Source	Interval from farrowing to AI[1]	Gestational age at slaughter	Pregnancy rate (%)	Embryo survival %
Gerritsen et al. (2007[b])	~21d[2]	23d	16/19 (76)	63±19
	Control, 28d		16/17 (94)	78±18
Gerritsen et al. (unpublished)	~21d[3]	30d	5/7 (71)	57 ± 8
	~35d[3]		9/12 (75)	62 ± 6
Gerritsen et al. (unpublished)	~21d[4]	30d	13/17 (76)	46±16
	~28d[4]		14/16 (88)	59±24
	Control, 28d		11/15 (73)	59±17

[1]In the limited nursing regimes listed, sows and piglets were separated for 12 h each day, commencing at 14 or 21 d of lactation. Consequently, sows were mated at ~21 and ~28 d of lactation. [2]These sows continued lactating until d23 of gestation, when they were slaughtered. [3]These sows were inseminated during the limited nursing regime, either at spontaneous oestrus (d21 post farrowing) or at PG600-induced oestrus (D35 post-farrowing), 14 d later. The sows were weaned one week after ovulation. [4]In this study, the limited nursing regime commenced on d14 or d21. Hence, oestrus and AI occurred around d21 or d28, respectively. Weaning occurred at ovulation for half of the sows and 20 d after ovulation for the other half.

From the studies listed in Table 19.2, the sows that did not ovulate either developed cystic follicles (7 out of 92) or did not grow any pre-ovulatory follicles (13 out of 92). The latter category of sows would grow follicles up to only 3 to 4 mm (ultrasound). The development of cystic follicles was probably due to the absence of an LH surge, because in the sows from which pre-ovulatory blood samples were assayed for LH (5 out of 7 cystic sows), hardly or no elevation in LH was observed around the time it was expected based on estimation of ovulation time in other sows using ultrasound. It is interesting to see that 4 out of 13 sows that were subjected to a two time 6-h separation regime (Gerritsen *et al.*, 2007[a]) developed cystic follicles, compared to 3 out of 79 sows in the 12-h separation regimes in all the other studies or treatments. Failure to grow pre-ovulatory size follicles was probably due to the lack of adequate pulsatile secretion of LH after the commencement of the limited nursing regime. In fact, the two sows that failed to grow pre-ovulatory follicles in the first study (Langendijk *et al.*, 2007; Table 2), did show an LH secretion pattern that was different from the sows that ovulated. Alternatively, at this stage of lactation, antral follicles that develop on the ovaries might not yet be responsive to an increase in LH secretion. Besides developing cystic follicles or not growing any pre-ovulatory follicles, there is a third form of deviant follicle development that has not been reported before. In a study with a Landrace×York line (TOPIGS20), only one-third of the sows showed oestrus and ovulated (Table 19.2). Surprisingly, the majority of the non-ovulating sows did show follicular development up to pre-ovulatory size, without showing any increase in peripheral oestradiol concentrations (Langendijk *et al.*, unpublished). LH pulsatile release did not seem to be limiting in these animals, and we therefore hypothesised that the antral follicle population that was on the ovaries at the time the limited nursing regime was imposed was not responsive to gonadotrophic input.

Follicle development and peri-ovulatory endocrine profiles

Follicle growth from around 2-3 mm up to pre-ovulatory size in sows that were subjected to the limited nursing regimes in our studies did show the same pattern as in weaned sows, at least for the TOPIGS40 line. Similarly, oestradiol secretion showed the same timing of increase and decline in the period before ovulation (Figure 19.5). This means that with the TOPIGS40 sow model oestrus and ovulation were established within a week after the start of the intermittent suckling regime, and ovulation occurred within a range of 3 days (sows ovulated between 4 and 7 d after the start of intermittent suckling). This was as synchronous as in conventionally weaned sows, even though average time of ovulation was one week earlier for the sows that ovulated during lactation, due to the fact that intermittent suckling commenced at d 14 and conventional weaning in control sows occurred at d 21 of lactation. Ovulation rate for the sows that ovulated during lactation was also similar to the sows that ovulated after a conventional lactation. In the older studies that are listed in Table 19.3, onset of oestrus relative to the start of the limited nursing regime was comparable to our more recent studies, at least for the sows that did show lactational oestrus. In the studies by Newton *et al.* (1987[a,b]) oestrus occurred 5.1 ± 0.3 days from the start of the limited nursing regime for sows nursing 3 h per day (Newton *et al.*, 1987[a]) and 4.5 ± 0.2 days (Newton *et al.*, 1987[a]) and 5.2 ± 0.5 days (Newton

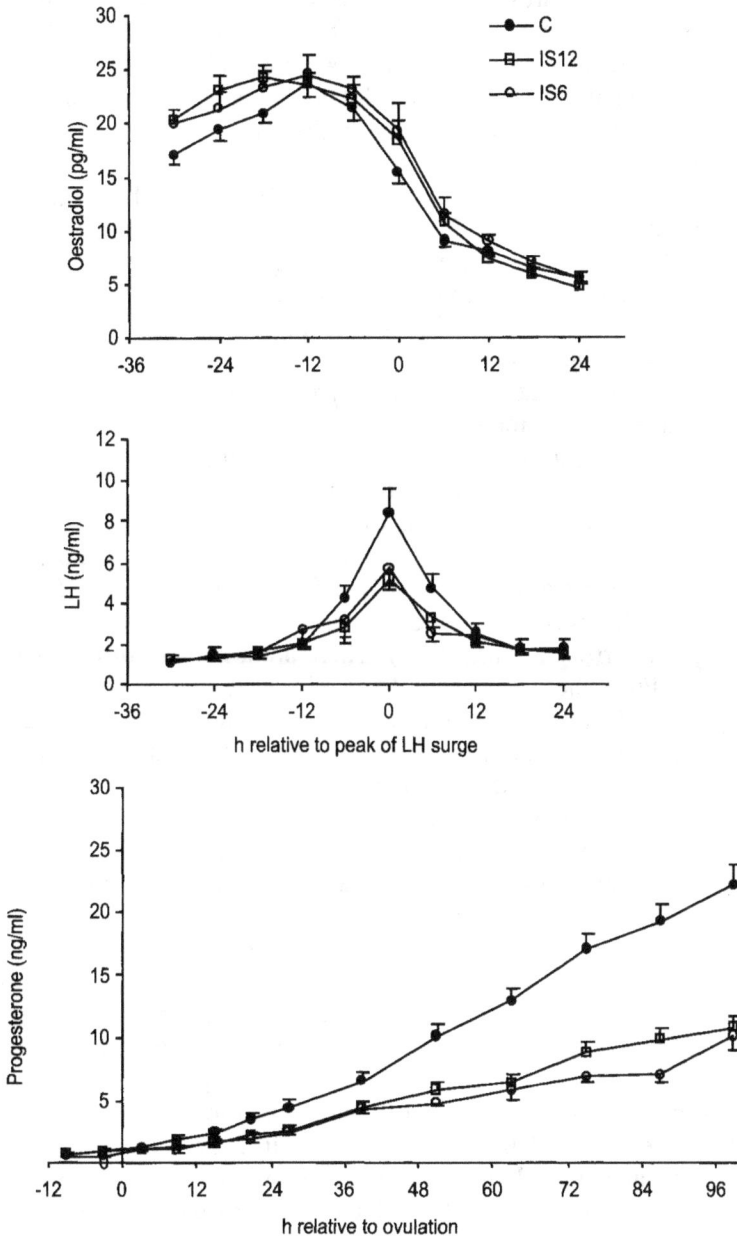

Figure 19.5 Peri-ovulatory concentrations of oestradiol-17ß, LH, and progesterone in serum of sows that showed oestrus and ovulated due to a limited nursing regime or after weaning following a conventional lactation (21d). Sows on a limited nursing regime were separated from their piglets during 12 h or twice during 6 h each day from d14 of lactation onwards. (From: Gerritsen et al, 2007[a]; courtesy of Reproduction in Domestic Animals, Blackwell Science)

et al., 1987[b]) from the start of limited nursing for sows nursing for 6 h per day. Stevenson and Davis (1984) recorded onset of oestrus at 6 days relative to the start of the limited nursing regime with a range of 4-8 days. However, the majority of the sows in the older studies did not show oestrus until after weaning. Including postweaning oestrus, the range in the time of oestrus for these studies is a few weeks.

The results with the TOPIGS40 line (Table 19.2) also show that within an intermittent suckling regime, follicle growth and follicle function (in terms of estradiol secretion) show a continuous development, and do not seem to be interrupted by periods of suckling. From Langendijk *et al.* (2007), it appears that after two days of the intermittent suckling regime LH secretion already becomes inhibited by increasing estradiol concentrations. It is therefore tempting to hypothesise that to initiate final follicle growth, only one or two episodes of inhibited suckling are sufficient, at least when sows are at the right stage of lactation and of the right genotype to respond to a limited nursing regime. For adaptation of piglets this is probably too short.

Although there were no indications of differences in follicular phase aspects between sows in a limited nursing regime and weaned sows, post-ovulatory progesterone concentrations were clearly affected (Figure 19.5). Progesterone increased at a much lower rate during the early luteal phase when sows ovulated during lactation (Gerritsen *et al.*, 2007). There are a number of possible reasons to explain this. First, the magnitude of the LH surge tended to be lower in the sows that ovulated during lactation as a consequence of the IS regime. The lower LH surge is not necessarily a consequence of the limited nursing regime, but might also be due to the earlier stage of lactation at which follicle development was initiated. The intermittent suckling regime commenced at d 14 of lactation, causing sows to ovulate around d21 of lactation, whereas continuously lactating sows were weaned at d 21 of lactation and consequently ovulated approximately one week later. Current studies do indicate that LH surge is indeed higher when the limited nursing regime commences at d 21 vs d 14. Another factor that might have caused the lower progesterone concentration is the fact that the sows were lactating. The high level of feed intake and metabolic rate in these sows might have caused a higher clearance rate of progesterone. It is generally accepted that a higher feeding level results in a higher metabolic clearance of progesterone (Miller *et al.*, 1999). Nevertheless, reduction of the feed allowance in a later experiment (4 kg vs 6.5 kg during early luteal phase) did not result in a higher progesterone concentration (Gerritsen *et al.*, 2006). There seem to be other factors involved that are related to the fact that the sows were lactating, although it is not clear what these are. There are a number of factors that have been reported to stimulate luteal function (see Schams *et al.*, 1999, for a review). IGF-1 for example has been shown to stimulate progesterone production by luteal cells in vitro (Webb *et al.*, 2002), and Langendijk *et al.* (2007[b]) have shown that there is an in vivo relation between IGF-1 and progesterone concentration.

IGF-1 and other metabolic factors could be limiting in a lactating sow and thus affect progesterone secretion.

In conclusion, a limited nursing regime can induce oestrus and ovulation during lactation, with timing of ovulation (synchrony) and percentage of animals ovulating being comparable to conventionally weaned sows, at least when a suitable sow model is used, and when the limited nursing regime is not imposed too early in lactation (beyond 3 weeks). Nevertheless, there is a striking difference in peripheral progesterone concentration between sows that ovulate during lactation and sows that ovulate after weaning. Consequences of the lower progesterone concentration for pregnancy are discussed in the next section. An important issue for the application of limited nursing regimes is genotype. There are probably large differences between genotypes in their response to a limited nursing regime in terms of follicle development and ovulation.

Establishment of pregnancy during lactation

In previous studies that have looked at pregnancy rates and embryo survival in limited nursing regimes, the regime was initiated shortly before weaning and ovulation occurred after weaning (Britt and Levis, 1982), or the regime was applied for a longer period but with the majority of ovulations occurring after weaning (e.g. Thompson *et al.*, 1981; Grinwich and McKay, 1985). In the latter studies no effects of limited nursing regimes are reported on conception rates in sows that ovulated during lactation (Britt and Levis, 1982; Thompson *et al.*, 1981) or embryo survival rate (Grinwich and McKay, 1985). However, conception and embryo development occurred after weaning in most of the sows in those studies, which makes it hard to draw any conclusion on how pregnancy and embryo survival are affected in a lactating sow.

In recent work from our group (see Table 19.3), pregnancy rates for multiparous sows that conceived during lactation and remained lactating for the major part of the embryonic phase, were between 71 and 76 % for the different studies (Table 3). Embryo survival for the pregnant sows was between 57 and 63%. For control sows, that conceived after a three-week lactation, pregnancy rates were 73 and 94%, and embryo survival 59 and 78% in two of these studies. Pregnancy rate and embryo survival seem to be affected when sows are lactating, although the differences were not significant in these studies. Because the number of animals in the different studies was low, and because progesterone was lower in sows that conceived in a limited nursing regime, effects on embryo development cannot simply be ruled out. There are two factors that can compromise embryo development when sows conceive during lactation: the time that has elapsed since farrowing and whether sows are lactating or not during the period of embryo development. The time that has elapsed since farrowing can influence involution and recovery of the uterus from the previous pregnancy, as well as the hypothalamus-pituitary axis. As highlighted earlier, the lower LH surge in lactating sows compared to weaned sows could in part have been a consequence of the fact that the lactating sows ovulated one week earlier relative to farrowing. The lower LH surge can affect luteinisation and subsequent function of the corpora lutea and thus affect embryonic development. The effect of time since farrowing is shown in preliminary data (Gerritsen *et al.*, unpublished) indicating that when sows are subjected to the limited nursing regime from d 21 instead of d 14 of lactation, pregnancy rates (88 vs 76%) and embryo survival (59 vs 46%) are higher (Table 19.3). Gaustad-Aas *et al.* (2004) also show that in sows that conceived after a spontaneous ovulation during a continuous lactation, pregnancy rate was seriously compromised when the sows were mated within 3 weeks after farrowing. They also show that weaned sows that were mated at a similar interval from farrowing, pregnancy rate was equally compromised (Table 19.4). This suggests that the time that has elapsed since the last farrowing is more important than whether the sows are lactating or not. However, preliminary data from our work indicate that sows that are mated at four weeks of lactation and weaned immediately after ovulation, have higher progesterone levels at 7d after ovulation (35 ± 5 vs 16 ± 3ng/ml; P<0.05) than sows that continue lactating after ovulation (Gerritsen *et al.*, 2007[c]). Furthermore, these sows have a numerically (but not significantly) higher pregnancy rate (100% (8 of 8) vs 75% (6 of 8) and embryo survival rate 62 vs 55% (Gerritsen, unpublished results). In sows that were mated three weeks post partum the differences were not as obvious.

Table 19.4 FARROWING RATES AND LITTER SIZE FOR SOWS THAT WERE MATED EITHER DURING (SPONTANEOUS) LACTATIONAL OESTRUS OR DURING OESTRUS AFTER WEANING, CATEGORISED ACCORDING TO TIME RELATIVE TO FARROWING. (AFTER GAUSTAD-AAS *et al.*, 2004)

Interval from farrowing to service[1]	Sows mated post-weaning		Sows mated during lactation	
	Farrowing rate, % (n)	Litter size (TNB)	Farrowing rate, % (n)	Litter size (TNB)
2nd wk	47 (283)	10.2 ± 0.30	50 (50)	10.8 ± 0.66
3rd wk	54 (179)	11.7 ± 0.32	60 (111)	11.1 ± 0.47
4th wk	71 (1274)	12.4 ± 0.11	75 (204)	12.7 ±0.30
5th wk	77 (12483)	13.1 ± 0.03	77 (209)	13.3 ± 0.28
6th wk	81 (41741)	13.7 ± 0.02		

[1]The interval from farrowing to service for sows that were mated postweaning was estimated based on stage of lactation at weaning and interval from weaning to service as presented in the original paper and are therefore rough indications.

Taken together, the available information suggests that post-ovulatory progesterone levels are lower in lactating sows compared to weaned sows, mainly due to the fact that they are lactating. This might affect pregnancy rate and embryo survival. However, in sows that are mated early after farrowing (around 3 weeks), involution of the uterus is an additional factor that can influence embryo survival and pregnancy. In the model used in our studies, the TOPIGS40 sow, it seems that when conception is established at least four weeks after farrowing, and sows are weaned after ovulation, progesterone concentration, embryo survival and pregnancy rate at d30 of gestation are as good as in conventionally weaned sows. Because in our studies sows were either weaned immediately after ovulation or around d 20 of gestation, it is not clear how embryo survival is affected when weaning occurs in between these times. Another question is whether there are differences between sows that conceive during a limited nursing regime and sows that conceive after weaning, in terms of eventual litter size. Based on the pregnancy rate and embryo survival that can be achieved in a limited nursing regime, at least when conception does not occur too early after farrowing and sows do not continue lactating too long during the embryonic period, the potential number of foetuses should be equal to sows that conceive after weaning. Studies are currently underway to establish the effects of a limited nursing regime on actual litter size at the end of gestation.

Concluding remarks

This review shows that limited nursing regimes offer the opportunity to gradually adapt piglets to a non-milk diet and definitive separation from the sow, without compromising the number of litters per year. By inducing oestrus and ovulation to occur during lactation, age of the piglets and the establishment of a next pregnancy are uncoupled, which enables the length of lactation to be extended. Extending lactation gives piglets even more time to adapt to the demands of weaning. Limited nursing regimes do not seem to negatively affect welfare of piglets. As opposed to studies in the 1980s and earlier, recent studies do indicate that an acceptable number of sows show oestrus and ovulate during a limited nursing regime. Pregnancy rates and embryo survival are also acceptable, provided that the limited nursing regime does not commence too early in lactation (before 3 weeks) and provided that sows do not remain lactating too long during early gestation. Genotype of the sow is also an important factor that determines the success of a limited nursing regime in terms of reproduction. Even though establishment of a subsequent pregnancy during lactation might seem an intensification of pig management, it should be realised that a limited nursing regime probably also improves the welfare of the sow. The extent of negative energy balance and mobilisation of body reserves are much reduced in a limited nursing regime (Gerritsen *et al.*, 2007[b]), which might also improve sow longevity in the long term. Furthermore, one could speculate that reducing the burden of continuous suckling improves the welfare of the sow. Separation from the piglets induces a short-lived increase in cortisol only the first two days of a limited nursing regime (Kluivers *et al.*, 2006). As such, limited nursing regimes offer opportunities for pig production in terms of improving welfare. By maintaining the number of litters per sow per year, limited nursing regimes are probably economically promising as well, although this has not yet been investigated. For organic farming systems, a limited nursing regime has the additional advantage of dramatically increasing the potential number of litters per year. Future research should focus more on the optimal stage of lactation to commence a limited nursing regime, as this influences both feed intake of the piglets and reproductive success of the sow. In terms of sow reproductive success, the optimal stage of lactation might depend on the genotype. In addition, the necessary duration of a limited nursing regime in terms of allowing piglets to develop adequate feed intake has to be explored, as continued lactation during pregnancy is a key issue when it comes to establishment of pregnancy and embryo survival.

References

Armstrong, J.D., Kraeling, R.R. and Britt, J.H. (1988) Effects of naloxone or transient weaning on secretion of LH and prolactin in lactating sows. *Journal of Reproduction and Fertility*, **83**, 301.

Berkeveld, M., Langendijk, P., Van Beers-Schreurs, H.M.G., Koets, A.P., Taverne, M.A.M. and Verheijden, J.H.M. (2007) Postweaning growth check in pigs is markedly reduced by intermittent suckling and extended lactation. *Journal of Animal Science*, **85**, 258–266.

Bevers, M.M., Willemse, A.H., Kruip, T.A. and Dieleman, S.J. (1985) The relation between prolactin and lactational anestrus in swine. *Tijdschrift Diergeneeskunde,* **110**, 107. Dutch.

Booman, P., Van der Wiel, D.F.M. and Jansen, A.A.M. (1982) Effect of exogenous prolactin on preipheral luteinizing hormone levels in the sow after weaning of piglets. *Rapport B-200 Instituut voor veeteeltkundig onderzoek Schoonoord.* Zeist, The Netherlands, p. 58. Dutch.

Brand, H. van den, Dieleman, S.J., Soede, N.M. and Kemp B. (2000) Dietary energy source at two feeding levels during lactation in primiparous sows. I Effects on glucose, insulin and LH and on follicle development, weaning-to-estrus interval and ovulation rate. *Journal of Animal Science*, **78**, 396-404.

Britt, J.H. and Levis, D.G. (1982) Effect of altered suckling intervals of early-weaned pigs on rebreeding performance of sows. *Theriogenology,* **18**, 201-207.

Britt, H.J. (1986) Improving sow productivity through management during gestation, lactation and after weaning. *Journal of Animal Science*, **63**, 1288.

Bruininx, E. M., Binnendijk, G.P., Van der Peet-Schwering, C.M.C., Schrama, J.W., Den Hartog, L.A., Everts, H. and Beynen, A.C. (2002). Effect of creep feed consumption on individual feed intake characteristics and performance of group-housed weanling pigs. *Journal of Animal Science*, **80**, 1413–1418.

Colson, V., Orgeur, P., Foury, A. and Mormede, P. (2006) Consequences of weaning piglets at 21 and 28 days on growth, behaviour, and hormonal responses. *Applied Animal Behaviour Science*, **98**, 70-88.

Crighton, D.B. (1970) Induction of pregnancy during lactation in the sow. *Journal of Reproduction and Fertility*, **22**, 223.

Duggan, R.T., Bryant, M.J. and Cunningham, F.J. (1982) Gonadotrophin, total oestrogen and progesterone concentrations in plasma of lactating sows with particular reference to lactational estrus. *Journal of Reproduction and Fertility*, **64**, 303–13.

Dybkjaer, L. (1992) The identification of behavioral indicators of 'stress' in early weaned pigs. *Applied Animal Behaviour Science*, **35**, 135-147.

English, P.R., Robb, C.M. and Dias, M.F.M. (1980) Evaluation of creep feeding using a highly-digestible diet for litters weaned at 4 weeks of age. *Animal Production* **30**, 496. (Abstr.)

Everts, H., Blok, M.C., Kemp, B., Van der Peet-Schwering, C.M.C. and Smits, C.H.M. (1995) Normen voor lacterende zeugen: uitgangspunten en factoriele afleiding van de behoefte aan energie en darmverteerbare aminozuren voor lacterende zeugen. *CVB-documentatierapport* **13**. ISSN 0928-0618.

Gaustad-Aas, A.H., Hofmo, P.O. and Karlberg, K. (2004) The importance of farrowing to service interval in sows served during lactation or after shorter lactation than 28 days. *Animal Reproduction Science*, **81**, 287-293.

Gerritsen, R., Soede, N.M., Langendijk, P., Laurenssen, B.F.A. and Kemp, B. (2006) Post-ovulatory feeding levels in sows with lactational ovulation: effects on progesterone and embryo survival. *Reproduction in Domestic Animals*, **41**, 369. Abst.

Gerritsen, R., Soede, N.M., Hazeleger, W., Dieleman, S.J., Langendijk, P. and Kemp, B. (2007[a]) Peri-estrus hormone profiles and follicle growth during lactation in sows under limited nursing (Intermittent Suckling) regimes. *Reproduction in Domestic Animals*, In Press.

Gerritsen, R., Soede, N.M., Langendijk, P., Taverne, M.A.M. and Kemp, B. (2007[b]) Early embryo survival and development in sows with lactational ovulation. *Reproduction in Domestic Animals*. In Press

Gerritsen, R., Soede, N.M., Langendijk, P., Dieleman, S.J., Hazeleger, W. and Kemp, B. (2007[c]) LH and progesterone profiles in sows with lactational oestrus induced by Intermittent Suckling at two different stages of lactation. European Society for Domestic Animal Reproduction, 2007. *Reproduction in Domestic Animals*. In Press.

Grinwich, D.L. and McKay, R.M. (1985) Effects of reduced suckling on days to estrus, conception during lactation and embryo survival in sows. *Theriogenology*, **23**, 449-459.

Guthrie, H.D., Bolt, D.J. and Cooper, B.S. (1990) Effects of gonadotropin treatment on ovarian follicle growth and granulosal cell aromatase activity in prepuberal gilts. *Journal of Animal Science*, **68**, 3719–26.

Hartke, J.L., Monaco, M.H., Wheeler, M.B. and Donovan, S.M. (2005) Effect of a short-term fast on intestinal disaccharidase activity and villus morphology of piglets suckling insulin-like growth factor-I transgenic sows. *Journal of Animal Science*, **83**, 2404–2413.

Henderson, R. and Hughes, P.E. (1984) The effects of partial weaning, movement and boar contact on the subsequent reproductive performance of lactating sows. *Animal Production*, **39**, 131-135

Kelly, D., Smyth, J.A. and McCracken, K.J. (1991[a]) Digestive development of the early-weaned pig. 1. Effect of continuous nutrient supply on the development of the digestive tract and on changes in digestive enzyme activity during the first week post-weaning. *British Journal of Nutrition*, **65**, 169–180.

Kelly, D., Smyth, J.A. and McCracken, K.J. (1991[b]) Digestive development of the early-weaned pig. 2. Effect of level of food intake on digestive enzyme activity during the immediate post-weaning period. *British Journal of Nutrition*, **65**, 181–188.

Kemp, B., Soede, N.M. and Langendijk, P. (2005) Effects of boar contact and housing conditions on estrus expression in sows. *Theriogenology*, **63**, 643-656.

Kirkwood, R.N., Lapwood, K.R., Smith, W.C. and Anderson, I.L. (1984) Plasma concentrations of LH, prolactin, oestradiol-17 beta and progesterone in sows weaned after lactation for 10 or 35 days. *Journal of Reproduction and Fertility*, **70**, 95.

Kluivers, M., Langendijk, P. and Van Nes, A. (2006) Cortisol profiles in sows submitted to an intermittent suckling regime compared to that of abruptly weaned sows. *Proceedings of the 19th International Pig Veterinary Society*, Copenhagen 2006 ,Vol 2 , p 606.

Koketsu, Y., Dial, G.D., Pettigrew, J.E., Xue, J., Yang, H. and Lucia, T. (1998) Influence of lactation length and feed intake on reproductive performance and blood concentrations of glucose, insulin and luteinizing hormone in primiparous sows. *Animal Reproduction Science*, **52**, 153.

Kunavongkrit, A., Einarsson, S. and Settergren, I. (1982) Follicular development in primiparous lactating sows. *Animal Reproduction Science*, **5**, 47.

Kuller, W.I., Soede, N.M., Van Beers-Schreurs, H.M.G., Langendijk, P., Taverne, M.A.M., Verheijden, J.H.M. and Kemp, B. (2004) Intermittent suckling:

Effects on piglet and sow performance before and after weaning. *Journal of Animal Science*, **82**, 405–413.

Langendijk, P., Dieleman, S.J., Van den Ham, C.M., Hazeleger, W., Soede, N.M. and Kemp, B. (2007[a]) LH pulsatile release patterns, follicular growth and function during repetitive periods of suckling and non-suckling in sows. *Theriogenology*, **67**, 1076-1086.

Langendijk, P., Van den Brand, H., Gerritsen, R., Quesnel, H., Soede, N.M. and Kemp, B. (2007[b]) Porcine luteal function in relation to IGF-1 levels following ovulation during lactation or after weaning *Reproduction in Domestic Animals*. In Press.

Marion, J., Biernat, M., Thomas, F., Savary, G., Le Breton, Y., Zabielski, R., Le Huerou-Luron, I. and Le Dividich, J. (2002) Small intestine growth and morphometry in piglets weaned at 7 days of age. Effects of level of energy intake. *Reproduction Nutrition and Development*, **42**, 339-54.

Matte, J.J., Pomar, C. and Close, W.H. (1992) The effect of interrupted suckling and split weaning on reproductive performance of sows: a review. *Livestock Production Science*, **30**, 195-212.

Mattioli, M., Galeati, G. and Seren, E. (1988) Control of LH and PRL secretion during lactational anestrus in the pig. In: *Eleventh International congress on animal reproduction*, 1988, p. 44.

Miller, H.M. and Slade, R.D. (2003) Digestive physiology of the weaned pig. In: Weaning the pig: Concept and consequences. Pluske, J.R., Le Dividich, J. and Verstegen, M.W.A. (ed.) Wageningen Academic Publishers, Wageningen, The Netherlands, p. 117-139.

Miller, H.M., Foxcroft, G.R., Squires, J. and Aherne, F.X. (1999) The effects of feed intake and body fatness on progesterone metabolism in ovariectomized gilts. *Journal of Animal Science* **77**, 3253-3261.

Nabuurs, M.J., Hoogendoorn, A. and Van Zijderveld-van Bemmel, A. (1996) Effect of supplementary feeding during the sucking period on net absorption from the small intestine of weaned pigs. *Research in Veterinary Science*, **61**, 72–77.

Nabuurs, M.J. (1998) Weaning piglets as a model for studying pathophysiology of diarrhea. *Veterinary Quarterly*, **20**(Suppl. 3), S42–S45.

Newton, E.A., Stevenson, J.S., Minton, J.E. and Davis D.L. (1987[a]) Endocrine changes before and after weaning in response to boar exposure and altered suckling in sows. *Journal of Reproduction and Fertility*, **81**, 599.

Newton, E.A., Stevenson, J.S. and Davis, D.L. (1987[b]) Influence of duration of litter separation and boar exposure on estrous expression of sows during and after lactation. *Journal of Animal Science*, 65, 1500-1506

Pajor, E.A., Fraser, D. and Kramer, D.L. (1991). Consumption of solid food by suckling pigs: Individual variation and relation to weight gain. *Applied Animal Behaviour Science*, **32**, 139–155.

Pajor, E.A., Weary, D.M., Caceres, C., Fraser, D. and Kramer, D.L. (2002). Alternative housing for sows and litters: Part 3. Effects of piglet diet quality and sow-controlled housing on performance and behaviour. *Applied Animal Behaviour Science*, **76**, 267-277.

Petchey, A.M. and Jolly, G.M. (1979). Sow service in lactation: An analysis of data from one herd. *Animal Production*, **29**, 183-191.

Pluske, J.R., Williams, I.H. and Aherne, F.X. (1996) Villous height and crypt depth in piglets in response to increase in intake of cow's milk after weaning. *Animal Science*, **62**, 145-158.

Pluske, J.R., Hampson, D.J. and Williams, I.H. (1997) Factors influencing the structure and function of the small intestine in the weaned pig: a review. *Livestock Production Science*, **51**, 215-236.

Quesnel, H. and Prunier, A. (1995) Endocrine bases of lactational anoestrus in the sow. *Reproduction Nutrition and Development*, **35**, 395.

Rantzer, D., Svendsen, J. and Westrom, B.R. (1995). Weaning of pigs raised in sow-controlled and in conventional housing systems. 2. Behaviour studies and cortisol levels. *Swedish Journal of Agricultural Research*, **25**, 61-71.

De Rensis, F., Hunter, M.G. and Foxcroft, G.R. (1993) Suckling-induced inhibition of luteinizing hormone secretion and follicular development in the early postpartum sow. *Biology of Reproduction*, **48**, 964.

Rowlinson, P. and Bryant, M.J. (1982) Lactational oestrus in the sow. 2. The influence of grouping, boar presence and feeding level upon the occurrence of oestrus in lactating sows. *Animal Production*, **32**, 283.

Schams, D., Berisha, B., Kosmann, M., Einspanier, R., Amselgruber, W.M. (1999) Possible role of growth hormone, IGFs, and IGF-binding proteins in the regulation of ovarian function in large farm animals. *Domestic Animal Endocrinology*, **17**, 279-85.

Sesti, L.A.C. and Britt, J.H. (1993[a]) Agonist-induced release of gonadotropin-releasing hormone, luteinizing hormone, and follicle-stimulating hormone and their associations with basal secretion of luteinizing hormone and follicle-stimulating hormone throughout lactation in sows. *Biology of Reproduction*, **49**, 332.

Sesti, L.A.C. and Britt, J.H. (1993[b]) Influence of stage of lactation, exogenous luteinizing hormone–releasing hormone, and suckling on estrus, positive feedback of luteinizing hormone, and ovulation in sow treated with estrogen. *Journal of Animal Science*, **71**, 989.

Smith, D.M. (1961) The effect of daily separation of sows from their litters upon milk yield, creep intake and energetic efficiency. *New Zealand Journal of Agricultural Research*, **4**, 232.

Soede, N.M, Quesnel, H., Prunier, A. and Kemp, B. (2000) Variation in Weaning-to-Oestrus intervals in sows: causes and consequences. *Reproduction in Domestic Animals*, **Supplement 6**, 111.

Stevenson, J.S. and Davis D.L. (1984) Influence of reduced litter size and daily litter separation on fertility of sows at 2 to 5 weeks postpartum. *Journal of Animal Science*, **59**, 284.

Thompson, L.H., Hanford, K.J. and Jensen, A.H. (1981) Estrus and fertility in lactating sows and piglet performance as influenced by limited nursing. *Journal of Animal Science*, **53**, 1419-23.

Van Beers-Schreurs, H.M.G., Vellenga, L., Wensing, T. and Breukink, H.J. (1992) The pathogenesis of the post-weaning syndrome in weaned piglets: A review. *Veterinary Quarterly*, **14**, 29–34.

Van Beers-Schreurs, H. M. G., Nabuurs, M.J., Vellenga, L., Kalsbeek-van der Valk, H.J., Wensing, T. and Breukink, H.J. (1998) Weaning and the weanling diet influence the villous height and crypt depth in the small intestine of pigs and alter the concentrations of short-chain fatty acids in the large intestine and blood. *Journal of Nutrition*, **128**, 947-953.

Varley, M.A. and Foxcroft, G.R. (1990) Endocrinology of the lactating and weaned sow. *Journal of Reproduction and Fertility*, **Suppl. 40**, 47.

Webb, R., Woad, K.J. and Armstrong, D.G. (2002) Corpus Luteum (CL) function: local control mechanisms. *Domestic Animal Endocrinology*, **23**, 277-85.

20

AD LIBITUM FEEDING OF GESTATING SOWS

PHILIP VAN DEN BRINK
Provimi B.V. Rotterdam, The Netherlands

Introduction

In practice the majority of pigs are fed ad libitumitum to obtain optimum performance results. This is the case for piglets, lactating sows and for most fatteners. The only exception is gestating sows, which are mainly fed restricted. In 1997 Provimi in The Netherlands initiated feeding ad libitumitum for gestating sows in practice. At this time there was no worldwide experience on how to optimize those diets and on how to create optimum pens. During the last decade Provimi have developed considerable experience with this system and more research was done on feeding high fibre diets and the consequences. This chapter will consider current know how on ad libitum feeding, and focus on the practical implementation of this knowledge and experience.

Group housing of sows

In the future there will be increased emphasis on group housing for gestating sows. The main drivers for this development are consumer demands and legislation. In The Netherlands, there has been an obligation to erect new buildings for gestating sows according to group housing systems since 1998.

There are several systems of group housing of sows and the most common are the following:

- Crates with free space behind the boxes
- Electronic sow feeders
- Long troughs
- Ad libitum feeding

Ad libitumitum feeding of gestating sows can be done in several ways. The most common current system are dynamic groups of sows housed on straw

385

bedding, stable groups on concrete floors or even outdoor systems. At the moment the system with ad libitumitum feeding in stable groups on concrete floors is the most popular. To make a success of this system it is very important to give attention to pen design and feed optimization.

Critical factors to make a success of ad libitum feeding

When gestating sows are fed ad libitum the most critical issue is to reduce feed intake. When using traditional gestating feed, intakes above 5 kg per day are quite normal and sows will then become seriously overweight. The two most critical factors to reduce feed intake are pen design and feed composition.

PEN DESIGN IN AD LIBITUM FEEDING SYSTEMS

EU regulations prescribe that the area provided should be2.25m² per sow. In groups with more than 40 sows this area can be reduced by 10%. The first factor to reduce ad libitum feed intake is by using separate feeders and drinkers. If these are placed at least 6 meter apart then this will lead to a reduction of feed intake. This also means that it is easier to make a success of ad libitum feeding for gestating sows in larger groups than in smaller groups. Figure 20.1 presents a basic pen design for a group of 40 sows.

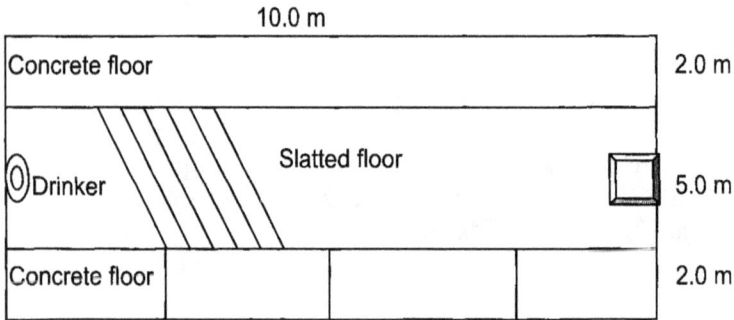

Figure 20.1 Basic pen design for a group of 40 gestating sows

The best feeders are specially developed for sows and are stronger and bigger than normal feeders that are used for fattening pigs. There are special drinkers on the market where no water is spilled. This is important to reduce slurry production.

FEED OPTIMIZATION OF AD LIBITUM FEED

In 1997 Provimi started with the inclusion of 550g sugar beet pulp/kg diet and it was noticed that this was really reducing feed intake to levels of 3.8 kg per sow per day. The problem at that time was that the resultant manure was too thin and lacked 'structure'. It was noticed that there are several kinds of sugar beet pulp on the market and that every factory produces products of different quality. The conclusion is that analysis of the beet pulp used in gestating diets is an important factor contributing to the success of this feeding system. Over the past 10 years the main challenge for Provimi was to reduce the amount of beet pulp in the feed and to find alternative raw material sources, which can have the same result as sugar beet pulp. The current formulations are mainly based on the use of a combination of sugar beet pulp, soyabean hulls, palmkernel expeller, sunflower expeller and alfalfa meal. The degree to which the the concept of ad libitum feeding can be applied will depend on the availability and the prices of these raw materials.

Ad libitum feeding and technical results

Previous sections have described that feed intake is a critical issue in ad libitum feeding and therefore the question now is the degree to which it can used for highly prolific farms. Table 20.1 below shows that it can be more economical to use the system on farms with good performance results.

Table 20.1 EXAMPLE OF THE EFFECT OF PERFORMANCE RESULTS ON THE QUANTITY OF SOW FEED (kg) USED PER PIGLET

Piglets/sow/year	20	25	30
Feed/sow/year (kg)	1100	1250	1350
Sow feed/piglet/year	55	50	45

Reproductive data in the ad libitum system

In 2000, Provimi started to gather data on farms where they had used the novel concept for at least two years. At that time it was seen that the performance results were satisfactory but the main question was whether these data where consustent. It was thought that, with heavier and fatter sows on the ad libitum system, production could decline after use of the concept on a long-term basis. Provimi investigated this through selecting 11 farms in The Netherlands and Germany where the litters over at least four parities could be followed (Table 20.2).

Table 20.2 PIGLETS BORN ALIVE PER LITTER ON SEVERAL FARMS

	Litters (n)	Gilts	Parity 2	Parity 3	Parity 4
Avg. of 11 farms	3789	10.85	11.55	12.2	12.24
Control 235 farms		10.2	10.5	11.3	11.6
Difference		+ 0.65	+1.05	+ 0.90	+ 0.64

Source: Provimi Field research 2000

The outcome was very impressive and even compared with the top 20% of farms in Holland the average results of the ad libitum concept was promising. The main question was at that time what the mode of action is behind these interesting results.

Vestergaard (1998) showed that blood insulin levels after feeding and over the whole day differed considerably between a cereal diet and a sugar beet pulp diet (Figure 20.2) with the former giving high insulin peaks immediately after feeding and low levels four hours after feeding whereas the latter gave much flatter levels of insulin but higher after 4 hours of feeding.

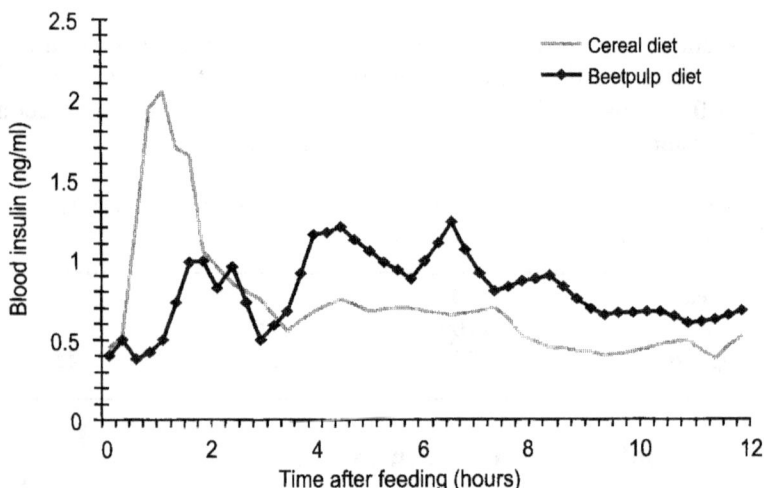

Figure 20.2 Insuline development after feeding 2 types of feed

The difference in insulin levels could be explained through enzymatic digestion of starch in the small intestine that gave high insulin peaks on the high starch cereal diet. In the sugar beet pulp diet, a second insulin "peak" was observed that could be explained by the fermentation of Fermentable Non Starch Polysacharides (VNSP) from the sugar beet pulp. This might be at least a partial of the explanation for the effect on numbers of pigs born alive on the ad libitum feed concept.

The theory is that if a high insulin peak is observed, the blood flow via the liver is higher and as a result progesterone clearance is higher. Progesterone is a key hormone in establishing and maintaining pregnancy and higher levels at the start of gestation can be beneficial for this and may also improve embryonic survival (Razdan, 2003).

A higher rate of survival of embryos was also shown by Ferguson (2003) in a trial where, during the period between weaning and mating, a VNSP-rich diet was compared with a starch-rich diet. After insemination all sows were fed the same gestation feed. On day 30 after insemination the embryonic survival on the latter 76.7% and on the former it was 85.2%. The mode of action behind this is not clear yet, but it is possible that the shape of the insulin curve has an influence on the survival rate of the embryos. In practice this research suggests that supplementing a gestation feed with a topdressing such as Profert (Fertility stimulation mix) can be interesting instead of feeding a lactation feed in the period between weaning and mating.

Another explanation for the beneficial effect on reproduction is given by De Leeuw et al. (2004) who measured the post-prandial plasma glucose levels of sows. Data were expressed as a proportion of the basal levels of glucose (figure 20.3). It was shown that the blood glucose was much more variable on the low VNSP diet (170 g/kg) compared to the higher (380 g/kg). The plasma glucose concentration was noticeably low in the hours immediately before the next meal.

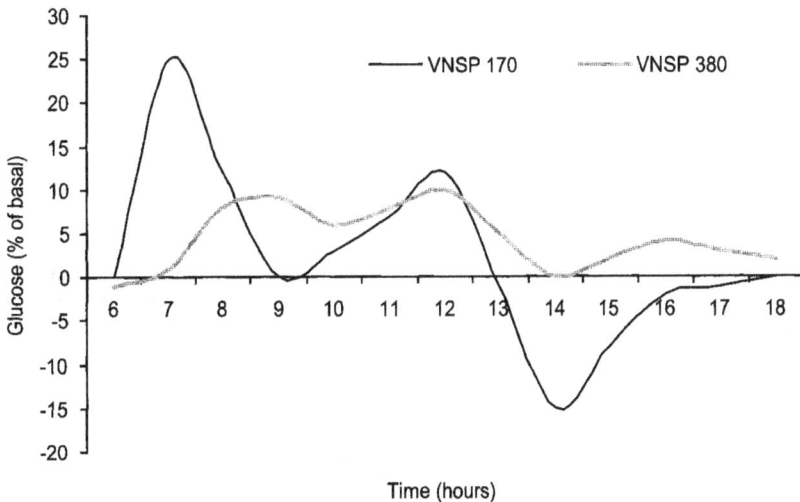

Figure 20.3

de Leeuw et al. (2004) also looked at the activity of the sows (figure 20.4) and found that positional changes in the sows on the low VNSP diet was about 60% higher in the hours before receiving a new meal. This shows that high

VNSP diets can lead to less restless and quieter sows, reduced stress and probably, as a result, better fertility.

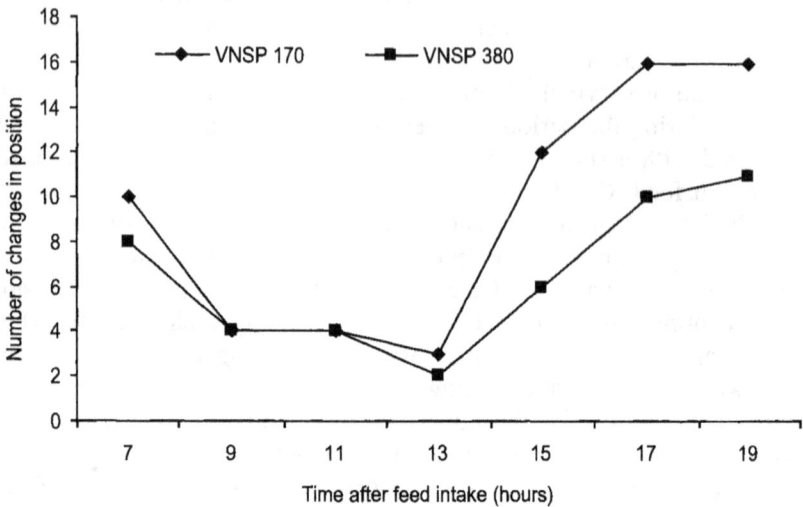

Figure 20.4

Data from van der Peet 2003 showed that feeding a VNSP-rich diet to gestating sows compared to starch-rich gestating feed gave the following benefits: .piglets born alive increased by 0.5 piglets per litter, the sows show less stereotypic behaviour and the feed intake during lactation was 0.4 kg higher per day.

Comparison between Electronic sow feeders and ad libitum feeding system

In collaboration with Wageningen University, Provimi and ABCTA a trial was undertaken at the experimental farm in Raalte, The Netherlands, by van der Peet (2003). where a "welfare feed" that was of high fibre content fed via electronic sow feeders (ESFs) was compared with an ad Libitum feed.

The ad libitum feed contained 450g sugar beet pulp and was fed unrestricted in dry sow feeders. 20 sows (2 pens of 10 sows) were fed in each group and they were followed for three parities. This was repeated three times with a total of 180 farroweings per treatment. The sows were housed in the group housing system immediately after mating.

Feed intake of the ad libitum feed was on average 4.2 kg per day vs 2.9 kg on the ESF. The feed intake curve (figure 20.5) was completely different between feeding systems.

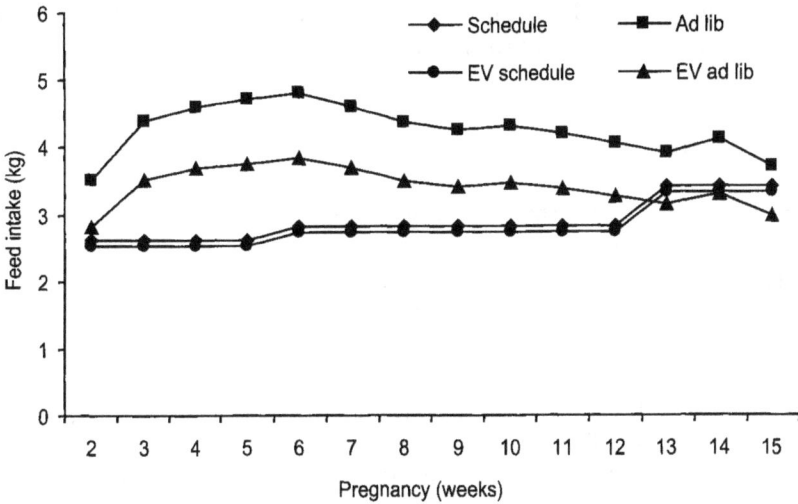

Figure 20.5

The energy density of the ad libitum feed is about 85% of the welfare feed so this should be taken into account if we want to look at nutrient requirement of sows. The effect of different feed intakes resulted in a difference in weight development and backfat development of the sows. At the third weaning of the sows (end of the trial) the ad libitum fed sows were on average 25.0 kg heavier and had 3 mm more backfat.

There were no significant differences in performance results (live piglets born per litter) but the means were good with around 11 weaned piglets per litter. In addition, litter birth weights was good in both groups reaching just over 19 kg for the live born piglets per litter. The litter weight of the weaned piglets with 95.1 kg for the ad libitum fed sows 4.1% higher then the restricted fed sows with 91.3 kg litter weight. This difference was also non significant.

Significant differences were recorded for the number of skin lesions of sows in different stages of gestation. This shows that sows are more quiet when they are fed on ad libitum systems instead of a restricted feeding system. However, the pens on the ad libitum system were dirtier than the pens of the sows which where on a restricted feeding system.

The overall conclusion is that in both restricted feeding and ad libitum feeding systems good performances are possible although the feed intake curves are completely different. The ad libitum feeding system is a simple that is beneficial for the welfare of the sows.

Economical evaluation

Although feed costs were higher in ad libitum systems, they can compete with the Electronic sow feeders. This is mainly due to lower investments, lower labour costs and extra income for the culled sows on ad libitum feeding systems.

Progeny studies

Provimi designed a trial investigating progeny from ad libitum fed sows. A group of gestating sows were divided, between day 30 and 80 of gestation, into 2 groups; half were fed ad libitum and the other half restricted, receiving 0.7 of the quantity consumed by the former, but using the same diet.. Subsequently sows were all mixed together and all fed ad libitum. The trial was repeated once. In the second trial, the progeny of the ad libitum-fed sows grew significantly faster in the last period of the fattening phase. In both trials, the lean meat content of the progeny of the ad libitum-fed sows was significantly higher then the restricted fed sows. This difference was caused by a significantly lower back fat thickness and a tendency for a higher muscle depth (+2 mm). The cause for this difference was initially investigated by Stickland ( ) effect who observed that sows fed on a higher nutritional level during gestation produced progeny with more secondary muscle fibres. This shows that the genetic potential of the piglet born is not only a question of breeding but also a result of feeding

Implementation in practice

On several farms in The Netherlands and Germany it has been demonstrated for years that ad libitum feeding can be possible in a technical way with excellent economic results. The best farms with ad libitum feeding are achieving performances of 30 weaned piglets per sow per year.

To find out if the system is applicable under different circumstances, it is important to check the availability of raw materials which are high in fermentable fibre. In addition the prices for these raw materials can be a handicap for the economic implementation of the ad libitum feeding concept. The ad libitum feeding system is simple but the two critical factors like pen design and feed formulation are the main factors to consider in making the system successful and to have a reasonable feed intake. One of the most important tools to lower the feed intake of ad libitum-fed gestating sows is to start with the system at 30 days of pregnancy. This means that, in the first part of gestation, sows stay in the mating centre and, after scanning for pregnancy, are moved to pens with the ad libitum system. In this way feed intake will be reduced by about 110 kg per sow per year compared to starting ad libitum feeding immediately after mating.

Conclusion

After 10 years of use of ad libitum feeding of gestating sows in practice it can be that the system has developed effectively. The advantages of ad libitum feeding are the simplicity of the system in combination with good performance technical results. Research has shown that reproductive data of ad libitum feeding are good and probably even better then on traditional feeds with low fermentable fibre. The know how of ad libitum feeding can be implemented in every feed for gestating sows all over the world. Through this there are possibilities for improved welfare and performance results in all systems of (group) housing of gestating sows.

21

FUNDAMENTAL ASPECTS OF MEAT SCIENCE - TENDERISATION

C. M. KEMP, P.L. SENSKY, S. J. GIBSON, R. G. BARDSLEY, P.J.
BUTTERY AND T. PARR.
*Division of Nutritional Sciences, School of Biosciences, University of
Nottingham, UK*

Introduction

An objective of the meat animal production industry is to achieve the efficient
conversion of dietary components nutrients into lean tissue growth and reduce
the partitioning of these components into adipose tissue.   However, the drive
for livestock production has increasingly moved from enhanced muscle growth
towards improving meat quality. Eating quality of meat depends on several
organoleptic properties including appearance, colour, fat content, flavour,
juiciness and tenderness.   Of all the meat traits, tenderness is considered to be
the most important (Miller *et al.*, 2001).   It has been demonstrated that
consumers are willing to pay higher prices for tender meat and there is an
added premium when meat is guaranteed to be tender (Lusk *et al.*, 2001).
However despite efforts to standardise variable factors in animal and meat
production, ensuring a consistently tender product still remains difficult to
control or predict.   The objective of the present chapter is to examine the
current understanding of the molecular factors that are involved in the
development of tender meat with particular reference to pork.

Postmortem degradation of muscle proteins and candidate proteolytic systems

The main determinant of ultimate tenderness appears to be the extent of
proteolysis of key target proteins within muscle fibres (Taylor *et al.*, 1995a).
During postmortem proteolysis specific myofibrillar and cytoskeletal proteins
are degraded. In tender meat specific myofibrillar (titin), Z line (desmin) and
costamere proteins (vinculin) are subjected to cleavage, but major myofibrillar
proteins such as actin and myosin and the major Z line protein α-actinin are
not (Fig 21.1; Goll *et al.*, 1992; Taylor *et al.*, 1995a).   There are several

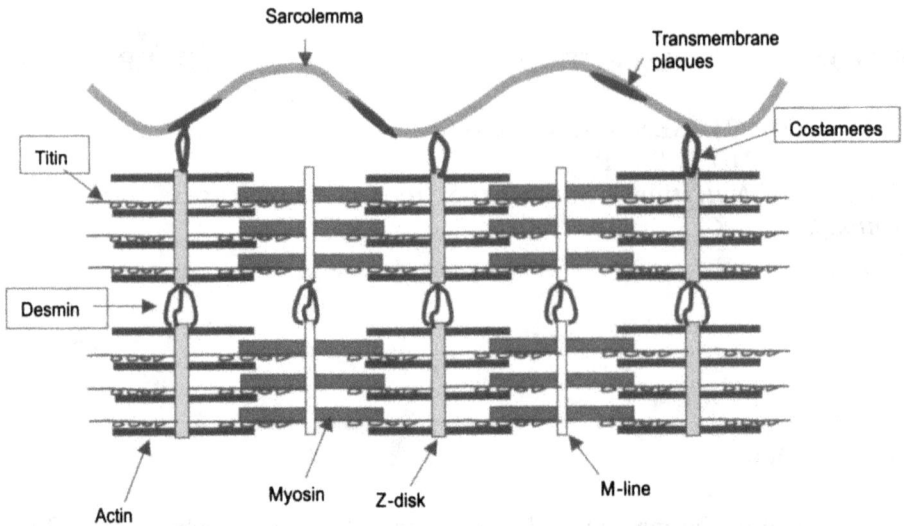

Figure 21.1 Schematic representation of muscle myofibrillar proteins showing the major components of the sarcromere. Boxes indicate the cytoskeletal structures and proteins susceptible to post-mortem cleavage (adapted from Taylor *et al.*, 1995a).

proteolytic systems in muscle which could have a role in postmortem proteolysis. The potential candidates for this role are the lysosomal cathepsins, the multicatalytic proteinase complex and the calpain system. To be considered as being responsible for postmortem proteolysis a system must be able to access the substrates cleaved postmortem and give the pattern of peptide fragments which is found in meat after tenderisation (Koohmaraie & Geesink, 2006). Previously both the lysosomal cathepsins and the multicatalytic proteasome have not been considered as strong candidates for the role as the principle instigators of postmortem proteolysis. Lysosomal cathepsins are contained within specific structures within the cell. It is suggested that they are unable to access the myofibril proteins during the critical stages of the proteolysis period as well as produce different patterns of myofibril protein degradation to those seen *in situ* (Koohmaraie, 1988). Although the multicatalytic proteasome is recognized as an important proteolytic enzyme system in muscle function, particularly during pathological states, it has previously been reported as being unable to cleave myofibrillar proteins in a similar pattern to that seen *in situ* during the meat conditioning period (Taylor *et al.*, 1995b). To date, research on these proteolytic systems indicates that calpains are the enzymes that are predominantly responsible for key peptide bond cleavage postmortem in all major livestock species (Koohmaraie, 1996).

The calpain system

The family of calpain isoforms is extensive, currently encoded by up to 14 genes (Goll *et al.*, 2003). To date, the few calpain proteins that have been isolated are characterized as being calcium-dependent cysteine proteinases. The two ubiquitously expressed isoforms are heterodimers consisting of a large and small subunit, the small subunit being common to both isoforms whilst the large subunit contains the catalytic site. These isoforms are active *in vitro* at low and high calcium ion concentration and are therefore given the names μ- and m- calpain respectively (Table 21.1). Associated with the calpain proteolytic enzyme family is the calpain-specific endogenous inhibitor, calpastatin (Wendt *et al.*, 2004). Calpastatin is a five-domain inhibitory protein (Takano *et al.*, 1988). The variable N terminal leader (L) domain does not appear to have any inhibitory activity, but may be involved in targeting or intracellular localisation, whilst the other domains (I – IV) are highly homologous and are each capable of inhibiting calpain. Calpastatin is expressed from a single gene in all tissues that contain calpains. The calpain system is highly sensitive to fluctuating levels of calcium ion, pH and temperature, all of which change rapidly in the immediate postmortem period. Calpains cleave substrates that are degraded *in situ* postmortem, but importantly not actin and myosin and the major Z line protein α-actinin (Goll *et al.*, 1992; Taylor *et al.*, 1995a). This has made the system a likely candidate having a major role in proteolysis associated with meat tenderisation. The calpain family also contains a number of tissue-specific calpain variants, which are products of different genes. The enzyme p94 (calpain 3) has been of particular interest, as it is expressed almost exclusively in skeletal muscle and its absence is associated with enhanced proteolysis in muscle. Sequence variants of the p94 gene are responsible for a muscular dystrophy and its associated muscle atrophy (Richard *et al.*, 1995). Consequently meat scientists have speculated on p94's role in meat conditioning.

Table 21.1 ESTIMATES OF THE CALCIUM CONCENTRATION (μM) REQUIRED FOR ACTIVATION, AUTOLYSIS AND INTERACTION WITH CALPASTATIN. THE FIGURES ARE THE CONCENTRATIONS REQUIRED FOR HALF MAXIMAL ACTIVITY, BINDING OR RATE OF AUTOLYSIS (ADAPTED FROM GOLL *et al.*, 2003).

Calpain property	*Autolysed μ-calpain*	*μ-calpain*	*Autolysed m-calpain*	*m-calpain*
Proteolytic activity	0.5-2	3-50	50-150	400-800
Calpastatin binding	0.042	40	25	250-500
Autolysis - phospholipids		50-150		550-800
Autolysis + phospholipids		0.8-50		90-400

Calpains and postmortem proteolysis

There is now considerable evidence linking the calpains to tenderisation in beef, lamb and pork. Correlations have shown that the different tenderisation rates between species (beef < lamb < pork) relate inversely to the ratio of calpastatin:calpain (beef >lamb > pork) (Koohmaraie *et al.*, 1991). Infusion of $CaCl_2$ in beef carcasses increases meat tenderisation, whilst infusion of the calpain inhibitor, $ZnCl_2$, reduces the rate of tenderisation in both beef and lamb (Geesink *et al.*, 1994; Koohmaraie, 1990). Hypertrophic effects of ß-adrenergic agonists on muscle are in part brought about by significant reductions in muscle protein degradation (Bohorov *et al.*, 1987). In meat producing species treatment with of ß-adrenergic agonists is associated with elevated activity and mRNA expression of the calpain-specific inhibitor calpastatin in skeletal muscle, suggesting that calpain activity is involved in muscle protein turnover (Higgins *et al.*, 1988; Bardsley *et al.*, 1992; Parr *et al.*, 1992; Killefer & Koohmaraie, 1994). Whilst ß-agonists induce muscle hypertrophy they generally produce a tougher meat product than in untreated animals (Dunshea *et al.*, 2005), indicating a central role of calpains in post-mortem proteolysis. Similarly the skeletal muscle hypertrophy which is present in callipyge lambs is associated with the high levels of calpastatin, greatly reduced postmortem proteolysis and significantly reduced meat tenderness (Geesink & Koohmaraie, 1999). Such studies indicate a central role of calpains in postmortem proteolysis. However both of the ubiquitous expressed µ- and m- calpain isoforms as well as the muscle specific p94 are found in skeletal muscle and therefore which calpain isoform is responsible for postmortem proteolysis has been area of considerable investigation.

p94 (calpain 3)

Several observations led meat science research groups to suggest that the muscle-specific calpain p94 could have a role in postmortem tenderisation. The enzyme is tightly bound to the giant elastic cytoskeletal protein titin in a region that corresponds to the sarcomeric N2 line (Labeit & Kolmerer, 1995) that is known to be susceptible to postmortem cleavage (Taylor *et al.*, 1995a). In addition it has been demonstrated that p94 will cleave a number of cytoskeletal proteins that are degraded in postmortem muscle (Taveau *et al.*, 2003). However, our investigations indicated the level of p94 at slaughter and its relative rate of postmortem decline in pig *longissimus dorsi*, taken as an index of the activation of p94 by autolysis, did not differ between tough and tender animals and showed no correlation with 8 d shear force values (Parr *et al.*, 1999b), (Table 21.2). However other studies showed a positive correlation between p94 mRNA expression and the rate of meat tenderization in sheep (Ilian *et al.*, 2001a; Ilian *et al.*, 2001b). In order to examine directly the effect on p94 muscle function, p94 knockout mice were used to examine the effects of the absence of p94 on postmortem muscle (Kramerova *et al.*, 2004). The effect of the absence of p94 on post-mortem muscle proteolysis was no change in the degradation of specific cytoskeletal proteins that are normally cleaved during the conditioning period, compared to wild type mice (Geesink *et al.*,

2005). Although these studies were carried out in mice, the body of experimental evidence available suggests that the muscle-specific calpain p94 does not appear to have a major role in postmortem proteolysis and the associated development of tenderness.

Table 21.2 THE P94 STABILITY RELATIVE TO α-ACTININ IN THE MYOFIBRILLAR FRACTION FROM LONGISSIMUS MUSCLE SAMPLED DURING 24 HOURS POSTMORTEM

Group	n	Shear force (kg)	Time postmortem (h)	p94/a-actinin ratio[a]	$t_{\frac{1}{2}}$ [b]
Tough	10	5.36[c]	0.7 ± 0.10[d]	0.666[c]	
			2.0 ± 0.00	0.628	
			3.7 ± 0.10	0.564	
			22.3 ± 0.30	0.255	
			-	-	13.8[c]
Tender	10	2.81	0.7 ± 0.10	0.609	
			2.0 ± 0.00	0.586	
			3.7 ± 0.10	0.554	
			22.3 ± 0.30	0.198	
			-	-	12.9
SED[e]		0.21		0.156	2.8
P-values					
Group		< 0.001		NS[f]	NS
Time		-		< 0.05	-
Interaction		-		NS	-

[a]based on densitometric measurements from immunoblots probed sequentially with p94 and a-actinin antisera.
[b]$t_{\frac{1}{2}}$ = half life of p94 decay relative to a-actinin (h).
[c]Mean values.
[d]Mean values ± SEM.
[e]Standard error of the difference of the means.
[f]NS = not statistically significant.

Micro- and milli-calpains

Since both μ- and m-calpain appear to be capable of cleaving the same target proteins, it is has been difficult to determine unequivocally which is the most important during the postmortem conditioning period. One of the major issues has been that the calcium concentration required to assay both isoforms *in vitro* is higher than that found *in situ*. However it appears that are various mechanisms by which the calcium requirement for activation can be reduced by interaction with activators, such as membranes, and the process of autolysis (Goll *et al.*, 2003; Suzuki & Sorimachi, 1998) (Table 21.1).   In general, the m-isoform persists longer than the less stable μ- in ageing muscle from all species studied, including the pig (Sensky *et al.*, 1996).   Following activation both these isoforms of calpain undergo autolysis and this characteristic allows researchers to determine which calpain isoforms have been active *in situ*. Examination of calpains in postmortem muscle have indicated that μ-calpain is activated early in postmortem (within 3d of slaughter), during the period where postmortem proteolysis of key myofibrillar proteins is known to take place.   In contrast, m-calpain is relatively stable during this period with no significant activation taking place (Veiseth *et al.*, 2001).   Such experiments have led workers in the field to suggest that it is μ-calpain that carries out proteolysis responsible for the development of tenderness.   Recently the evidence for significant role of μ-calpain in post-mortem proteolysis has further been strengthened from observations made in μ-calpain knockout mice (Geesink *et al.*, 2006). In these animals postmortem proteolysis was significantly inhibited. Although these studies were carried out in mice, the experimental evidence available suggests that μ-calpain is the enzyme component of the calpain system that has the most significant role in postmortem proteolysis and the development of tenderness. The importance of μ-calpain in the tenderisation process is also strengthened by the discovery in cattle of single nucleotide polymorphisms (SNP) in the *CAPN1* gene, encoding m-calpain, which are associated with tender meat.   Two bovine *CAPN1* gene SNPs, which introduce amino acid substitutions (non-synonymous SNPs) in μ-calpain, are associated with tenderness (Page *et al.*, 2002), forming part of a commercial test for tenderness in cattle (GeneSTAR ®Tenderness, Genetic Solutions Pty. Ltd.). However in pigs, to date, no relationship between polymorphisms in *CAPN1* and meat quality have yet been reported.

Relationship between calpastatin and meat toughness

Calpastatin is the specific endogenous calpain inhibitor and its variable expression has consistently been shown to be associated with variability in meat quality. The single calpastatin gene contains multiple promoters that generate several different transcripts that are also alternatively spliced into multiple mRNAs (Takano *et al.*, 1999; 2000), resulting in multiple protein isoforms being derived from a single gene (Parr *et al.*, 2000; 2001; 2004; Reynard *et al.*, 2005). The micro-heterogeneity of calpastatin in different cells and tissues (Parr *et al.*, 2000) may determine its intracellular localisation and its physiological role, being a potential mechanism by which the single calpastatin gene can regulate the activity of the products from multiple calpain genes. Calpastatin inhibits both μ- and m- calpain and this process requires calcium concentrations that are close to or below those that are required to activate calpain (Goll *et al.*, 2003), (Table 1). However calpastatin is itself susceptible to proteolysis but the resulting fragments retain inhibitory activity. The role of this complex calpastatin micro-heterogeneity in regulating calpain activity is not full understood.

However, as indicated above, one of the consistent observations of the calpain system's involvement in tenderness is that high levels of calpastatin are associated with poor quality meat. An explanation of this association is that high levels of calpastatin reduce the activity of calpain thereby reducing the proteolysis that is essential for the development of tender meat. In ruminant species there is relationship between calpastatin activity in the muscle 24h after slaughter and the degree of tenderisation achieved after conditioning, with differences in calpastatin accounting for 0.40 of the variation in tenderness (Shackelford *et al.*, 1994). Our studies on a random selection of commercially slaughtered pigs have shown that a high level of calpastatin (both activity and protein levels) in the first few hours after slaughter are associated with an increased incidence of toughness (Sensky *et al.*, 1998; Parr *et al.*, 1999a), (Figure 21.2). By monitoring calpastatin at these times it should therefore be possible to predict whether or not any given carcass will tenderise to an acceptable degree. This provides an opportunity to grade carcasses at a much earlier time postmortem than is currently possible.

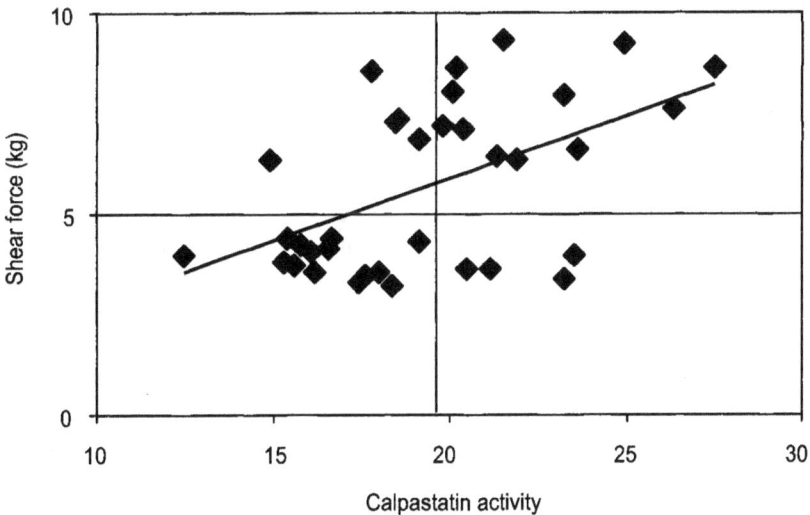

Figure 21.2 Correlation between slaughter calpastatin activity ($\times 10^7$ fluorescence units/kg) and 8 day shear force in porcine *longissmus dorsi* ($p < 0.05$).

The use of antibody-based methods of calpastatin detection is complicated by the fact that there is considerable micro-heterogeneity in calpastatin forms that are generated by alternative splicing and phosphorylation (Parr *et al.*, 2000; 2001). The function of such variability may be to control the inhibitory activity of calpastatin, and therefore the identification of these isoforms and the mRNA species from which they originate may lead to an even closer predictor of toughness/tenderness. Studies have established the heritability of the link between calpastatin level and beef toughness (Shackelford *et al.*, 1994). Given the potential this has for marker-assisted breeding programmes, a number of groups have now identified calpastatin gene polymorphisms and have shown that some of these are predictive of carcass quality in cattle and in pigs (Barendse, 2002; Ciobanu *et al.*, 2004). Currently there are markers within the calpastatin and μ-calpain genes that are able to identifying beef cattle with the genetic potential to produce tender meat (Casas *et al.*, 2006) and these are commercially available as a genetic test (GeneSTAR ®Tenderness, Genetic Solutions Pty. Ltd.).

In pigs, five genetic polymorphisms have been identified that are missense mutations leading to amino acid substitutions within subdomains of the calpastatin protein (Ciobanu *et al.*, 2004). Furthermore a specific combination of some of these SNPs forms a haplotype that has been shown to have a significant association with improved pork eating quality. Recently in our laboratory we have generated novel calpastatin genomic DNA sequence in order to measure the distribution of these polymorphisms by direct sequencing of PCR amplicons. In addition, rapid throughput assays for these calpastatin SNPs based on primer extension technology (SNuPE) have been developed and tested on DNA both from blood samples and from fresh and processed meat products. A preliminary analysis has shown that the distribution of favourable alleles and predicted haplotypes differs markedly between four authenticated pure-bred pig breeds.

Variability in calpastatin and its inhibition of calpain mediated-proteolysis is likely to be affected by environmental effects. For example elevated plasma adrenaline increases calpastatin activity and expression in pigs, implying that the link between stress and meat toughness may indeed be partially mediated via the calpain system (Sensky *et al.*, 1996; Parr *et al.*, 2000). The hypothesis that environmental factors influence calpastatin function and subsequently that this affects meat quality is further strengthened by the observation that the gene's sequence variation, that specifies the haplotype linked to pig meat quality, alters peptide consensus sequences predicted to be phosphorylated by cyclic 3', 5'-adenosine monophosphate-dependent protein kinase (PKA). This kinase is part of the signalling pathway stimulated by adrenergic agonists. Porcine calpastatin can be phosphorylated by PKA (Parr *et al.*, 2000) and in other species PKA-mediated phosphorylation alters calpastatin's solubility and cell localisation (Averna *et al.*, 2001). In addition it has been shown that phosphorylation reduces calpastatin activity (Salamino *et al.*, 1997). It therefore appears that any effect of calpastatin on pig meat quality is influenced by gene-environment interaction. The effect of genetic variability on the expression of calpastatin and its relationship to meat quality was illustrated by our observations in pigs of different breeds. In the *longissimus dorsi* of Duroc,

Large White (LW) and LW x Duroc (LWxD) pigs, several bands of calpastatin immunoreactivity were identified in extracts, equivalent to approximately 135, 120, 90 and 80 kDa (Figure 21.3).   This range of calpastatin protein bands is typical of the heterogeneity of calpastatin in muscle (Parr *et al.*, 2000; 2001). These bands are thought to be generated *de novo* from different mRNAs or by post-translational processing.   The intensities of the three heavier bands did not differ between the 3 breeds (Table 21.3). The mean intensity of the 80 kDa calpastatin band was higher in the Duroc breed ($P < 0.01$), with the band being virtually undetectable in the majority of animals in the other breeds (Figure 21.3). Considering all individual animals independent of breed, the intensity of the 80 kDa calpastatin band 2 h after slaughter was found to correlate negatively with 8d shear force values ($r^2 = 0.221$, $p < 0.05$). Micro-calpain activity was highest in Duroc ($P < 0.05$), whilst m-calpain activity did not differ between breeds (Table 21.3). Micro-calpain activity correlated positively with 80 kDa band intensity ($r^2 = 0.540$, $P < 0.01$) and also correlated negatively with SF ($r^2 = 0.253$, $P < 0.05$). This suggests that either the 80 kDa calpastatin fragment is less effective in inhibiting tenderisation than the 135 kDa fragment, or that the fragmentation of calpastatin permits the calpain-mediated tenderization of skeletal muscle to proceed more effectively.

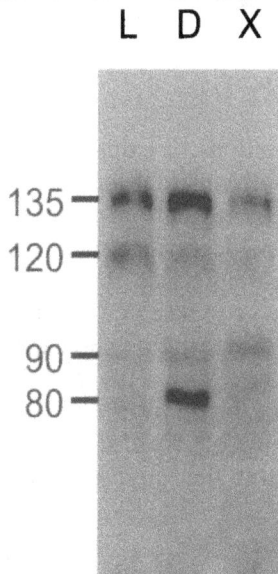

Figure 21.3 Representative gel showing ECL visualisation of bands immunoreactive to a specific calpastatin antibody in extracts of *longissmus dorsi* taken from Large White (L), Duroc (D) and Large White x Duroc (X) pigs within the first 90 min of slaughter. Molecular weight (kDa) of the immunoreactive bands is indicated.

Table 21.3 DIFFERENCES IN CARCASS CHARACTERISTICS, CALPAIN ACTIVITY (X10[7] FLUORESCENCE UNITS/KG) AND IMMUNOREACTIVITY OF CALPASTATIN BANDS (ARBITRARY DENSITOMETRIC UNITS) IN EXTRACTED SAMPLES OF LD TAKEN IN THE EARLY POSTMORTEM PERIOD FROM THREE DIFFERENT PIG BREEDS: DUROC (D), LARGE WHITE (LW) AND LW X DUROC (LWXD)

		D	*LW*	*LW x D*	*SED*[a]	*P*
	n	*11*	*11*	*11*		
Carcass	Carcass weight	63[b]	65	72	2.7	0.008
characteristic	P_2	12.2	9.4	12.4	0.94	0.005
	pH_{45}	6.32	6.29	6.44	0.095	NS[c]
	pH_u	5.62	5.58	5.76	0.054	0.001
	Shear Force (N)	39	50	48	2.9	0.004
Activity	μ-calpain	13	4	6	3.6	0.041
	m-calpain	14	14	13	2.7	NS
Calpastatin	135kDa	0.35	0.26	0.32	0.063	NS
	120kDa	0.05	0.06	0.06	0.024	NS
	90kDa	0.04	0.03	0.04	0.010	NS
	80kDa	0.11	0.02	0.01	0.03	0.005

[a]Standard error of the differences of the means.
[b]Mean values.
[c]Not statistically significant.

It therefore appears that variations in calpastatin that relate to meat toughness/ tenderness are subject to environmental and genetic regulation. Understanding the mechanism by which such regulation can control calpastatin function is of paramount importance to enable further meat quality improvement and may yield specific genetic or protein based markers that are indicative of tender meat. Our current studies have focused on determining the factors that regulate calpastatin gene expression. Calpastatin gene expression is regulated via several promoters each of which is responsible for the expression of a different calpastatin isoform. These promoters each have a different response to secondary messenger pathways (Parr *et al.*, 2004; Sensky *et al.*, 2006) which mediate the response to factors that stimulate growth or a stress response.

The role of novel proteolytic systems in postmortem proteolysis

Although there is considerable experimental evidence to suggest that the activity of the calpain system early in the conditioning period influences the ultimate tenderness, it has been suggested that it is not the sole proteolytic determinant of meat quality (Sentandreu *et al.*, 2002; Herrera-Mendez *et al.*, 2006, Ouali *et al.*, 2006). In addition to the calpain system, the other proteolytic systems that are present in muscle and have been relatively extensively studied are the cathepsin system and the proteasome or multicatalytic protease. The cathepsin system's involvement in postmortem proteolysis has been the subject of a number of studies and there appear to be no significant experimental observations to suggest that it explains the variability in tenderness. Digestion of myofibrils with cathepsins was reported as yielding a pattern of peptide fragments that was not similar to that seen *in situ* (Koohmarie, 1988). Compared to cathepsins the multicatalytic proteasome system was discovered more recently (Wilk & Orlowski, 1980) and therefore the number of studies that have examined its role in postmortem proteolysis are relatively few. Although Koohmaraie and Geesink (2006) in a recent review excluded the role of this system as a significant contributor to the tenderisation, Herrera-Mendez *et al.* (2006) have indicated that it could potentially contribute to tenderisation. However, unlike μ-calpain, there is limited evidence supporting its role in the kind of proteolysis that influences meat quality. Although some studies have indicated that, *in vitro*, the multicatalytic proteasome will degrade myofibril structures, this pattern of degradation does not mimic that seen *in situ* (Taylor *et al.*, 1995b). The role of multicatalytic proteasome in tenderisation currently is an area of ongoing research.

Recently, reviews examining the role of proteolytic enzymes in postmortem proteolysis have put forward the hypothesis that other proteases, such as caspases, could be activated postmortem and influence the tenderisation process (Sentandreu *et al.*, 2002; Herrera-Mendez *et al.*, 2006, Ouali *et al.*, 2006). The caspases are a family of cysteine-proteases that are predominantly associated with apoptosis, which is the ordered biological process where a cell's internal structures are disrupted through carefully choreographed proteolysis resulting in programmed cell death. This process is distinct from necrotic cell death were the cell membrane is disrupted and the release of the cell contents causes an inflammatory response (Buja *et al.*, 1993). There are currently 14 members of the caspase family so far identified (Fuentes-Prior & Salvesen, (2004) and these can be broadly divided into two groups (Thornberry & Lazebnik, 1999):

The *effector caspases*, such as 3, 6 and 7, are involved in the direct cleavage of key structural proteins resulting in cell disassembly. The *initiator caspases*, such as 8, 9, 10 and 12, are involved in initiating this disassembly in response to pro-apoptotic signals by cleaving the effector caspases, thereby activating them.

In addition the caspase system includes specific inhibitors that regulated caspase activity, such as the Apoptosis Repressor with Caspase recruitment domain (ARC) (Koseki *et al.*, 1998).

Activation of caspases is via a variety of signalling pathways that are outlined in Figure 21.4. The cell death receptor mediated pathway is stimulated by ligands which then activate initiating caspases such as caspase 8, which in turn activate effector caspases such as caspase 3. There is also an intrinsic activation

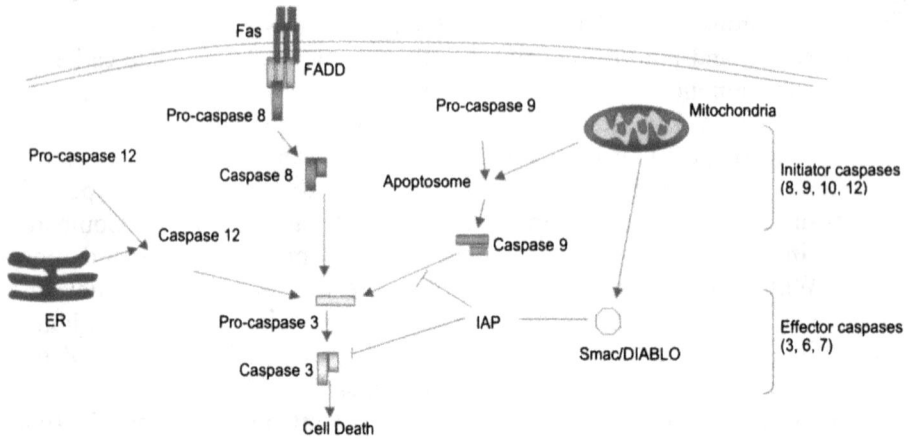

Figure 21.4 Schematic diagram of the intrinsic, extrinsic and ER-mediated apoptosis pathways showing the caspases involved in each pathway. FADD – Fas associated death domain, IAP- inhibitor of apoptosis, Smac - secondary mitochondrial activator of caspases, DIABLO - direct IAP-binding protein with low pI (adapted from Holcik, 2002).

pathway that involves signalling stimuli that originate from the mitochondria when these are subjected to a variety of stressors, including oxidation. This pathway activates the initiator caspase, caspase 9, which in turn activates the effector caspases. In addition there are signalling pathways that originate from a stress stimuli response of the endoplasmic reticulum (ER), which activates caspase 12 leading to the activation of effector caspases (Keiko *et al.*, 2005). The intricacies of these pathways is beyond the scope of this review and readers are referred to the relevant reviews (Herrera-Mendez *et al.*, 2006; Fuentes-Prior & Salvesen, 2004).

There are a variety of stimuli that can activate caspases. However it is clear that caspases are activated early in pathological events associated with hypoxia/ ischemia (Gustafsson & Gottlieb, 2003), which is not dissimilar to the hypoxic conditions in muscle after slaughter. In meat animals the process of exsanguination occurs after slaughter, depriving all cells and tissues of nutrients and oxygen. It has recently suggested that this process will inevitably cause muscle cells to engage the processes of cell death resulting in the activation of the caspase system (Herrera-Mendez *et al.*, 2006). The most likely mechanisms by which this process is activated are by the intrinsic systems involving either the response of mitochondria or the ER to cell stress. During the process of cell death in living orgainsms, the activated effector caspases target and specifically cleave a number of substrates including cytoskeletal proteins such as vimentin, desmin and spectrin that have an important structural role (Wang, 2000). These proteins are amongst those cleaved during the postmortem conditioning of meat. Caspases carry out selective initiating cleavage events targeting a similar spectrum of proteins as does calpain and often it is difficult distinguish between the effects of the two proteolytic systems (Wang, 2000). Therefore the caspase system might contribute to proteolysis early postmortem and thereby, in part, be responsible for the process of tenderisation. Our recent research work has focused on trying to determine whether caspases are active in post-mortem muscle and whether they cleave proteins found within the muscle myofibril structures. According to criteria set by Koohmaraie (1996), in order for proteases to contribute to meat tenderisation they must be endogenous to skeletal muscle cells, have the ability to replicate myofibrillar changes *in vitro* and have access to myofibrils in muscle tissue.

Recent studies have shown that caspases are involved in skeletal muscle development and remodelling, with casapase expression being essential for normal muscle differentiation during myogenesis (Fernando *et al.*, 2002). Caspases are up-regulated in muscles in conditions such as in sarcopenia (Dupont-Versteegden, 2005), muscular dystrophies (Sandri *et al.*, 2001) and cachexia (Belizario *et al.*, 2001). Our work in pigs has shown that components of the caspase system are expressed and are active in varying proportions in different skeletal muscles (Kemp *et al.*, 2006a). Subsequently we examined whether caspases were active in postmortem pig *longissimus dorsi* muscles which were conditioned for period of up to 8d (Kemp *et al.*, 2006b). As shown in Table 21.4 the activity of both caspase 3/7 (the assay used could not distinguish between these two effector caspases) and 9 declined over the postmortem period. In these an animals, specific degradation products of alpha II spectrin and poly (ADP-ribose) polymerase (PARP), which are known indicators of caspase activity and apoptosis were detected on immunoblots of muscle samples taken over the postmortem period. Protein levels of alpha II spectrin cleavage products across the conditioning period were found to correlate positively to caspase 3/7 activity ($P < 0.01$) and caspase 9 activity ($P < 0.05$), indicating that caspase- mediated cleavage was occurring *in situ*. There was a negative relationship between shear force and the 0:32 h ratio of caspase 3/7 ($P = 0.053$) and caspase 9 activities ($P < 0.05$) (Figure 21.5). In addition there was also a negative relationship between shear force and the level of the caspase generated alpha II spectrin 120 kDa degradation product ($P < 0.05$) (Figure 21.5). The findings of this preliminary study indicate that changes in caspase activity and caspase-mediated cleavage take place in muscle during the conditioning period and this could be associated with the development of tender meat.

Subsequent to this study we investigated whether caspase 3 was directly capable of degrading pig muscle myofibril proteins. A full-length human recombinant caspase 3 (rC3) was expressed in *E.coli* and purified to homogeneity. The rC3 was incubated with myofibrillar proteins prepared from porcine *longissimus dorsi* in buffer which was designed to simulate the postmortem conditions in the muscle according to Winger and Pope (1981). In this *in vitro* system there was a visible increase in myofibril degradation with the detection of protein degradation products at approximately 32, 28

Table 21.4 CHANGES IN CASPASE 3/7 AND 9 ACTIVITIES AND PROTEIN LEVELS OF POLY (ADP-RIBOSE) POLYMERASE (PARP 89KDA) AND ALPHA II SPECTRIN 120 KDA DEGRADATION PRODUCT (SBDP120), THE CASPASE-MEDIATED PROTEOLYSIS BREAKDOWN PRODUCTS, OVER TIME IN PORCINE *LONGISSMUS DORSI* .

Time h	n	Caspase 3/7 activity fluorescence/ μg protein	Caspase 9 activity luminescence/ μg protein	PARP 89 kDa units[1]/mg protein	SBDP120 units[1]/mg protein
0	10	110	55	22	2.6
2	10	130	59	11	3.0
4	10	116	57	20	2.8
8	10	111	40	19	1.2
16	10	121	48	ND	1.0
32	10	73	26	ND	1.0
192	10	5	3	ND	1.1
SED[2]		10.3	5.2	5.1	0.57
P value		P<0.001	P<0.001	P<0.001	P<0.001

ND = not detectable
[1] arbitrary densitometry units
[2] Standard error of the differences of the means

and 18 kDa with increasing concentrations of rC3 (Figure 21.6A) as determined by SDS-PAGE. The clearly discernible degradation products were analysed using MALDI-TOF mass spectrometry and identified as the products of the proteolysis of actin, troponin T and myosin light chain. The appearance of these degradation products was inhibited by the caspase-specific inhibitor Ac-DEVD-CHO (data not shown). In incubation conditions that mimicked those at which carcasses are stored over the conditioning period (4°C for 8 days) there was visible degradation of a number of myofibril proteins including desmin and troponin I (Figure 21.6B). This study has shown that rC3 was capable of causing myofibril degradation, hydrolysing myofibrillar proteins under conditions that are similar to those found in muscle in the postmortem conditioning period.

From our current investigations and those of others it is clear that caspases are expressed in muscle. We have shown that caspase activity can be detected in pig muscles *in situ* during the postmortem condition period, over a similar early postmortem time course as m-calpain. These investigations have indicated that the caspase proteases appear to fulfil some of the criteria that were stipulated by Koohmaraie (1996) as a requirement for a protease system to influence meat quality. Firstly caspases are endogenous to skeletal muscle fibres, secondly it appears that they have the ability to degrade myofibrillar proteins. Our investigations are seeking to build on our observations to investigate whether caspases do contribute to the process of meat tenderisation *in situ* and whether variability within this system or in its response to stimuli at slaughter could influence meat quality.

(a)

(b)

(c)

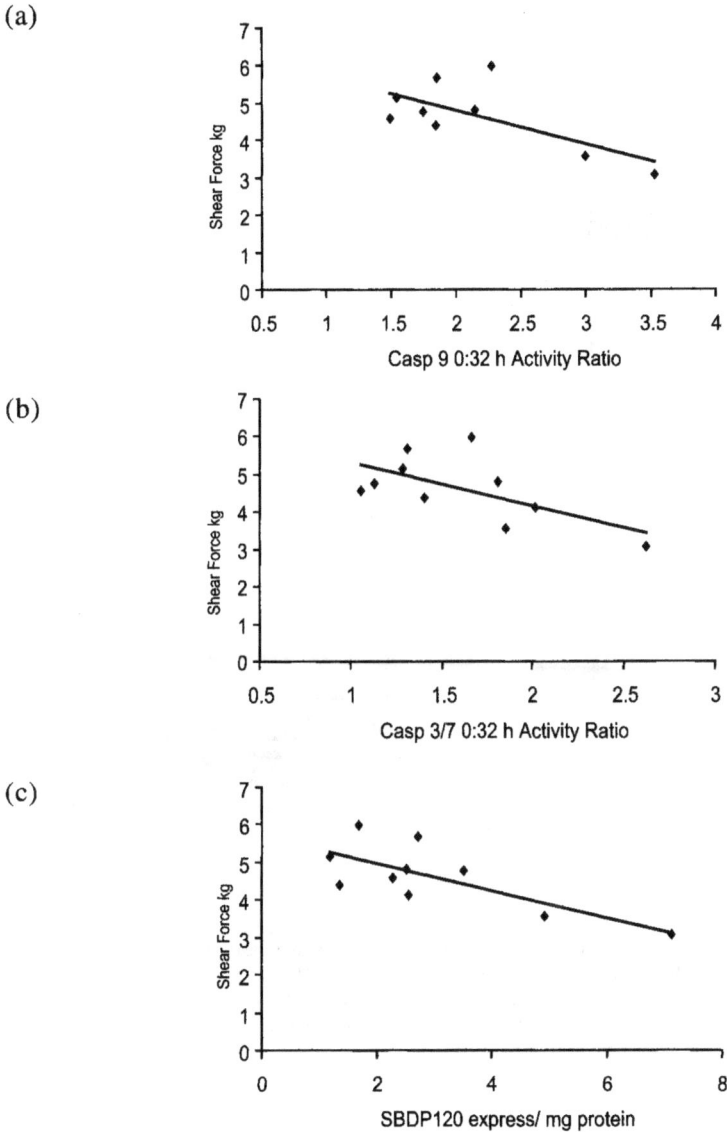

Figure 21.5 Relationships observed between porcine *longissimus dorsi* 8 day shear force and a) caspase 9 0:32 h activity ratio; b) caspase 3/7 0:32 h activity ratio, where the ratio of activities at 0 and 32 h being taken as an approximation of the change in caspase activity during the early postmortem period and c) protein levels of the specific caspase 3 spectrin 120 kDa degradation product at 2 h post mortem, n = 10.

Conclusion

The role of proteolysis in the development of tender meat has been shown to

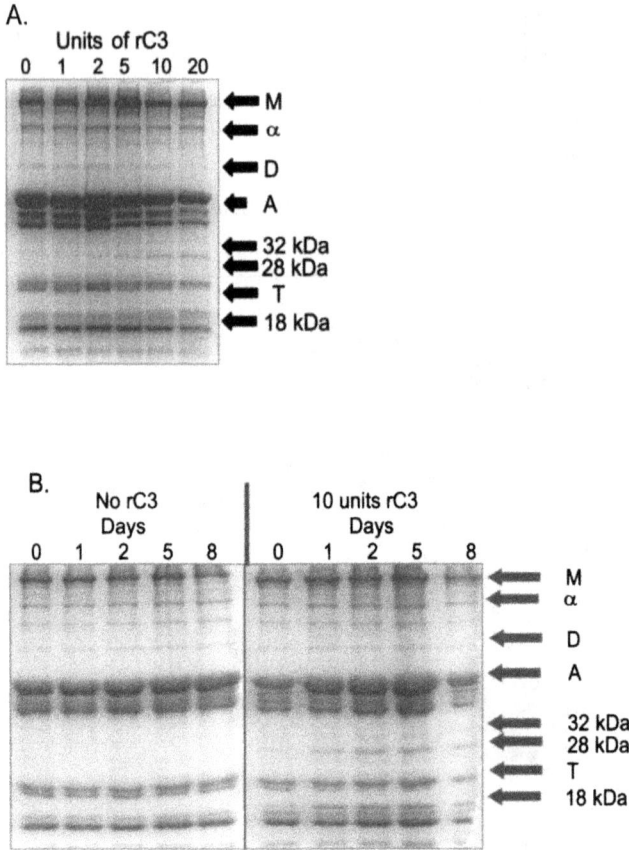

Figure 21.6 The effects of incubating porcine myofibrils with recombinant caspase 3 (rC3). A. The effect of incubating myofibrils with increasing concentrations of rC3 at 37°C for 24 hours. B. The effect of incubating myofibrils for 8 days at 4°C without or with 10 units of rC3.   In both figures the major degradation products generated by caspase-mediated proteolysis are indicated by their molecular weights. Abbreviations: M, myosin heavy chain; α, α-actinin; D, desmin; A, actin; T, troponin-I.

be an important factor in the production of meat that is valued by consumers. Specific components of the calpain system have been shown to influence postmortem proteolysis and their endogenous inhibitor calpastatin has an important role in influencing tenderisation and acts as a marker for meat quality. However it must be remembered that other novel proteolytic systems may contribute to postmortem proteolysis and thereby influence meat quality. The full extent of their contribution has yet to be determined.

References

Averna, M., De Tullio, R., Passalacqua, M., Salamino, F., Pontremoli, S. & Melloni, E. (2001). Changes in intracellular calpastatin localization are mediated by reversible phosphorylation. *Biochem. J.* 354: 25–30.

Bardsley, R.G., Allock, S.M.J., Dawson, J.M., Dumelow, N.W., Higgins, J.A., Lasslett, Y.V., Lockley, A.K., Parr, T. & Buttery, P.J. (1992). Effects of Beta-Agonists on the Expression of Calpain and Calpastatin Activity in Skeletal Muscle. *Biochimie* 74; 267-273.

Barendse, W.J. (2002). DNA markers for meat tenderness. International patent application PCT/AU02/00122. International patent publication WO 02/064820 A1.

Belizario, J.E., Lorite, M.J. & Tisdale, M.J. (2001). Cleavage of caspases-1, -3, -6, -8 and -9 substrates by proteases in skeletal muscles from mice undergoing cancer cachexia. *Brit.J.Cancer* 84: 1135-1140.

Bohorov, 0., Buttery, P. J., Correia, J. H. R. D. & Soar, J. B. (1987). The effect of the b-2-adrenergic agonist clenbuterol or implantation with oestradiol plus trenbolone acetate on protein metabolism in wether lambs. *Brit.J.Nut.* 57: 99-107.

Buja, L.M., Eigenbrodt, M.L. & Eigenbrodt, E.H. (1993). Apoptosis and necrosis. Basic types and mechanisms of cell death. *Arch. Path. Lab. Med.* 117: 1208–1214.

Casas, E., White, S.N., Wheeler, T.L., Shackelford, S.D., Koohmaraie, M., Riley, D.G., Chase, C.C. Jr., Johnson, D.D. & Smith, T.P. (2006). Effects of calpastatin and u-calpain markers in beef cattle on tenderness traits. *J.Anim.Sci.* 84: 520-525.

Ciobanu, D.C., Bastiaansen, J.W.M., Lonergan, S.M., Thomsen, H., Dekkers, J.C.M., Plastow, G.S. & Rothschild, M.F. (2004). New alleles in calpastatin gene are associated with meat quality traits in pigs. *J.Anim.Sci.* 82: 2829-2839.

Dunshea, F. R., D'Souza, D. N., Pethick D. W., Harper, G. S. & Warner, R. D. (2005). Effects of dietary factors and other metabolic modifiers on quality and nutritional value of meat. *Meat Science*, 71, 8-38.

Dupont-Versteegden, E.E. (2005). Apoptosis in muscle atrophy: Relevance to sarcopenia. *Exp. Gerontol.* 40: 473-481.

Fernando, P., Kelly, J.F., Balazsi, K., Slack, R.S. & Megeney, L.A. (2002). Caspase 3 activity is required for skeletal muscle differentiation. *Proc.Natl.Acad.Sci.(USA)* 99: 11025-11030.

Fuentes-Prior, P. & Salvesen, G.S. (2004). The protein structures that shape caspase-activity, specificity, activation, and inhibition. *Biochem.J.* 384: 201–232.

Geesink, G.H. & Koohmaraie, M. (1999). Postmortem proteolysis and calpain/calpastatin activity in callipyge and normal lamb biceps femoris during extended postmortem storage. *J.Anim.Sci.* 77: 1490–1501.

Geesink, G.H., Kuchay, S., Chishti, A.H. & Koohmaraie, M. (2006). m-Calpain is essential for postmortem proteolysis of muscle proteins. *J.Anim.Sci.* 284: 2834-2840

Geesink, G.H., Smulders, F.J.M. & Van Laack, R. (1994). The effects of calcium-, sodium- and zinc-chlorides treatment on the quality of beef. *Sci Aliments* 14:485–502,

Geesink, G.H., Taylor, R.G. & Koohmaraie, M. (2005). Calpain 3/p94 is not involved in postmortem proteolysis. *J.Anim.Sci.* 83: 1646–1652.

GeneSTAR ®Tenderness, Genetic Solutions Pty. Ltd., *http:// www.geneticsolutions.com.au/content/v4_standard.asp?* (accessed April 2007).

Goll, D.E., Thompson, V.F. Li, H., Wei, W. & Cong, J. (2003). The Calpain System. *Physiol. Rev.* 83: 731– 801.

Goll, D.E., Thompson, V.F., Taylor, R.G. & Christiansen, J.A. (1992). Role of the calpain system in muscle growth. *Biochimie* 74: 225-237

Gustafsson, A.B. & Gottlieb, R.A. (2003). Mechanisms of apoptosis in the heart. *J.Clin.Imm. 23:* 447-459.

Herrera-Mendez, C.H., Becila, S., Boudjellal, A. & Ouali, A. (2006). Meat ageing: Reconsideration of the current concept. *Trends in Food Sci. & Technol* 17: 394-405

Higgins, J.A., Lasslett, Y.V., Bardsley, R.G. & Buttery, P.J. (1988). The relation between dietary restriction or clenbuterol (a selective b-2 agonist) treatment on muscle growth and calpain proteinase (*EC* 3.4.22.17) and calpastatin activities in lambs. *Brit.J.Nut.* 60: 645-652

Ilian, M. A., Morton, J. D., Bekhit, A., Roberts, N., Palmer, B., Sorimachi, H. & Bickerstaffe R.. (2001a). Effect of preslaughter feed withdrawal period on longissimus tenderness and the expression of calpains in the ovine. *J. Agric. Food Chem.* 49:1990–1998.

Ilian, M. A., Morton, J. D., Kent, M. P., Le Couteur, C. E., Hickford, J., Cowley, R. & Bickerstaffe, R. (2001b). Intermuscular variation in tenderness: Association with the ubiquitous and muscle-specific calpains. *J.Anim.Sci.* 79:122–132.

Keiko, N., Sudo, T. & Morishima, N. (2005). Endoplasmic reticulum stress signaling transmitted by ATF6 mediates apoptosis during muscle development. *J. Cell Biol.* 169: 555–560

Kemp, C.M., Parr,T., Bardsley, R.G. & Buttery, P.J. (2006a). Comparison of the relative expression of caspase isoforms in different porcine skeletal muscles. *Meat Science*, 73, 426-431.

Kemp, C.M., Bardsley, R.G. & Parr, T., (2006b). Changes in caspase activity during the post mortem conditioning period and its relationship to shear force in porcine longissimus muscle *J.Anim.Sci.* 84: 2841-2846.

Killefer, J. & Koohmaraie, M. (1994). Bovine skeletal muscle calpastatin: cloning, sequence analysis, and steady-state mRNA expression. *J.Anim.Sci* 72: 606-614.

Koohmaraie, M. (1988). The role of endogenous proteases in meat tenderness. In Proceedings of 41st annual reciprocal meat conference, Wyoming, USA (pp. 89–100).

Koohmaraie, M. (1990). Inhibition of postmortem tenderization in ovine carcasses through infusion of zinc. *J.Anim.Sci.* 68: 1476–1483.

Koohmaraie, M., (1996). Biochemical factors regulating the toughening and tenderization processes of meat. *Meat Science,* 43, S193-S201.

Koohmaraie, M. & Geesink G.H. (2006). Contribution of postmortem muscle biochemistry to the delivery of consistent meat quality with particular focus on the calpain system *Meat Science* 74: 34–43.

Koohmaraie, M., Killefer, J., Bishop, M.D., Shackelford, S.D., Wheeler, T.L. & Arbona, J.R. (1995). Calpastatin-based methods for predicting meat tenderness. In: *Expression of Tissue Proteinases and Regulation of Protein Degradation as Related to Meat Quality* (A. Ouali, D. Demeyer and F. Smulders, eds.) pp 395-411. ECCEAMST, Utrecht, The Netherlands.

Koohmaraie, M., Whipple, G., Kretchmar, D.H., Crouse J.D., & Mersmann, H.J. (1991). Postmortem proteolysis in longissimus muscle from beef, lamb and pork carcasses. *J.Anim.Sci.* 69: 617-624.

Koseki, T., Inohara, N., Chen, S., & Nunez, G. (1998). ARC, an inhibitor of apoptosis expressed in skeletal muscle and heart that interacts selectively with caspases. *Proc.Natl.Acad.Sci.(USA),* 95: 5156-5160.

Kramerova, I., Kudryashova, E., Tidball, J.G., & Spencer, M. J. (2004). Null mutation of calpain 3 (p94) in mice causes abnormal sarcomere formation in vivo and in vitro. *Hum.Molec.Genetics* 13: 1373–1388

Labeit, S. & Kolmerer, B. (1995). Titins, giant proteins in charge of muscle ultrastructure and elasticity. *Science* 270: 293-296.

Lusk, J. L., Fox, J. A., Schroeder, T. C., Mintert, J. & Koohmaraie, M. (2001). In-store valuation of steak tenderness. *Am.J.Agric.Econ.* 83:539–550.

Miller, M. F., Carr, M. A., Ramsey, C. B., Crockett, K. L. & Hoover, L. C. (2001). Consumer thresholds for establishing the value of beef tenderness. *J.Anim.Sci.* 79: 3062-3068.

Ouali, A., Herrera-Mendez, C.H., Coulis, G., Becila, S., Boudjellal, A., Aubry, L., & Sentandreu, M.A. (2006). Revisiting the conversion of muscle into meat and the underlying mechanisms *Meat Science,* 74, 44-58.

Page, B. T., Casas, E., Heaton, M. P., Cullen, N. G., Hyndman, D. L., Morris, C. A., Crawford, A. M., Wheeler, T. L., Koohmaraie, M., Keele, J. W. and Smith, T. P. (2002). Evaluation of single-nucleotide polymorphisms in CAPN1 for association with meat tenderness in cattle. *J.Anim.Sci.* 80:3077–3085.

Parr, T., Bardsley, R.G., Gilmour, R.S. & Buttery, P.J. (1992). Changes in Calpain and Calpastatin mRNA Induced by Beta-adrenergic Stimulation of Bovine Skeletal Muscle. *Eur.J.Biochem.* 208: 333-339.

Parr, T., Jewell, K.K., Sensky, P.L., Brameld, J.M., Bardsley, R.G. & Buttery, P.J (2004). Expression of calpastatin isoforms in muscle and functionality of multiple calpastatin promoters *Arch.Biochem.Biophys. 427: 8-15.*

Parr, T., Sensky, P. L., Arnold, M. K., Bardsley, R. G. & Buttery, P. J. (2000). Effects of Ephinephrine Infusion on Expression of Calpastatin in Porcine Cardiac and Skeletal Muscle. *Arch.Biochem.Biophys. 374: 299-305.*

Parr, T., Sensky, P. L., Bardsley, R. G. & Buttery, P. J. (2001). Calpastatin Expression in Cardiac and Skeletal Muscle and Partial Gene Structure. *Arch.Biochem.Biophys. 395: 1-13.*

Parr, T., Sensky, P. L., Scothern, G., Bardsley, R. G., Buttery, P. J., Wood, J. D. & Warkup, C. C. (1999a). Immunochemical study of the calpain system in porcine longissimus muscle with high and low shear force values. *J.Anim.Sci. 77 (suppl 1): 164.*

Parr, T., Sensky, P. L., Scothern, G. P., Bardsley, R. G., Buttery, P. J., Wood, J. D. & Warkup, C. (1999b). Skeletal muscle-specific calpain and variable postmortem tenderization in porcine longissimus muscle. *J.Anim.Sci. 77: 661-668.*

Raynaud, P., Gillard, M., Parr, T., Bardsley, R., Amarger, V. & Leveziel, H. (2005). Correlation between bovine calpastatin mRNA transcripts and protein isoforms. *Arch.Biochem.Biophys.* 440: 46–53.

Richard, I., Broux, O., Allamand, V., Fougerousse, F., Chiannilkulchai, N., Bourg, N., Brenguier, L., Devaud, C., Pasturaud, P., Roudaut, C., Hillaire, D., Passos-Bueno, M.R., Zatz, M., Tischfield, J.A., Fardeau, M., Jackson, C.E., Cohen, D. & Beckmann, J.S. (1995). Mutations in the proteolytic enzyme calpain 3 cause limb-girdle muscular dystrophy type 2A. Cell 81:27– 40

Salamino, F., Averna, M., Tedesco, I., DeTillio, R., Melloni, E. & Pontremoli, S. (1997). Modulation of rat brain calpastatin by post-translational modifications. *FEBS Lett.* 412:433–438

Sandri, M., El Meslemani, A.H., Sandri, C., Schjerling, P., Vissing, K., Andersen, J.L., Rossini, K., Carraro, U. & Angelini, C. (2001). Caspase 3 expression correlates with skeletal muscle apoptosis in Duchenne and facioscapulo human muscular dystrophy. A potential target for pharmacological treatment? *J.Neuropath. Expt Neurol. 60:* 302-312.

Sentandreu, M.A., Coulis, G. & Ouali. A., (2002). Role of muscle endopeptidases and their inhibitors in meat tenderness. *Trends in Food Sci. & Technol.* 13: 400-421.

Sensky, P. L., Jewell, K. K., Ryan, K. J. P., Parr, T., Bardsley, R .G. & Buttery, P. J. (2006). Effect of anabolic agents on calpastatin promoters in porcine skeletal muscle and their responsiveness to cAMP- and Ca^{2+}-related stimuli. *J.Anim.Sci.* 84: 2973-82

Sensky, P.L., Parr, T., Bardsley, R.G. & Buttery, P.J. (1996). The relationship between plasma epinephrine concentration and the activity of the calpain enzyme-system in porcine longissimus muscle. *J.Anim.Sci. 74, 380-387.*

Sensky, P.L., Parr, T., Scothern, G.P., Perry, A., Bardsley, R.G., Buttery, P.J., Wood, J.D. & Warkup, C. (1998). Differences in the calpain enzyme system in tough and tender samples of porcine longissimus dorsi. *Proc.Brit.Soc.Anim.Sci. (1998)*, 14.

Shackelford, S. D., Koohmaraie, M., Cundiff, L. V., Gregory, K. E., Rohrer, G. A. & Savell, J. W. (1994). Heritabilities and phenotypic and genetic correlations for bovine postrigor calpastatin activity, intramuscular fat content, Warner-Bratzler shear force, retail product yield, and growth rate *J.Anim.Sci.* 72: 857-863.

Suzuki, K. & Sorimachi, H. (1998). A novel aspect of calpain activation. *FEBS Lett.* 433 1-4.

Suzuki, K., Sorimachi, H., Yoshizawa, T., Kinbara, K. & Ishiura, S. (1995). Calpain: Novel family members, activation, and physiological function. *Biol.Chem. Hoppe-Seyler* 376:523-529.

Takano, J., Kawamura, T., Murase, M., Hitomi, K. & Maki, M. (1999). Structure of mouse calpastatin isoforms: implications of species-common and species-specific alternative splicing. *Bioch.Biophys.Res.Comm.* 260: 339-345;

Takano, E., Maki, M., Mori, H., Hatanaka, M., Marti, T., Titani, K., Kannagi, R., Ooi, T. & Murachi, T. (1988). Pig heart calpastatin: identification of repetitive domain structures and anomalous behaviour in polyacrylamide gel electrophoresis. *Biochem.* 27: 1964-1972.

Takano, J., Watanabe, M., Kiyotaka, H. & Maki, M. (2000). Four Types of Calpastatin Isoforms with Distinct Amino-Terminal Sequences Are Specified by Alternative First Exons and Differentially Expressed in Mouse Tissues. *J.Biochem.* 128: 83-92;

Taveau, M., Bourg, N., Sillon, G., Roudaut, C., Bartoli, M. & Richard, I. (2003). Calpain 3 Is Activated through Autolysis within the Active Site and Lyses Sarcomeric and Sarcolemmal Components. *Mole.Cell.Biol.* 23:9127-9135

Taylor, R. G., Geesink, G. H., Thompson, V. F., Koohmaraie, M. & Goll, D. E. (1995a). Is Z-disk degradation responsible for postmortem tenderization? *J.Anim.Sci.* 21: 1351–1367.

Taylor, R. G., Tassy, C., Briand, M., Robert, N., Briand, Y., & Ouali, A. (1995b). Proteolytic activity of proteasome on myofibrillar structures. *Molec.Biol.Reports* 21, 71–73.

Thornberry, N.A. & Lazebnik, Y. (1998). Caspases: enemies within. *Science*, **281**, 1312-1316.

Veiseth, E., Shackelford, S. D., Wheeler, T. L. & Koohmaraie, M. (2001). Effect of post-mortem storage on m-calpain and m-calpain in ovine skeletal muscle. *J.Anim.Sci.* 70: 3035–3043.

Wang, K.K. (2000). Calpain and caspase: can you tell the difference? *Trends Neurosci*, **23,** 20-26.

Wendt, A., Thompson, V. F. & Goll, D. E. (2004). Interaction of calpastatin with calpain: a review. *Biol. Chem.* 385: 465–472.

Wilk, S. & Orlowski, M. (1980). Cation-sensitive neutral endopeptidase: Isolation and specificity of the bovine pituitary enzyme. Journal of Neurochemistry,

35(5), 1172–1182.

Winger, R.J. & Pope, C.G. (1981). Osmotic properties of post-rigor beef muscle. *Meat Science*, **5,** 355-369.

22

COMPENSATORY GROWTH IN PIGS: EFFECTS ON PERFORMANCE, PROTEIN TURNOVER AND MEAT QUALITY

NIELS OKSBJERG AND MARGRETHE THERKILDSEN
University of Aarhus, Faculty of Agricultural Sciences, P.O. Box 50, DK-8830 Tjele

Introduction

Pig research has previously focused on performance and carcass quality. However, in recent years eating quality and especially meat tenderness have become increasingly important. Several factors in the chain from conception to consumption of pigmeat affect various meat quality traits. These factors may encompass, for example,. maternal nutrition, genotype, sex, rearing space, feeding strategy and feed composition. Based on several studies a hypothesis has been proposed suggesting a relationship between the rate of muscle protein degradation *in vivo* at the time of slaughter and the rate and extent of *post mortem* tenderness development. In this contribution, initially there will be a short introduction to basic principles of muscle growth, and secondly, several pieces of evidence suggesting that the rate of protein degradation at the time of slaughter is related to the rate and extent of tenderness development, and that the calpain system is probably the link between these two traits will be presented. Finally, feeding strategy involving compensatory growth leading to increased protein degradation rate will be considered.

Basic principles of muscle growth

Muscle growth is the major determinant of the performance of meat-producing animals. The amount of meat produced is related to the number of muscle fibres and the growth of the individual muscle fibre. Muscle fibres are formed during the embryonic and foetal stages and, in most mammals, the number is fixed around birth. Thus, postnatal growth is related to growth in the cross-sectional area (hypertrophy) and to length of the fibres by adding additional sarcomeres. Postnatal growth is determined by the difference between two dynamic processes; i) the rate of protein synthesis and ii) the rate of protein degradation, referred to as the protein turnover. Protein turnover is in many situations directly related to division and incorporation of the satellite cells. Thus, during postnatal muscle growth, the rate of synthesis exceeds the rate of degradation. However, during growth these processes decline and approach each other, and in adult animals they become equal, while in aging animals the rate of protein degradation exceeds the rate of protein synthesis. The rate of protein turnover is also related to the fibre type frequency being higher in fast-moving fibres than in slow-moving fibres (for review, see Oksbjerg *et al.* 2004).

Relationship between muscle protein degradation *in vivo* and tenderisation *post mortem*

The growth rate of farm animals is related to muscle protein turnover. Thus the more positive the muscle protein balance, the better the growth performance, the efficiency of growth, and the lean composition; thus making this trait economically essential in meat production. Moreover, the efficiency is of critical importance in terms of the environmental load of nitrogen and phosphorus during production. The proteolytic potential in the muscle at the time of slaughter has long-been regarded as an important factor in the tenderisation process in meat, which claims high muscle protein turnover in healthy animals at the time of slaughter. Consequently, management of muscle protein turnover may enable control of the three meat quality attributes - *price, tenderness* and *sustainability.*

Several reports suggest that a relationship exists between the rate of muscle protein degradation and the rate and extent of tenderness development in meat (Table 22.1). Thus, in situations where the rate of protein degradation is decreased, this may lead to increased muscle growth, but decreased tenderness, e.g. *i)* treatment with ß-adrenergic agonists (for review, see Beermann, 1993), *ii)* restricted feeding (Therkildsen *et al.,* 2004), *iii)* bulls versus steers (Morgan *et al.,* 1993) and *iv)* animals possessing the callipyge gene (Lorenzen *et al.,* 2000). In contrast, treatment with porcine growth hormone results in increased muscle growth by stimulating the rates of both synthesis and degradation (Sevé *et al.,* 1993) without change in tenderness (Oksbjerg *et al.,* 1995). Finally, short-term fasting for 5 days leads to increased rate of muscle protein degradation resulting in increased tenderness in lambs (McDonagh *et al.,* 1999). The link between the rate of muscle protein degradation and tenderness development may be coupled to the calpain system, that is known to be the rate-limiting proteolytic system disassembling the myofibrillar proteins to their individual constitutive proteins (Goll *et al.,* 1992; Koohmaraie *et al.,* 2002). Thus, in cattle the inhibitor of the calpains (μ-calpain to be involved in disassembly of myofibrillar proteins), calpastatin, is inversely related to the rate of muscle protein degradation and positively correlated with the shear force of the meat. Having established a link between the rate of protein degradation and *post mortem* tenderisation, the challenge becomes to implement this into a feeding strategy, which will result in increased tenderness without compromising performance. The feeding strategy used in this project is called compensatory growth or catch-up growth.

Table 22.1 FACTORS AFFECTING PROTEIN DEGRADATION PARAMETERS AND SHEAR FORCE IN MUSCLE AND MEAT, RESPECTIVELY, OF MEAT PRODUCING MAMMALS

	Muscle growth	FSR[a]	FDR[b]	Calpastatin activity	Shear force
ß-agonist	⇑	⇔,⇑	⇓	⇑	⇑
pGH[c]	⇑	⇑	⇑		⇔
Bull *vs* steer	⇑	⇔	⇓	⇑	⇑
Fasting (short time)	⇓	⇓	⇑	⇓	⇓
Fasting (long time)	⇓	⇓	⇓	⇑	⇑
Callipyge gene	⇑	⇔	⇓	⇑	⇑

[a]Fractional synthesis rate, [b]Fractional degradation rate; [c]porcine Growth Hormone

Compensatory growth

For decades it has been recognised that pigs (McMeekan, 1940), whose growth rate has been slowed by nutritional deprivation, may exhibit an enhanced rate of growth when re-alimented. If this exceeds the maximal rate of gain when adequate nutrition has been provided, the animal is said to have undergone compensatory or catch-up growth. In pigs the total growth period can be divided into the weaning, starter, grower and the finishing period. Thus, various research groups have imposed diet restriction or amino acid/crude protein deficiency to cause retarded growth in either the weaning, starter or grower period or in a combination of periods to study compensatory growth in the following periods until slaughter. Sarkar *et al.* (1983) restricted dietary energy of pigs during the weaning period until d 35. They found that, although restricted pigs had a significantly lower body weight at d 35 and d 70, they were not significantly different from control pigs at d 166 due to compensatory growth between d 70 and d 166. Others have demonstrated compensatory growth in the growing/ finishing period following restrictive feeding (Nielsen, 1964; Campbell *et al.*, 1983) or amino acid deficiency (Zimmermann and Kharajaren, 1973; Thaler *et al.*, 1986; Campbell and Biden, 1983; Kyriazakis *et al.*, 1991; Pond *et al.*, 1980; Hogberg and Zimmermann, 1978; Valaja *et al.*, 1992) during the starter period (5-20 kg body weight).

Compensatory growth was also observed in the finishing period following diet restriction (Prince *et al.*, 1983; Donker *et al.*, 1986; Oksbjerg *et al.*, 2002; Therkildsen *et al.*, 2004) or following a period of feeding an amino acid deficient diet (Wahlstrom and Libal, 1983; de Greef *et al.*, 1992; Critser *et al.*, 1995; Chiba, 1994; Chiba *et al.*, 1999; Chiba *et al.*, 2002; Fabian *et al.*, 2002) in the grower period (25-50 kg body weight). Using diets deficient in amino acids to slow the rate of growth results in fatter carcasses at the end of the restricted period. However, compensatory growth is supported by improved efficiency rather than increased feed intake, and the carcass quality is normally unchanged. Following a period with diet restriction, the rate of both muscle and fat deposition is decreased, but the efficiency is improved. Following this period, compensatory growth is supported by improved efficiency, but in a few cases increased feed intake was noted. Thus, using diet restriction may improve the overall feed efficiency. At slaughter the carcass lean is unchanged or slightly improved. Compensatory growth may be related to both internal organs and the carcass. Recently, Skiba *et al.*, (2004) showed that the rate of protein accretion in the carcass increased during compensatory growth.

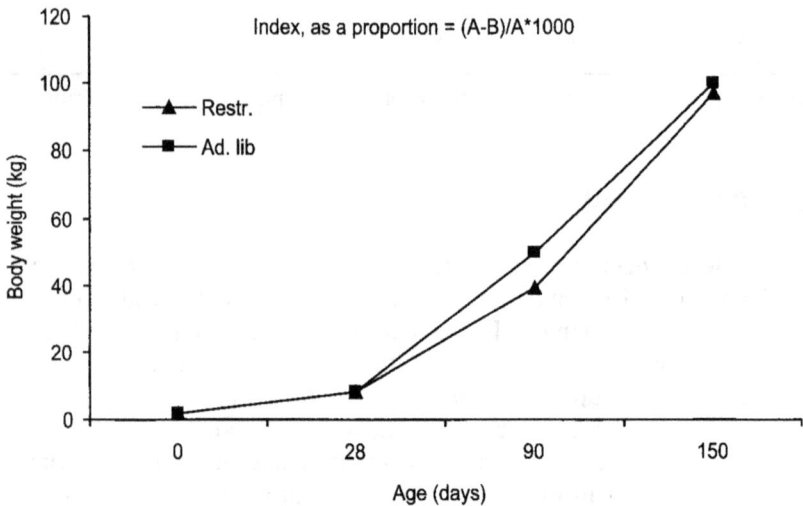

Figure 22.1 Calculation of compensatory growth in pigs (Oksbjerg *et al.*, 2002) where A is the difference in weight between feed restricted animals and *ad libitum*-fed animals following the restriction period, and B is the difference between *ad libitum*-fed animals and compensatory animals at the end (at slaughter) of the period as defined by Hornick *et al.*, 2000.

To be used in the production, a compensatory growth strategy may be of economic interest meaning that such animals must more or less achieve normal weight for their age, improved feed conversion ratio or exhibit better carcass quality.

   The degree of compensatory growth or the index is calculated as in figure 22.1. The degree of compensatory index may depend on:

•   Restricted period:

- Length
- Developmental stage
- Severity
- Stage of development
• Realimentation period:
- Length
- Level of feed intake
- Diet compostion
• Genotype and sex

All these factors should be taken into consideration to obtain full compensation. From experiments where the nutritive restriction has been imposed during the grower period, it seems as if the compensatory growth response increases up to approximately 80-90 g/day, when the growth is restricted up to 20%. A further restriction beyond 20% increases a further compensatory growth only slightly. Generally, a reduction of growth beyond 20% in the restriction period will only result in partial compensation. Assume a finishing period of 60 days with a compensatory growth response of 80 g/day, then the pigs will compensate with 4800 g. This means that compensation will be complete only when pigs are restricted thus that the body weight will be 4800 g behind at the end of the restriction period. Assuming the grower period to be 52 days (8 – 40 kg), this means that the pigs should be restricted thus that their growth rate is reduced by 4800/52=92 g/d (see Therkildsen *et al.*, 2004; Figure 22.2).

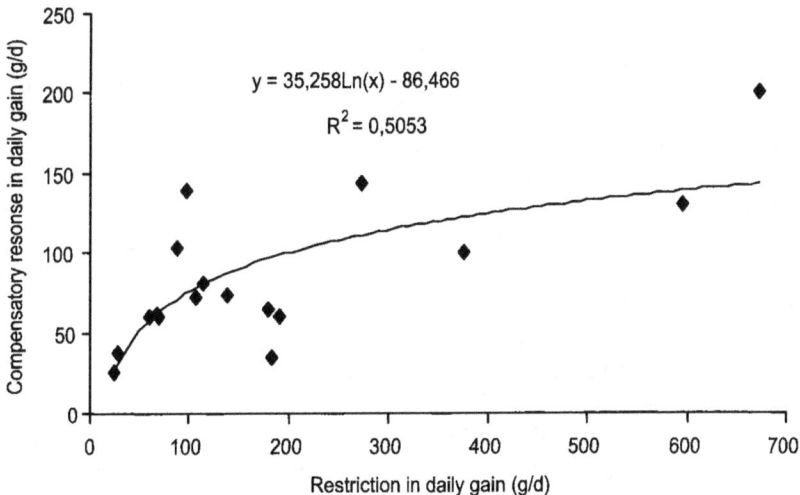

Figure 22.2 The influence of the degree of restriction on the subsequent compensatory growth response (Therkildsen *et al.*, 2004).

Danish experiments

THE COMPENSATORY GROWTH STRATEGY

For reviews covering earlier experiments on compensatory growth, see Andersen *et al.*.2005a,b. During compensatory growth, the rates of both protein synthesis and degradation are elevated according to findings in the rat (Milward *et al.*, 1975; Schradereit *et al.*, 1995) and cattle (Jones *et al.*, 1990). Consequently, implementation of a compensatory growth strategy in production of meat-producing animals could be a means to improve tenderness of meat taking the production economy into consideration. However, the increase in protein turnover during compensatory growth is dynamic. Thus, initially during compensatory growth, increased protein synthesis is evident, while protein degradation remains low as a consequence of the former restricted feeding regime. Later on also the rate of protein degradation increases gradually and eventually exceeds the rate of protein degradation of control *ad libitum*-fed animals.

Thus, one of the goals to implement successfully a compensatory feeding approach in the production of meat of high quality is to establish the length of the compensatory period, which results in highest muscle protein degradation potential at the time of slaughter. This was examined by Therkildsen *et al.*, (2002), and together with results published by Kristensen *et al.* (2002) using markers for protein synthesis (concentrations of RNA and elongation factor-2) and protein degradation (activity of μ-calpain and myofibrillar fragmentation index, MFI), it was suggested that until 48 days of compensatory growth, the rate of protein turnover increased steadily, but the difference between protein synthesis and degradation diminished with every day on compensatory growth. Beyond 48 days of compensatory growth, the muscle protein turnover exceeded the turnover of control *ad lib*-fed animals.

Consequently in the compensatory growth model, pigs were fed pigs restrictively (0.60-0.70 of *ad lib* intake) for 52-62 days from day 28 until they were 80 or 90 days of age and subsequently ad lib until 140 days of age (50-60 days). Feed uptake in the restriction period was reduced by 60-70%, daily gain by 20-25% and kg feed:kg gain was reduced by 20%. At the end of the restriction period the cross-sectional area of the LD muscle and the fat layer were measured by ultra sonic equipment. In restrictively fed pigs, the CSA of LD was reduced by 15% and the fat layer by 26%. During re-alimentation, feed uptake was unchanged, while daily gain was significantly increased by 70-80 g/day due to an improved kg feed:kg gain of 6%. During the entire growth period from day 28 to day 140 of age, compensatory growth resulted in similar daily gain, carcass weight and meat percentage, while the kg feed:kg gain ratio was reduced. The significance of the latter is that slaughter pigs on compensatory growth strategy can be produced on 15 kg less feed per pig (Oksbjerg *et al.*, 2002; Therkildsen *et al.*, 2004, Kristensen *et al.*, 2002).

These experiments were carried out in a conventional production system, and the pigs reared in individual pens. However, when pigs were reared on pasture and fed restrictively until transfer to indoor facilities with access to an outdoor concrete area at 40 kg live weight and fed ad lib until slaughter, compensatory growth was also found (Oksbjerg *et al.*, 2005).

Muscle protein turnover

Using markers for protein turnover revealed that protein synthesis (concentration of muscle RNA and eEF-2) was increased at slaughter (Oksbjerg *et al.*, 2002, Therkildsen *et al.*, 2004) or at an earlier stage of growth (day 105, Kristensen *et al.*, 2002). In the latter study, no indication of increased protein synthesis was observed, probably because the length of the compensatory growth period was 75 days compared with 50-60 days in the other studies (Therkildsen *et al.*, 2004) Additionally, results on MFI were not consistent, because in one study the MFI was increased was nether consistent. However, data on markers for protein degradation were not consistent. Thus, the activity μ-calpain was not changed. On the other hand, protein degradation rate was also measured on muscle strips form M. semitendinosus *in vitro* by release of tyrosine in the medium, and using muscle proteome after compensation in female pigs also indicates high proteolytic activity because of increased amounts of MLC (myosin light chain) II and III (Lametsch *et al.*, 2006). These data showed a trend towards an increased protein degradation in pigs on the compensatory growth strategy.

Meat quality and compensatory growth

The technological meat quality (pH, drip loss, pigment and colour) was not affected by compensatory growth (Oksbjerg *et al.*, 2002; Kristensen *et al.* 2002), although the pigment concentration was reduced after compensatory growth in partly outdoor reared pigs, probably due to higher intake of roughage (Oksbjerg *et al.* 2005).

Consistently, the shear force of LD muscle was reduced, indicating better tenderness (Kristensen *et al.*, 2002), or tenderness measured by a sensory panel increased by compensatory growth (Kristensen, 2004). This was true in female pigs, but not in castrated male pigs. The reason for this may be that compensatory growth reduced intramuscular fat (IMF) in castrated male pigs, but not in female pigs. IMF increased tenderness linearly in the range up to 25g/kg. Consequently, decreased IMF may counteract the beneficial effect of compensatory growth on tenderness.

In conclusion

Studies on compensatory growth show that this strategy reduces the kg feed:kg gain efficiency. At the same time, the overall daily gain was unaltered, as were also the meat content in both castrated male and female pigs. Moreover, the technological meat quality was unaltered, but the eating quality (tenderness) was increased in female pigs, but not in castrated male pigs.

Literature cited

Andersen, H.J., Oksbjerg, N., and Therkildsen, M. 2005a. Potential quality control tools in the production of fresh pork, beef and lamb demanded by the European society. Livest. Prod. Sci. 94:105-124.

Andersen, H.J., Oksbjerg, N., Young, J.F., and Therkildsen, M. 2005. Feeding and meat quality – a future. Meat Sci. 70:543-554.

Beermann, D.H. (1993). b-adrenergic agonists and growth. In *The Endocrinology of Growth, Development, and Metabolism in Vertebrate*, pp 345-366. Academic Press.

Campbell, R.G. and Biden, R.S. 1978. The effect of protein nutrition between 5.5 and 20 kg live weight on the subsequent performance and carcass quality of pigs. Anim. Prod. 27:223-228.

Campbell, R. G., Taverner, M. R., & Curic, D. M. (1983). Effects of feeding level from 20 to 45 kg on the performance and carcass composition of pigs grown to 90 kg live weight. Livest. Prod. Sci., 10, 265-272.

Critser, D. J., Miller, P. S., & Lewis, A. J. (1995). The effects of dietary protein concentration on compensatory growth in barrows and gilts. J. Anim. Sci., 73, 3376-3383.

Donker, R. A., Den Hartog, L. A., Brascamp, E. W., Merks, J. W. M., Noordewier, G.J., & Buiting, G. A. J. (1986). Restriction of feed intake to optimize the overall performance and composition of pigs. Livest. Prod. Sci., 15, 353-365.

Goll, D. E., Thompson, V. F., Taylor, R. G., & Ouali, A. (1998). The calpain system and skeletal muscle growth. Canadian Journal of Animal Science, 78, 503-512.

Hogberg, M.G., and Zimmerman, D.R. 1978. Compensatory responses to dietary protein, length of starter period and strain of pig. J. Anim. Sci. 47:893-899.

Hornick, J. L., Van Eenaeme, C., Gerard, O., Dufrasne, I., & Istasse, L. (2000). Mechanisms of reduced and compensatory growth. Domestic Animal Endocrinology, 19 (2), 121-132.

Jones, S. J., Starkey, D. L., Calkins, C. R., & Crouse, J. D. (1990). Myofibrillar protein turnover in feed-restricted and realimented beef cattle. J. Anim. Sci., 68, 2707-2715.

Koohmaraie, M., Kent, M. P., Shackelford, S. D., Veiseth, E., & Wheeler, T. L. (2002). Meat tenderness and muscle growth: is there any relationship? Meat Science, 62, 345-352.

Kristensen, L., Therkildsen, M., Riis, B., Sørensen, M. T., Oksbjerg, N., Purslow, P. P., & Ertbjerg, P. (2002). Dietary induced changes of muscle growth rate in pigs: Effects on *in vivo* and post-mortem muscle proteolysis and meat quality. J. Anim. Sci. 80, 2862-2871.

Kristensen, L., Therkildsen, M., Aaslyng, M.D., Oksbjerg, N., and Ertbjerg, P. 2004. Compensatory growth improves meat tenderness in gilts but not in barrows. J. Anim. Sci. 82:3617-3624.e

Kyriazakis, I., Stamataris, C., Emmans, G.C., and Whittemore, C.T. 1991. The effects of food protein content on the performance of pigs previously given foods with low or moderate protein contents. Anim. Prod. 52:165-173.

Lametsch, R., Kristensen, L., Larsen, M.R., Therkildsen, M., Oksbjerg, N., and Ertbjerg, P. 2006. Changes in the muscle proteome after compensatory growth in pigs. J. Anim. Sci. 84:918-924.

Lorenzen, C.L., Koohmaraie, M., Shackelford, S.D., Jahoor, F., Freetly, H.C., Wheeler, T.L., Sawell, J.W., & Fiorotto, M.L. (2000). Protein kinetics in callipyge lambs. J. Anim. Sci., 78, 78-87.

McDonagh, M. B., Fernandez, C., & Oddy, V. H. (1999). Hind-limb protein metabolism and calpain system activity influence post-mortem change in meat quality in lamb. Meat Science, 52, 9-18.

McMeekan, C.P. (1940). Growth and development in the pig, with special references to carcass quality characters. III. Effects of plane of nutrition on the form and composition of the bacon pig. J. Agric. Sci. (Camb.), 30, 511-569.

Millward, D. J., Garlick, P. J., Stewart, J. C., & Nnanyelugo, D. O. (1975). Skeletal-muscle growth and protein turnover. Biochem. J., 150, 235-243.

Morgan, J. B., Wheeler, T. L., Koohmaraie, M., Crouse, J. D., & Sawell, J. W. (1993). Effect of castration on myofibrillar protein turnover, endogenous proteinase activities, and muscle growth in bovine skeletal muscle. J. Anim. Sci., 71, 408-414.

Nielsen, H. E. (1964). Effects in bacon pigs of differing levels of nutrition to 20 kg body weight. Anim. Prod., 6, 301-308.

Oksbjerg, N., Petersen, J. S., Sorensen, M. T., Henckel, P., Agergaard, N., Bejerholm, C., & Erlandsen, E. (1995) The influence of porcine growth hormone on muscle fibre characteristics, metabolic potential and meat quality. Meat Science, 39, 375-385.

Oksbjerg, N., Sørensen, M. T., & Vestergaard, M. (2002). Compensatory growth and its effect on muscularity and technogicla meat quality in growing pigs. Acta Agric. Scand., Sect. A, Animal Sci., 52, 85-90.

Oksbjerg, N., Gondret, F., & Vestergaard, M. (2004). Basic principles of muscle development and growth in meat-producing mammals as affected by the insulin-like growth factor (IGF) system. Domest. Anim. Endocrinol. 27:219-240.

Oksbjerg, N., Strudsholm, K., Lindahl, G., Hermansen, J.E. 2005. Meat quality of fully or partly outdoor reared pigs in organic production. Acta Agric. Scan. Sect. A, 55: 106-112.

Prince, T. J., Jungst, S. B., & Kuhlers, D. L. (1983). Compensatory responses to short-term feed restriction during the growing period. J. Anim. Sci., 56, 846-852.

Sakar, N.K., Lodge, G.A., Williams, C.J., and Elliot, J.I. 1983. The effects of undernutrition of suckled pigs on subsequent growth, and body composition after nutritional rehabilitation. J. Anim. Sci. 57:34-42.

Schradereit, R., Klein, M., Rehfeldt, C., Kreienbring,, F., Krawielitzki, K. 1995. Influence of nutrient restriction and realimentation on protein and energy metabolim, organ weights and muscle structure in growing rats. J.Anim. Physiol., Anim., Nutr. 74:253-262.

Sève, B., Ballèvre, Ganier, P., Noblet, J., Prugnaud, J., & Obled, C. (1993). Recombinant porcine somatotropin and dietary protein enhance protein synthesis in growing pigs. J. Nutr., 123, 529-540.

Thaler, R.C., Libal, G.W., and Wahlstrom R.C. 1986. Effect of lysine levels in pjig starter diets on performance to 20 kg and on subsequent performance and carcass characteristics. J. Anim. Sci. 63:139-144.

Therkildsen, M., Riis, B., Karlsson, A., Kristensen, L., Ertbjerg, P., Purslow, P. P., Dall Aaslyng, M. & Oksbjerg, N. (2002). Compensatory growth response in pigs, muscle protein turn-over and meat texture: effects of restriction/realimentation period. Anim. Sci. 75, 367-377.

Therkildsen, M., Vestergaard, M., Busk, H., Jensen M. T., Riis, B., Karlsson, A. H., Kristensen, L., Ertbjerg, P., & Oksbjerg, N. (2004). Compensatory growth in slaughter pigs-in vitro muscle protein turnover at slaughter, circulating IGF-I, performance and carcass quality. Livest. Prod. Sci., 88, 63-74.

Valaja, J., Alaviuhkola, T., Suomi, K., and Immonen, I. 1992. Compensatory growth after feed restriction during the rearing period in pigs. Agric. Sci. Finl. 1:15-20

Zimmerman, D.R., and Khajarern, S. 1973. Starter protein nutrition and compensatory response in swine. J. Anim. Sci. 36:189-194.

23

SELECTION FOR CARCASE QUALITY

SAM HOSTE
Quantech Solutions

Introduction

Selection for pig carcase quality has been underway since pigs were first kept. However, the nature of "selection" and meaning of "quality" have changed. This chapter will discuss quality in a general way as it relates to pig production and concentrate on selection in pig populations for carcase quality. There has been a renaissance of interest in selection for eating quality through the use of molecular information; generally however the industry's lack of interest in meat quality is caused by a failure of market signals in the pork chain. The involvement and dialogue between players in the whole pork meat chain needs to be encouraged to engender genetic change in carcase quality.

What is carcase quality? This is not a trivial issue as meat is processed and used in many different ways (de Vries *et al.* 1998).

Quality is rather like the 6 blind men and the elephant in the poem by John Godfrey Sax

> And so these men of Indostan
> Disputed loud and long,
> Each in his own opinion
> Exceeding stiff and strong,
> Though each was partly in the right,
> And all were in the wrong!
>
> So oft in theological wars,
> The disputants, I ween,
> Rail on in utter ignorance
> Of what each other mean,
> And prate about an Elephant
> Not one of them has seen!

Quality has been examined in many different ways and from many different perspectives, and therefore seen different things. These definitions complement and add richness and elements to our understanding of what quality means to geneticists, producers, processors, retailers and customers. For example consumers are interested in tenderness, juiciness, flavour and colour; retailers in drip loss, shelf life, colour and fatness, and processors in yield, colour and lack of drip. However such lists display a considerable information loss in that quality surely is defined and relative to culture and social aspects, i.e. context. They even display a lack of information for the industry context as different quality criteria may apply if the product is fresh, cooked, smoked, dried or processed.

Despite the problems that there are with defining quality, the definition is if anything expanding to incorporate various environmental, welfare, safety, traceability and trust aspects (Grunert 2006).

The above description details some of the landscape of quality which is affected by the actions and interactions of events by the breeding company, on-farm and during processing (Table 23.1).

Table 23.1 BREEDING COMPANY, PRODUCER AND PROCESSING COMPANY FACTORS AFFECTING CARCASS AND MEAT QUALITY

Breeding company	On-farm	Processing
Between breed choice	Feed	Pre-slaughter treatment
Within line selection criteria	Management & handling	Slaughter conditions
	Stocking density	Cooling profile
	Transport	Ageing
	Slaughter age & weight	Hanging method
	Pre-slaughter treatment	Cooking

For most breeding companies that are marketing in a number of countries the situation is further complicated by regional or country specific requirements. Two brief examples will be given to illustrate. The Hal gene (or the Ryanodine receptor gene) has two alleles: N being normal and dominant and n being the mutant and recessive. The nn genotype relative to the NN genotype has higher frequency of stress susceptibility, Pale Soft and Exudative (PSE) meat and a higher lean proportion. Thus the nn genotype was not attractive to the UK fresh pork market due to the colour and drip, but was beneficial in the German market where further processing was important. The story is similar with the Rendement Napole (RN) or Hampshire, acid meat gene. The RN gene has two alleles: rn+ which is normal and recessive whilst RN⁻ is dominant and gives rise to a large increase in muscle glycogen, leading to pH decrease, i.e. acid meat and a marked decrease in cooked ham processing yield. As with the Hal gene there are a number of pleiotrophic affects of the RN⁻allele. RN⁻ meat is pale, has a higher meat flavour and higher odour intensity whilst there is some controversy over the effect on tenderness. Swedish researchers have found RN meat to be more tender (Jonsall *et al.* 2001) and others tougher (Lundstrom *et al.* 1996; Le Roy *et al.* 2000). These apparently different results may be a consequence of the cooking temperature used (65^{0}C in Sweden and 80^{0}C in France; Monin 2003). One of the main problems of the RN⁻ gene in the French market is the slicing losses in hams cooked for *jambon superieur.* Thus the Swedish view RN⁻ as favourable for flavour and tenderness whereas in the French market RN⁻ is a problem for processing: in different markets different alleles may be desirable.

The sources of variation in carcase quality

The sources of variation in meat quality unfortunately are infrequently reported. The series of reports for Scottish Executive Environment and Rural Affairs Department (SEERAD) "Meat Eating Quality – A Whole Chain Approach" does detail the sources of variation for sheep and beef, but not pigs (Table 23.2). Generally individual animal effects account for approximately 0.30 to 0.50 of the variation. This between-animal variation remains poorly understood. Farm and farm protocols, abattoir and slaughter batch have relatively little effect, with the exception of slaughter batch for abnormal flavour. The processing protocol (electrical stimulation and hip suspension) was a significant source of variation for texture and flavour (in beef only). "Other" is the unexplained variation that includes measurement error and small effects of interactions between processing and farm protocols, and between processing protocol and abattoir but varies between 0.11 and 0.62. Both the "between animal" and "other" or unexplained sources of variation would appear to be useful areas to investigate further by pork chain researchers.

Examples of successful "Whole Chain" marketing in pork products

Currently in the UK there are a number of successful collaborations between producers, processors and retailers in marketing pork products based on individual breeding company lines.

- George Adams and sons with PIC (Genus) have the Link Pork scheme.
- George Adams and sons have a link with JSR: "Titan Pork Partners".
- Cranswick and JSR for Sainsbury's "Taste the difference" range.
- JSR have a link with producers to produce pork in collaboration with Grampian for Marks & Spencers in their Muirden range of products.
- PIC and British Quality Pigs (BQP) have a link to supply Waitrose.

Table 23.2 SOURCES OF VARIATION FOR TEXTURE, JUICINESS, FLAVOUR AND ABNORMAL FLAVOUR IN BEEF AND SHEEP (DATA AS PERCENTAGES)

	Source of variation	*Beef*	*Sheep*
Texture	Animal	27	31
	Farm	6	} 0
	Farm protocol	0	
	Slaughter batch	7	2
	Abbattoir	0	0
	Processing protocol	38	56
	Season	NA	0
	Other	22	11
Juiceness	Animal	19	33
	Farm	4	} 4
	Farm protocol	0	
	Slaughter batch	4	6
	Abbattoir	19	4
	Processing protocol	0	1
	Season	NA	0
	Other	54	52
Flavour	Animal	31	54
	Farm	10	} 0
	Farm protocol	1	
	Slaughter batch	9	0
	Abbattoir	0	2
	Processing protocol	10	2
	Season	NA	0
	Other	39	40
Abnormal flavour	Animal	7	34
	Farm	1	} 0
	Farm protocol	0	
	Slaughter batch	30	0
	Abbattoir	0	0
	Processing protocol	0	0
	Season	NA	35
	Other	62	31

These collaborations have all achieved the link up in the pork chain gaining premium prices and sharing of the added value. They have all been based around utilising different lines from breeding companies, although there is not an un-ending supply of "new" lines and breeds. There is however great scope for within line selection for meat and carcase quality traits and also the use of on-farm and processing. The way that organisations compete is changing: there are examples of supply chains competing as in Professor Martin Christopher 's often quoted: "supply chains compete, not companies" (Christopher & Peck 2003). Importantly, although these schemes are based around different breeding company lines, they also have a series of "standard operating procedures" for the farming system used, processor, retailer and cooking information for the consumer.

Genetic selection for carcase quality

Genetic improvement comprises the utilisation of between breed (or specific company line) differences, crossing lines and breeds, within line selection and subsequent dissemination of the improved genotypes. There will a brief discussion of between-breed effects, crossbreeding and a fuller discussion of within breed selection.

SELECTION BETWEEN BREED

Until recently between-breed selection has been utilised to exploit lines with particular reproductive, growth or carcass merit that are not too far from the average performance required by the market. For example lines such as the Berkshire that offer advantages in meat quality are excessively fat for the UK market. Even though it may be possible to achieve a premium price on loins there remains the rest of the carcase to sell possibly at a severe discount. Similarly the incorporation of the Chinese Meishan line in female (reproductive) lines has only been achieved successfully by NPD with the advantage of an extremely lean female line with which to cross the pure Meishan.

However, what would have been described as "rare" breeds are experiencing a come-back as people become more interested in a range of food issues. In the UK, animal welfare, outdoor rearing and growing of pigs, the concept of local food, food miles and reducing the carbon footprint, farmers markets are all topical issues. Alongside this there has been considerable media (e.g. Jimmy's farm) and celebrity chef (Jamie Oliver) interest.

Obviously this is a niche market, albeit with considerable media attention and regional differences. Around 0.53 of Londoners surveyed in a recent IGD report (Padbury 2006) "buy local food and wish to buy more" compared with an average of only 0.39 for the rest of Britain (England, Wales and Scotland; Table 23.3). Londoners also show, a not too surprising, preference for "food produced in adjoining counties to where I live". This contrast between London and the rest of Britain is also shown for organic food.

This is a whole system concept (small producer, few animals, different feed, outdoor, local, etc.) rather than just a different line or breed of animal, but it does provide a market for number of animals produced and slaughtered locally with what people often say will be "tasty" meat (Wood & Lutwyche 2007).

Table 23.3 REGIONAL DIFFERENCES IN THE PERCENTAGE OF SHOPPER WHO THINK ORGANIC IS IMPORTANT

Proportion of shoppers who think organic is important – by region	
London	0.71
South	0.46
Midlands	0.45
North	0.35
Wales	0.45
Scotland	0.30

Source: www.igd.com/CIR.asp?menuid=34&cirid=2057 , Accessed 29/04/2007

Some examples of between breed differences are shown in Table 23.4.

Table 23.4 LEAST SQUARES MEANS FOR GROWTH, CARCASS AND SENSORY BETWEEN BREED AND WITHIN MUSCLE

Trait	*Breed*				*SED*	*Signif.*
	Berkshire	*Duroc*	*LW*	*Tamworth*		
Cold carcass weight(kg)	48.4	63.4	64.8	48.2	2.322[A]	*[1]
Average daily gain (g/d)	424.7	659.4	658	435.4	22.979[A]	*[1]
P2 fat thickness (mm)	15.1	9.1	7.8	14.7	0.662[A]	<0.001
Subcutaneous fat	32.1	13	12	25.9	1.043[A]	<0.001
Tenderness (in LD[*2])	4.16	3.96	4.35	3.93	0.127[B]	<0.01
Juiciness (in LD)	4.29	4.18	4.18	4.05	0.103[B]	n.s.
Pork flavour (in LD)	3.60	3.83	3.35	3.53	0.092[B]	<0.05
Overall liking (in LD)	3.71	3.47	3.60	3.60	0.109[B]	n.s.

Source: (Wood *et al.* 2004)
[A] SED, standard error of the difference, average sample size of 22.5 pigs per breed and diet group
[B] SED, standard error of the difference, average sample size of 44.3 pigs per breed for each muscle
*[1] Significant breed * diet interaction
*[2] longissimus dorsi muscle
n.s. not significant
Tenderness, juiciness, pork flavour and overall liking scored on a 1 to 8 categorical scale

The two relatively un-improved breeds, the Berkshire and Tamworth, were considerably lighter after the 12 week feeding period and were significantly fatter (Table 23.4). Tenderness was higher in the Large White compared with the Duroc and Tamworth. Pork flavour tended to be higher in the two traditional breeds, whilst abnormal flavour higher in the two modern breeds. Flavour liking was higher in the two traditional breeds. Berkshire had a higher overall liking than Duroc. However all these breed effects were small and inconsistent.

To utilise the more traditional, unselected and fatter breeds there seems to be great scope for creating synthetic lines between modern breeds and traditional breeds thereby removing some of the adverse affects of fatness.

Within breed selection

ON FARM TESTING

Within-breed selection for growth and carcase traits has been a major activity of pig breeding companies since the 1960s. The types of on-farm test have varied in detail but have included some measure of weight, back fat and possibly food intake, often with performance being recorded on individual animals. Specific details of testing are often considered proprietary information which has led to a lack of discussion in the literature of these aspects.

Backfat and muscle depth and muscle area

Back fat depth has been measured at one or several reference points along the back of the live animal, often offset from the centre line. A variety of equipment has been used such as A or B mode ultrasonic scanners. Commercial examples of machines include the Meritronics (Merit, Lowson and French Ltd, England), SFK (SFK Technology, Denmark), Renco (Renco Corporation, Minneapolis, USA), Sonalyser and Medata. There have been various comparisons of A-mode machines (Greer *et al.* 1987; Magowan & McCann 2006) and several historical reviews of ultrasound technology(Moeller 2001 [Moeller provides a partial list of evaluations of ultrasound accuracy]; Stouffer 2004) . A-mode scanners provide a relatively in-expensive method of measuring backfat.

Many pig breeding companies have moved to using B-mode machines such as the Aloka where a real time screen picture is obtained and images can be saved and used for measurement of fat depth and muscle depth. Loin eye muscle area can also be estimated by using a transverse rather than a longitudinal scan.

Correct use of the equipment is probably a matter of greater concern than differences between the better machines. Using an easily recognisable anatomical reference point (such as the last rib) and how consistently the reference location is found, consistency of transducer angle to the body and animal movement whilst measurements are taking place. Subsequently the image quality and correct interpretation by the assessor are additional factors influencing bias and precision. Additionally there is variation between and within operators. This necessitates regular training and benchmarking amongst assessors and also calibration between the animals live and as carcases. ACMC has undertaken a significant number of trials evaluating the repeatability of weight, fat depth, muscle depth and muscle area measurements in addition to regular staff benchmarking. The repeatability (Table 23.5) or intra-class correlation is a measure of the similarity between repeat observations within a person which are due to measurement error and also to general environmental error: $\sigma^2_B/\sigma^2_B + \sigma^2_w$.

Table 23.5 REPEATABILITY OF VARIOUS TRAITS AT END OF PERFORMANCE TEST

Trait	Repeatability
Weight at end of test (~100kg)	0.95
Fat depth (mm)	0.76
Muscle depth (mm)	0.54
Muscle area (cm²)	0.54

Source (ACMC, unpublished)

The lower relative repeatability for the measurements of muscle depth and muscle area indicates that there is benefit in accuracy to measuring these variables a number of times. Animals off test are run through the off-test facilities twice, on consecutive days, and weighed and scanned.

Repeatability of an operator tracing round saved images of muscle area is similarly important. Saved Aloka images from a trial consisting of 177 males and females from three breeds were used to test the accuracy of the operator tracing the muscle area on consecutive days. The results are shown in Figure 23.1. If the results were identical the regression slope would be 1, however it lay between 0.93 and 0.96.

New developments in assessment

VISUAL IMAGE ANALYSIS

There has been considerable analysis of the Osborne VIA system which is beneficial in providing a sequential growth curve for an animal. Obtaining sequential growth measurements is otherwise time consuming and weigh platforms do not work reliably in straw based systems such as in the UK. The equipment also allows the possibility of recording body length, shoulder and ham width which could be utilised in selection.

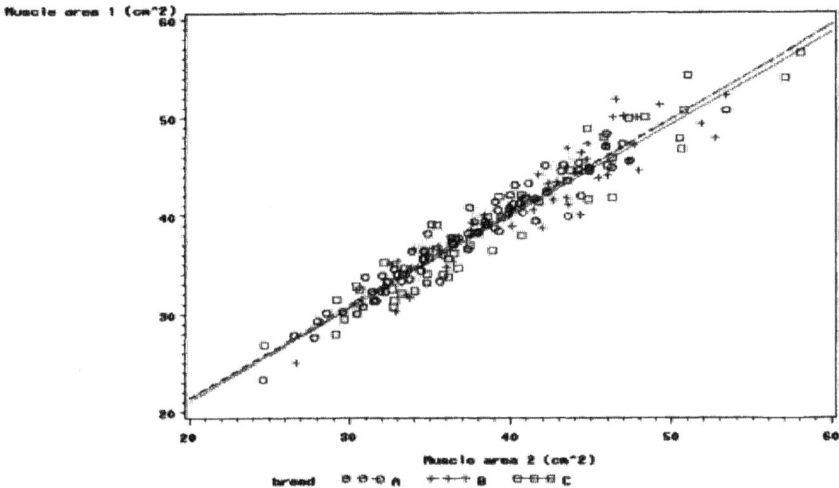

Figure 23.1 Relationship between repeat measures of loin eye muscle area (Source ACMC).

ULTRASOUND

The Aloka 500 scanner has been the mainstay in animal agriculture for some considerable time. However the Aloka machine is relatively large and delicate in comparison with newer models from Aloka (www.aloka-europe.com/), GE Healthcare (www.gehealthcare.com/usen/ultrasound/) or Sonosite (www.sonosite.com). Potentially there is scope for future development in the use of ultrasound. From the sales and demonstration material there appears to have been significant improvements in the image quality which may make it easier to measure muscle area but also extend the capability to measure intramuscular fat.

COMPUTER TOMOGRAPHY (CT)

CT scanning is being used by groups in Edinburgh, Scotland and in New Zealand to assess carcase composition predominantly in sheep, but also deer and pigs (Table 23.6). CT scanning is now used routinely in UK terminal sheep sire selection. CT scanning in pigs does lead to higher accuracies as shown in Table 23.7.

Table 23.6 IMPROVED ACCURACIES FROM USING
CT SCANNING IN PIGS (COURTESY OF JSR GENETICS)

Trait	With CT scan	Without CT scan
Muscle	94%	88.3%
Fat	94.0%	89.0%
Bone	86.1%	74.9%

However, the benefit from CT scanning probably lies in utilising the other factors that potentially can flow from its use. In sheep, CT scanning provides good measures of muscularity and carcase composition (Jones *et al.* 2002a;Jones *et al.* 2002b;Jones *et al.* 2004). More recently CT scanning has been used to investigate genetic parameters for carcase composition (Karamichou *et al.* 2007a;Karamichou *et al.* 2007b) with medium heritability estimates. The CT scan has been used to estimate muscle density (as indicated by the relative darkness of the scan) which was consistently related to meat quality traits. "Genetic correlations of muscle density with live weight, fat class, subcutaneous fat score, dry matter proportion, juiciness, flavour and overall liking were all moderately to strongly negative, and significantly different from zero" (Karamichou *et al.* 2006; Table 23.7). This opens up the ability to select for meat quality characteristics using live animals. This provides considerable additionality to the usefulness of CT scanning.

Table 23.7 HERITABILITY IN SHEEP FOR SLAUGHTER AND MEAT QUALITY TRAITS AND GENETIC CORRELATION WITH AVERAGE MUSCLE DENSITY. SOURCE: (KARAMICHOU *et al.* 2006)

Trait	Heritability	Genetic correlation
Cold weight (kg)	0.47	-0.3
Conformation value (units)	0.52	0.38
Subcutaneous fat (g/kg)	0.34	-0.62
Shear force (kg)	0.39	-0.49
Hue angle (0^0 =red, 90^0=yellow)	0.3	-0.31
pH_{45}	0.54	0.3
pH ultimate	0.21	0.41
Intramuscular fat (mg/100g muscle)	0.32	-0.67
Juiciness (units)	0.21	-0.71
Overall liking (units)	0.22	-0.8
Flavour (units)	0.11	-0.73

Selection within breed for meat quality traits

Until recently the Halothane gene has affected meat quality in lines with the mutation, causing PSE meat. Secondly, the Renedemont Napole gene has also had an effect in Hampshire lines. There are examples of breeding companies and national schemes that have undertaken selection to improve meat quality: Danish (http://www.danavl.dk/); Swiss, SUISAG (http://www.suisag.ch/suisag/); Austrian national scheme (http://www.schweine.at) and PIC (Genus PLC). The weighting in an index has generally been small (<10%) with the exception of Austria (Table 23.8).

Table 23.8 EXAMPLES OF WITHIN-LINE SELECTION FOR MEAT QUALITY FROM NATIONAL SCHEMES AND PIC

Organisation or company	Trait or marker	Breeds, % weighting in index
Danish national scheme	Ultimate pH (Stopped ~2003)	sire line, 5; Dam line 5
Austrian national scheme	Colour, pH, electrical conductance	Pietran, 20; Landrace 15
	Intramuscular fat	Pietran, 10; Landrace 15
Swiss national scheme	Intramuscular fat	14 to 21
	pH_{1h} *post mortem* (pm)	7
	Meat reflectance 30h pm	13
PIC	Genetic markers	Unknown

USING pH AS A SELECTION CRITERIA

Selection for an increase in the ultimate pH value was seen as a useful method to improve meat quality. The Halothane gene was viewed as responsible for accelerated pH decline, low ultimate pH, low water holding capacity and high purge(Huff-Lonergan & Lonergan 2005). Since the removal of the Halothane gene and the RN⁻ gene "ultimate pH may not be the ideal indicator of meat quality" (Rosenvold & Andersen 2003). Similarly "recent studies have shown that increasing the ultimate pH in the carcass does not have any effect on the meat quality including the processing yield" (Danish National Committee for Pig Production 2003, p. 8). Rosenvold refers to the paper by Schäfer *et al* (Schaefer *et al.* 2001) who reported that pH24 only accounted for 0.04 of the variation in water holding capacity (WHC) whereas pH_{1h} *post mortem* explained 0.72 of the variation in WHC. However, there is need for cautious interpretation of the data. Firstly the results are phenotypic correlations and not genetic and secondly Schäfer is using an experimental model "exercising immediately prior to stunning on a treadmill till exhaustion" (Schaefer *et al.* 2002) to invoke a range of values which may not represent the range of values in a low stress abattoir environment (It may still represent the range of values found in practice).

A review (Sellier 1998) provides estimates of genetic parameters for meat quality and carcase traits, however it is not always clear whether the populations concerned are affected by the Halothane gene. Subsequent papers have also sometimes been poor at documenting the Halothane status of the animals, together with relatively few reports of genetic parameters (Table 23.9).

Table 23.9 HERITABILITY OF PH MEASURES FROM SELLIER (1998) AND AFTER WITH THEIR HALOTHANE STATUS, IF KNOWN

Trait	Halothane status	Muscle	Average h^2	Author
ph1	Mixed		0.16	(Sellier 1998)
pHu	Mixed		0.21	(Sellier 1998)
pH45m	Unclear	?	0.19, 0.14 & 0.37	(Knapp *et al.* 1997)
pH30mins	Presumably clear	*longissimus*	0.32	(Larzul *et al.* 1997)
pH24h	Presumably clear	*longissimus*	0.13	(Larzul *et al.* 1997)
pHu	Halothane and RN- free	*longissimus*	0.3	(Oksbjerg *et al.* 2004)
pHu	Halothane and RN- free	Ham	0.39	(Oksbjerg *et al.* 2004)
pH45m	Unclear, some Hal prob	*longissimus*	0.15	(Hermesch *et al.* 2000)
pH24h	Unclear, some Hal prob	*longissimus*	0.14	(Hermesch *et al.* 2000)
pHu	Free of Halothane	*longissimus*	0.11	(van Wijk *et al.* 2005)

The halothane free heritability estimates of pH_u range from 0.11 to 0.39 with a pH_{24h} of 0.13 (Larzul *et al.* 1997). Genetic correlations between the various pH measures from (Sellier 1998, derived from table 16.5 & presumably of mixed Halothane status) are shown in Table 23.10.

Table 23.10 GENETIC CORRELATIONS OF MEAT QUALITY TRAITS WITH EARLY AND ULTIMATE pH

Trait	pH_1	pHu
Drip loss	-0.27	-0.71
Water holding capacity	-0.65	0.45
Cooking loss	-0.14	-0.68
Napole yield		0.7
Reflectance	-0.38	-0.53
Tenderness	0.27	0.49
Overall acceptability		0.59

The only genetic correlation from more recent work and known to be free of the Halothane gene is between pH_u with drip, -0.86 and with purge, -0.92 (van Wijk *et al.* 2005) which still indicates the usefulness of utilising pH to improve meat quality.

WATER HOLDING CAPACITY (WHC)

There appears to be heightened cognisant of drip loss, WHC and purge by processors in North America in comparison with Europe. This difference in attitude may relate to a number of factors such as distance to abattoir, extremes in temperature and humidity, older handling facilities, peri-slaughter handling and cooling facilities. Whatever the reason this has led to an increase in measuring pH in North America.

The use of temperature and early pH to predict WHC as suggested by (Schaefer *et al.* 2002) may be worthwhile considering. The genetic variance of ultimate pH is low: 0.005 (Hermesch *et al.* 2000, derived from table 23.6), 0.002 (van Wijk *et al.* 2005, given in table 23.2) limiting possible improvement in ultimate pH and related traits. Genetic variance of pH_{45m} was considerably higher 0.0195 relative to pHu (0.005) (Hermesch *et al.* 2000, derived from table 6) and the standard deviation was higher: pH_{45m}, 0.46 and pH_{24h}, 0.26. Similarly the standard deviation was higher in Figure 2a (Schaefer *et al.* 2002) early post mortem compared with the standard deviation of ultimate pH. Warriss and Brown (1987) reported standard deviations of pH_{45m}, 0.299 and pH_u 0.106 (Warriss & Brown 1987, derived from table 1) and similarly more recently on ACMC animals pH_{45m}, 0.23 and pH_{24h} 0.048 (Hoste, unpublished). pH measurements early post mortem show favourable genetic correlations with meat quality traits and have relatively greater variation compared with the variation in ultimate pH suggesting that early post mortem pH may still be useful to improve meat quality.

COLLECTING MEAT QUALITY DATA

Traditionally it has always been maintained that obtaining slaughter information from sibs of selected animals is too difficult and too late for selection decisions. Given the will to do so collecting individual information from the abattoir is often achievable. Selection also tends to be considered as selecting for some improvement on one tail of the distribution curve for some trait breeding value. However, it is equally possible to consider the removal of animals (such as AI boars) that have a breeding value at the other end of the distribution, thereby removing the worst offenders. Processors are often interested in reducing variation , i.e. increasing uniformity (Heuven *et al.* 2003), although they also may be utilising the variability they have to satisfy differing markets. However, it is still useful to return to consider that the optimum approach depends on the slaughter conditions and type of processing (de Vries *et al.* 1998).

Genomics

Genomics selection offers large benefits particularly for meat and carcass quality as potential breeding animals do not have to be slaughtered to evaluate their genetic merit and results can be known soon after birth. Table 23.11 shows the known candidate genes, their effect and industry usage.

Table 23.11 KNOWN GENES AFFECTING CARCASS AND MEAT QUALITY AND COMPOSITION AND INDUSTRY USAGE

Candidate genes	Traits	Industry use	Author
HAL	meat quality, leanness, stress	Extensive	(Fujii *et al.* 1991)
MC4R	Growth & fatness	Exclusive	(Kim *et al.* 2000)
RN	Meat quality	Extensive	(Milan *et al.* 2000)
PRKAG3	Meat quality	Exclusive	(Ciobanu *et al.* 2001)
AFABP, HFABP	Intra-muscular fat	Unknown	
CAST	Tenderness	Exclusive	(Ciobanu *et al.* 2004)
IGF2	carcass composition	Exclusive	(Jeon *et al.* 1999), (Nezer *et al.* 1999)

The Halothane gene and RN gene are used across breeding companies whilst others are used exclusively by various breeding companies.   The PRKAG3 and CAST alleles have been utilised by PIC and are being used in conjunction with processors and retailers to improve meat quality (Westfleisch and EDEKA in Germany).

Information capture and near market data

There has been increased interest in both obtaining information from lower in a breeding pyramid and also utilising this in selection decisions at nucleus (Casey *et al.* 2006). Slice information (Heuven *et al.* 2003) illustrates a scheme to have a portion of the pyramid individually recorded all the way through to slaughter. Ideally nucleus boars can be used on commercial sow farms to obtain sire information at this level. Capturing near market information can obviously go beyond the abattoir to the retailer. Using a market research approach of conjoint analysis would allow an alternative approach to finding what customers want. Consumers are asked to assign preferences to products which differ in key attributes of interest. Consumers can be tested on multiple sets of competing variants of the same type of product. Analysis of the results reveals the implicit relative ranking or preference placed on each key attribute of interest (Amer 2006). The big advantage is that such an analysis reveals the tacit or intangible aspects that consumers are interested in rather than our perception or perspective of what they are interested in. Such analyses are common in consumer market research and could be a useful method to help with meat quality selection. Such a use has been described in the pricing of colour in Salmon (Alfnes *et al.* 2006).

Conclusion

The current state of affairs with regard to selection for carcass quality is mixed. For many organisations meat quality does not appear high on the agenda. There is an on-going concern of the lack of market signals and reward for producing meat of high quality. This is a whole chain problem as quality is affected by the breeding company, producer and processor. Building trust and collaboration and understanding of each other is vital if we are to better utilise the resources we have. Within breed selection for carcass quality could be achieved with a greater degree of involvement by both the breeding and processing companies. The opportunities range from greater use of pH measurements to utilising CT scanning in live animals to predict meat quality. In the first case this is eminently possible and in the second potentially interesting. Utilising genomic information is also available now. Linking information (and importantly also people involved in the meat chain) through a pyramid slice and also market research type analyses could build a coherent and rewarding link for all players.

Reference

Alfnes, F., Guttormsen, A. G., Steine, G., & Kolstad, K. (2006). Consumers' willingness to pay for the color of salmon: A choice experiment with real economic incentives. *American Journal Of Agricultural Economics* **88**(4), 1050-1061.

Amer, P. R. (2006). Approaches to formulating breeding values. Belo Horizonte, Mg Brazil.

Casey, D., Perez, M., Mclaren, D., & Short, T. (2006). Crossbreeding values: Selecting for commercial performance in pigs. Belo Horizonte, Mg Brazil.

Christopher, M. & Peck, H. (2003). *Marketing Logistics*. Oxford: Elsevier Butterworth-Heinemann.

Ciobanu, D. C., Bastiaansen, J. W. M., Lonergan, S. M., Thomsen, H., Dekkers, J. C. M., Plastow, G. S., & Rothschild, M. F. (2004). New alleles in calpastatin gene are associated with meat quality traits in pigs. *Journal Of Animal Science* **82**(10), 2829-2839.

Ciobanu, D., Bastiaansen, J., Malek, M., Helm, J., Woollard, J., Plastow, G., & Rothschild, M. (2001). Evidence for new alleles in the protein kinase adenosine monophosphate-activated {gamma}3-subunit gene associated with low glycogen content in pig skeletal muscle and improved meat quality. *Genetics* **159**(3), 1151-1162.

Danish National Committee For Pig Production. The National Committee For Pig Production Annual Report. 2003. Copenhagen. Ref Type: Report

De Vries, A. G., Sosnicki, A., Garnier, J. P., & Plastow, G. S. (1998). The role of major genes and DNA technology in selection for meat quality in pigs. *Meat Science* **49**(Supplement 1), S245-S255.

Fujii, J., Otsu, K., Zorzato, F., Leon, S. D., Khanna, V. K., Weiler, J. E., O'brien, P. J., & Maclennan, D. H. (1991). Identification of a mutation in porcine ryanodine receptor associated with malignant hyperthermia. *Science* **253**(5018), 448-451.

Greer, E. B., Mort, P. C., Lowe, T. W., & Giles, L. R. (1987). Accuracy of ultrasonic backfat testers in predicting carcass P2 fat depth from live pig measurement and the effect on accuracy of mislocating the P2 site on the live pig. *Australian Journal Of Experimental Agriculture* **27**(1), 27-34.

Grunert, K. G. (2006). Future trends and consumer lifestyles with regard to meat consumption. *Meat Science* **74**(1), 149-160.

Hermesch, S., Luxford, B. G., & Graser, H.-U. (2000). Genetic parameters for lean meat yield, meat quality, reproduction and feed efficiency traits for Australian pigs: 1. Description of traits and heritability estimates. *Livestock Production Science* **65**(3), 239-248.

Heuven, H. C. M., Van Wijk, H. J., & Van Arendonk, J. A. M. (2003). Combining traditional breeding and genomics to improve pork quality. *Outlook On Agriculture* **32**, 235-239.

Huff-Lonergan, E. & Lonergan, S. M. (2005). Mechanisms of water-holding capacity of meat: the role of postmortem biochemical and structural changes. *Meat Science* **71**(1), 194-204.

Jeon, J. T., Carlborg, O., Tornsten, A., Giuffra, E., Amarger, V., Chardon, P., Ndersson-Eklund, L., Andersson, K., Hansson, I., Lundstrom, K., & Andersson, L. (1999). A paternally expressed qtl affecting skeletal and cardiac muscle mass in pigs maps to the Igf2 locus. *Nat Genet* **21**(2), 157-158.

Jones, H. E., Lewis, R. M., Young, M. J., & Simm, G. (2004). Genetic parameters for carcass composition and muscularity in sheep measured by X-ray computer tomography, ultrasound and dissection. *Livestock Production Science* **90**(2-3), 167-179.

Jones, H. E., Lewis, R. M., Young, M. J., & Wolf, B. T. (2002a). The use of X-ray computer tomography for measuring the muscularity of live sheep. *Animal Science* **75**, 387-399.

Jones, H. E., Lewis, R. M., Young, M. J., Wolf, B. T., & Warkup, C. C. (2002b). Changes in muscularity with growth and its relationship with other carcass traits in three terminal sire breeds of sheep. *Animal Science* **74**, 265-275.

Jonsall, A., Johansson, L., & Lundstrom, K. (2001). Sensory quality and cooking loss of ham muscle (M. Biceps Femoris) from pigs reared indoors and outdoors. *Meat Science* **57**(3), 245-250.

Karamichou, E., Richardson, R. I., Nute, G. R., Mclean, K. A., & Bishop, S. C. (2006). Genetic analyses of carcass composition, as assessed by X-ray computer tomography, and meat quality traits in Scottish Blackface sheep. *Animal Science* **82**, 151-162.

Karamichou, E., Richardson, R. I., Nute, G. R., Mclean, K. A., & Bishop, S. C. (2007a). A partial genome scan to map quantitative trait loci for carcass composition, as assessed by X-ray computer tomography, and meat quality traits in Scottish Blackface sheep. *Animal Science* **82**(03), 301-309.

Karamichou, E., Richardson, R. I., Nute, G. R., Mclean, K. A., & Bishop, S. C. (2007b). Genetic analyses of carcass composition, as assessed by X-ray computer tomography, and meat quality traits in Scottish Blackface sheep. *Animal Science* **82**(02), 151-162.

Kim, K. S., Larsen, N., Short, T., Plastow, G., & Rothschild, M. F. (2000). A missense variant of the porcine melanocortin-4 receptor (Mc4r) gene is associated with fatness, growth, and feed intake traits. *Mammalian Genome* **11**(2), 131-135.

Knapp, P., Willam, A., & Solkner, J. (1997). Genetic parameters for lean meat content and meat quality traits in different pig breeds. *Livestock Production Science* **52**(1), 69-73.

Larzul, C., Lefaucheur, L., Ecolan, P., Gogue, J., Talmant, A., Sellier, P., Le Roy, P., & Monin, G. (1997). Phenotypic and genetic parameters for longissimus muscle fiber characteristics in relation to growth, carcass, and meat quality traits in large white pigs. *Journal Of Animal Science* **75**(12), 3126-3137.

Le Roy, P., Elsen, J.-M., Caritez, J. C., Talmant, A., Juin, H., Sellier, P., & Monin, G. (2000). Comparison between the three porcine RN genotypes for growth, carcass composition and meat quality traits. *Genetics Selection Evolution* **32**, 165-186.

Lundstrom, K., Andersson, A., & Hansson, I. (1996). Effect of the RN gene on technological and sensory meat quality in crossbred pigs with Hampshire as terminal sire. *Meat Science* **42**(2), 145-153.

Magowan, E. & Mccann, M. E. E. (2006). A comparison of pig backfat measurements using ultrasonic and optical instruments. *Livestock Science* **103**(1-2), 116-123.

Milan, D., Jeon, J. T., Looft, C., Amarger, V., Robic, A., Thelander, M., Rogel-Gaillard, C., Paul, S., Iannuccelli, N., Rask, L., Ronne, H., Lundstrom, K., Reinsch, N., Gellin, J., Kalm, E., Roy, P. L., Chardon, P., & Andersson, L. (2000). A mutation in PRKAG3 associated with excess glycogen content in pig skeletal muscle. *Science* **288**(5469), 1248-1251.

Moeller, S. J. (2001). Evolution and use of ultrasonic technology in the swine industry.

Monin, G. (2003). Genomics: improving qualitative characteristics and value of meat from pigs. *Outlook On Agriculture* **32**, 227-233.

Nezer, C., Moreau, L., Brouwers, B., Coppieters, W., Detilleux, J., Hanset, R., Karim, L., Kvasz, A., Leroy, P., & Georges, M. (1999). An imprinted QTL with major effect on muscle mass and fat deposition maps to the IGF2 locus in pigs. *Nat Genet* **21**(2), 155-156.

Oksbjerg, N., Henckel, P., Andersen, S., Pedersen, B., & Nielsen, B. (2004). Genetic variation of *in vivo* muscle glycerol, glycogen, and pigment in Danish purebred pigs. *Acta Agriculturae Scandinavica, Section A - Animal Sciences* **54**(4), 187-192.

Padbury, G. (2006). *Retail And Foodservice Opportunities For Local Food*. Institute Of Grocery Distribution.

Rosenvold, K. & Andersen, H. J. (2003). Factors of significance for pork quality—a review. *Meat Science* **64**(3), 219-237.

Schaefer, A., Rosenvold, K., Purslow, P. P., Andersen, H. J., & Henckel, P. (2001). Critical post mortem pH and temperature values in relation to drip loss in pork. Pp. 206-207. Krakow, Poland.

Schaefer, A., Rosenvold, K., Purslow, P. P., Andersen, H. J., & Henckel, P. (2002). Physiological and structural events post mortem of importance for drip loss in pork. *Meat Science* **61**(4), 355-366.

Sellier, P. (1998). Genetics of meat and carcass traits. In *The Genetics Of The Pig* (Eds M. F. Rothschild & A. Ruvinsky), P. 463. Wallingford, Uk: Cab International.

Stouffer, J. R. (2004). History of ultrasound in animal science. *Journal Of Ultrasound In Medicine* **23**(5), 577-584.

Van Wijk, H. J., Arts, D. J. G., Matthews, J. O., Webster, M., Ducro, B. J., & Knol, E. F. (2005). Genetic parameters for carcass composition and pork quality estimated in a commercial production chain. *Journal Of Animal Science* **83**(2), 324-333.

Warriss, P. D. & Brown, S. N. (1987). The relationships between initial pH, reflectance and exudation in pig muscle. *Meat Science* **20**(1), 65-74.

Wood, J. D. & Lutwyche, R. H. L. (2007). Niche market opportunities for rare livestock breeds. P. 254.

Wood, J. D., Nute, G. R., Richardson, R. I., Whittington, F. M., Southwood,

O., Plastow, G., Mansbridge, R., Da Costa, N., & Chang, K. C. (2004). Effects of breed, diet and muscle on fat deposition and eating quality in pigs. *Meat Science* **67**(4), 651-667.

24

FARMINGNET, THE BEST WAY TO A HIGHER YIELD

HAROLD THEUNISSEN
Chain Manager Farming, VION Food Group, The Netherlands

The structure of the VION Food Group is shown in Figure 24.1.

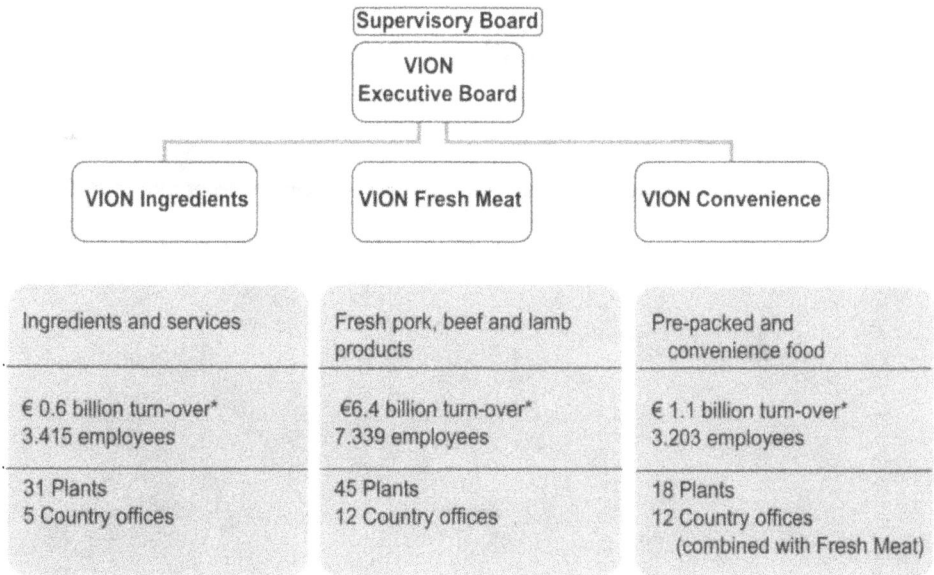

Figure 24.1 Schematic diagram of the VION Food Group

VION's supply organization for pigs in the Netherlands is based on 75 employees and 5 Field offices. Annually, it deals with 8.4 million slaughter pigs of which 4.4 million and handled directly and 4.0 million indirectly. It also handles 2.8 million piglets and110000 cull sows, and has 3400 farmers supplying animals.

VION's marketing approach is based on passion for better food. The overall yield of its operation is centered around carcass value, concept value and management value. The Farmingnet-system is internet-based and an important tool for VION Yield Approach. There is detailed feedback of slaughter data that includes classification, weight, concept value, health of the pigs, classification of males and females, and feedback on gut contents. All these data are analysed and presented in subsequent tables and figures.

A typical table for a middle man from a number of farms is presented in figure 24.2.

Vorige 1-15 van 57 ▼ Volgende 15																			
CLASSIFICATIE							SALDO tov BASIS(ct/kg)						SLACHTBEVINDINGEN						
aantal varkens	gem gewicht	% vlees	spekdikte	spierdikte	%type AA	%type A	%type B	% concept	gew korting	vlees toeslag	type toeslag	concept toeslag	TOTAAL	% gezond	% pleuritis	% Lever	% Longen	gezondheid ct/kg	% dpv
	80,9	56,3	15,3	51,3	2	86	73	16	-1,1	-1,0	-0,6	0,0	-8,0	91	2	1	2	0,3	57
	90,1	56,8	15,9	58,5	14	77	9	53	-1,0	-1,4	-0,2	2,8	0,1	74	13	4	14	-0,7	44
50	91,6	56,7	16,7	62,1	14	76	10	60	-0,6	-1,8	-0,2	3,1	0,5	30	68	6	22	-2,8	
71	93,0	56,6	16,4	59,8	8	87	4	52	-0,8	-1,7	0,0	2,7	0,1	90	7	1	4	0,1	
80	84,2	56,8	14,9	53,2	4	81	15	34	-4,1	-1,1	-0,7	1,9	-3,9	43	36	2	41	-3,2	
138	90,2	55,8	16,6	54,9	4	85	12	39	-0,5	-1,6	-0,5	2,1	-0,7	52	35	8	28	-2,2	
100	89,9	57,0	15,8	59,8	6	79	16	67	-0,6	-2,2	-0,7	3,5	0,0	63	17	7	35	-2,1	
70	87,1	55,8	17,1	58,0	4	79	17	36	-0,9	-2,2	-0,8	1,9	-2,1	67	23	2	10	-1,0	
101	89,1	57,1	15,6	58,9	9	80	11	60	-1,0	-1,5	-0,4	3,2	0,0	80	2	4	11	-0,3	
115	90,3	57,3	15,5	60,6	4	93	3	70	-0,8	-0,8	0,0	3,7	2,0	88	8	6	5	0,1	
92	90,9	56,4	16,5	59,3	13	78	9	55	-0,7	-2,3	-0,1	2,9	-0,3	83	8	0	11	-0,3	
60	90,2	56,5	16,5	60,1	8	82	10	47	-1,3	-2,5	-0,4	2,5	-1,9	90	2	0	7	0,2	
56	86,4	58,7	13,3	57,9	11	84	5	54	-4,4	-0,7	0,0	2,9	-2,1	84	9	7	9	-0,2	
	79,7	57,8	13,4	52,3	5	91	4	80	-1,7	-0,2	-0,1	4,8	2,8	74	10	6	13	-0,7	35
57	80,5	58,0	13,2	52,4	6	95	0	93	-0,8	-0,3	0,1	5,5	4,6	74	11	7	14	-0,5	
Vorige 1-15 van 57 ▼ Volgende 15																			

Figure 24.2 Daily supply data for the company. In this screen, the average classification and health status per delivy per supplier is shown in the rows. Successively, the number of pigs, the average slaughter weight, the average lean meat content, the average muscle depth, the average fat depth, the rating of the carcasses, the percentage of pigs, who meet the specifications of the concept, the average deduction for weight, lean meat content and rating of the carcasses, the concept bonus, the total yield, the percentage of healthy pigs, the percentage of pigs with pleuritis, liver damage, lung damage, the health yield of the delivery and the average filling of the gut is represented. Red numbers indicate deviant values. The green rows show the average values for the concept.

Data feedback to farmers is presented in figure 24.3.

Leverdatum:							03-01-2006	Slachtplaats:				VION HELMOND BV		
Aantal geleverd:							179	Gemiddeld gewicht:				90,1		
Aantal geslacht:							179	Gemiddeld vlees %:				56,8		
Aantal binnen kwaliteitsconcept:							104	Gemiddelde spierdikte:				59,1		
Percentage binnen kwaliteitsconcept:							58,1	Gemiddelde spekdikte:				16,2		
Aantal overliggers:							0	Aantal geclassificeerd				179		

Kg	Aantal	%	VI%	Spek	Spier		Type	Aantal	%	Vlees	Aantal	%		Afdeling	Aantal	Kg
<70	2	1,1	59,8	11,6	55,2		AA	13	7,3	<54	15	8,4		1623	55	87,0
70-75	9	5,0	58,4	13,1	53,8		A	151	84,4	54-58	115	64,2		1613	38	87,7
76-79	5	2,8	59,8	11,4	53,7		B	15	8,4	>58	49	27,4		1513	43	93,0
80-85	23	12,8	58,1	14,1	56,8		C							1023	43	93,1
86-92	67	37,4	56,3	16,8	59,1											
93-98	57	31,8	56,3	17,2	60,8											
99-103	14	7,8	56,8	16,6	61,6											
104-110	2	1,1	55,0	18,4	59,2											
>110																
Gem. 90.1	179		56,8	16,2	59,1											

Saldo t.o.v. basis (ct/kg)

Vlees %	Type	Gewicht	Concept	Sekse	Totaal
-1,3	-0,3	-1,1	3,1	-0,3	0,1

Figure 24.3 Slaughter information summary for the farmer. Data describe the delivery date, the number of delivered pigs, the number of slaughtered pigs, the number of pigs that meet the concept specifications, the number of pigs that stayed overnight in the lairage (first column), the slaughter location, the average weight, the average lean meat content, the average muscle depth, the average fat depth and the number of classified pigs (second column) is shown. In the tables in the lower part of the screen, the segmentation of the slaughter weight (with, for each weight category, the number of pigs, the percentage of the total, the average lean meat content, the average fat and muscle depth) is shown on the left, in the second left table the rating of the carcasses is shown, in the middle right table the segmentation for lean meat content is shown, in the far right table the number of pigs and average weight per compartment is shown. In the bottom table, the composition of the yield is shown.

An example the management information on classification that can be passed back to producers is presented in figure 24.4.

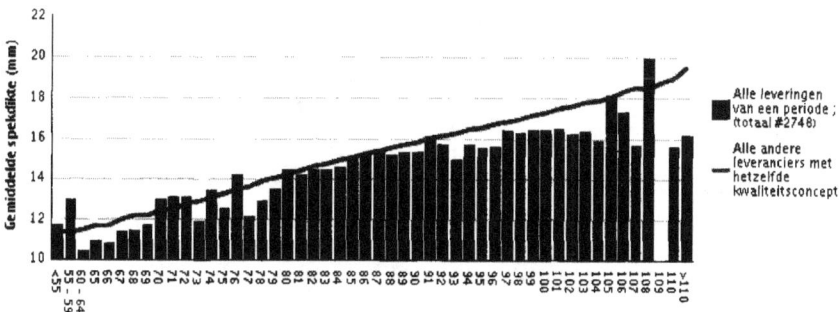

Figure 24.4 Relationship between slaughter weight and fat depth, compared to the average results of the total group. On the x-axis the weight is represented, on the y-axis the fat depth. The columns represent the pigs delivered by the specified supplier for a set period, the line represents the VION average for this period

As described elsewhere in these proceedings (e.g. Chapter 3) herd health is a crucial aspect of pig production. Accordingly it is very valuable for producers to be given information on health status. An example is given in Figure 24.5.

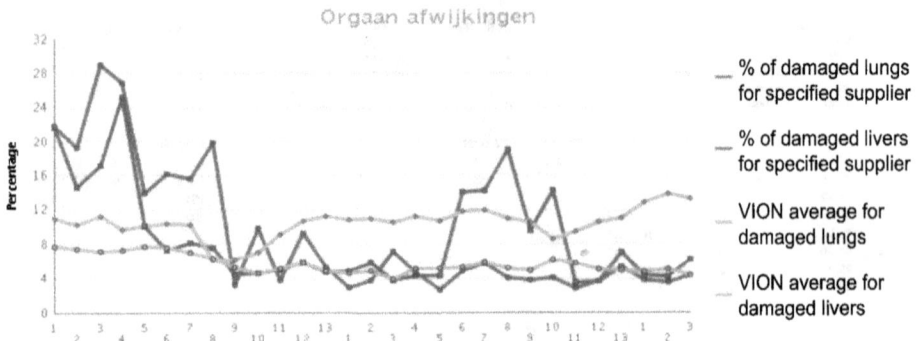

Figure 24.5 Developments in herd health per period of four weeks.

The success of the system can be seen in figure 24.6 that presents the use of Farmingnet, expressed as the number of pigs (% of total supply).

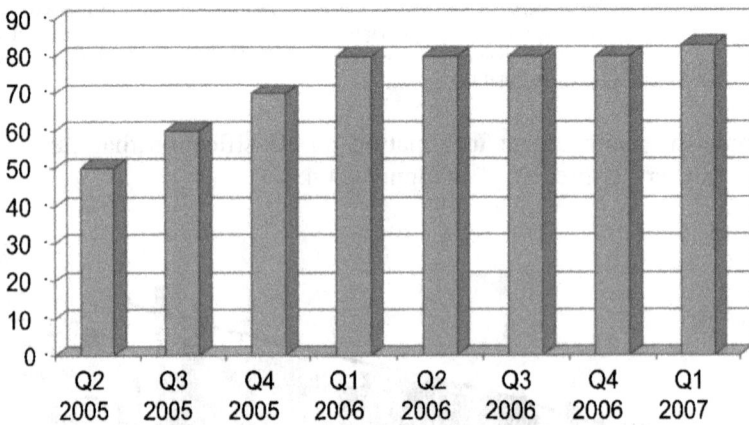

Figure 24.6 Development of participation in Farminget (% of total supply to VION)

Observations on the applicability of the system are that farmers and middle men need training on how to work with the application. Of crucial importance is that information should be quickly available and reliable. This will lead to effective information exchange within the chain. In daily practice, the system is a valuable tool to communicate with the industry, to include feed supplies and veterinarians.

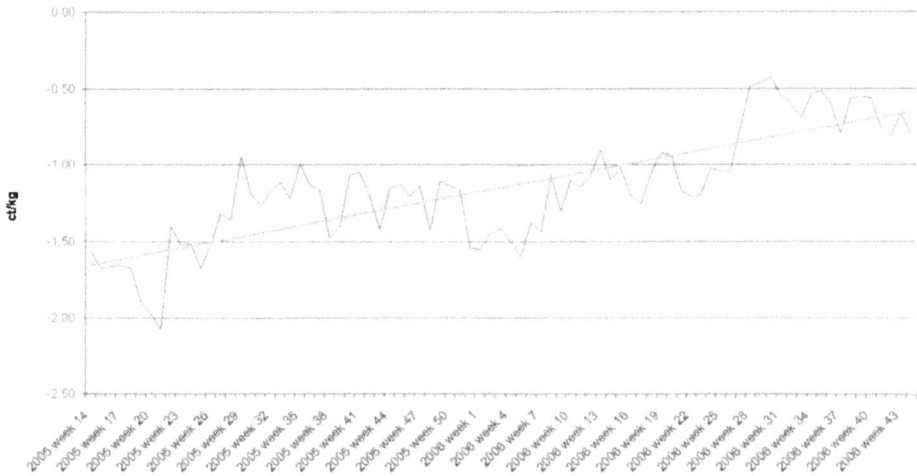

Figure 24.7 Average yield development to base over time. The graph shows that the average yield has increased by € 0.01 per kg. slaughter weight since the introduction of Farmingnet.

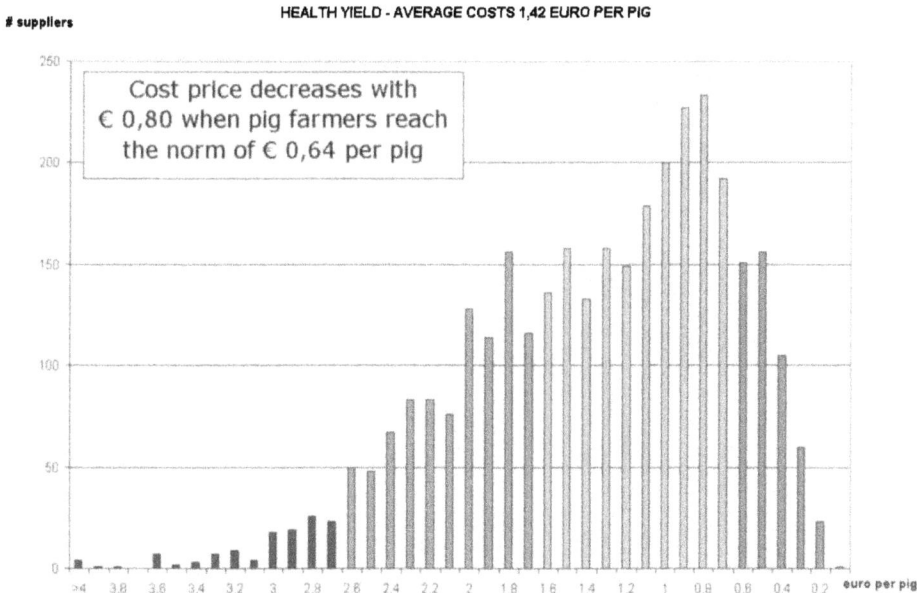

Figure 24.8 Room for improvement on health status. This graph shows the variation in health yield for the suppliers to VION. When all suppliers reach the average of € 0.64 of health yield per pig, the average cost price for VION suppliers decreases by € 0.80.

Finally, it is important to consider the future developments of Farmingnet. These include electronic planning of slaughter pigs, food safety information, digital settlement reports and information exchange tools to chain partners.

25

COMMODITY FUTURES

H. R. BURTON
Raw Material Manager, ABN, ABN House, Oundle Road, Peterborough, PE2 9PW, UK

Introduction

Initially, commodities in the context of this chapter will be defined as those agricultural commodities that are commonly associated with the feeding of pigs. The outlook detailed reflects the factors prevailing at the time (March 2007) but as will be discussed, commodity markets can change very quickly.

Commodity markets historically have been influenced primarily by supply and demand. The volatility of prices has attracted speculative investors' money into commodity futures exchanges for many years. More recently commodity indices such as the Goldman Sachs and Commodity Research Bureau have become popular with investors. These Indices have a portfolio of commodity investments ranging from crude oil to metals to corn.

The heavy level of investment in the commodity futures markets tends to distort pricing and cause confusion around the long term direction of markets for those whose livelihoods are more directly linked to individual or small baskets of commodities. Price risk management has become an important feature in the thinking of successful pig producers globally in relation to feed costs.

This comes at a time when the world has also become very concerned about the generation of greenhouse gases and using renewable energy sources. Thus in addition to the continually growing demand for commodities for direct human consumption and livestock production, there is new legislation being implemented across the world, that is creating a huge demand to use commodities as a raw material for energy production. The implication of the alternative use for energy production is that the growth of global crop production must start to increase more rapidly to keep pace with demand.

Figure 25.1, that shows Chicago futures market soya meal prices 2004-2007 ranging from $150-330 per short ton, is a good example of volatile prices and hence the price risk to which pig producers are exposed.

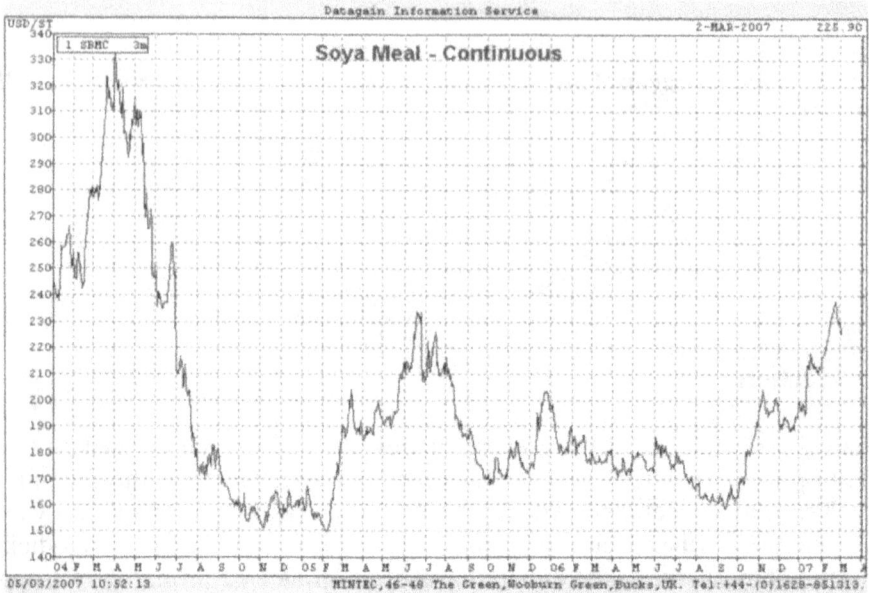

Figure 25.1 Chicago futures market soya meal prices 2004-2007

Supply

Global production of , wheat and soya continue to grow but there are some changes developing in terms of what is grown where.

CEREALS

World wheat production has been largely static over the last ten years. Figure 25.2 shows that there was a steady decline in wheat production between 1997/98 and 2003/04 with the highest ever global production of 628.84 million tonnes in 2004/05. Figure 25.3 illustrates the trends in major wheat producing countries that have a trend that can be identified. In China and the United States the volume of wheat as a proportion of world production appears to be declining where as in the E.U. it appears to be increasing. In China it is widely reported that there is a definite trend away from growing commodity crops; farmers are instead preferring to grow fruit and vegetable crops for which they can more readily get paid in cash. This is illustrated in Figure 25.4.

World maize production was quite stable between 1997/98 and 2003/04 but has increased by around 100 million tonnes in the last three years (Figure 25.5). Further increases are expected with additional demand from the bio-fuel sector.

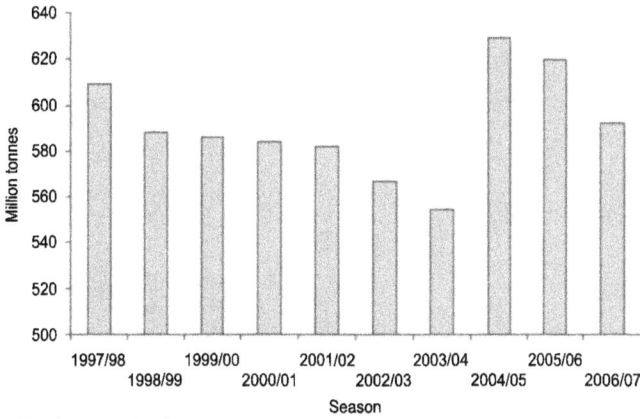

Figure 25.2 World wheat production

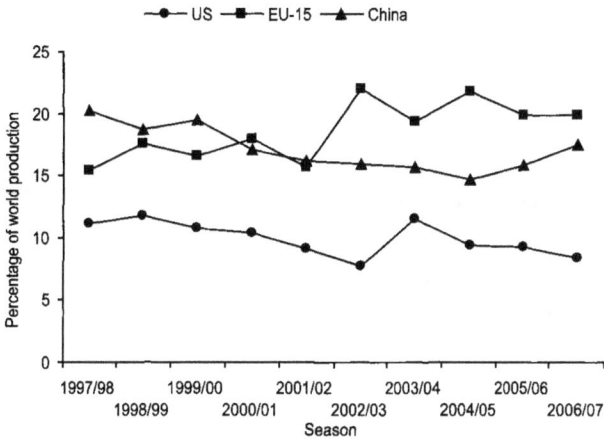

Figure 25.3 Percentage of world wheat production

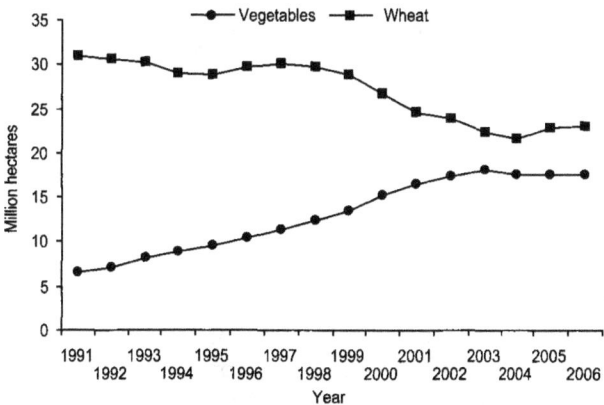

Figure 25.4 China planted area wheat *vs* vegetables

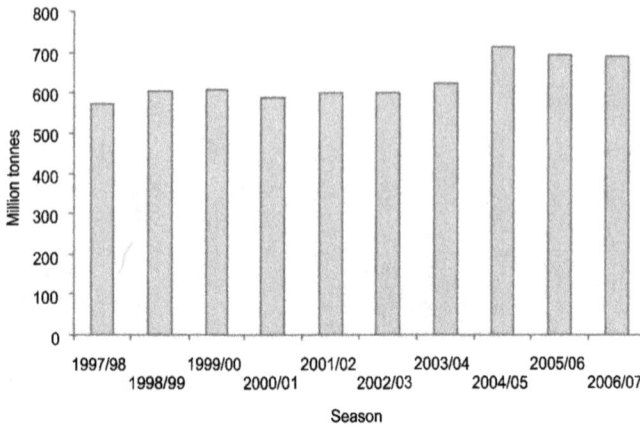

Figure 25.5 World corn production

The countries growing maize have not changed significantly in the last ten years with the United States being by far the largest producer at 0.40 of the World Production followed by China at 0.20. It looks very likely that the United States will be even more dominant in terms of world maize production in the future. This is explored further in relation to bio-fuels but Figures 25.6 and 25.7 show a projected change in U.S. planted acreage between 2006 and 2007.

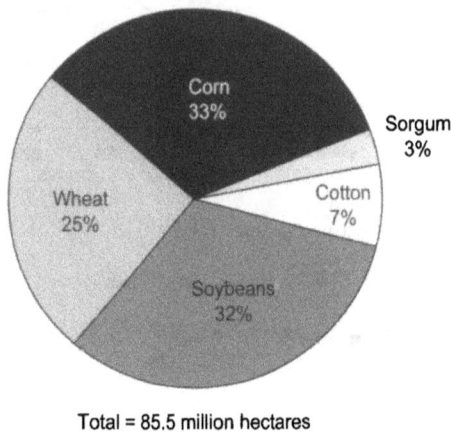

Total = 85.5 million hectares

Figure 25.6 US planted area 2006

SOYA

Global soya production has increased from 158 million tonnes in the 1997/98 season to a forecast of 227 million tonnes for 2006/07 (Figure 25.8). Figure 25.9 shows that there has been a significant increase in the proportion of soya grown in South America in the last ten years. Brazil has nearly doubled its production in that time and now equates to 0.25 of world production.

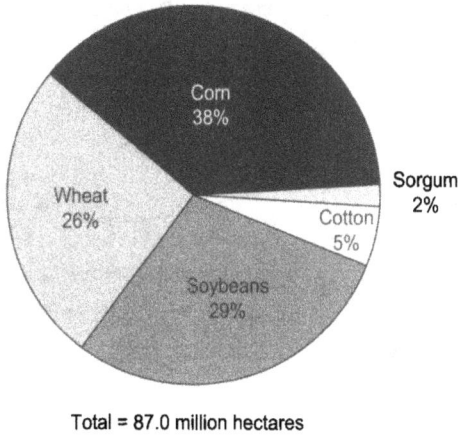

Total = 87.0 million hectares

Figure 25.7 US planted area 2007 projection

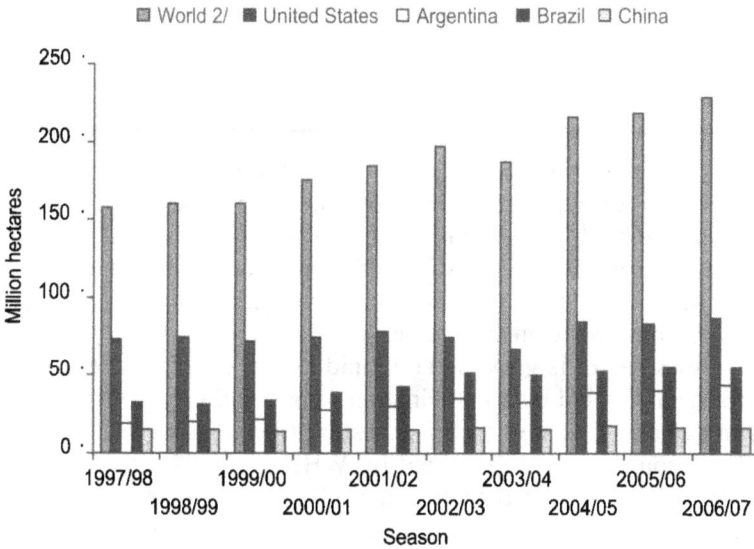

Figure 25.8 World soya production

Argentina has more than doubled production and now accounts for 0.19 of the global figures. Whilst production in the United States has increased, especially in the last three years, their proportion of global production has declined. The picture in China is similar to the United States albeit on a smaller scale. It is expected that there will be an acreage shift away from soya into maize in the United States for the 2007/08 season and that this trend may continue for the next few years.

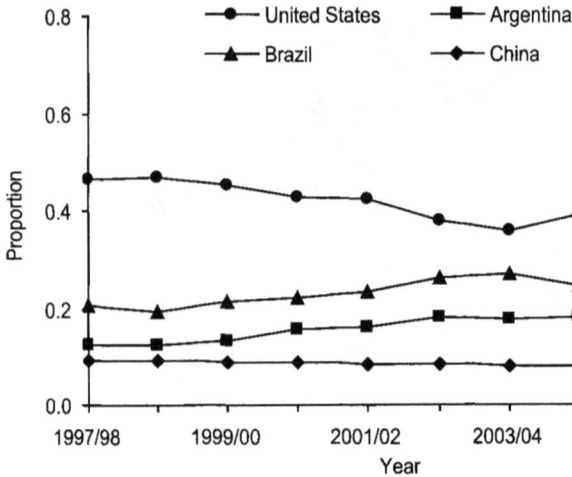

Figure 25.9 Soya production: proportion of world total

Demand

POPULATION

The world's population is increasing by around 80 million people every year.

Demand for food and feed will continue to grow as population expands and diets change with increased wealth. In Asia, in particular, people have tended to move from the countryside to the cities and raised their requirements for living standards. This combined with higher incomes obtained from industrial jobs has led to more consumption of meat and diary products.

The livestock sector is growing at a rapid rate in order to meet this increased demand for meat created by a combination of population growth, rising incomes and urbanization. The per capita consumption of livestock products for industrialised countries was estimated by WHO to be 88.2 kg per year for the period 1997-1999 compared to less than 10 kg per year in South and East Asia. Figure 25.10 makes some very useful comparisons between meat consumption in different countries as of 2006. If one was to make the broad assumption that the consumption in Hong Kong is where China might be some time in the future then it could be concluded that demand for meat in China will more than double. It then becomes no surprise that world annual meat production is forecast to increase from 218 million tonnes in 1997-1999 to 376 million tonnes by 2030.

Demand for food and feed is expected to double in the next 25 years if these trends continue. In China the demand for food is believed to be growing at 9-10% per year. The increased demand for pork is particularly strong (see Figure 25.11). There is some debate about whether China's production capacity is able to keep pace with this rapid increase in food demand. Certainly imports

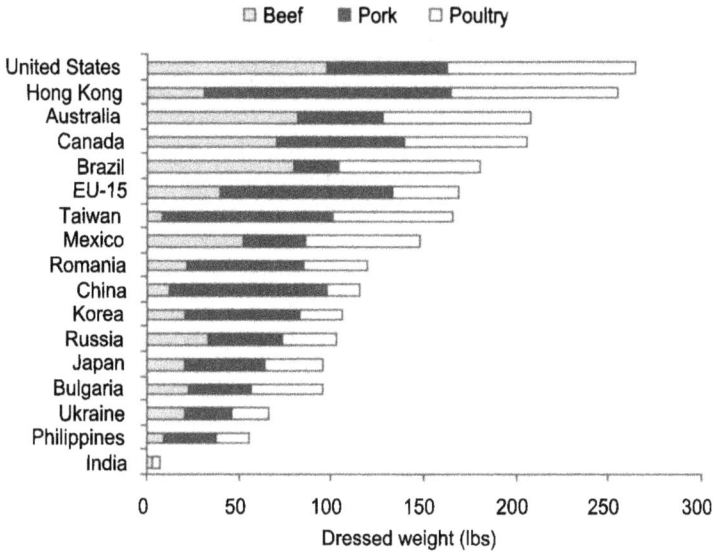

Figure 25.10 2006 per capita protein consumption by country

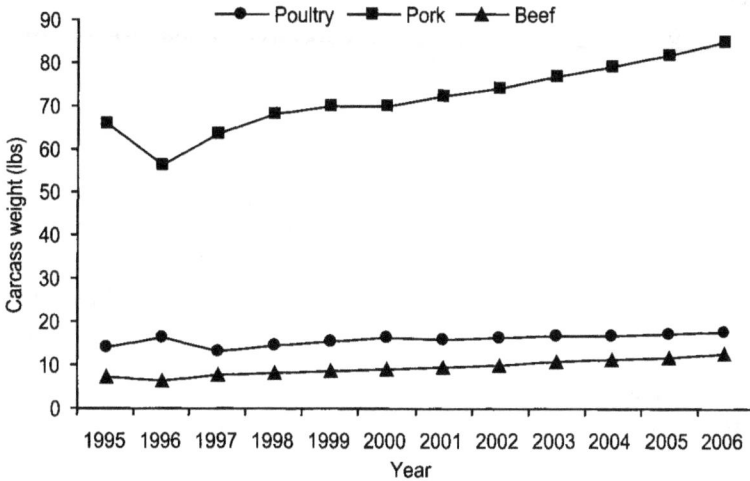

Figure 25.11 China per capital consumtpion of meat

of soya beans into China have increased dramatically in the last 10 years from 2.94 million tonnes in 1997/98 season to a projected 32 million tonnes in 2006/07. China is currently a small exporter of maize, shipping out 4 million tonnes per annum. With internal demand increasing it could be that China may become a net importer of maize within a few years.

BIO-FUELS

For much the same reasons as with food, demand for energy has also increased, but at the same time the world has become increasingly aware of the need to reduce greenhouse gas emissions.

The Kyoto Protocol, which came into force in February 2005, assigns mandatory targets for the reduction of greenhouse gas emissions to those nations that are signatories to the United Nations framework convention on climate change. Kyoto has influenced participating countries to set targets on the use of bio-fuels. Some of the main bio-fuels being considered can be produced from crops currently grown for use in human foods and animal feeds. The main bio-fuels being discussed at this point in time are biodiesel and bio-ethanol.

Biodiesel

Biodiesel is a fuel derived from vegetable oil, animal fat or used cooking oils that can be used in unmodified diesel engines. Biodiesel is biodegradable and typically produces 60% less net carbon dioxide emissions than petroleum based diesel. Whilst biodiesel can be produced from a wide variety of vegetable oils, the main crops being considered commercially appear to be soya beans, rapeseed and palm. In the United States soya bean oil is typically used where as in Canada and the E.U. rapeseed oil predominates.

About 3.4 kg of oil are required to produce one gallon of biodiesel. The raw oil is processed by combining with alcohol to produce biodiesel (mono-alkyl ester) and crude glycerol. The by-product glycerol (also known as glycerine) constitutes about 0.10 of the output.

The National Biodiesel Board in the U.S. reported that, at the end of January 2007 there were 105 active biodiesel plants with an annual production capacity of 864 million gallons. A further 77 new plants were under construction with 8 existing plants being expanded. If all this expansion were completed then it would add a further 1.7 billion gallons per year of biodiesel production capacity.

If all the existing plants were at capacity they could use 2.94 million tonnes of oil. If this all were to come from soya oil then it equates to 0.32 of 2006/07 forecast U.S. production. Soya beans typically yield around 190g oil/kg so it would take 15.5 million tonnes of soya beans to produce the required oil or 0.18 of the 2006/07 production.

The additional plants under construction/expansion are scheduled to be completed by the end of 2008. If all these plants were at maximum capacity by late 2008 and assuming that they all used soya oil then a total of 46 million tonnes of soya beans would be required for biodiesel production alone. It does not follow that all the biodiesel will come from soya oil so this is very much theoretical maximum. Additionally, whilst there are U.S. tax incentives to produce biodiesel, unlike bio-ethanol in the U.S. there is not currently any mandatory minimum usage for biodiesel so it will need to compete economically to get any where near this level of production. The American Soybean Association reported that monthly soya bean oil used in Biodiesel production peaked at 95,710 tonnes in August 2006 and is now declining due to eroding margins.

Figure 25.12, courtesy of Rosenthal Collins Group, illustrates the profitability table for biodiesel derived from soya oil and shows that during a snapshot during February 2007 it was unprofitable.

We could extrapolate an annual U.S. usage to around one million tonnes, currently, and with the plant expansion that is going on this could possibly double to two million tonnes by 2008. This would mean that the volume of soya beans required would also double from 5.3 to 10.6 million tonnes. It is not possible to be precise with these figures but unless the U.S. is able to increase significantly production of soya then there will be less soya beans available for export, although more soya meal.

Brazil may also export less soya beans if their voluntary 0.02 biodiesel blending becomes mandatory in 2008 as has been suggested by the Association of Soybean Producers of Matto Grosso. It is believed that Brazil plans to replace 0.05 of diesel usage with biodiesel by 2013. One estimate suggests that Brazil would need to use over 2 million tonnes of vegetable oil per annum in order to do this. Undoubtabely this would substantially reduce Brazil's soya oil exports also which currently stand at around this figure. Argentina is reported to have adopted a 0.02 biodiesel mandate with plans to increase to 0.05. Current export tax structure favors the exports of biodiesel rather than soya oil.

Malaysia and Indonesia are also reported to be making significant investments in the biodiesel sector. Malaysia may have 5 million tonnes of capacity by the end of 2007 with Indonesia about one year behind them. It is suggested that this production will be exported to the E.U. and U.S. This would reduce the volume of palm oil available for other countries to make biodiesel.

The E.U. directive 2003/30/EC on the promotion of the use of biofuels or other renewable fuels for transport has stimulated a significant increase in biodiesel production. The EU25 produced just over 3.1 million tonnes in 2005 with Germany being by far the largest producer at just over 1.6 million tonnes. According to the European Biodiesel Board the 2006 EU25 capacity was around 6 million tonnes.

The directive states that member states should ensure that a minimum proportion of biofuels and other renewable fuel is placed on their markets. The reference values for biofuel targets given were on an energy basis as 0.02 of petrol and diesel by 31st December 2005 and 0.0575 by 31st December 2010. As the energy value of biodiesel is lower than diesel, a volume inclusion of 0.06-0.065 biodiesel will be needed by 2010.

In order to meet the targets by 2010 the E.U. will need use at least 18 million tonnes of biofuels. The proportion of this total that biodiesel will form is not defined but if 0.50 is assumed then biodiesel production would need to

Theoretical US Bio-diesel Margin Grid - Spot.

NY Harbor WHLSL Diesel	CBOT Soybean Oil Futures - cents per lb										
	22.45	24.45	26.45	28.45	30.45	32.45	34.45	36.45	38.45	40.45	42.45
$1.04	-$0.32	-$0.47	-$0.62	-$0.77	-$0.92	-$1.07	-$1.22	-$1.37	-$1.53	-$1.68	-$1.79
$1.14	-$0.22	-$0.37	-$0.52	-$0.67	-$0.82	-$0.97	-$1.12	-$1.27	-$1.43	-$1.58	-$1.69
$1.24	-$0.12	-$0.27	-$0.42	-$0.57	-$0.72	-$0.87	-$1.02	-$1.17	-$1.33	-$1.48	-$1.59
$1.34	-$0.02	-$0.17	-$0.32	-$0.47	-$0.62	-$0.77	-$0.92	-$1.07	-$1.23	-$1.38	-$1.49
$1.44	$0.08	-$0.07	-$0.22	-$0.37	-$0.52	-$0.67	-$0.82	-$0.97	-$1.13	-$1.28	-$1.39
$1.54	$0.18	$0.03	-$0.12	-$0.27	-$0.42	-$0.57	-$0.72	-$0.87	-$1.03	-$1.18	-$1.29
$1.64	$0.28	$0.13	-$0.02	-$0.17	-$0.32	-$0.47	-$0.62	-$0.77	-$0.93	-$1.08	-$1.19
$1.74	$0.38	$0.23	$0.08	-$0.07	-$0.22	-$0.37	-$0.52	-$0.67	-$0.83	-$0.98	-$1.09
$1.84	$0.48	$0.33	$0.18	$0.03	-$0.12	-$0.27	-$0.42	-$0.57	-$0.73	-$0.88	-$0.99
$1.94	$0.58	$0.43	$0.28	$0.13	-$0.02	-$0.17	-$0.32	-$0.47	-$0.63	-$0.78	-$0.89
$2.04	$0.68	$0.53	$0.38	$0.23	$0.08	-$0.07	-$0.22	-$0.37	-$0.53	-$0.68	-$0.79
$2.14	$0.78	$0.63	$0.48	$0.33	$0.18	$0.03	-$0.12	-$0.27	-$0.43	-$0.58	-$0.69
$2.24	$0.88	$0.73	$0.58	$0.43	$0.28	$0.13	-$0.02	-$0.17	-$0.33	-$0.48	-$0.59
$2.34	$0.98	$0.83	$0.68	$0.53	$0.38	$0.23	$0.08	-$0.07	-$0.23	-$0.38	-$0.49
$2.44	$1.08	$0.93	$0.78	$0.63	$0.48	$0.33	$0.18	$0.03	-$0.13	-$0.28	-$0.39
$2.54	$1.18	$1.03	$0.88	$0.73	$0.58	$0.43	$0.28	$0.13	-$0.03	-$0.18	-$0.29
$2.64	$1.28	$1.13	$0.98	$0.83	$0.68	$0.53	$0.38	$0.23	$0.07	-$0.08	-$0.19
$2.74	$1.38	$1.23	$1.08	$0.93	$0.78	$0.63	$0.48	$0.33	$0.17	$0.02	-$0.09

Feb 08, 2007

Profit margin in $ per gallon

Assumptions:
$1.00 per gallon tax subsidy fully captured by biodiesel manufacturer
Crude soybean oil basis 300 points under futures (rough average east to west)
Soybean oil refining cost 270 points, plus 80 point transportation costs to delivery to biodiesel plant
Conversion rate of 7.55 lbs soyoil per gallon of biodiesel
Total conversion costs of 63 cents/gallon - split evenly between chemical inputs and manufacturing

Figure 25.12 Biodiesel profitability grid: wholesale diesel fuel *vs* SBO future prices

treble from the 2005 production level. If this was the case and it was all produced from rapeseed then, at a 0.40 oil extraction rate, 22.5 million tonnes of rapeseed would be required. This compares to the current EU25 production level of 16 million tonnes of rapeseed. In reality rapeseed oil is unlikely to be the sole source in the E.U. with used cooking oil, soya oil and palm oil the likely alternatives. It is fair to say that the volume of rapeseed grown and crushed for its oil is likely to increase significantly, that will lead to a very good availability of rapeseed meal for feeding to pigs and other species in the E.U.

It is already evident that the volume of glycerol now being produced as a co-product of E.U. biodiesel production is exceeding the requirements of traditional uses such as cosmetics. Glycerol prices are therefore falling and it has already found its way into animal feeds as a molasses substitute. As the volume of biodiesel production continues to increase further uses for glycerol will be required.

Bio-ethanol

Bio-ethanol is an alternative to petrol and is produced by the fermentation of carbohydrate derived from crops. This is as opposed to the generic ethanol, which can be produced by other means such as hydration of ethylene from petroleum.

Up to now production of bio-ethanol is primarily from sugarcane, maize and sugar beet. Brazil, Colombia, China and the United States have already developed bio-ethanol fuel programs and many others including the UK have plans to go down this route.

In the United States around 55 million tonnes of the maize crop is expected to be destined for bio-ethanol production in the 2006/07 season (this is 0.20 of the crop). In the United States, the Energy Policy Act of 2005 (EPACT) initiated a raft of biofuel incentives which included the Renewable Fuels Standard (RFS). The RFS dictates a scale of increase for the volume of the U.S. fuel supplies, which must come from renewable fuels. This started at four billion gallons in 2006 and rises to 7.5 billion gallons by 2012. This requirement provides a baseline calculation for the minimum U.S. production of biofuels. Beyond this the price relationship with crude oil prices will have a large influence on ethanol production.

Figure 25.13, courtesy of Rosenthal Collins Group, illustrates their view on the relationship between maize prices and ethanol prices in terms of profitability of ethanol production. The market price of ethanol is, in turn, influenced largely by the crude oil price. This snapshot taken 22nd February, 2007 shows a good profit margin despite the recent rises in maize prices.

With current technology one bushel of maize produces about 2.8 gallons of bio-ethanol or one tonne of maize produces 110 gallons of bio-ethanol. The U.S. Renewable Fuels Association (RFA) estimate that the U.S. capacity as of February 2007 was 5.6 billion gallons of bio-ethanol with further plants under

Theoretical US Ethanol Profit Margin Grid - 4th Qtr ...

Ethanol $ gallon	$3.10	$3.20	$3.50	$3.70	$3.90	$4.10	$4.30	$4.50	$4.70	$4.90	$5.10
						Cash Corn Price - $ per bushel					
$1.05	-$0.82	-$1.02	-$1.22	-$1.42	-$1.62	-$1.82	-$2.02	-$2.22	-$2.42	-$2.62	-$2.82
$1.15	-$0.55	-$0.75	-$0.95	-$1.15	-$1.35	-$1.55	-$1.75	-$1.95	-$2.15	-$2.35	-$2.55
$1.25	-$0.27	-$0.47	-$0.67	-$0.87	-$1.07	-$1.27	-$1.47	-$1.67	-$1.87	-$2.07	-$2.27
$1.35	$0.00	-$0.20	-$0.40	-$0.60	-$0.80	-$1.00	-$1.20	-$1.40	-$1.60	-$1.80	-$2.00
$1.45	$0.28	$0.08	-$0.12	-$0.32	-$0.52	-$0.72	-$0.92	-$1.12	-$1.32	-$1.52	-$1.72
$1.55	$0.55	$0.35	$0.15	-$0.05	-$0.25	-$0.45	-$0.65	-$0.85	-$1.05	-$1.25	-$1.45
$1.65	$0.83	$0.63	$0.43	$0.23	$0.03	-$0.17	-$0.37	-$0.57	-$0.77	-$0.97	-$1.17
$1.75	$1.10	$0.90	$0.70	$0.50	$0.30	$0.10	-$0.10	-$0.30	-$0.50	-$0.70	-$0.90
$1.85	$1.38	$1.18	$0.98	$0.78	$0.58	$0.38	$0.18	-$0.02	-$0.22	-$0.42	-$0.62
$1.95	$1.65	$1.45	$1.25	$1.05	$0.85	$0.65	$0.45	$0.25	$0.05	-$0.15	-$0.35
$2.05	$1.93	$1.73	$1.53	$1.33	$1.13	$0.93	$0.73	$0.53	$0.33	$0.13	-$0.07
$2.15	$2.20	$2.00	$1.80	$1.60	$1.40	$1.20	$1.00	$0.80	$0.60	$0.40	$0.20
$2.25	$2.48	$2.28	$2.08	$1.88	$1.68	$1.48	$1.28	$1.08	$0.88	$0.68	$0.48
$2.35	$2.75	$2.55	$2.35	$2.15	$1.95	$1.75	$1.55	$1.35	$1.15	$0.98	$0.75
$2.45	$3.03	$2.83	$2.63	$2.43	$2.23	$2.03	$1.83	$1.63	$1.43	$1.23	$1.03
$2.55	$3.30	$3.10	$2.90	$2.70	$2.50	$2.30	$2.10	$1.90	$1.70	$1.50	$1.30
$2.56	$3.58	$3.38	$3.18	$2.98	$2.78	$2.58	$2.38	$2.18	$1.98	$1.78	$1.58
$2.75	$3.85	$3.65	$3.45	$3.25	$3.05	$2.85	$2.65	$2.45	$2.25	$2.05	$1.85

Profit margin in $ per bushel

Feb 22, 2007

Assumptions:
Above grid assumes "zero" subsidy to the ethanol producer
All-in operational and conversion costs of $1.70 per bushel
Breakdown costs - 70 cents/bu energy, 38 chemicals and 62 misc (includes labor, maintenance, etc.)
Conversion rate of 2.75 gallons of ethanol per bushel of corn
Using a DDG price of $128 per short-ton

LaSalle Group
ROSENTHAL COLLINS GROUP

Figure 25.13 Even with corn prices in excess of 4.00/bushel, spot ethanol margins are near $1.00/bushel

construction which could take the capacity to nearly 12 billion gallons well before 2012. This enhanced capacity could use up to 109 million tonnes of corn! Estimates for 2007/08 season predict usages of between 70 and 82.5 million tonnes of maize, which will represent around one quarter of the U.S. crop! This huge demand for maize has led to a significant increase in prices of maize, with a knock on effect on wheat and soya prices also.

The main by product of bio-ethanol production is distillers grains which may be produced in a wet or dried form. It is estimated that a maize grind of 70 million tonnes would produce 22 million tonnes of distillers grains. The quality of distillers grains produced from modern bio-ethanol plants is widely believed to be good and is allowing inclusions in pig finisher and sow rations of up to 150g/kg. This inclusion is reducing the reliance on soya meal in these rations.

The U.S. legislation also requires that beginning 2013, a minimum of 250 million gallons per annum of cellulosic-derived ethanol be included in the RFS. Cellulosic ethanol may in theory be derived from plants such as switchgrass but will require further technological breakthroughs before commercial production could be contemplated.

Whilst the U.S. is forging ahead with biofuel production, the Chinese government, by contrast has put a halt to ethanol plant construction due to the threat it poses to the country's food sector.

In the E.U. it is expected that bio-ethanol will play a significant part in reaching the 0.0575 incorporation of biofuels by 2010. Toepfer estimate that 22 million tonnes of grain (predominantly wheat) will be required for ethanol production and that 7 million tonnes of distillers grains per annum will be produced as by-product. However it is not even clear that all this bio-ethanol will be produced in the E.U. as it could well be imported. Brazil is widely considered a very economic source of bio-ethanol produced from sugarcane and even Russia and Kazakhstan have plans to build large plants to produce bio-ethanol from wheat and export to the E.U. It is clear, though, that demand for grain is going to be very strong on a global basis.

Combustion (or co-firing)

In the UK the Renewables Obligation requires power suppliers to derive from renewables a specified proportion of the electricity they supply to their customers. It started at 0.03 in 2003, and rises gradually to 0.15 by 2015.

Eligible renewable generators receive Renewables Obligation Certificates (ROCs) for each MWh of electricity generated. These certificates can then be sold to suppliers in order to fulfil their obligation. Suppliers can either present enough certificates to cover the required proportion of their output, or they can pay a 'buyout' price of currently £34.30 /MWh for any shortfall in 2007/ 08. All proceeds from buyout payments are recycled to suppliers in proportion to the number of ROCs they present.

So far the renewables that have been successfully used include palm kernel expeller, sheanut, olive pulp, wheatfeed and wood. The legislation limits the amount of ROCs from co-firing, non-energy crops to 0.10 so that any co-fired ROCs above this are worthless unless they are produced from energy crops. Co-firing currently ceases to be eligible for NIROCs after 31 March 2016. There are further amendments to the Obligation expected which may define banding of technologies that could penalise co-firing in favour of more expensive technologies such as offshore wind and photoelectric cell technology.

The probable energy crops that will be co-fired in the future are miscanthus grass, coppice (poplar or willow) and possibly cereals. Whichever energy crops are grown there will be competition for arable acres. Other E.U. countries have similar polices to encourage combustion of renewable biomass.

Land use

It is obvious that competition for acres to grow crops will be fierce in the years to come with demand for food, feed, biofuels and combustion. Set-a-side in the E.U. and the Conservation Reserve Programme (CRP) in the U.S. will surely be phased out as non productive land will not be an option if all these requirements are to be met. Additionally uncultivated areas of land in other countries such as South America, Eastern Europe, the Former Soviet States or even Africa may be brought into production. Minority crops which can not be used in biofuel production or command a premium for direct human uses are likely to decline. The need for crop rotation can not be ignored, but pulses such as field beans and feed peas are likely to decline. Whilst the headlines in the U.S. are about a switch out of soya beans and into maize, a number of other crops are also reducing acreage to make way for maize, such as sorghum, rice and cotton.

World stocks

It is perhaps no surprise that having detailed the growing demand for commodities for both food and fuel that the world is struggling to maintain adequate stocks.Figure 25.14 shows world grain stocks over the last 10 years. Grain stocks have declined dramatically in the last eight years and are just below the levels of 10 years ago. It is a concern that world grain production has only exceeded world usage twice in the last 8 years. Whilst stocks are only slightly lower than 10 years ago in total, when expressed as a ratio of total use as, in Figure 25.15, it is possible to appreciate that world grain stocks are relatively very low.

It is the very low forecast stocks that have ultimately influenced prices to move higher, in order that stocks are conserved where use is not essential. This will also potentially mean that grain prices will become even more volatile

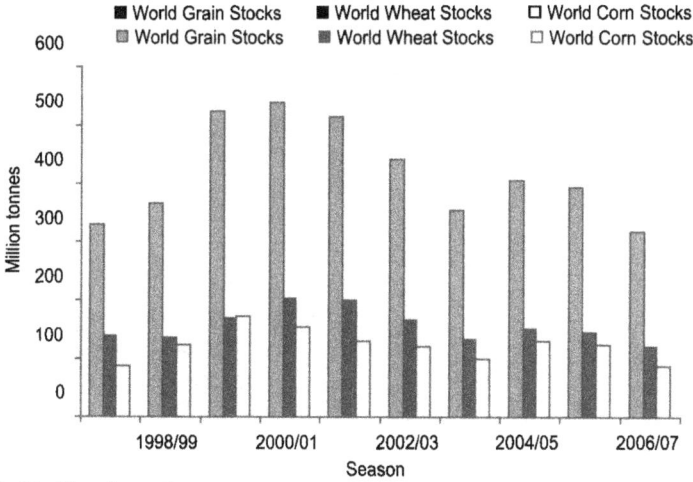

Figure 25.14 World grain stocks

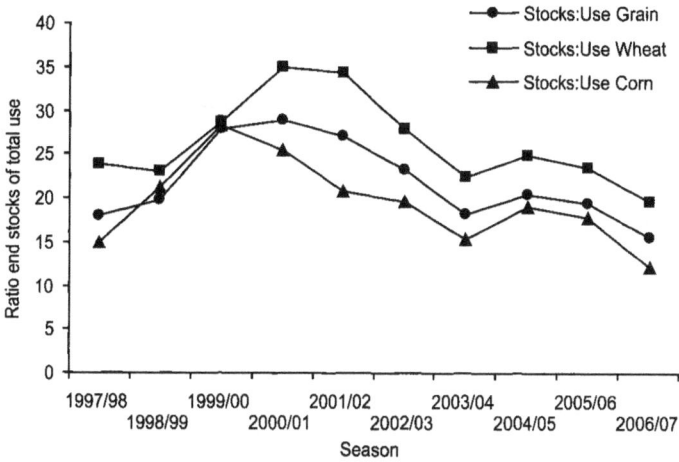

Figure 25.15 World grain stocks:usage ratio

because crop yields, good or bad, have more potential impact when stocks are so precariously positioned, as do any changes in the demand picture.

Due to a succession of good harvests in North and South America, world soya stocks are at all time highs. Figure 25.16 shows how world soya stocks have grown in the last 10 years and this is still the case when expressed as a stock to usage ratio, as in Figure 25.17. Logically one would have expected soya prices to have declined, in order to stimulate increased usage of soya but the market dynamics are more complex than this. The Chicago Board of Trade Futures markets have a large influence on world prices. The market talk has recently been very focused on the battle for acres between soya and maize in

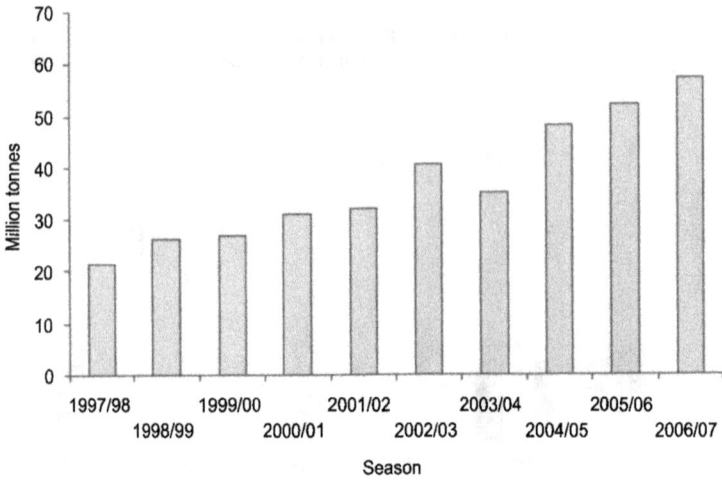

Figure 25.16 World soya stocks

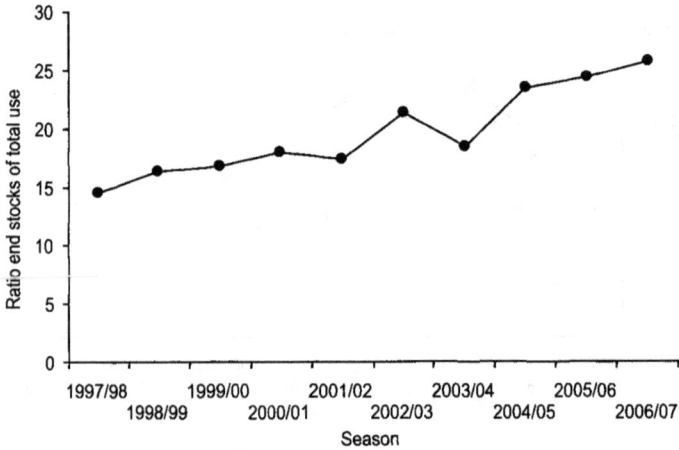

Figure 25.17 World soya stocks:usage ratio

the U.S. for the planting period April/May 2007. A switch in acreage out of soya and into maize is expected compared to the previous season. Estimates of this acreage move vary from 4 to 12 million acres. Soya prices appear to have been held up fundamentally by the need to prevent too big a move from soya to corn and also the longer term outlook that if soya acreage reduces then stocks will eventually be eroded. The situation has been influenced further by the level of investment and speculation, in commodities, in the Chicago markets.

The funds

The price volatility in commodity prices has attracted speculative investment for many years, particularly in the Chicago futures market. It has become accepted that investment funds play a major role in Chicago futures trades. Investment funds will buy or sell commodities depending on their investment policies, which often do not relate directly to supply or demand fundamentals of world crops. Increasingly computer models are used to buy or sell on a semi- automated basis. As a snapshot example for Chicago soya meal on 26th February, 2007, the speculative funds owned 67,100 contracts (each contract is 100 short tons). This long holding represented 0.293 of the open interest in the soya meal market, which is one of the highest recorded positions held by the speculative funds. This perhaps suggests that commodity markets are becoming increasingly interesting to speculators.

2005 saw the rise of the index funds. Index funds such as the Goldman Sachs Index or the Commodity Research Bureau (CRB) Index have existed for many years in the U.S. but 2005 saw a strong trend for investors to put money into commodity indexes. Commodity indexes are made up of a basket of commodities for example the Goldman Sachs index includes wheat, maize, soya beans, coffee, sugar, pigs, crude oil, copper, gold and silver.

Investors buy the index and their money is distributed across the basket of commodities at set percentages. Money entering the market creates a long holding in the relevant futures markets and only when investors withdraw their money are those long holdings sold back. So index funds tend to be in for the long term and would not sell the market short, although they will reposition long holdings further into the future, before they are in any danger of being called for physical delivery. The index funds typically represent a smaller proportion of the market than speculative funds.

There is also a category of funds termed hedge funds that have more complex and longer term investment strategies than the largely short term speculative funds. The volume that hedge funds represent is not reported but it would be quite possible for the three categories of funds in total on hold positions, which represented 0.50 of open interest in soya beans, maize or wheat. With such large positions held by investors, who have no direct interest in producing or consuming the commodities, price movements are often distorted or exaggerated away from the pure fundamentals of supply and demand. The increasing influence of the various funds will surely create further market volatility.

Outlook

So far this chapter has indicated that the following outcomes look likely over the next five years.

1.   In the U.S. there will be a shift in planted acreage away from soya and into maize.
2.   Chinese meat consumption will continue to increase and China may become a net importer of maize.
3.   Countries which have traditionally been big exporters of soya beans, such as U.S., Brazil and Argentina, will export a lower proportion of their crop as beans or oil, in order to produce biodiesel. They may, however, increase soya meal exports.
4.   A lower proportion of palm oil will be exported from Malaysia and Indonesia, as they will also begin biodiesel production.
5.   Availability of oilseed meals such as soya meal and rapeseed meal will be good where these crops are being grown for biodiesel production.
6.   Glycerol will be available in much greater quantities as a co-product of biodiesel production and may have a role to play in the feeding of pigs.
7.   Demand for cereal grains looks set to be very strong, primarily due to additional requirements for bio-ethanol production.
8.   There will be good availability of distillers grains from bio-ethanol plants that will have an increasing influence on pig nutrition as a partial substitute for soya meal.
9.   Competition for arable acres will be very strong leading to a phasing out of set-a-side and CRP. Other areas of land may need to be brought into production.
10.   Volatility in pricing of agricultural commodities looks set to increase.
11.   There will be increased trade (exports) of biofuels.

COMMODITY PRICE OUTLOOK

The emphasis over the next five years appears to be about the strongly growing demand for food, feed and bioenergy, with supply struggling to keep up. The impact of any crop failures, especially in the cereal sector, will probably be very severe with world stocks already very low. Thus prices for cereals and oilseeds look set to trend higher but at a level that is not possible to predict, although it is obvious that the price of crude oil will have a big impact on agricultural commodity prices, as it will determine the economics of producing biofuels from them. Where biofuel usage is mandatory food and feed prices will be forced to take the burden of any crop failures, so prices will move perhaps higher than have been seen before.

Availability of oilseed meals, distillers grains and glycerol for use in pig feeds look set to improve so prices relative to cereals, whole oilseeds and vegetable oil should fall. Please note the word relative. It should be the case that it will be more economic to feed the above mentioned by-products in pig feeds over the next five years.

The key question is whether prices will keep increasing forever. Of course, nothing is forever. Producing maize, in particular, is very energy intensive when the energy required to produce the fertiliser, pesticide and herbicide required to grow the crop are taken into account. There is considerable debate as to whether it takes more energy to produce bio-ethanol than it provides. Ethanol is difficult to move because it absorbs water and corrodes pipes, so it uses a lot of energy to transport it.

It appears that, when the animal feed co-products are taken into account, the bio-ethanol production process does release more energy than it consumes. There is also further debate as to whether growing cereal crops on the same land over and over again is sustainable as soil fertility will be depleted. On the positive side, it is cheaper to build ethanol refineries than crude oil refineries, and cereal crops are renewable and also carbon neutral.

Producing ethanol from cereals is not the perfect answer, it is just the best so far.

Producing ethanol from cellulose promises much greater progress in reducing greenhouse gas emissions but the process needs to be perfected and currently it remains uneconomic. If and when cellulosic ethanol production technology is improved, world cereal prices will fall dramatically.

If cellulosic ethanol production proves not to be the answer then other solutions will be sought. The current plans for biofuel production are creating international controversy over the concerns about rising food prices and environmentally-contentious land use. Legislation can change at any time and will have a big influence on commodity futures markets in the years to come.

PRICE RISK

Pig producers will have tough decisions to make over the next few years when agreeing feed contracts as prices will become higher. Use of futures market options or maximum price feed contracts may become more common place. This allows producers the confidence that, if committing to long term feed prices at higher levels than they have done historically, they have some insurance if prices subsequently fall. Of course there is an additional cost with any type of insurance and options are no different in this respect, it is just that they give more peace of mind when prices are high.

NUTRITION

Nutritionists will also be forced into new territory and will be asked to produce diets to utilise more co-products and reduce reliance on cereals, whole oilseeds and vegetable oils. Where bio-ethanol plants produce wet distillers grains, rather than a dry pellet, then new feeding systems will also be required.

Summary

Global cereal and oilseed production is going to be driven to expand at a rapid rate over the next five years. Higher prices for these commodities are necessary in order to stimulate increased land area to be brought into production. The world's quest to reduce greenhouse gas emissions and use sustainable sources of energy is leading to a massive potential increase in demand for these commodities. This coincides with increasing demand from a growing world population with evolving dietary requirements.

Increased speculation and investment in commodity markets and low world cereal stocks will increase price volatility in commodities. Pig producers will need clearly defined strategies to manage the price risk of their feedingstuffs.

Global pig production will need to align itself, to best utilise the higher volume of oilseed meals and distillers grains that will become available as by-products of biodiesel and bio-ethanol production.

26

NUTRITIONAL VALUE OF CO-PRODUCTS FROM VEGETABLE FOOD INDUSTRY

A. SERENA, H. JØRGENSEN, AND K. E. BACH KNUDSEN
Aarhus University, Faculty of Agricultural Sciences, Department of Animal Health, Welfare and Nutrition, PO Box 50, DK-8830 Tjele Denmark

Introduction

The demand for high-energy refined plant foods for human consumption and, more recently, as raw materials for the production of bio-energy has promoted the interest in the use of co-products from the food industry as raw materials in the feed industry. In Europe, a number of co-products are available (Crawshaw, 2003) the most important being those deriving from the grain processing industries as well as from the production of sugar, starch, beer and pectin. While most of the co-products in the past have been used primarily for the feeding of ruminants, recent research has focused on the use of fibre rich co-products for improved health and welfare in pigs.

The main purpose of the present chaper is to give an overview of the chemical composition and physico-chemical properties of common co-products available for the feeding of pigs, and the nutritive value thereof. The co-products considered will be: sugar beet pulp, potato pulp, pectin residue and brewers spent grain, all of which are wet when leaving the industrial process line, and dry co-products such as hulls, brans, and seed residue. Whole grain wheat and barley are used as reference materials.

Types of co-products from the vegetable food industry in Europe

Co-products from the vegetable food industry represent a very heterogeneous group of plant residues derived from different plant families and botanical origin (cereals, tubers, roots, fruits, culms, shells and hulls). During the processing steps they are exposed to a wide variety of different physical and chemical treatments for the extraction of the economically important component. The residue will consequently have a different matrix with plant cell walls in the form of non-starch polysaccharides (NSP) and lignin representing the major part of the dry weight. Pea hull, cereal bran and seed residues are only exposed to mechanical forces during production – sieving for rye grass, abrasive separation and sieving for pea hull, and roller milling and sieving for wheat bran. The majority of brewers spent grain provided by the breweries in Europe is from beer production, which is based predominantly on either barley or wheat. In this process the malt is finely divided and mashed in temperatures ranging from 56 to 78 °C to inactivate the enzymes in the malt. Thereafter the water is removed with an additional rinsing with 78 °C hot water to remove the remaining sugar and finally the residue is transported to silos. Pectin residue derives from the production of pectin for the food industry from citrus pulp, which is a co-product consisting of peel, membranes, juice vesicles and seed after the extraction of juice (Van Heerden *et al.*, 2002). In commercial pectin production, the pectin polysaccharides are dissolved by weak acid, followed by filtration in order to separate the pectin from the remaining plant materials. In the filtration step it is common practice to add wood cellulose to improve the porosity of the filter cakes (Rolin *et al.*, 1998). The production of potato starch from potatoes and of sugar from sugar beet roots involves washing, grating and extraction of starch or sugar from the plant material. The most noticeable difference between the two methods is found in the extraction procedure where potato starch is extracted by cold water (Eriksen and Hedegaard, 2006), whereas sugar is extracted in hot water at 70 °C (Danisco, 2003).

Chemical composition and physicochemical properties of co-products

The chemical composition of cereals (wheat, barley, and oats), and of bran and hulls is shown in Table 26.1 and of co-products in Table 26.2. A common feature of the co-products is high content of dietary fibre (DF, NSP + lignin); the concentration of starch, oil and protein for most co-products is lower than that of cereals. The exception, however, is brewers spent grain with a relatively high content of oil and protein.

The composition of the DF fraction of co-products depends largely on the botanical origin of the plants in combination with the type of processing applied. In monocotyledonous plants (cereals), the main polysaccharides of the NSP fraction are cellulose, arabinoxylanes and ß-glucan, while cellulose, xyloglucanes and pectic polysaccharides are the main cell-wall constituents in dicotyledonous plants (Selvendran, 1984; Raven *et al.*, 1992). These differences are also reflected in the composition of the DF fraction of the co-products such as sugar beet pulp, potato pulp, pectin residue and pea hull, all of which can be characterised as being relatively high in soluble DF, whereas brewers spent grain and seed residue have a high content of insoluble DF in the form of insoluble NSP and lignin. In several respects the DF characteristics of seed residues and brewers spent grain have much in common with straw, bran and barley hulls. The high content of lignin in pectin residues is caused by the inclusion of wood cellulose in the filtering step and not is a remnant of the citrus pulp.

Table 26.1 CHEMICAL COMPOSITION OF DIFFERENT FEED FOR PIGS[1]

	Wheat	Barley	Oat	Wheat bran	Barley hull
g/kg dry matter					
Ash	17	22	24	52	57
Crude Protein	108	111	91	170	101
EE (fat)	27	31	58	57	45
Starch	673	594	468	222	174
Total Fibre	128	217	379	348	577
Soluble NSP	37	52	42	27	61
Insoluble NSP	80	131	247	268	399
Lignin	11	35	90	53	117
Gross Energy MJ/kg	18.56	18.53	19.34	19.22	18.97

NSP, Non-starch polysaccharides
[1]Data from (Bach Knudsen, 1997; Just *et al.*, 1983b; Serena and Bach Knudsen, 2007).

Table 26.2 CHEMICAL COMPOSITION OF CO-PRODUCTS[1]

	Potato pulp[2]	Seed residue Rye grass	Pectin residue[2]	Brewers spent grain[2]	Sugar beet pulp[2]	Pea Hulls	Straw
g/kg dry matter							
Ash	36	104	66	48	54	31	42
Crude Protein	51	91	82	215	106	116	36
EE (fat)	9	28	26	117	26	18	21
Starch	249	15	4	60	5	88	33
Total Fibre	612	555	873	589	737	730	753
Soluble NSP	280	21	117	36	290	121	39
Insoluble NSP	297	409	605	427	410	600	556
Lignin	35	124	130	126	37	9	158
Gross Energy MJ/kg	17.18	18.29	18.83	21.74	17.40	18.00	18.32

NSP, non-starch polysaccharides
[1]Data from (Bach Knudsen *et al.*, 1997; Just *et al.*, 1983b; Serena and Bach Knudsen, 2007) and Bach Knudsen (unpublished).
[2] Heat dried and pelleted.

The main factors influencing the physico-chemical properties – swelling and water binding capacity (WBC) - are the polysaccharide composition of the plant cell wall, the intermolecular organisation of the polysaccharides within the cell wall and the degree of lignification. Secondary lignified cell walls usually tend to have lower swelling and WBC than the primary cell wall (Chen et al., 1984). This is also the general picture of Figures 26.1 and 26.2, which demonstrate that WBC and swelling in general are higher in the pectin rich co-products – sugar beet pulp, potato pulp and pectin residue – than in cereals, cereal bran and hull, and the other co-products. Processing of the co-products, however, also influences the physio-chemical properties. Although pectin residue and pea hulls in terms of NSP composition have much in common, the former co-product both as is and after freeze-drying has a much higher swelling capacity and WBC than the latter. The most likely reason for this difference is the acid treatment and heating applied during the extraction of pectin polysaccharides that dissociates the cell wall matrix and the addition of wood cellulose with a high WBC and swelling itself. For comparison, the cell walls in pea hulls have not been treated either by heat or chemicals, and the cell walls will consequently retain their original form. The impact of insoluble DF and lignification on swelling and WBC can also be seen as the low swelling and WBC of brewers spent grain and seed residues, both having lower swelling and WBC than vegetable fibre as also reported by McConnell et al. (1974). A further factor that may have an impact on swelling and WBC is the physical form of the plant material. Auffret et al. (1994) studied the influence of grinding on the WBC and found that the WBC decreased in response to decreased particle size. This phenomenon is seen with pea hulls, but not in seed residues presumably caused by a reflection of the lignification of the cell walls that makes the cell walls of the seed residues more rigid. The significantly lower WBC in freeze-dried, heat-dried and ground materials compared with wet co-products is due to the grinding process but also to the irreversible changes in the polysaccharides after drying. It can therefore be concluded that, after drying and milling, the plant cells are no longer capable of binding water and swelling to the same extent as they were in the fresh material.

Digestion and dietary energy value of co-products in growing pigs and sows

As shown in Table 26.1 and 26.2, the co-products from the vegetable food industry have in general a high dietary DF content. This will inevitably have a negative impact on the total tract digestibility of components as the DF portion is resistant to the endogenous enzymes present in the small intestine. Consequently a larger proportion of the dietary dry matter is subject to microbial fermentation in the large intestine. Because of the chemical organisation of the DF fraction and degree of lignification (cross-linkages between cell wall

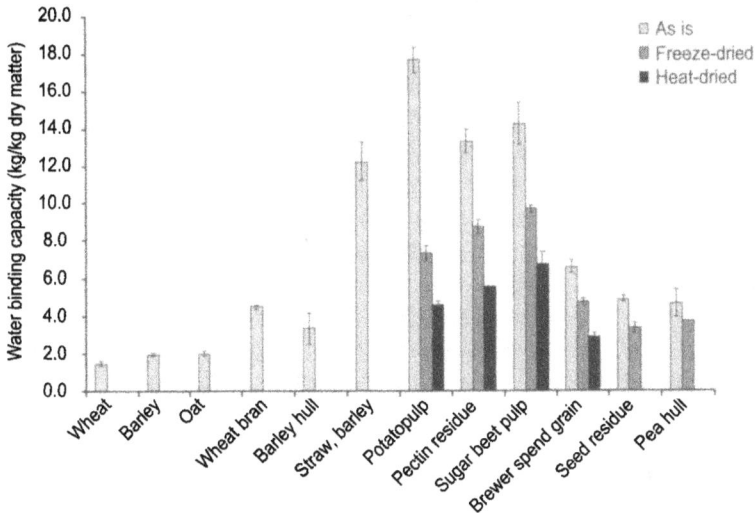

Figure 26.1 Water binding capacity of cereals and co-products. Data from Serena and Bach Knudsen (2007) and Serena and Bach Knudsen (unpublished data).

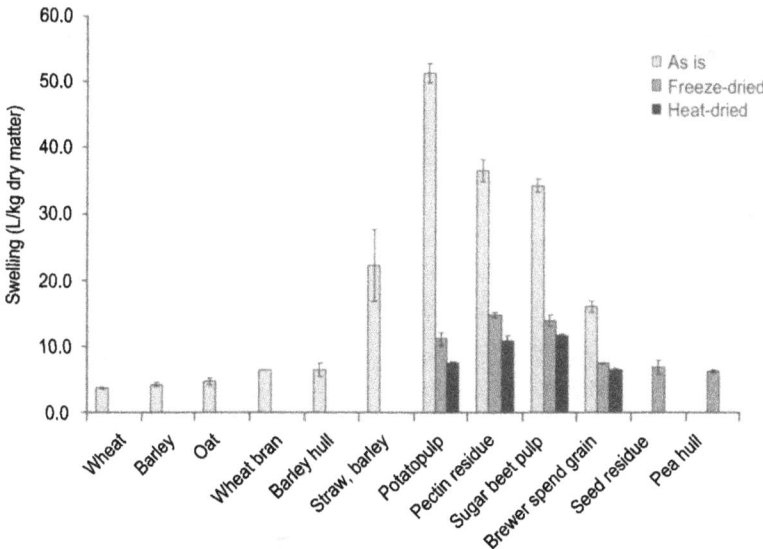

Figure 26.2 Swelling of cereals and co-products. Data from Serena and Bach Knudsen (2007) and Serena and Bach Knudsen (unpublished data).

polysaccharides), a variable proportion of the DF polysaccharides will be resistant to microbial degradation in contrast to starch, oligosaccharides and sugars where there is no structural hindrance for the hydrolysis of the glycosidic linkages either by endogenous or by microbial enzymes. Moreover DF, in particular those types characterised as soluble, will influence the digestibility of protein and fat as well as the endogenous secretion.

The impact of DF of co-products with different botanical origin on digestibility (Le Goff *et al.*, 2002; Fernández *et al.*, 1986) is clearly seen in Tables 26.3 and 26.4. The DF sources with high pectin content and low degree of lignification - sugar beet pulp, potato pulp or pea hulls - are digested to a much greater degree in growing pigs compared with wheat straw, pectin pulp and brewers spent grain that all have a relatively high lignin content. The same has been reported in growing pigs and sows by Graham et al. (1986), Longland et al. (1993), Andersson and Lindberg (1997a; 1997b), Serena (2005). In all cases, the highest digestibilities are found with co-products high in pectin (Guillon *et al.*, 1998) and/or low lignin and/or a high proportion of soluble DF, and the lowest with high lignin and high cellulose levels in DF (Noblet and Le Goff, 2001; Bach Knudsen and Hansen, 1991).

Table 26.3 DIGESTIBILITY AND ALLOWANCE OF CEREALS IN GROWING PIGS AND SOWS[1]

	Wheat	*Barley*	*Oat*	*Wheat bran*	*Barley hulls*
Coefficient of total tract apparent digestibility					
(a) growing pigs					
Protein	0.81	0.74	0.79	0.52	0.56
Fat	0.35	0.46	0.69	0.42	0.55
Fibre	0.43	0.37	0.15	0.32	0.21
(b) adult sows					
Protein	0.85	0.80	0.82	0.71	
Fat	0.41	0.51	0.71	0.69	
Fibre	0.47	0.43	0.17	0.42	
Digestible energy, pig MJ/kg DM	15.93	14.84	13.47	9.90	9.11
Digestible energy, sow MJ/kg DM	16.26	15.35	13.82	11.62	NA
Maximum allowance (g/kg diet)					
(a) pigs	700	700	500	200	NA
(b) sows	800	900	800	200-300	NA

[1]Data from Just *et al.* (1983b), Fernández *et al.* (1986), Jørgensen *et al* (unpublished), Crawshaw (2003) and Dansk Svineproduktion (2007).
NA – not available.

Several factors favour a more extensive degradation of fibrous components in sows or adult animals compared to piglets and growing pigs. Adult animals will usually have a lower feed intake per unit of body weight, a slower digesta transit, a greater intestinal volume and a higher cellulolytic activity. Varel et al.

Table 26.4 DIGESTIBILITY AND ALLOWANCE OF CO-PRODUCTS IN GROWING PIGS AND SOWS[1]

Coefficient of total tract apparent digestibility	Potato pulp	Seed residue rye grass	Pectin residue	Brewers spent grain[2]	Sugar beet pulp[2]	Pea hulls	Straw
(a) growing pigs							
Protein	-0.41	-0.01	-0.82	0.58	-0.04	0.51	-0.72
Fat	-0.92	0.19	0.45	0.49	-0.35	0.38	0.38
Fibre	0.95	0.19	0.60	0.34	0.80	0.65	0.06
(b) adult sows							
Protein	0.17	0.02	-0.02	0.65	0.43	0.73	-0.46
Fat	NA	-0.25	0.20	0.55	-0.16	0.36	0.34
Fibre	0.91	0.16	0.60	0.38	0.81	0.87	0.20
Digestible energy, pig MJ/kg DM	13.92	4.28	8.17	9.66	11.31	11.49	0.53
Digestible energy, sow MJ/kg DM	13.86	3.42	8.91	10.97	12.58	14.98	2.78
Maximum allowance (g/kg diet)							
(a) pigs	NA	NA	50	NA	200-300	NA	NA
(b) sows	NA	NA	900	NA	800	NA	NA

[1]Data from Fernández et al. (1986), Jørgensen et al (unpublished), Crawshaw (2003) and Dansk Svineproduktion (2007).
[2]Mean of two samples one wet and one dry.
NA – not available.

(1987) reported that adult animals had 6.7 times more cellulolytic bacteria in the colon than growing pigs, and trials have shown a generally higher digestibility of crude fibre and higher content of metabolisable energy with sows than in growing pigs (Table 26.5) (Fernández *et al.*, 1986; Nobelet and Shi, 1993). The most significant differences between the two groups of animals were seen for some cereal by-products and most roughages (Fernández *et al.*, 1986). A study of Jørgensen et al. (1978) used three diets with increasing levels of insoluble fibre fed to pigs weighing 20, 90 and 225 kg, respectively. An increased total tract digestibility of energy in response to age was seen for the medium and high fibre diets but not for the diet with a low level of fibre. In a study of Noblet and Bach Knudsen (1997) it was also found that the digestibility of cellulose and non-cellulosic polysaccharides was consistently higher in sows than in growing pigs.

The physico-chemical properties of the DF fraction can also influence digestibility. The high swelling and WBC of for example sugar beet pulp (Figure

Table 26.5 COMPARATIVE DIGESTIBILITY FROM GROWING PIGS AND SOWS

Coefficient of total tract apparent digestibility	Mean of 72 diets[1]		Mean of 14 diets[2]	
	Growing pigs	Sows	Growing pigs	Sows
Crude protein	0.50	0.60	0.75	0.85
Crude fat or EE	0.38	0.42	0.55	0.69
Fibre	0.49	0.55	0.38	0.64
ME (MJ/kg DM)	12.23	13.15	13.49	14.43

EE, Ether extracts; ME, metabolisable energy.
[1] Data from Fernández et al. (1986) and Jørgensen et al. (unpublished).
[2] Data from Noblet and Shi (1993).

26.1 and 26.2) increases the surface area, which encourages easy colonization of the residue. These conditions are responsible for the high degradation in general of soluble DF seen for ß-glucans, arabinoxylans and pectin polysaccharides with growing pigs (Bach Knudsen *et al.*, 1993; Canibe and Bach Knudsen, 1997; Glitso *et al.*, 1999; Petkevicius *et al.*, 1997) and sows (Noblet and Le Goff, 2001; Noblet and Bach Knudsen, 1997; Serena, 2005). Lignified co-products such as bran, hull and straw are, on the other hand, not easy to digest, independent of the age of the animal, as the lignin cross-linkages make the cell wall very rigid and difficult to penetrate by microbial enzymes.

It is clear from Tables 26.3 and 26.4 that the digestible energy (DE) of the co-products is much more variable compared with that of wheat. The highest level is found in sugar beet pulp, potato pulp and pea hulls; a medium level is found in pectin pulp, brewers spent grain and wheat bran whereas the lowest value for DE compared to wheat was found in seed residues and straw. For all co-products, the value for DE is higher in sows compared to growing pigs which, as discussed above, presumably is due to factors including retention time and cellulolytic activity.

As shown above it is clear that digestibility varies with the botanical origin of DF but also with the age of the animal. The increased digestibility of DF with age is however also dependent on the botanical origin of DF. This means that DF does not have a constant nutritional effect (Noblet and Le Goff, 2001; Fernández and Jørgensen, 1986).

Inclusion of co-products must be done with caution, and not all co-products can be included in the diet in equal amounts. Co-products with a high proportion of soluble DF are usually easier to incorporate than co-products high in insoluble DF. The co-products can generally be incorporated at a higher level in diets for sows than for growing pigs. In sows a study by O'Sullivan et al. (2003) has shown that inclusion of citrus pulp could be included in amounts up to 200g/kg diet without any negative effect on reproductive performance. In a study by Serena (2005) it was found that up to 420g/kg diet could consist of co-products that were characterised to be high in soluble DF, whereas the inclusion of co-products high in insoluble DF at the same level caused problems in maintaining the weight of the sows. The maximum inclusion level is usually somewhat lower in growing pigs.

A study by Antuszewiez and Swiech (2006) showed that an inclusion of 50g apple pulp, apple pectin, sugar beet pulp, potato pulp and artichoke offal did not affect apparent ileal digestibility of protein and amino acids compared to the control diet containing similar amounts of cellulose. Moreover, a study of Hedemann et al. (Mette Skou Hedemann, Aarhus University, Faculty of Agricultural Sciences, Denmark; unpublished data) did not show any negative effects on growth parameters by including up to 200g/kg diet, with co-products that were characterised as soluble and insoluble, respectively, when proper corrections were done for the differences in net energy content.

Absorption of nutrients from co-products in sows

Co-products that are characterised to be high in soluble DF and with high WBC and swelling have been shown to influence the rate by which glucose is absorbed into the body and the apparent insulin production. This is illustrated in Figures 26.3 and 26.4 that show the results of a study where 6 sows were fed three experimental diets – a low fibre (LF) control diet and two high fibre diets with added soluble DF from sugar beet pulp, potato pulp and pectin residues (HF1) or the three fibre sources diluted by insoluble DF from brewers spent grain, pea hulls or seed residue (HF2) (Serena, 2005). While the insoluble DF reduces the uptake of glucose due to the substitution of DF for starch, this did not influence absorption patterns following feeding the diet high in soluble DF that not only reduces the uptake of glucose but also delayed the uptake. A similar pattern has been seen when adding highly viscous guar gum to a diet for growing pigs (Ellis *et al.*, 1995). The differences observed in the uptake of glucose are associated with the apparent production of insulin (Figure 26.4) as have been found in the concentration of insulin in the peripheral blood (de Leeuw *et al.*, 2004; Ellis *et al.*, 1995; Vestergaard, 1997).

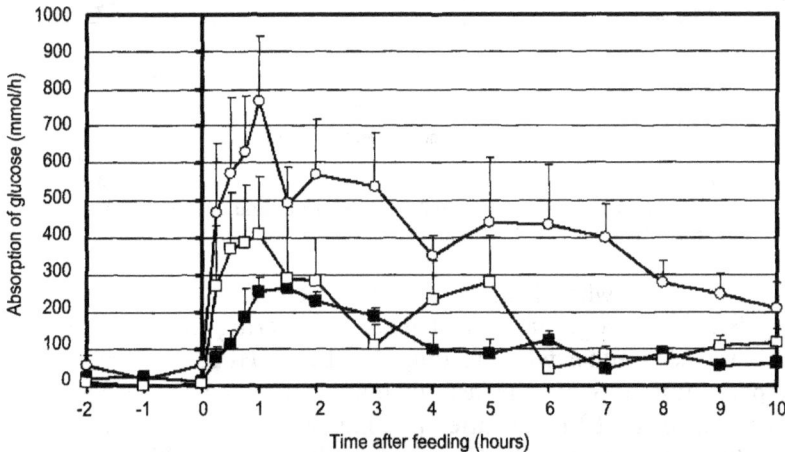

Figure 26.3. Absorption of glucose in sows fed diet low fibre (LF), high fibre - soluble (HF1) or high fibre - insoluble (HF2) (difference between partial vein and mesenteric artery).Values are means ± SEM, *n* = 6. -○-, LF; -■-, HF1; -□-, HF2. Data from Serena (2005).

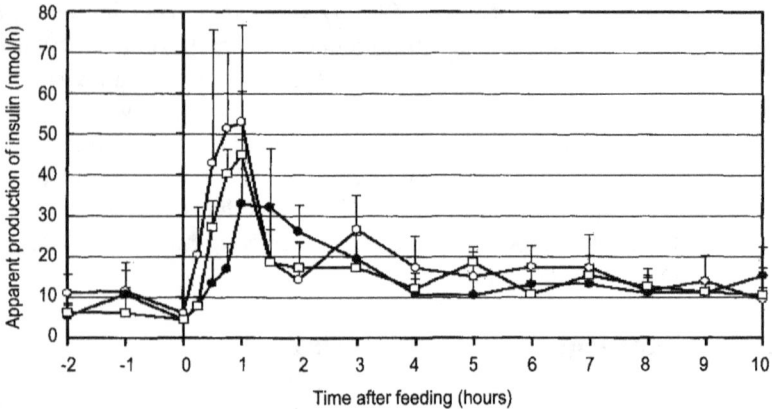

Figure 26.4 Apparent production of insulin in sows fed diet low fibre (LF), high fibre – soluble (HF1) or high fibre - insoluble (HF2) (difference between partial vein and mesenteric artery). Values are means ± SEM, *n* = 6. -O-, LF; -■-, HF1; -□-, HF2. Data from Serena (2005).

While soluble DF delayed the uptake of glucose, the same diet (diet HF1) resulted in a substantial rise in the absorption of short chain fatty acids (SCFA) compared with the low DF diet as well as the high fibre diet with high level of insoluble DF (Figure 26.5). From Figure 26.5 it can also be seen that the

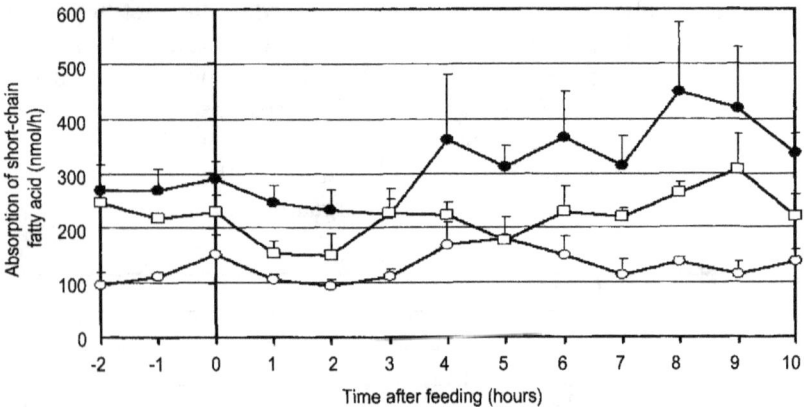

Figure 26.5 Absorption of total SCFA in sows fed diet low fibre (LF), high fibre - soluble (HF1) or high fibre - insoluble (HF2) (difference between partial vein and mesenteric artery). Values are means ± SEM, *n* = 6. -O-, LF; -■-, HF1; -□-, HF2. Data from Serena (2005).

absorption of SCFA when feeding diet HF1 increased four hours after feeding probably because a large amount of material entering the large intestine. The same was not seen either when feeding diet LF or HF2. It is further noticeable that the portal flux of SCFA the last two hours before feeding was approximately at the same level as 10 hours after feeding thus confirming the constant and high supply of energy from SCFA when feeding the two high DF diets.

Taken as a whole, the accumulated apparent uptake of energy 0 - 10 h post-feeding from carbohydrate derived-components amounted to 18.2 MJ, 9.9 MJ and 10.4 MJ in diets LF, HF1 and HF2, respectively (Table 26.6). Although the intake of digestible energy was practically the same for diet LF and HF1, the accumulated uptake of glucose was only one third during the 10 hours post feeding period when feeding diet HF1 compared to diet LF while the energy from SCFA was more than twice as high with the former as compared to the latter diet.

Table 26.6 APPARENT ENERGY ABSORPTION 0 TO 10 HOURS POST-FEEDING IN SOWS FED LOW FIBRE (LF), HIGH FIBRE (SOLUBLE) (HF1) OR HIGH FIBRE (INSOLUBLE) (HF2) (MEAN VALUES, N = 6)

Item	*LF*		*HF1*		*HF2*	
Glucose* (MJ)	15.9[a]	(0.87)	5.3[b]	(0.54)	7.2[b]	(0.69)
LA* (MJ)	0.68	(0.04)	0.61	(0.06)	0.53	(0.05)
SCFA* (MJ)	1.68[c]	(0.09)	3.95[a]	(0.40)	2.64[b]	(0.26)
Total AE (MJ)	18.2[a]		9.9[b]		10.9[b]	
AE / DE	0.69		0.38		0.49	

LA, lactate; SCFA, short-chain fatty acid; DE, digestible energy; AE, absorbed energy.
*Values in parentheses are proportion of total AE.
a, b, c: significant difference between diets ($p < 0.05$).
Data from Serena (2005).

Utilization of absorbed nutrients in growing pigs and sows

When feeding high fibre co-products, especially insoluble DF, an increased amount of organic matter is fermented in the large intestine (Jørgensen *et al.*, 2007). For a high fermentation of fibrous material in the large intestine, an adequate supply of nitrogen is required by the colonic flora (Mosenthin *et al.*, 1990). Normally adequate levels of nitrogen are provided by residual feed protein and urea recycling to the gut (Mosenthin *et al.*, 1992). The fermentation not only occurs in the large intestine. In a review by Bach Knudsen and Jørgensen (2001) the average digestibility of NSP up to the end of the small intestine was found to be 0.24, with large variations between experiments. The large variation was not only influenced by soluble proportions of NSP. However, feeding level, sampling technique, marker type etc undoubtedly also was responsible for the variations. The colonic fermentation of digesta results in a lower energetic utilization than for carbohydrates which are digested and absorbed from the small intestine. From the average of several experiments with growing finishing pigs, the efficiency of utilisation of energy fermented in the hindgut was estimated to be 0.69 of energy absorbed from the small intestine (Table 26.7). The difference is due to additional losses of H_2, CH_4, increased loss as fermentation heat and also a lower utilization of SCFA in the intermediary metabolism of the organism. However, in experiments with adult sows the efficiency seems somewhat higher at 0.90 in spite of a higher loss of CH_4 from the older animals.

Table 26.7 UTILIZATION OF ENERGY FERMENTED IN THE HINDGUT

Diet component	Body weight range, kg	Energy fermented hind-gut, (proportion of DE)	Efficiency of utilisation (RE*/ME**)	Author
Potato starch, cellulose	60-90	0.18-0.33	0.51	(Just *et al.*, 1983a)
Potato, sugar beet, grass, Lucerne meal	90-180	0.09-0.40	0.66	(Hoffmann *et al.*, 1990)
Maize starch, cellulose, soya hull	30-105	0.13-0.27	0.63	(Bakker *et al.*, 1994)
Beet pulp, corn distillers grain, sunflower meal, etc	38-47	0.03-0.27	0.82	(Noblet *et al.*, 1994)
Pea fibre, pectin	40-125	0.07-0.29	0.73	(Jørgensen *et al.*, 1996)
Barley straw, barley hulls, wheat bran, potato fibre, soya fibre	50-70	0.04-0.29	0.76	(Jørgensen, unpublished)
Wheat bran, sugar beet pulp, seed residues, brewers spent grain, pea hulls, potato pulp,	46-125	0.08-0.35	0.69	(Jørgensen, unpublished)
pectin residue, sugar beet pulp	160-243	0.05-0.40	0.90	(Jørgensen, unpublished)

* Efficiency of utilisation of energy obtained from the hindgut
** Metabolisable energy

The production of CH_4 is not only dependent on the animals weight/age but also the amount and botanical origin of the fibres in the diet shown by Jørgensen (2007) when summarising several experiments on growing pigs as well as adult sows. The study showed that sows lost a higher proportion of DE per g fermented DF as methane than growing pigs. Furthermore the study showed that of the co-products especially pea hull results in a high production of CH_4, and that pectin reduces the CH_4 production (Jørgensen *et al.*, 2007).

When comparing infusion of SCFA into the hindgut with calculated amount of fermented material from a high fibre diet a higher efficiency of 0.82 was found (Jørgensen *et al.*, 1997) compared to 0.73 from fermentation (Jørgensen *et al.*, 1996). Other experiments have shown lower efficiency (Table 26.8) where the SCFA either were given orally (Jentsch *et al.*, 1968) or infused into the caecum of sows or growing pigs.

Table 26.8 UTILIZATION OF INFUSED SHORT CHAIN FATTY ACIDS INTO THE HINDGUT

Diet type	Body weight range, kg	Infusate	Utilization, proportion of ME	Author
Barley/Fishmeal	140-180	Ethanol Lactic acid Acetic acid	Ethanol: 0.72 Lactic acid: 0.75 Acetic acid: 0.60	(Jentsch *et al.*, 1968)
Barley/Soybean meal	160-200	Acetic acid Propionic acid	Acetic acid: 0.79 Propionic acid: 0.75	(Roth *et al.*, 1988)
Grain/Oat meal by-product/ Wheat bran, Soybean meal/ Fish meal	55-120	Acetic acid, Propionic acid Butyric acid	Acetic acid: 0.65 Propionic acid: 0.71 Butyric acid: 0.67	(Gädeken *et al.*, 1989)
Barley/Soybean meal	179±17	Mixture of: Acetic acid + propionic acid	Mixture: 0.70	(Müller *et al.*, 1991)
Barley/Wheat starch/Fish meal/ Casein	60-120	Mixture of: Acetic acid + propionic acid + butyric acid	Mixture: 0.82	(Jørgensen *et al.*, 1997)

Increased fermentation from co-products might have beneficial effect on ammonia emission, because nitrogen (N) is excreted in the more stable structure of bacterial protein (Kirchgessner *et al.*, 1994; Sørensen and Fernández, 2003).

Use of co-products to influence welfare of sows

Dietary fibre is found to contribute to welfare of the animals kept in intense conditions and cause reduction in stereotypic behaviour (Meunier-Salaun, 2001). Satiety signals consist of several elements incurred by stretch- and chemoreceptors in the stomach and duodenum, which operate in the short term and metabolic signals brought about by fluctuations in plasma glucose, free fatty acids, SCFA, amino acids, lactic acid (LA) and insulin that operate over a much longer timeframe (Read *et al.*, 1994).

Several studies have investigated the incidence of aggressiveness, stress and/or stereotypic behaviour in sows caused by hunger (Bergeron *et al.*, 2000; Brouns *et al.*, 1997; Danielsen and Vestergaard, 2001; de Leeuw *et al.*, 2004; de Leeuw *et al.*, 2005b; de Leeuw *et al.*, 2005a; Ramonet *et al.*, 1999), and it is expected that diets that cause modest variation in the energy delivery from the feed over a longer period (Grieshop *et al.*, 2001) and stabilize glucose (de Leeuw *et al.*, 2004; de Leeuw *et al.*, 2005b) could prevent some of this undesirable behaviour.

Feeding DF-rich co-products to sows results in higher amounts of materials in the whole gastrointestinal tract compared to a cereal base low fibre diet (Serena, 2005), which may increase satiety. In addition the absorption of energy as SCFA before feeding is higher when consuming a high fibre diet compared with a low fibre diet. Sows consuming a high DF diets can thus be expected to have a longer-lasting feeding of satiety due to physical as well as metabolic regulation. Feeding co-products with a high content of soluble fibres will probably in higher degree influence satiety by a regulation of metabolic pathways than indigestible DF, which may influence satiety by physical mechanisms.

Conclusion

The traditional way looking at nutritive values of feed is through digestible or net energy, and chemical composition. However, the physicochemical properties also have a great influence on digestibility and absorption, and welfare; in this way nutritive value is also influenced. The nutritive value of co-products is also very dependent on the age of the animal. The inclusion of co-products in the feed for growing pigs and sows can, in the right amounts, be undertaken without any negative effects and also with some beneficial effects in particular for sows. Therefore there are good reasons for increased attention on the inclusion of co-products for improving welfare in pig production.

References

Andersson,C. and Lindberg,J.E. (1997a) Forages in diets for growing pigs. 1. Nutrient apparent digestibilities and partition of nutrient digestion in barley-based diets including lucerne and white-clover meal. *Animal Science*, **65**, 483-491.

Andersson,C. and Lindberg,J.E. (1997b) Forages in diets for growing pigs. 2. Nutrient apparent digestibilities and partition of nutrient digestion in barley-based diets including red-clover and perennial ryegrass meal. *Animal Science*, **65**, 493-500.

Antuszewiez,A. and Swiech,E. (2006) Apparent ileal digestibility of amino acids in pig diets containing various sources of dietary fibre. *Journal of Animal and Feed Sciences*, **15**, 53-56.

Auffret,A., Ralet,M.C., Guillon,F., Barry,J.L. and Thibault,J.F. (1994) Effect of grinding and experimental conditions on the measurement of hydration properties of dietary fibres. *Lebensmittel-Wissenschaft und-Technologie*, **27**, 166-172.

Bach Knudsen,K.E. (1997) Carbohydrate and lignin contents of plant materials used in animal feeding. *Animal Feed Science and Technology*, **67**, 319-338.

Bach Knudsen,K.E. and Hansen,I. (1991) Gastrointestinal implications in pigs of wheat and oat fractions. 1. Digestibility and bulking properties of polysaccharides and other major constituens. *British Journal of Nutrition*, **65**, 217-232.

Bach Knudsen,K.E., Jensen,B.B. and Hansen,I. (1993) Digestion of polysaccharides and other major components in the small and large intestine of pigs fed on diets consisting of oat fractions rich in ß-D-glucan. *British Journal of Nutrition*, **70**, 537-556.

Bach Knudsen,K.E., Johansen,H.N. and Glitsø,V. (1997) Methods for analysis of dietary fibre - advantage and limitations. *Journal of Animal and Feed Sciences*, **6**, 185-206.

Bach Knudsen,K.E. and Jørgensen,H. (2001) Intestinal degradation of dietary carbohydrates - from birth to maturity. *Digestive Physiology of Pigs* (ed. by J.E.Lindberg and B.Ogle), pp. 109-120. CABI Publishing, Wallingford.

Bakker,G.C.M., Dekker,R.A., Jongbloed,R. and Jongbloed,A.W. (1994) The effect of starch, fat and non-starch polysaccharides on net energy and on the proportion of digestible organic matter or energy that disappeared in the hindgut. *Energy Metabolism of Farm Animals* (ed. by J.F.Aguilera), pp. 163-166. CSIB Publishing Service, Madrid.

Bergeron,R., Bolduc,J., Ramonet,Y., Meunier-Salaün,M.C. and Robert,S. (2000) Feeding motivation and stereotypies in pregnant sows fed increasing levels of fibre and/or food. *Applied Animal Behaviour Science*, **70**, 27-40.

Brouns,F., Edwards,S.A. and English,P.R. (1997) The effect of dietary inclusion of sugar-beet pulp on the feeding behaviour of dry sows. *Animal Science*, **65**, 129-133.

Canibe,N. and Bach Knudsen,K.E. (1997) Digestibility of dried and toasted peas in pigs. 1. Ileal and total tract digestibilities of carbohydrates. *Animal Feed Science and Technology*, **64**, 293-310.

Chen,J.Y., Piva,M. and Labuza,T.P. (1984) Evaluation of water binding capacity (WBC) of food fiber sources. *Journal of Food Science*, **49**, 59-63.

Crawshaw,R. (2003) *Co-Product Feeds. Animal Feeds from the Food and Drinks Industries*. Nottingham University Press, United Kingdom.

Danielsen,V. and Vestergaard,E.-M. (2001) Dietary fibre for pregnant sows: effect on performance and behaviour. *Animal Feed Science and Technology*, **90**, 71-80.

Danisco .(2003) From sun to sugar. http://www.danisco.com/cms/resources/file/eb28740c5656bde/from_sun_to_sugar_english.pdf Copyright:, Danisco Sugar.

Dansk Svineproduktion .(2007) INFO SVIN http://www.infosvin.dk/_.

Copyright:, Dansk Svineproduktion.

de Leeuw,J.A., Jongbloed,A.W., Spoolder,H.A.M. and Verstegen,M.W.A. (2005a) Effects of hindgut fermentation of non-starch polysaccharides on the stability of blood glucose and insulin levels and physical activity in empty sows. *Livestock Production Science*, **96**, 165-174.

de Leeuw,J.A., Jongbloed,A.W. and Verstegen,M.W.A. (2004) Dietary fiber stabilizes blood glucose and insulin levels and reduces physical activity in sows (Sus scrofa). *Journal of Nutrition*, **134**, 1481-1486.

de Leeuw,J.A., Zonderland,J.J., Altena,H., Spoolder,H.A.M., Jongbloed,A.W. and Verstegen,M.W.A. (2005b) Effects of levels and sources of dietary fermentable non-starch polysaccharides on blood glucose stability and behaviour of group-housed pregnant gilts. *Applied Animal Behaviour Science*, **94**, 15-29.

Ellis,P.R., Roberts,F.G., Low,A.G. and Morgan,L.M. (1995) The effect of high-molecular-weight guar gum on net apparent glucose absorption and net apparent insulin and gastric inhibitory polypeptide production in the growing pig: relationship to rheological changes in jejunal digesta. *British Journal of Nutrition*, **74**, 539-556.

Eriksen,A. and Hedegaard,U. (2006) Fra kartofler til ren, hvid stivelse. www.kartoffel-info.dk/Kend_Kartoflen/Kartoffelkulen/Kartoffelmel/ Kartoffelmel_tekst.htm.

Fernández,J.A., Jørgensen,H. and Just,A. (1986) Comparative digestibility experiments with growing pigs and adult sows. *Animal Production*, **43**, 127-132.

Fernández,J.A. and Jørgensen,J.N. (1986) Digestibility and absorption of nutrients as affected by fibre content in the diet of the pig. Quantitative aspects. *Livestock Production Science*, **15**, 53-71.

Gädeken,D., Breves,G. and Oslage,H.J. (1989) Efficiency of energy utilization of intracaecally infused volatile fatty acids in pigs. *Proceedings of the 11th symposium on Energy metabolism of farm animals*, **EAAP Publ. no.43**, 115-118.

Glitso,L.V., Gruppen,H., Schols,H.A., Hojsgaard,S., Sandström,B. and Bach Knudsen,K.E. (1999) Degradation of rye arabinoxylans in the large intestine of pigs. *Journal of the Science of Food and Agriculture*, **79**, 961-969.

Graham,H., Hesselman,K. and Åman,P. (1986) The influence of wheat bran and sugar-beet pulp on the digestibility of dietary components in a cereal-based pig diet. *Journal of Nutrition*, **116**, 242-251.

Grieshop,C.M., Reese,D.E. and Fahey,G.C. (2001) Nonstarch polysaccharides and oligosaccharides in swine nutrition. *Swine Nutrition* (ed. by A.J.Lewis and L.L.Southern), pp. 107-130. CRC Press, New York.

Guillon,F., Auffret,A., Robertson,J.A., Thibault,J.F. and Barry,J.L. (1998) Relationships between physical characteristics of sugar-beet fibre and its fermentability by human faecal flora. *Carbohydrate Polymers*, **37**,

185-197.

Hoffmann,L., Jentsch,W. and Schiemann,R. (1990) Energieumsatzmessungen am adulten Schwein bei Verfütterung von Rationen mit Kartoffelstärke, Kartoffeln, Rüben, Pressschitzeln und Grobfuttermitteln als Zulagen zu einer Grundrationen. 1.Energieumsatz und Energieverwertung. *Archives of Animal Nutrition,* **40,** 191-207.

Jentsch,W., Schiemann,R. and Hoffmann,L. (1968) Modellversuche mit Schweinen zur Bestimmung der energetischen Verwertung von Alkohol, Essig- und Milchsäure. *Archiv für Tierernährung,* **18,** 352-357.

Jørgensen,H., Just,A. and Fekadu,M. (1978) Fodermidlernes værdi til svin. 9. Formalingsgradens og træstofkoncentrationens indflydelse på foderets fordøjelighed hos svin af forskellig alder (vægt). *Meddelelse fra Statens Husdyrbrugsforsøg,* **230,** 4pp.

Jørgensen,H., Larsen,T., Zhao,X.Q. and Eggum,B.O. (1997) The energy value of short-chain fatty acids infused into the caecum of pigs. *British Journal of Nutrition,* **77,** 745-756.

Jørgensen,H., Zhao,X.Q. and Eggum,B.O. (1996) The influence of dietary fibre and environmental temperature on the development of the gastrointestinal tract, digestibility, degree of fermentation in the hind-gut and energy metabolism in pigs. *British Journal of Nutrition,* **75,** 365-378.

Jørgensen,H. (2007) Methane emission by growing pigs and adult sows as influenced by fermentation. *Livestock Science,* **109,** 216-219.

Jørgensen,H., Serena,A., Hedemann,M.S. and Bach Knudsen,K.E. (2007) The fermentative capacity of growing pigs and adult sows fed diets with contrasting type and level of dietary fibre. *Livestock Science,* **109,** 111-114.

Just,A., Fernández,J.A. and Jørgensen,H. (1983a) The net energy value of diets for growth in pigs in relation to the fermentative processes in the digestive tract and the site of absorption of the nutrients. *Livestock Production Science,* **10,** 171-186.

Just,A., Jørgensen,H., Fernández,J.A., Bech-Andersen,S. and Hansen,N.E. (1983b) *The Chemical Composition, Digestibility, Energy and Protein Value of Different Feedstuffs for Pigs.* 556. Bert., 99pp. National Institute of Animal Science, Copenhagen.

Kirchgessner,M., Kreuzer,M., Machmüller,A. and Roth-Maier,D.A. (1994) Evidence for a high efficiency of bacterial protein synthesis in the digestive tract of adult sows fed supplements of fibrous feedstuffs. *Animal Feed Science and Technology,* **46,** 293-306.

Le Goff,G., Le Groumellec,L., van Milgen,J., Dubois,S. and Noblet,J. (2002) Digestibility and metabolic utilisation of dietary energy in adult sows: influence of addition and origin of dietary fibre. *British Journal of Nutrition,* **87,** 325-335.

Longland,A.C., Low,A.G., Quelch,D.B. and Bray,S.P. (1993) Adaptation to the digestion of non-starch polysaccharide in growing pigs fed on cereal or semi-purified basal diets. *British Journal of Nutrition*, **70**, 557-566.

McConnell,A.A., Eastwood,M.A. and Mitchell,W.D. (1974) Physical characteristics of vegetable foodstuffs that could influence bowel function. *Journal of the Science of Food and Agriculture*, **25**, 1457-1464.

Meunier-Salaun,M.C. (2001) Fibre in diets of sows. (ed. by P.C.Garnsworthy and J.Wiseman), pp. 323-339. Nottingham University Press, Nottingham.

Mosenthin,R., Sauer,W.C., Henkel,H., Ahrens,F. and de Lange,C.F.M. (1992) Tracer studies of urea kinetics in growing pigs. II. The effect of starch infusion at the distal ileum on urea recycling and bacterial nitrogen excretion. *Journal of Animal Science*, **70**, 3467-3472.

Mosenthin,R., Sauer,W.C., Völker,L. and Frigg,M. (1990) Synthesis and absorption of biotin in the large intestine of pigs. *Livestock Production Science*, **25**, 95-103.

Müller,H.L., Kirchgessner,M. and Roth,F.X. (1991) Energetic Efficiency of a Mixture of Acetic and Propionic Acid in Sows. *Journal of Animal Physiology and Animal Nutrition*, **65**, 140-145.

Noblet,J. and Bach Knudsen,K.E. (1997) Comparative digestibility of wheat, maize and sugar beet pulp non-starch polysaccharides in adult sows and growing pigs. *Energy Metabolism of Farm Animals* (ed. by K.J.McCracken, E.F.Unsworth and A.R.G.Wylie), pp. 371-374. CAB International, Wallingford.

Noblet,J., Fortune,H., Shi,X.S. and Dubois,S. (1994) Prediction of net energy value of feeds for growing pigs. *Journal of Animal Science*, **72**, 344-354.

Noblet,J. and Le Goff,G. (2001) Effect of dietary fibre on the energy value of feeds for pigs. *Animal Feed Science and Technology*, **90**, 35-52.

Noblet,J. and Shi,X.S. (1993) Comparative digestibility of energy and nutrients in growing pigs fed ad libitum and adults sows fed at maintenance. *Livestock Production Science*, **34**, 137-152.

O'Sullivan,T.C., Lynch,P.B., Morrissey,P.A. and O'Grady,J.F. (2003) Evaluation of citrus pulp in diets for sows and growing pigs. *Irish Journal of Agricultural and Food Research*, **42**, 243-253.

Petkevicius,S., Knudsen,K.E.B., Nansen,P., Roepstorff,A., Skjoth,F. and Jensen,K. (1997) The impact of diets varying in carbohydrates resistant to endogenous enzymes and lignin on populations of *Ascaris suum* and *Oesophagostomum dentatum* in pigs. *Parasitology*, **114**, 555-568.

Ramonet,Y., Meunier-Salaün,M.C. and Dourmad,J.Y. (1999) High-fiber diets in pregnant sows: digestive utilization and effects on the behavior of the animals. *Journal of Animal Science*, **77**, 591-599.

Raven,P.H., Evert,R.F. and Eichhorn,S.E. (1992) *Biology of Plants*. Worth Publishers, New York.

Read,N., French,S. and Cunningham,K. (1994) The role of the gut in regulating food intake in man. *Nutrition Reviews*, **52**, 1-10.

Rolin,C., Nielsen,B.U. and Glahn,P.E. (1998) Pectin. *Polysaccharides. Structural Diversity and Functional Versatility* (ed. by S.Dumitriu), pp. 377-431. Marcel Dekker, Inc., New York; Basel; Hong Kong.

Roth,F.X., Kirchgessner,M. and Müller,H.L. (1988) Energetische Verwertung von intracaecal infundierter Essig- und Propionsäure bei Sauen. *Journal of Animal Physiology and Animal Nutrition*, **59**, 211-217.

Selvendran,R.R. (1984) The plant cell wall as a source of dietary fiber: chemistry and structure. *American Journal of Clinical Nutrition*, **39**, 320-337.

Serena,A. (2005) *Physiological Properties of Dietary Carbohydrates for Sows - PhD Thesis*. Danish Institute of Agricultural Sciences, Department of Animal Health, Welfare and Nutrition, Research Centre Foulum, Tjele.

Serena,A. and Bach Knudsen,K.E. (2007) Chemical and physicochemical characterisation of co-products from the vegetable food and agro industries. *Animal Feed Science and Technology*, **in press**.

Sørensen,P. and Fernández,J.A. (2003) Dietary effects on the composition of pig slurry and on the plant utilization of pig slurry nitrogen. *Journal of Agricultural Science*, **140**, 343-355.

Van Heerden,I., Cronj,,C., Swart,S.H. and Kotz,,J.M. (2002) Microbial, chemical and physical aspects of citrus waste composting. *Bioresource Technology*, **81**, 71-76.

Varel,V.H., Robinson,I.M. and Jung,H.J.G. (1987) Influence of dietary fiber on xylanolytic and cellulotic bacteria of adult pigs. *Applied and Environmental Microbiology*, **53**, 22-26.

Vestergaard,E.M. (1997) *The Effect of Dietary Fibre on Welfare and Productivity of Sows - PhD Thesis*. Danish Institute of Agricultural Science, Tjele and The Royal Veterinary and Agricultural University, Copenhagen, Foulum.

27

FOOD INTAKE AND PERFORMANCE OF PIGS DURING HEALTH, DISEASE AND RECOVERY

ILIAS KYRIAZAKIS[1, 2] AND JOS G M HOUDIJK[1]
[1]*Animal Nutrition and Health Department, SAC, West Mains Road, Edinburgh, EH9 3JG*
[2]*Faculty of Veterinary Medicine, University of Thessaly, PO Box 199, 43100, Karditsa, Greece*

Introduction

In the preceding conference of the series, Kyriazakis (2003) presented a framework that was able to account for the performance and food intake of healthy pigs. The framework, summarised on Figure 27.1, makes the genotype and state of the pig the defining force of maximum pig performance, which is achieved through food intake. Whether maximum performance is achieved depends on the composition of the food and the environment in which the pig is kept. This is because the environment imposes constraints on the ability of the pig to consume adequate food of a given composition. Such environmental constraints include the ambient temperature, competition amongst pigs and space allowance. At the same time constraints on food intake and hence performance may arise from the capacity of the pig to deal with the required food. If, for example, the dietary energy contcentration is low, the pig will need to consume more food than from a high dietary energy concentration in order to achieve the same amount dietary energy intake. Whether the pig is able to achieve this will depend on its capacity to deal with the extra bulkiness and nutrients that result from the higher food intake. If the pig is unable to overcome such constraints, then its food intake will be reduced and the *actual* performance of the pig will be below its *maximum.*

The above framework has been successfully applied to predict the performance of healthy pigs kept under different environments (eg Wellock *et al.*, 2003a,b). At the same time it has been used to define the environmental and food composition requirements of pigs of different genotypes. As genetic selection changes the rate and composition of pig growth, then the composition of the food and the environment in which the pig is kept will also need to change, if the maximum performance of the new genotype is to be achieved (Emmans and Kyriazakis, 2001). Until recently, however, the framework had not been applied to account for the performance of pigs exposed to infectious

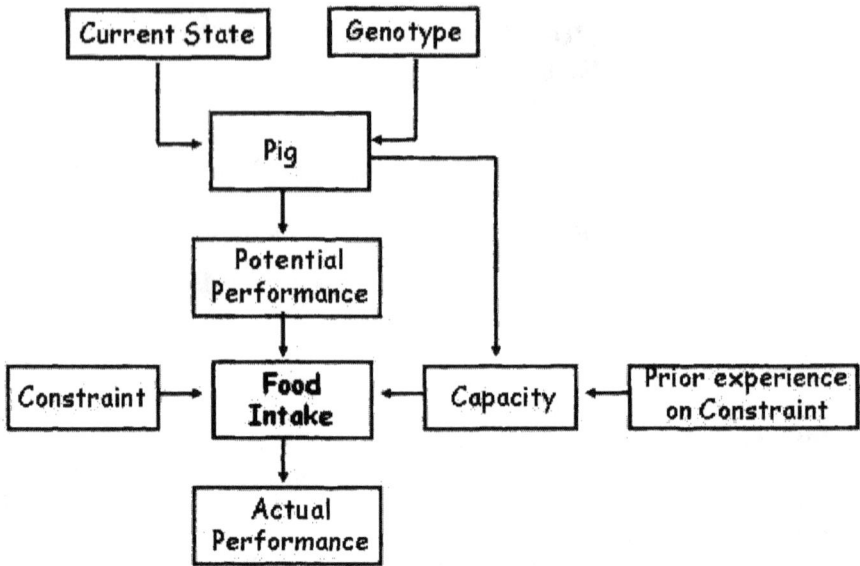

Figure 27.1 A schematic description of the framework developed to predict the food intake and performance of healthy pigs. For details see text.

pathogens. This is because there has been no general agreement over the issues that need to be taken into account in order to expand the framework in this dimension. An added complexity to the problem has arisen from the fact that recent trends in genetic improvement are increasingly taking into account traits that characterise the ability of the pigs to cope with infectious pathogens and the resultant disease, alongside performance traits.

The aim of the current chapter is to characterise changes in food intake and performance of pigs during exposure to pathogens. By doing so, it is planned to develop a framework that is able to account for the effects of the infectious environment on different pig genotypes kept in different environments, including access to foods of different composition. Through the actions of the immune response pigs are eventually able to contain pathogens and minimise their consequences on performance, although this may come with a cost associated with immunopathology. The issue then is to account for the performance of such pigs during recovery from the consequences of infection. This constitutes the focus of the latter part of the chapter.

The consequences of exposure to pathogens on food intake

The starting point is a pig that has no prior exposure to a specific pathogen, i.e. it is immunologically naïve. The consistent feature of the exposure to pathogens in hosts without prior experience of a pathogen is a voluntary reduction in the food intake of the animal (Figure 27.2). This temporary reduction in food intake, henceforth called *anorexia,* is also apparent when the immune system of the animal is stimulated by non-pathogenic antigens, such as though lipo-polysaccharide (LPS) stimulation. This implies that there is a link between the occurrence of anorexia and the immune system (see below). The characteristics of anorexia are: (1) its rate of development, i.e. how quickly it manifests following exposure, (2) its extent or degree, i.e. by how much food intake reduces, compared to a similar pig that has not been challenged, (3) its duration and (4) the rate of recovery. These characteristics are greatly influenced by the type of pathogen and its amount (dose), and it is possible that they may be influenced by the genotype of the pig and the environment it is kept in. Below we examine these influences on the first three characteristics of anorexia by giving emphasis on subclinical disease. This implies that we deal mainly with relatively small reductions in the food intake of the pigs. These reductions are at the same time responsible for the reductions in the performance of pigs seen in commercial enterprises. The last characteristic will be addressed in the latter parts of the chapter.

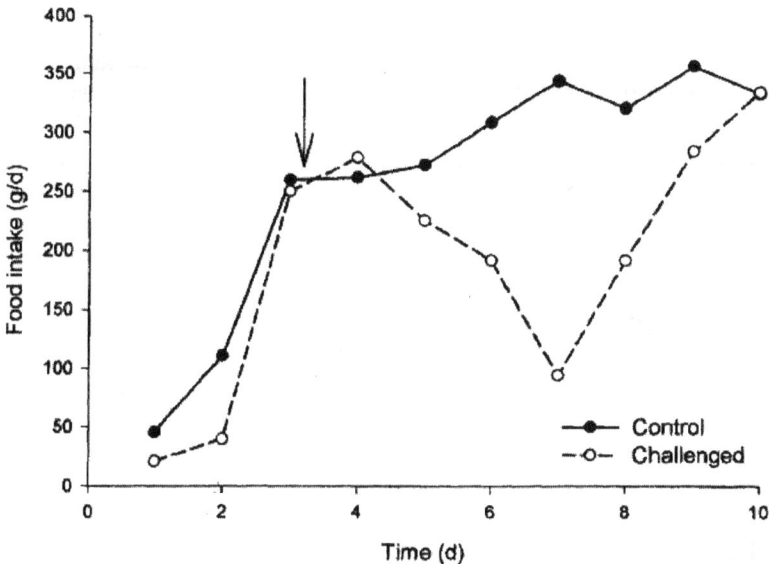

Figure 27.2 The effect of a challenge with 10^8 colony forming units of enterotoxigenic *Escherichia coli* on the food intake of control and challenged pigs over a 10 day time period post weaning (from Houdijk *et al.,* 2007). The arrow indicates the point of infection.

EFFECT OF PATHOGEN TYPE AND DOSE

The rate of development of anorexia is greatly dependent upon the type of pathogen to which the pig is exposed. This ranges from a few hours or days for bacterial and viral challenges (Houdijk *et al.*, 2007; Greiner *et al.*, 2000), to several days or weeks for macroparasitic challenges, such as gastrointestinal helminths (Hale, 1985). A possible explanation for these differences may be that greater amounts of antigen are presented from bacterial and viral pathogens, due to their greater rates of proliferation, than are presented by non-proliferating macroparasites that accumulate in the host over a period of time (May and Nowak, 1995). On the other hand, it may be due to recognition being different, with macroparasites avoiding detection. The above suggestions imply that not only pathogen type, but also its dose will affect the rate of anorexia development. This has indeed been observed by both Houdijk *et al.* (2007) and Fraser *et al.* (2007) who have shown that low doses of bacterial pathogens such as *Escherichia coli* and *Salmonella enterica* either do not result in anorexia, or pigs need repeated doses of these pathogens for a reduction in food intake to be observed. Currently it is unknown whether pathogen virulence may also affect the rate of anorexia development. Here, it is hypothesized that different serovars of the same pathogen that lead to different consequences on the host (eg local vs systemic consequences) may also affect the rate of anorexia development through, for example, their rate of recognition by the immune system.

Similarly the extent of anorexia is greatly affected by pathogen type and its dose. Higher pathogen doses lead to very dramatic decreases in the food intake and even complete cessation of feeding (Black *et al.*, 1999). However, there is a wide range of pathogen doses, which are associated with subclinical disease, over which food intake is reduced to 0.75-0.80 of the expected voluntary food intake. For this reason, Sandberg *et al.* (2006) have suggested that for all intents and purposes, the extent of anorexia during subclinical disease should be considered independent of pathogen type, and a constant reduction of 0.25 to the food intake of non-infected counterparts, provided they are not offered limiting foods (see below), should be considered as the outcome of exposure to pathogens. This suggestion is schematically shown in Figure 27.3 that is reproduced from Kyriazakis *et al.* (1998). However, it should be emphasised that the extent of anorexia will be substantially affected both by pathogen type and dose, beyond the threshold (dose) that leads to clinical disease.

The duration of anorexia may reflect the time taken by the immune system to begin controlling and subsequently eliminating the pathogens. It is thus seen as a reflection of the rate of acquisition of acquired immunity and needs to be seen as being specific to the type of pathogen. Duration of anorexia can be very short, as in the example shown in Figure 2, and in the cases of exposure to *Salmonella typhimurium* (Turner *et al.*, 2002) and the porcine reproductive and respiratory syndrome (PRRS) virus (Greiner *et al.*, 2000). On the other hand, duration of anorexia can last for several days or weeks for gastrointestinal parasite challenges (Hale, 1985). Given also that the rate of acquisition of immunity appears to be independent of both the pathogen dose and its virulence

(Sandberg *et al.*, 2006), it is suggested that anorexia duration will be independent of these factors.

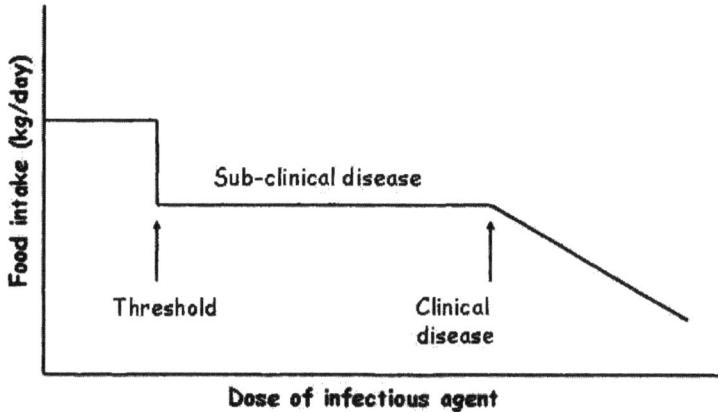

Figure 27.3 A schematic description of the effect of dose of an infectious agent on the daily rate of food intakes of pigs that have not been previously exposed to the infectious agent (from Kyriazakis *et al.*, 1998).

EFFECT OF HOST GENOTYPE

Currently there is no information to link host genotype to the features of anorexia during exposure to pathogens in pigs. This is despite the fact that pigs have been selected for increased cellular and humoral immune responses to a wide range of pathogens (Wilkie and Mallard, 1998). Above it was suggested that there was a link between anorexia and the stimulation of the (acquired) immune response of the host by pathogens. This suggestion is further supported by Greer *et al.* (2005) who found that immuno-suppression during the phase of acquisition of immunity, prevented immunologically naïve sheep challenged with gastrointestinal nematodes to develop any anorexia. It is possible, therefore, that differences in the ability of pigs to cope with pathogens also lead to differences in the rate of development, extent and duration of anorexia.

A genotype that is more capable of coping with a pathogen challenge is here defined as a genetically resistant genotype. This resistance may manifest as anorexia being observed at a higher pathogen dose and its duration being shorter. This is because in resistant genotypes the innate immune system may be more capable of dealing with a certain pathogen dose (Wakelin, 2000) and thus the acquired immune response may not be stimulated. In a similar manner, resistant genotypes may be able to acquire immunity towards the pathogen at a faster rate and as a consequence reduce their pathogen load. This in turn may result in a shorter duration of anorexia.

Similarly there is no information in the literature on whether pigs selected for productive traits (e.g. faster growth) differ in their anorexia responses to pathogens. With other things being equal, the expectation is that such genetic selection should not affect the dynamics of anorexia following exposure to pathogens.

EFFECT OF FOOD COMPOSITION

There is some evidence in the literature that the composition of the food offered to pigs exposed to pathogens may affect the characteristics of anorexia. This comes from the experiments of Williams *et al.* (1997a) and is shown in Table 27.1. In these experiments pigs were offered *ad libitum* foods of difference protein (lysine) contents in environments intended to have either high or low levels of "stimulation of the immune system". This was achieved by exposure to a "dirty environment", in the absence of dietary antimicrobials and through repeated vaccination. The data suggest that as the lysine content of the food was reduced the difference in the food intake between the "control" and "immune stimulated" pigs was reduced; the food intake of these two groups of pigs was identical at the lowest level of lysine. This effect can be accounted for by the fact that at the lowest level of lysine the food intake of the "control" pigs was reduced, compared to the intake of the control pigs on the highest lysine level. Wellock *et al.* (2003 b) have suggested that the inability of the pigs to increase their food intake on the low lysine food is due to their inability to cope with the excess energy intake that would accompany this situation. Food intake of pigs on the lowest lysine level was thus constrained. Stimulation of the immune system would not reduce any further the already constrained actual food intake.

Table 27.1 THE FOOD INTAKE AND LIVEWEIGHT GAIN OF PIGS THAT WERE KEPT EITHER IN A CLEAN ENVIRONMENT (C) OR A HIGH IMMUNE SYSTEM ACTIVATION ENVIRONMENT (IS) AND OFFERED FOODS OF DIFFERENT DIETARY LYSINE CONTENT FROM 6 TO 27 kg BODY WEIGHT (WILLIAMS *et al.*, 1997 a).

	Immune Activation	*Dietary Lysine (g/kg)*			
		6	*9*	*12*	*15*
Food Intake (g/d)	C	896	1025	1052	1002
	IS	889	954	889	911
	Ratio IS: C	0.99	0.93	0.85	0.91
Gain (g/d)	C	400	556	644	663
	IS	357	495	510	504
	Ratio IS: C	0.89	0.89	0.79	0.76

There is also some evidence from other species (sheep, Kyriazakis *et al.*, 1996; Zaralis *et al.*, 2007) to suggest that the nutrient content of the food offered to animals exposed to pathogens affects the duration, but not the extent, of anorexia. This can be accounted for by the fact that the immune response may be affected by the level of host nutrition (Coop and Kyriazakis, 1999; see below) and hosts on a higher nutrient content food may eliminate pathogens at a faster rate. In this instance the duration of anorexia would also be expected to be shorter.

Different conclusions on the effect of food composition on the extent and duration of anorexia may be drawn when comparisons between control (uninfected) and infected pigs are made at a weight or a time basis. This can be accounted for by the different growth trajectories of the control and infected pigs, which in turn may have direct effects on their food intake. Thus, for a fair comparison, control and infected pigs may best be compared at either the same body weight, or through scaling food intake relative to body weight.

MECHANISMS OF ANOREXIA DEVELOPMENT

Several factors have been proposed as the specific cause of reduction in food intake during pathogen challenges (Johnson, 1998; Broussard *et al.*, 2001). These factors include members of the cytokine family (Langhans, 2000) and leptin (Grunfeld *et al.*, 1996). These physiological parameters may have multiple effects on the host, as well effects on food intake. They are unlikely, therefore, to provide an answer how anorexia during exposure to pathogens comes about and, therefore, advance our predictive framework.

Sandberg *et al.* (2006) have proposed that there are at least two functional mechanisms that may lead to a reduction in food intake during exposure to pathogens: (i) food intake is reduced because the potential for protein retention of the challenged animals is reduced, and (ii) the reduction in food intake is a direct consequence of the exposure to pathogens. Black *et al.* (1999) have argued that both mechanisms may be acting simultaneously. Mechanism (i) assumes a direct relationship between pathogen load and the maximum growth rate of the animal. It is consistent with the approach of Wellock *et al.* (2003b), who suggested that all environmental stressors lower the maximum growth in the pig. This in turn leads to a decrease in the maximum daily gain the pig is able to achieve and hence a reduction in its required food intake. Mechanism (ii) has been favoured by Sandberg *et al.* (2006), who suggested that the reduction in the voluntary food intake of the challenged animal was made a function of the pig's pathogen load. Appropriately designed experiments can distinguish between the two mechanisms and lead to consistent predictions of the effect of pathogen challenges on the food intake of pigs. This will be discussed in further detail subsequently when dealing with the consequences of exposure to pathogens on the maximum capacity for growth.

The consequences of exposure to pathogens on performance

The experiment of Williams *et al.* (1997a), described above, is a typical example of the consequences of exposure to pathogens on pig performance. The performances of the "control" and "immune stimulated" pigs given access to foods of different protein (lysine) content have been plotted on Figure 27.4.

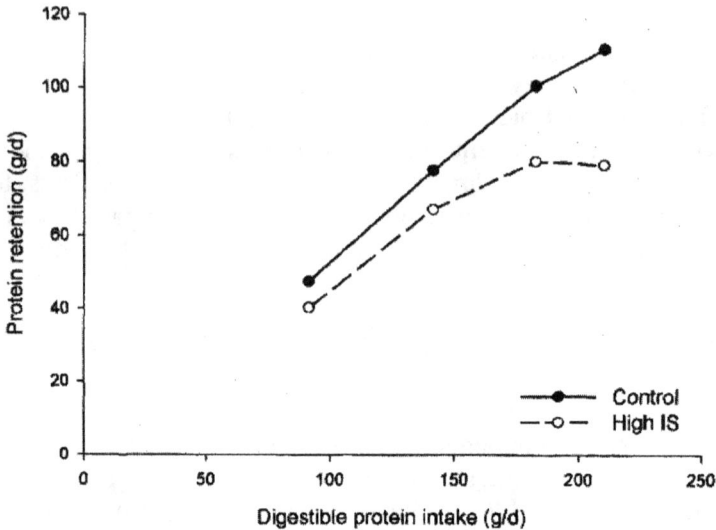

Figure 27.4 The response in protein retention to digestible protein intake for pigs that were kept either in a clean environment (Control) or a high immune system activation environment (High IS) from 6 to 27kg body weight (from Williams *et al.*, 1997a).

These consequences can be summarised as follows:

1.   An increase in the level of lysine intake was required to achieve a certain level of protein retention (PR) below maximum, in challenged pigs. This would be equivalent to saying that the maintenance requirements of the challenged pigs were increased, in this case by a factor of ~1.6, as estimated from extrapolation of the data to zero protein retention.
2.   No changes in the marginal response of PR to protein intake, i.e. the slope of the response, was observed between challenged and unchallenged pigs.
3.   Challenged pigs did not achieve the same maximum rate of PR. In the case of the above experiment this was estimated as 0.74 of the "control" pigs, which coincides very well with the aforementioned 0.25 expected reduction in food intake for anorexia as a general rule.

The causes of these effects will be considered in turn.

EFFECTS OF PATHOGEN CHALLENGE ON MAINTENANCE

In the experiment of Williams *et al.* (1997a) the associated increase in maintenance requirements for protein of challenged pigs was approximately twice as much as that of their controls. The same magnitude of increase was seen in the experiment of Williams *et al.* (1997b) on bigger pigs whose immune system was challenged through the same methodology. Experiments on poultry (Webel *et al.*, 1998a,b) suggest that chicks challenged with the antigenic stimulant LPS show a small increase in maintenance requirements for certain amino acids (e.g. lysine), but not for others (e.g. arginine). These experiments strongly suggest that it might be necessary to consider the maintenance requirements for individual amino acids to account fully for reductions in growth during pathogen challenges.

Various experiments have reported an increase in heat production of pigs challenged with different pathogens (e.g. van Diemen *et al.*, 1995). The experiments of Fagbemi *et al.* (1991) and Otesile *et al.* (1991) measured the energy retention of pigs challenged with the same dose of *Trypanosoma brucei* at two different liveweights. The 10 kg pigs had a greater increase in maintenance requirements for energy (2.2 times of the healthy controls),than the 100 kg pigs, that had maintenance requirements 1.7 times of the healthy controls. This suggests that pigs of different sizes may have different changes in maintenance requirements at the same challenge dose, although these differences may be accounted for, at least to some extent, by scaling to body weight.

Causes of increased requirements during pathogen challenges

Pathogen challenges may cause increased requirements for protein due to the requirements for the immune response, and the repair and replenishment of damaged tissues. Given that the stimulation of the effector mechanisms of the immune response will depend on the nature of the pathogen challenge, it is reasonable to suggest that both pathogen type and dose will affect maintenance protein requirements. The experiments of Webel *et al.* (1998a,b), where chicks were exposed to LPS challenge, would have mainly stimulated the innate immune response (e.g. the production of acute phase proteins) and for this reason the increases in maintenance requirements were relatively small. Differences between hosts in their ability to mount an immune response also open up the possibility that host genotype may affect such maintenance requirements when pigs are challenged by the same pathogen.

Pathogens may cause damage to a host's tissues (e.g. gut wall) or specific cells (e.g. red blood cells) and cause body fluids to leave their natural compartments, such as plasma leaking into the gastrointestinal tract. The pig would need to repair such damage, or replace lost fluids to maintain normal function that is, thus, a direct cost to the animal (Berendt *et al.*, 1977). Pigs have been challenged by different parasitic helminths that affected the different internal organs of the stomach, small intestine and kidneys (Hale, 1985). The consequences of parasitism on N metabolism depended on the organ affected. The kidney parasite was not associated with any measurable effects on N metabolism, whereas the intestinal parasites were associated with increased N in the faeces resulting from the associated damaged tissues and endogenous secretions. At a given level of N intake pigs thus parasitised had a lower level of N retention than their respective controls. Therefore, the type of pathogen and by extension the organ(s) affected need to be taken into account when considering the costs of repair and replacement of damaged tissues.

Literature evidence also suggests that the extent of the costs associated with damage is not only dependant on pathogen type, but also on the level of pathogen challenge. Powanda *et al.* (1975) have suggested that the relationship is of an exponential type. This may indicate that a host that has already suffered a certain amount of damage, may be less capable to deal with additional pathogen challenges.

Increases in energy requirements during a pathogen challenge may arise from the functioning of the immune system, the cost associated with damage and repair (such as additional N processing) and expression of fever. It has been proposed that the energetic costs of the former two processes are relatively small (Sandberg *et al.*, 2007). In the case of gastrointestinal parasites, the most significant energetic cost appears to be associated with damaged tissues resulting in additional amounts of N appearing in the urine (MacRae *et al.*, 1982). Such costs have not been estimated directly on pigs, but would be expected to contribute to a reduced efficiency of food use. Appropriately designed pair-feeding experiments could shed light on this.

The most significant increases in energy requirements due to exposure to pathogens appear to be associated with the expression of fever. A febrile response accompanies most bacterial, viral and parasitic infections of pigs, although it appears to be absent in infections where pathogens are localised. For example, van Diemen *et al.* (1995) did not record any increase in body temperature in pigs that were suffering from atrophic rhinitis. Similarly, gastrointestinal parasitism (Hale, 1985) and sub-clinical post weaning colibacillosis (Houdijk *et al.* 2007) did not seem to be accompanied by fever. Sandberg *et al.* (2007) linked the febrile response to acquired immunity and saw the duration of fever as a function of pathogen load. There appear to be parallels in the febrile response and the anorexia due to exposure to pathogens in terms of their relationship to pathogen dose.

Experiments with both "artificial" antigenic challenges and pathogens have shown increases in maintenance requirements for energy that ranged from 1.05 – 1.35 to that of unchallenged animals. These estimates not only involved the energetic costs of the immune response and tissue repair, but they have been taken in the presence of anorexia and in situations where challenged animals with fever were showing behavioural coping responses (eg huddling).

They may, therefore, be underestimates of the direct increases in energy requirements due to fever. The previously reported experiments of Fagbemi *et al.* (1991) and Otesile *et al.* (1991) have suggested significant greater increases in maintenance requirements for energy. In these experiments these requirements were measured through marginal responses.

EFFECTS OF PATHOGENS CHALLENGES ON THE MARGINAL RESPONSES TO NUTRIENT INTAKE

The earlier described experiments of Williams *et al.* (1997a,b) on pigs and Webel *et al.* (1998a,b) on chicks found no effect on the marginal responses in PR to protein intake. However, there are other experiments in the literature that suggest that marginal responses in growth to protein or amino acid intake are affected by exposure to pathogens. Willis and Baker (1981a,b), for example, challenged chicks with different doses of the protozoan *Eimeria acervulina*, whilst offering them different amounts of amino acid intake. The slope of the response in liveweight gain was only marginally affected by the low doses of oocytes, in comparison to the response of the unchallenged chicks (10% reduction, which was not significantly different). However, the high oocyte dose reduced significantly (by 22%) the marginal response (i.e. the slope of the response). These results suggest that the marginal response to nutrient intake is affected by pathogen type and dose, and may be more pronounced at times of clinical disease compared to sub-clinical disease.

Differences in the slope of the marginal response to nutrient intake between challenged and uninfected pigs would be expected to arise when the following three conditions are met: (1) the requirement of the immune response towards a pathogen is significant, (2) the nutrient (e.g. amino acid) composition of the immune response is different from that of growth and (3) there is a competition between growth and immune functions. There is now sufficient evidence in the literature in support of (1) and (2), as has been extensively reviewed by Sandberg *et al.* (2007). The acquired immune response has both cellular and humoral components, that also contain large amounts of protein; as a consequence their production would contribute towards an additional requirement for protein or amino acids during pathogen challenges. In addition there are very marked differences in terms of amino acid composition between the immune proteins and the whole body protein of pigs.

There is also more circumstantial evidence in the literature in support of condition (3). One or more components of the effector arms of the immune response appear to respond to intake of nutrient intake above maintenance during exposure to pathogens (van Heugten *et al.*, 1995; Spurlock *et al.*, 1997; Johansen *et al.*, 1997). It would be fair to say that this response does not seem to be systematic or repeatable. An example of such a response is given on Figure 27.5. Bhargava *et al.* (1970) found that both growth rates and antibody titres increased in chicks challenged with the Newcastle virus, when they were given increasingly intakes of valine. Interestingly, the response in antibody titer increased further, when the chicks had reached a plateau in their growth response. This response suggests that there may be a partitioning of scarce resources between growth and the immune response in animals challenged by pathogen. Moreover, the pattern of response of growth and immune response to stepwise increments of valine would suggest a higher priority for scarce nutrient allocation to growth functions over immune response, as has been hypothesized in a nutrient partitioning framework (Coop and Kyriazakis, 1999).

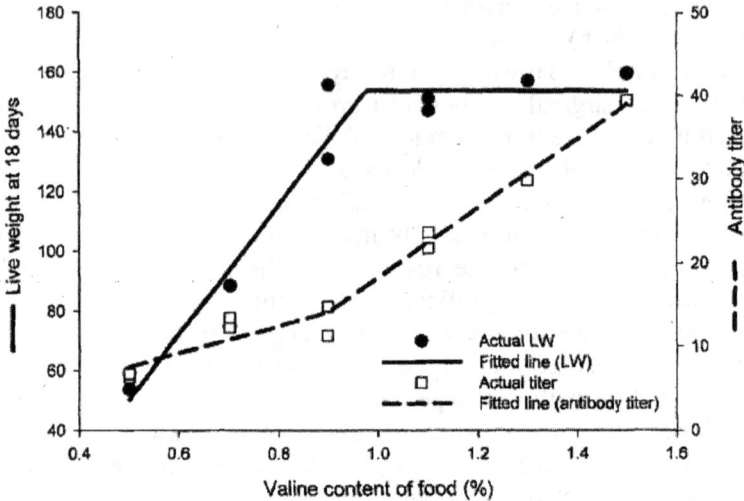

Figure 27.5 The responses in live weight over 18 days (LW) and antibody titers (AT) to increasing valine contents (V) of a food for chicks challenged with the Newcastle virus (from Bhargava *et al.*, 1970). Linear plateau response was approximated as a linear function of valine content (LW=217.V-58.2) until the plateau of 153.6 was reached. The linear regressions of antibody titer against valine content until maximum LW was reached was AT=16.37.V-0.6325, whereas for the subsequence phase was AT=41.35.V-23.04.

The question is how the findings of the experiments that find an effect of pathogen challenges on the marginal responses to nutrients with those that do not can be reconciled In the experiments of Williams *et al.* (1997a,b) and Webel *et al.* (1998a,b) the animals were stimulated by "mild" antigenic stimulants (a dirty environment and LPS challenge respectively). It is possible that such a stimulation of the immune system did not sufficiently increase the requirements for the immune response. As a consequence, any changes in the marginal responses to amino acid and intakes were either absent or non detectable. Therefore, such differences may be ascribed to different pathogens and/or doses uses across the difference experiments.

The idea that immune and growth functions compete for scarce nutrient resources above maintenance is contrary to the traditional concepts of nutrient partitioning that consider that requirements for the immune response need to be met before any growth to occur. As a consequence any increases in requirements for the function of the immune response would be expected to be seen only as increases in maintenance requirement. Some authors have proposed that pathogen challenges *always* lead to a release in amino acids from body protein stores such as muscle (Klasing *et al.*, 1987). This may be the case in the case of sub-maintenance intakes, where there is some logic for such an animal to maintain a degree of immunity. However, it is unlikely that an animal would preferentially utilise body protein when dietary resources are available. Here, it is concluded that some properly designed experiments to measure the effects of nutrient intake above maintenance on components of the immune response will further advance understanding of the issue addressed in this section. Recent advances in the detection and quantification of the immune response to pathogens make this now possible.

EFFECTS OF PATHOGEN CHALLENGE ON THE UPPER-LIMIT FOR GROWTH

It would appear that, during pathogen challenges, animals cannot achieve the same upper limit for growth as their healthy controls (Figure 27.4 and Table 27.1). This has led authors to conclude that the maximum or upper limit for growth of challenged animals is reduced (Escobar *et al.*, 2004). In the first instance, the results of Williams *et al.* (1997a,b) agree with this interpretation, as increases in protein intake did not lead to any changes in PR beyond a certain level. However, this response can also arise for other reasons.

In the experiments of Williams *et al.* (1997a,b) the increase in protein intake was achieved by changing the ratio of the energy to protein content in the food. It is well established that a plateau in PR in pigs can be achieved when there is insufficient amount of energy in relation to the protein intake (Kyriazakis and Emmans, 1992). The highest protein to energy ratio food used by Williams *et al.* (1997a,b) to achieve the higher level of protein intake in challenged pigs, could have led to such a situation.

On the other hand a number of experiments on pathogen-challenged animals are conducted over a period of time, rather than over a weight range. A transient decrease (Figure 27.2) in the food intake of the challenged animal *per se* will have direct consequences on growth and hence the size of the pig. Thus considering the upper limit for growth over a period of time could include both the phase of anorexia and recovery, over which the size of the pig is reduced. A smaller sized pig during recovery will also have a smaller upper limit for growth at a time.

Whether pathogen challenges affect directly the upper limit for growth is still a matter of debate. Carefully designed experiments may contribute towards our understanding of how growth is reduced during pathogen challenges.

The performance of pigs during recovery from disease

The immune response enables the animal to eventually eliminate the pathogens from and contain their consequences on its system. Alternatively, a pharmaceutical intervention may lead to the same outcome, but the animal might not have had the opportunity to have developed its immune response in full. This naturally or artificially-induced point will be termed as "the start of recovery". On the basis of the framework presented on Figure 27.1 and developed briefly in the Introduction, the growth of a pig previously exposed to pathogens will depend on its state at this point in time.

Because the major consequence of exposure to pathogens is a reduction in the intake of and resources from a given food, a pig exposed to pathogens will be delayed in its growth. In other words, it will be smaller in size than its uninfected control at the start of recovery. Its body composition may also be different from the control pig, but this will depend on the composition of the food the pigs have had access to during the period of exposure (Stamataris *et al.*, 1991). This view concurs with the results of Bassaganya–Riera *et al.* (2001) and Escobar *et al.* (2004) who exposed pigs to either a "dirty environment" or to PRRS virus respectively. In both cases pigs were delayed in their growth and had a lower lipid to protein in their bodies compared to unchallenged pigs.

The question is whether the pig will attempt to correct any of the above differences from its controls, when pathogens are removed from its body naturally or artificially. For the purposes of the current chapter, it has been assumed that exposure to pathogens has had no irreversible damage on any of the organs and their functioning in the pig, although it is appreciated that this might not always be the case. The latter may be the outcome of prolonged clinical disease.

THE GROWTH OF PIGS DURING RECOVERY FROM EXPOSURE TO PATHOGENS

Because of the paucity of information below, we rely on the literature of the growth of pigs following a period of food restriction. Such restriction leads to similar body composition changes to pigs whose food intake has been reduced due to pathogen induced anorexia (Bassaganya-Riera *et al.*, 2001; Escobar *et al.*, 2004). In such instances the pigs will be expected to attempt to restore their fatness to a level determined by their genotype; in other words they would show compensatory fattening (Stamataris *et al.*, 1991). Providing that the period of recovery is sufficiently long, this compensatory fattening should result in the same end lipid weights in the carcass (Kyriazakis and Emmans, 1992).

The issue of whether pigs will also show compensatory growth in protein gain and thus reduce some of the time lost during the period of food restriction is more contentious. Kyriazakis and Emmans (1992) have summarised an extensive body of literature on growth following re-alimentation in pigs and concluded that compensatory growth in protein of skeletal muscle does not seem to exist. In other words time lost in the growth of muscle protein can never be regained.

A compensatory gain in lipid and no compensatory gain in body protein will manifest as compensatory liveweight gain during recovery until body fatness is restored. On the other hand, if the level of fatness has been undisturbed by the exposure to pathogens, then liveweight gain will not show any compensatory gain. This seems to have been the case in the experiment of Hein (1968), whose findings are reproduced on Figure 27.6. In this experiment chicks were infected with different doses of *Eimeria acervulina* oocysts. The higher number of oocysts reduced liveweight gain to a greater extent that the lower numbers, as expected. Following recovery, the growth of animals appeared to be parallel to that of the animals and was dictated by the extent to which the healthy and challenged animals differed in size after they had overcome the infection.

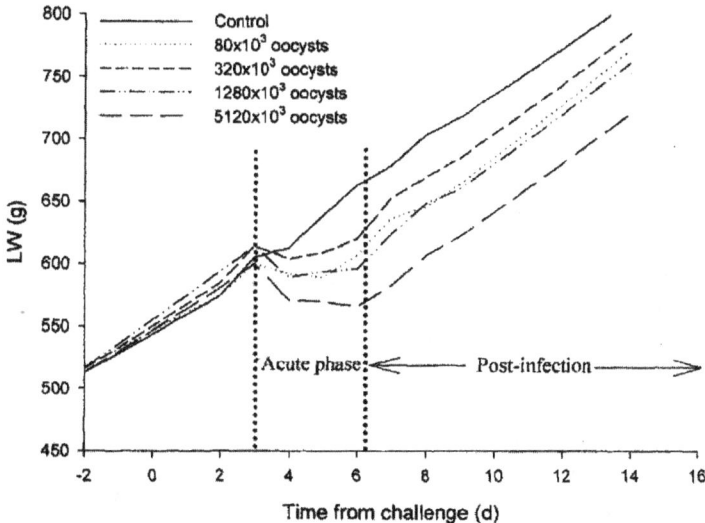

Figure 27.6 The effect of different single challenge doses of *Eimeria acervulina* on the live weight, *LW* g, of chicks over the time course of an experiment, including the infection (acute phase) and recovery (post infection) phases (from Hein, 1968).

The above conclusions have been drawn in the absence of suitable experiments that observed the growth of pigs during recovery from exposure to pathogens.

Properly designed experiments that record this and dissect the composition of the growth of the pigs during recovery would be welcomed.

THE FOOD INTAKE OF PIGS DURING RECOVERY FROM EXPOSURE TO PATHOGENS

In order to restore any differences in its composition at the point of recovery from the uninfected controls, the pig will need to modify its food intake. If the pig attempts to show compensatory fattening, as suggested above, then the expectation is that the pig will also increase the rate of its food intake. The extent of the increase will depend on the level of fatness at the point of recovery (Stamataris *et al.*, 1991). It has not been possible to identify suitably designed experiments that have measured food intake during recovery in pigs previously exposed to pathogens. This is because, in most cases, the point of recovery can not be clearly identified and the food intake of animals is reported over a period of time that includes both anorexia and recovery.

The rate of food intake that enables restoration of the body composition will depend on the composition of the food, the digestive capacity of the pig and the environment it is kept in (Stamataris *et al.*, 1991, see also Figure 1). An increase in food intake will also increase the amount of heat production by the pig and for this reason the ambient temperature may define the extent of this increase (Wellock *et al.*, 2003a). In the literature of compensatory growth, the conditions of re-alimentation are considered responsible for the confusion that exists over the issue of if and how quickly pigs overcome the abnormalities in their body composition (Kyriazakis and Emmans, 1992). Here it is suggested that environmental requirements of a pig that is recovering from exposure to pathogens may be very different from an uninfected pig.

A special case to consider is the food intake of the pig that continues to be exposed to pathogens following recovery. This is a realistic scenario, as exposure to most pathogens is a constant feature of commercial practice. On the basis of the arguments developed in previous sections, this stimulation of the immune system will be associated with an increased nutrient requirement, even for a fully immune pig. This increase in requirements will be expected to be manifested as an increase in food intake. The extent of the increase will depend on the pathogen type and its dose, as discussed above. It has not been possible to identify relevant experiments to support this suggestion. It would be useful to investigate this issue, as it may affect the requirements of a pig continuously exposed to pathogens.

Conclusions

A framework that is able to account for the performance of pigs during exposure to pathogens has been developed. By doing so a number of issues have been identified where research effort can be usefully directed to in order to enhance the predictive value of the framework. As many characteristics of the effects of pathogens on pig performance are common across pathogen types and doses (eg degree of anorexia during subclinical disease), progress in certain components of the developed framework can be achieved relatively quickly.

The description of the genetic ability of the pig to cope with pathogens is central to the developed framework. The definition and quantification of the traits that characterise the pig in this way will further enhance the predictive value of the developed framework. This can be achieved by a closer communication between pig breeders and other animal scientists, such as nutritionists. Similarly, the characterisation of the pig's immune response has been a consistent limitation in our understanding of how animals deal with pathogens and has had implications upon our ability to enhance them. For example, under what circumstances can the immune response of a pig be enhanced through nutrition? Recent advances in the characterisation and quantification of the immune response should enable progress on this subject to be made.

As reliance to chemoprophylaxis to control farm animal pathogens is decreasing, due for example to consumer concerns, legislation and pathogen resistance to drugs (Olesen *et al.*, 2000), interest in the understanding of the performance of animals in the presence of pathogens will increase. A framework that accounts for the performance of pigs during exposure to pathogens may then have a value in the developments of strategies, including nutritional and breeding ones, to deal with this challenge.

References

Bassaganya-Riera, J., Hontecillas-Magarzo, R., Bregendahl, K., Wannesmuchier, M. J. and Zimmerman, D. R. (2001). Effects of dietary conjugated conjugated linopeic acid in nursery pigs of dirty and clean environments on growth, empty body composition and immune competence. *Journal of Animal Science,* **79:** 714-721.

Berendt, R. F., Long, G. G., Abeles, F. B. Canonico, P. G., Elwell, M. R. and Powanda, M. C. (1997). Pathogensis of respiratory *Klebsiella pneumoniae* infection in rats: bacterial and histological findings and metabolic alterations. *Infection and Immunity,* **15:** 586-593.

Bhargava, K. K., Hanson, R. P. and Sunde, M. L. (1970). Effects of methionine and valine on antibody production in chicks infected with Newcastle disease virus. *Journal of Nutrition,* **100:** 241-248.

Black, J. L., Bray, H. J. and Giles, L. R. (1999). The thermal and infectious environment. In *A Quantitative Biology of the Pig,* pp 71-97. Edited by I. Kyriazakis). CAB International, Wallingford.

Broussard, S. R., Zhou, J. H., Venters, H. D., Bluthe, R. M., Freund, G. G.,

Johnson, R. W., Dantzer, R. and Kelley, K. W. (2001). At the interface of environment-immune interactions: cytokine and growth-factor receptors. *Journal of Animal Science,* **79**(Suppl.E): E268-E284.

Coop, R. L. and Kyriazakis, I. (1999). Nutrition-parasite interaction. *Veterinary Parasitology,* **84**: 187-204.

Emmans, G. C. and Kyriazakis, I. (2001). Consequences of genetic change in farm animals on feed intake and feeding behaviour. *Proceedings of the Nutrition Society,* **60**: 115-125.

Escobar, J., Van Alstine, W. G., Baker, D. H. and Johnson, R. W. (2004). Decreased protein accretion in pigs with viral and bacterial pneumonia is associated with increased myostatin expression in muscle. *Journal of Nutrition,* **134**: 3047-3053.

Fagbemi, B. O., Otesile, E. B., Makinde M. O. and Akinboade O. A. (1990). The relationship between dietary energy levels and the severity of *Trypanosoma brucei* infection in growing pigs. *Veterinary Parasitology,* **35**: 29-42.

Fraser, J. N., Davis, B. L., Skjolaas, K. A., Burkey, T. E, Dritz, S. S., Johnson, B. J. and Minton, J. E. (2007). Effects of feeding *Salmonella enterica* serovar Typhimurium or serovar Cholerasuis on growth performance and circulating insulin-like growth factor-I, tumor necrosis factor-α, and interleukin–ß in weaned pigs. *Journal of Animal Science,* **85**: 1161-1167.

Greer, A. W., Stankiewicz, M., Jay, N. P., McNulty R. W. and Sykes, A. R. (2005). The effect of concurrent corticosteroid induced immuno-suppression and infection with the intestinal parasite *Trichostrongylus colubriformis* on feed intake and utilisation in both immunologically naïve and competent sheep. *Animal Science,* **80**: 89-99.

Greiner, L. L., Stahly, T. S. and Stabel, T. J. (2000). Quantitative relationship of systemic virus concentration on growth and immune response in pigs. *Journal of Animal Science,* **78**: 2690-2695.

Grunfeld, C., Zhao C., Fuller J., Pollock A., Moser, A., Friedman, J. and Feingold, K. R. (1996). Endotoxin and cytokines induce expression of leptin, the Ob Gene product, in hamsters: a role for leptin in the anorexia of infection. *Journal of Clinical Investigation,* **97**: 2152-2157.

Hale, O. M. (1985). The influence of internal parasite infections on the performance of growing finishing swine. *Research Bulletin, Georgia Agricultural Experiment Stations,* **341**: 1-5.

Hein, H. (1968). The pathogenic effects of *Eimeria acervulina* in young chicks. *Experimental Parasitology* **22**: 1-11.

Houdijk, J. G. M., Campbell, F. M., Fortomaris, P. D., Eckersall, P. D. and Kyriazakis, I (2007). Effects of sub-clinical post-weaning colibacillosis and dietary protein on acute phase proteins in weaner pigs. *Livestock Science,* **108**: 182-185.

Johnson, R. W. (1998). Immune and endocrine regulation of feed intake in sick animals. *Domestic Animal Endocrinology,* **15**: 309-319.

Klassing, K. C., Laurin, D. E., Peng, R. K. and Fry, M. (1987). Immunologically mediated growth depression in chicks: Influence of feed intake, corticosterone and interleukin-1. *Journal of Nutrition,* **117**: 1629-1637.

Kyriazakis, I. (2003). The control and prediction of food intake in sickness and in health. In *Perspectives in Pig Science,* pp 381-403. Edited by J. Wiseman, M. A. Varley and B. Kemp. Nottingham University Press, Nottingham.

Kyriazakis, I. and Emmans, G. C. (1992). The effects of varying protein and energy intakes on the growth and body-composition of pigs. 1. The effects of energy-intake at constant, high protein-intake. *British Journal of Nutrition,* **68**: 603-613.

Kyriazakis, I. and Emmans, G. C. (1992). The growth of mammals following a period of nutritional limitation. *Journal of Theoretical Biology,* **156**: 485-498.

Kyriazakis, I., Anderson, D. H., Oldham, J. D., Coop, R. L. and Jackson F. (1996). Long-term subclinical infection with *Trichostrongylus colubriformis:* effects on feed intake, diet selection and performance of growing lambs. *Veterinary Parasitology,* **61**: 297-313.

Kyriazakis, I., Tolkamp, B. J. and Hutchings, M. R. (1998). Towards a functional explanation for the occurrence of anorexia during parasitic infection. *Animal Behaviour,* **56**: 265-274.

Langhans, W. (2000). Anorexia of infection: current prospects. *Nutrition,* **16**: 996-1005.

MacRae, J. C., Smith, J. S., Sharman, G. A. M. and Corrigal, W. (1982). Energy metabolism of lambs infected with *Trichostrongylus colubriformis.* In *Energy Metabolism of Farm Animals, EAAP publication no 29,* pp 112-115. Edited by A. Ekern and R. F. Sundstol. The Agricultural University of Norway, Aas.

May, R. M., and Nowak, M. A. (1995). Coinfection and the evolution of parasite virulence. *Proceedings of the Royal Society, London Series B: Biological Sciences,* **261**: 209-215.

Olesen, I., Groen, A. F. and Gjerde, B. (2000). Definition of animal breeding goals for sustainable production systems. *Journal of Animal Science,* **78**: 570-582.

Otesile, E. B., Fagbemi, B. O. and Adeyemo, O. (1991). The effect of *Trypanosoma brucei* infection on serum biochemical parameters in boars on different planes of dietary energy. *Veterinary Parasitology,* **40**: 207-216.

Powanda, M. C., Cockerall, G. L., Moe, J. B., Abeles, F. B., Pekarek, R. S. and Canonico, P. G. (1975). Induced metabolic sequelae of tularaemia in the rat: correlation with tissue damage. *American Journal of Physiology,* **229**: 479-483.

Sandberg, F. B., Emmans, G. C. and Kyriazakis, I. (2006). A model for predicting food intake of growing animals during exposure to pathogens. *Journal of Animal Science,* **84:** 1552-1566.

Sandberg, F.B., Emmans, G. C. and Kyriazakis, I. (2007). The effects of pathogen challenges on the performance of naïve and immune animals: the problem of prediction, *Animal* **1:** 67-86.

Spurlock, M. E., Frank, G. R., Willis, G. M., Kuske, J. L. and Cornelius, S. G. (1997). Effect of dietary energy source and immunological challenge on growth performance and immunological variables in growing pigs. *Journal of Animal Science,* **75:** 720-726.

Stamataris, C., Kyriazakis, I. and Emmans, G. C. (1991). The performance and body-composition of young pigs following a period of growth retardation by food restriction. *Animal Production,* **53:** 373-381.

Turner, J. L. Dritz, S. S., Higgins, J. J., Herkelman, K. L. and Minton, J. E. (2002). Effect of a *Quillaja saponaria* extract on growth performance and immune function of weanling pigs challenged with *Salmonella typhimurium. Journal of Animal Science,* **80:** 1939-1946.

van Diemen, P. M., Henken, A. M., Schrama, J. W., Brandsma, H. A. and Verstegen M. W. A. (1995). Effects of atrophic rhinitis by *Pasteurella multocide* toxin on heat production and activity of pigs kept under different climatic conditions. *Journal of Animal Science,* **73:** 1658-1665.

van Heugten E, Coffey, M. T. and Spears, J. W. (1996). Effects of immune challenge, dietary energy density, and source of energy on performance and immunity in weanling pigs. *Journal of Animal Science,* **74:** 2431-2440.

Wakelin, D. (2000). Rodent models of genetic resistance to parasitic infections. In *Breeding for Disease Resistance in Farm Animals,* pp 107-126. Edited by R. F. E. Axford, S. C. Bishop, F. W. Nicholas and J. B. Owen. CABI, Wallingford.

Webel, D. M., Johnson, R. W. and Baker, D. H. (1998a). Lipopolysaccharide-induced reductions in body weight gain and feed intake do not reduce the efficiency of arginine utilization for whole-body protein accretion in the chick. *Poultry Science,* **77:** 1893-1898.

Webel, D. M., Johnson, R. W. and Baker, D. H. (1998b). Lipopolysaccharide-induced reductions in food intake do not decrease the efficiency of lysine and threonine utilization for protein accretion in chickens. *Journal of Nutrition,* **128:** 1760-1766.

Wellock, I. J., Emmans, G. C., and Kyriazakis, I. (2003a). Modelling the effects of thermal environment and dietary composition on pig performance: model logic and concepts. *Animal Science,* **77:** 255-266.

Wellock, I. J., Emmans, G. C. and Kyriazakis, I. (2003b). Predicting the consequences of social stressors on pig feed intake and performance. *Journal of Animal Science,* **81:** 2995-3007.

Wilkie, B. N. and Mallard, B. A. (1998). Multi-trait selection for immune response: A possible alternative strategy for enhanced livestock health and productivity. In *Progress in Pig Science*, pp 29-38. Edited by J. Wiseman. Nottingham University Press, Nottingham.

Williams, N. H., Stahly, T. S. and Zimmerman, D. R. (1997a). Effect of chronic immune system activation on the rate, efficiency and composition of growth and lysine needs of pigs fed from 6 to 27 kg. *Journal of Animal Science,* **75:** 2463-2471.

Williams, N. H., Stahly, T. S. and Zimmerman, D. R. (1997b). Effect of chronic immune system activation on body nitrogen retention, partial efficiency of lysine utilization, and lysine needs of pigs. *Journal of Animal Science,* **75:** 2472-2480.

Willis, G. M. and Baker, D. H. (1981a). *Eimeria acervulina* infection in the chicken: a model system for estimating nutrient requirements during coccidiosis. *Poultry Science,* **60:** 1884-1891.

Willis, G. M. and Baker, D. H. (1981b). *Eimeria acervulina* infection in the chicken: sulphur amino acid requirement of the chick during acute coccidiosis. *Poultry Science,* **60:** 1892-1897.

Zaralis, K., Tolkamp, B. J., Houdijk, J. G. M. and Kyriazakis, I. (2007). Protein supplementation consequences on anorexia and expression of immunity of parasitized ewes of two breeds. In: *12ᵗʰ Seminar on Sheep and Goat Nutrition: Nutritional and foraging ecology of sheep and goats.* Thessaloniki, Greece (in press).

28

IN VITRO CHARACTERIZATION OF STARCH DIGESTION AND ITS IMPLICATIONS FOR PIGS

THEO VAN KEMPEN[1,3], SERENA PUJOL[1,4], SIMON TIBBLE[2], AND AITOR BALFAGON[1]
[1]*Provimi Research and Technology Centre, Brussels, Belgium;* [2]*SCA Iberica, Mequinenza, Spain;* [3]*North Carolina State University, Raleigh, NC, USA;* [4]*IRTA, Reus, Spain*

Introduction

Pigs and poultry are typically fed diets based on cereals (or their co-products) and oilseed meals. The main component in cereal is starch, which is approximately 0.40 by weight for oats to 0.70 by weight for rice. In complete feeds it is thus common to find starch levels in the range of 300 to 500g/kg.

Starch consists of two polymers of glucose: amylose and amylopectin (highly branched). These polymers are stacked inside granules in both crystalline and amorphous forms, while the granules themselves are packaged in a protein matrix. Upon digestion, starch is converted into glucose which can be readily absorbed by animals and used for energy-yielding purposes (see Buléon *et al.*, 1998, or Vandeputte and Delcour, 2004, for more comprehensive reviews of starch structure).

The nature of the starch, including the amylose/amylopectin ratio, the crystallinity, the protein matrix in which the granules are embedded, and the link between the protein and the starch granules have a strong effect on the rate and extent of digestion in the animal (e.g., Stevnebø *et al.*, 2006). This rate of digestion then affects blood glucose levels, which is monitored by the pancreas. In response, the pancreas releases insulin (Björck *et al.*, 2000). Insulin, as well as the availability of glucose, are key metabolic signals for the health and growth of animals.

Despite the importance of starch, little is known about the intricacies of its digestion and the effects this has on the growth and health of the animal. The objective of current starch research was to obtain a better understanding of the digestion of starch in commercial feed materials and the impact this has on animal performance. For this purpose an *in vitro* assay was employed to characterize starch digestion, and animal trials were used to test the effect of different starch digestion profiles on subsequent animal performance.

In vitro methodology

The *in vitro* assay used for characterizing starch digestion was modified after Englyst (e.g., Englyst *et al.* 2000, 2003). In brief, a sample of a cereal is ground through a 1 mm screen. One gram of ground material is then incubated at pH 2 in a buffer containing pepsin for a period of 45 minutes. Subsequently, the sample is neutralized and pancreatin, invertase, and amyloglucosidase are added. At time 0, 20, 60, 120, and 240 min. an aliquot of this incubation medium is analyzed for glucose.

A sample of expanded maize was included as a reference in each run of the assay, and between-assay variation was corrected based on variation between runs observed in this reference. For this reference, it was observed that the between-assay coefficient of variation (CV) for the digestible starch content was 1.8%. For the rate of digestion, the CV was substantially higher reaching 14%. The latter CV is higher than desired, but given the type of assay it is nevertheless acceptable. Given that sample data were corrected for assay variability based on this reference it can be expected that assay error is smaller than that for the reference.

Using this assay, approximately 350 samples of both raw and processed cereals have been analyzed. These materials were collected from feed mills around the world. Materials were typically routine samples of material entering the respective feed mills. Processing techniques include expanding, extruding, flaking, and roasting material. As such, this database gives a good picture of the characteristics of feed material available in the field.

In vitro results

Englyst (Englyst *et al.* 2000, 2003) modeled glucose release data using the following model:

Glucose release = Plateau x (1-exp(-K x time))

In which:

Glucose release = glucose release as a % of sample weight (expressed as starch-equivalent)

Plateau = maximal glucose release as a % of sample weight (at time infinity and expressed as starch equivalent). Practically, the plateau value can be equated to metabolizable energy.

K = rate of glucose release (unit of glucose/unit of time).

Englyst *et al.* (2003) also divided the starch degradation curve into rapidly degradable starch (RDS), or the starch that is hydrolyzed in the first 20 minutes, slowly degradable starch (SDS), or the starch that is hydrolyzed between 20 and 120 minutes, and resistant starch (RS) starch not hydrolyzed in 120 minutes. Resistant starch is comprised of four chemically or physically distinct starches: RS1 which is starch inaccessible to digestive enzymes because the starch is trapped in an indigestible matrix, RS2 which is starch with an indigestible crystalline nature, RS3 which is starch which is retrograded, and RS4 which is chemically modified indigestible starch.

Current data showed that this mathematical model had two shortcomings. Given the very good repeatability of the *in vitro* work it was observed that *in vitro* digestion was sigmoidal for some feedstuffs, e.g., wheat. This sigmoidal feature could be incorporated into the model by raising it to the power c, as per the Chapman-Richards model (e.g., Fekedulegn *et al.*, 1999). This model could typically be fitted to the *in vitro* data such that over 99.7% of variation was explained. Assigning sigmoidal degradation patterns, though, has a strong impact on the classification of Englyst (RDS, SDS, RS) which deserves further study.

Data also showed that the original Englyst assay is affected by the amount of starch present in the incubation, giving higher rates of digestion for samples low in starch (likely due to a higher enzyme to substrate ratio). Testing the same sample at different concentrations (not shown) showed that this problem could be corrected by dividing K by plateau and by dividing c by plateau and adding 1. The final model used is as follows:

$$\text{Glucose release} = \text{Plateau} \times (1-\exp(-\text{Kplateau/plateau} \times \text{time}))^{(\text{Cplateau/plateau}+1)}$$

In which:

Glucose release = glucose release as a % of sample weight (expressed on a starch equivalent base).

Plateau = maximal glucose (expressed on a starch equivalent base) release as a % of sample weight (at time infinity). Practically, the plateau value can be equated to metabolizable energy.

Kplateau = rate of glucose release corrected for plateau effects

Cplateau = sigmoidal glucose release corrected for plateau effects. It indicates the presence of a lag time before glucose release; the higher the C, the bigger the lag time.

A typical degradation curve based on duplicate analysis is shown in Figure 28.1 (wheat). An insignificant amount of free glucose is observed at time 0 (start of pancreatin incubation), followed by a sigmoidal increase in hydrolyzed starch over time.

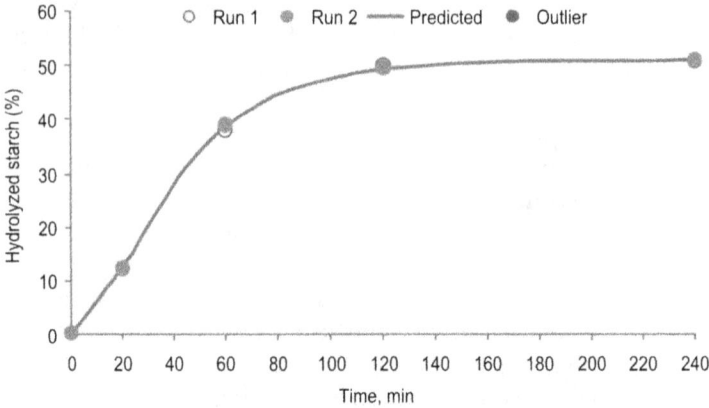

Figure 28.1 Starch digestion curve as observed in a wheat sample. The assay is carried out in duplicate (Run 1 and Run 2; outliers are automatically highlighted when present) and glucose release is plotted over time.

Research studies reviewed for the current programme have not described starch kinetics using a sigmoidal curve. Performing the above assay with additional sampling points early in the assay confirmed that for some materials glucose release is undeniably sigmoidal (data not shown). Additionally, it was observed that there was a firm relationship between Kplateau and Cplateau that depended on the nature of the material (Fig. 28.2). This also suggests that the sigmoidal nature is a feature of the materials tested rather than the assay itself. Figure 28.2 does show that some materials appear to be identified incorrectly. Given the nature of this database (field samples which include mixed materials from cereal processing plants), errors like this can be expected.

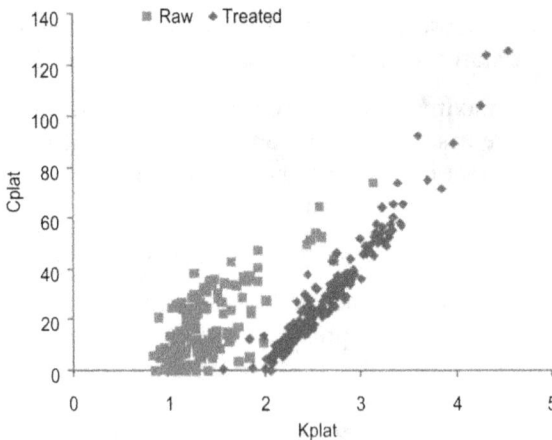

Figure 28.2 Relationship between Kplateau and Cplateau in raw and heat-treated cereal samples. Within heat-treated cereals a high correlation ($r^2=0.94$) between these two parameters was observed. In raw material, this relationship was less clear although the r^2 was still a very respectable 0.65.

A possible explanation for a sigmoidal starch degradation is that, typically, starch granules are digested from the 'inside-out'. Thus, digestive enzymes first need to enter into the center of the starch granule (through existing pores or through channels formed by amylase). At the centre, the enzymes can then initiate the digestion of amylopectin. Given the branched nature of amylopectin it follows from this that substrate availability increases when digestion proceeds (Gallant *et al.*, 1997, Zhang *et al.*, 2006). Another possible explanation is that the inside-out digested starch-granule acts as a reservoir for digested glucose which is abruptly released when the granule is digested to the surface. Processing the cereal should alter this behavior as starch can turn liquid at high temperatures which can result in a physical reorganization of the material, especially if mechanical stresses are applied at the same time. However, during some types of processing, the granule organization returns to a similar layout after processing; and thus the inside-out mechanism may be preserved.

It should also be noted that an important feature for the rate of starch digestion is how the starch granules break during particle-size reduction. In some materials, the fractures are typically at the protein matrix surrounding the granules; while in others the protein forms such strong links between granules that the granules are broken apart during processing (Buléon, personal communication). It is easy to imagine that this has a strong influence on the rate of starch digestion, but what role it plays in practice remains unclear. Other factors that affect the rate of starch digestion include the amylose content, granule size, crystallinity, and their interactions (Stevnebø *et al.*, 2006, Svihus *et al.*, 2005).

A sigmoidal degradation curve has interesting implications for glucose release. For example if the Cplateau value is around 0 as in maize and rice, then glucose release commences immediately upon mixing digestive enzymes with the test material. In vivo, these feedstuffs are thus more likely to result in glucose release in the mouth, possibly causing a sweet taste, and also in glucose release in the duodenum. In contrast, feedstuffs like wheat which is distinctly sigmoidal will only slowly release glucose upon mixing with digestive enzymes (salivary amylase is destroyed in the stomach and thus should have only a minimal impact on overall starch digestion). An implication of this may be that wheat tastes less sweet to an animal. Wheat may also have different effects on blood glucose since glucose release becomes more important further down the intestinal tract, which is not the case for maize and rice. There appear to be no studies that have looked at the impact of this *in vivo*. Trials in cooperation with Lucta (Barcelona, Spain) and IRTA (Reus, Spain) have shown that the in vitro glycemic index affects palatability, explaining approximately 36% of the variation in palatability (manuscript in preparation).

Typical starch degradation curves for maize, rice, and wheat are shown in Fig. 28.3 that shows that rice has a high plateau value; it has a high amount of digestible starch. This figure also shows that wheat has a stronger sigmoidal degradation pattern which translates to a higher Cplateau value.

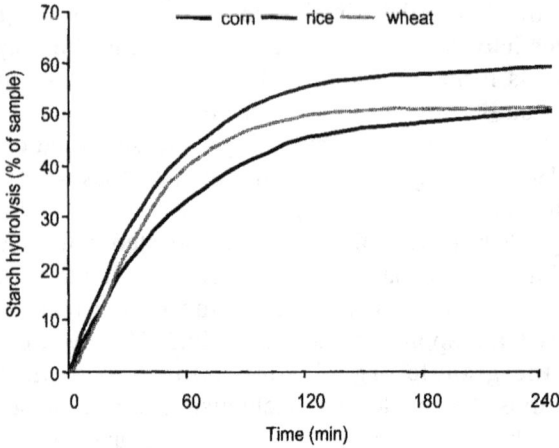

Figure 28.3 Typical in vitro starch hydrolysis patterns for maize, rice, and wheat.

Instead of looking at absolute starch hydrolysis, it is probably more relevant to look at glucose release over time (Fig. 28.4) as this corresponds closer to what the animal is experiencing. A high rate of glucose release should result in a high level of blood glucose. These levels of blood glucose then would affect insulin, which affects feed intake patterns and efficiency of nutrient utilization, for example high levels of insulin stimulate the conversion of glucose to fat resulting in possibly fatter, less efficiently growing animals. Thus, based on the *in vitro* starch hydrolysis profiles, it may be possible to estimate what happens with blood glucose and insulin in an animal and, thus, the efficiency of nutrient utilization, keeping in mind that the time course observed *in vitro* is likely accelerated relative to what is observed *in vivo*.

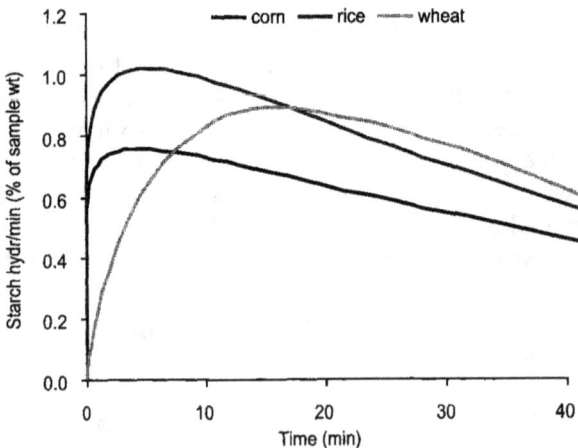

Figure 28.4 Glucose release per minute for maize, rice, and wheat upon *in vitro* digestion.

Figure 28.5 shows average plateau values for various raw and heat-treated cereals. This graph shows that oats are low in digestible starch while rice is very high. Both maize and wheat are intermediate. The effect of heat treatment is interesting as there seems to be an interaction between the treatment technique and the cereal being treated; for example, in oats, a large increase in plateau value is observed with rolling, extrusion, and flaking. In contrast, in wheat, extrusion and flaking appear to have no positive effect on plateau value. For maize, expansion seems without effect on plateau while extrusion and flaking do have positive effects. Given that samples were typically not matched (there were few samples of the same material before and after heat treatment), it is important to be careful in drawing general conclusions from these data with respect to the efficacy of heat treatment techniques.

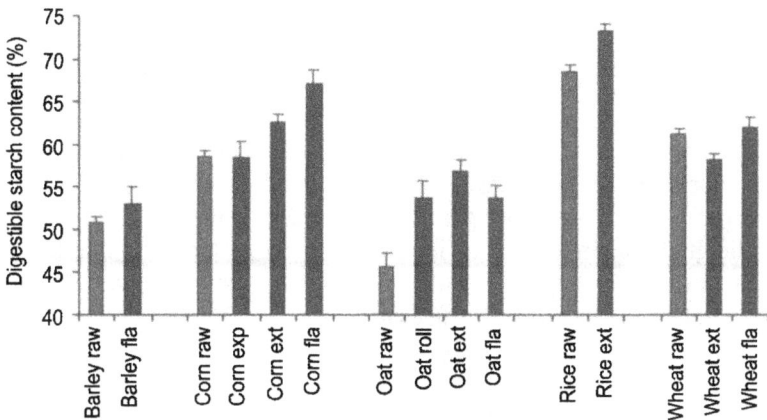

Figure 28.5 Plateau values, or the content of digestible starch in various raw and processed cereals as measured *in vitro*.

Like for plateau value, there are large differences in the initial rate of starch digestion between the different samples (Fig. 28.6). Raw maize has the slowest rate (Kplateau) of starch digestion while raw oats have the fastest rate. Heat treatment of cereals increases the rate of digestion by a factor of 2 for maize and brings the rate of digestion for all cereals much closer together.

In human nutrition, starch-rich foods are commonly evaluated based on glycemic index (GI) that is the blood glucose profile integrated over the first 2 hours after ingesting a known amount (e.g., 50 g) of a food-borne glucose source (Jenkins *et al.*, 1981). Numerous studies since the original work from Jenkins have confirmed the value of the glycemic index and its relation to insulin release (e.g., Björck *et al.*, 2000).

Based on the *in vitro* glucose release curves, calculated maximal glucose release has been calculated which was used as a measure of the glycemic index of feed materials. Figure 28.7 summarizes this glycemic index relative

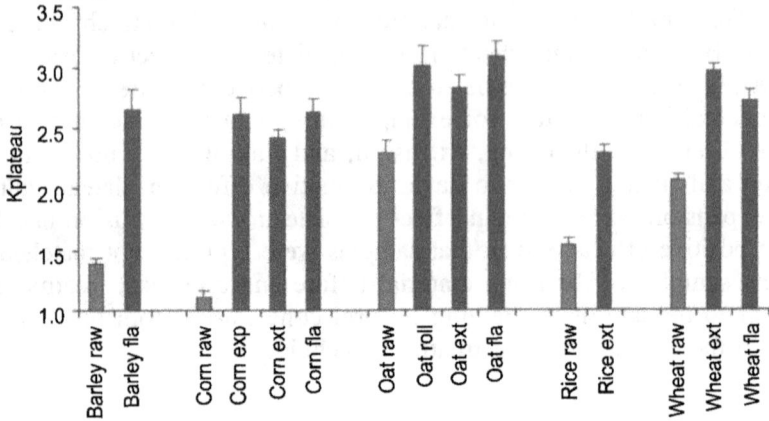

Figure 28.6 Initial rate of starch digestion corrected for plateau effects (Kplateau). (Fla=flaked, exp=expanded, ext=extruded, roll=rolled).

to the value observed in raw maize and shows that raw oats have a higher glycemic index than raw maize despite a lower digestible starch content. Heat treatment results in comparable glycemic indexes for the different cereals, close to twice the glycemic index of raw maize. It is again interesting that the variation within categories is extremely small. A reason for this may be that only feed-grade material was used in *in vitro* assays.

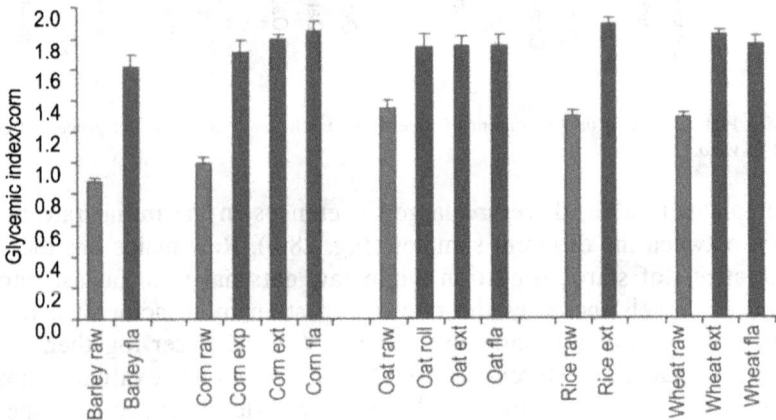

Figure 28.7 Calculated glycemic index (maximal glucose release/min) relative to that of maize.

Animal studies

The biggest question, of course, is if this is relevant to animals. Englyst has validated the *in vitro* assay against glycemic index studies with humans and has shown that the *in vitro* assay corresponds well (Englyst *et al.*, 2003). These findings were corroborated by various other studies, for example Garsetti *et al*, (2005).

Similarly, we worked with Dr. Gonzalo Mateos from the University of Madrid who performed glycaemic index studies in young pigs using feedstuffs characterized with the *in vitro* assay. His observation agreed with Englyst in that there was a good correlation between the calculated *in vitro* glycaemic index and the observed *in vivo* glycaemic index (manuscript in preparation).

A more practical evaluation was performed at the Piglet Evaluation Centre of SCA Iberica, Spain. For this trial, 6 diets were formulated that each included 480g pure rice starch/kg; rice was derived from 3 different genetic varieties of rice, either raw or heat-treated Pure starches were chosen so that any effect observed could be related to the nature of the starch rather than to effects on protein matrix, fibre, etc.

In vitro studies of these rice starches showed that there were large differences in digestion kinetics and, thus, glycemic index. The slow raw rice starch had a plateau value of 58% and a glycemic index of 1.1, while the fast, gelatinized starch had a plateau value of 91% and a glycemic index of 1.8.

For the animal portion of the trial, approximately 900 nursery piglets 25 days old (4 days after weaning) were used. Animals were blocked by weight (small, medium, large) and sex (gilts and barrows), and 12 pens housing 12 to 13 animals each were assigned per treatment. Feed intake and growth were measured weekly over the course of the experiment (3 weeks). In addition, blood glucose was measured using Bayer Ascensia blood glucometers on day 19 mid-morning from 4 randomly selected pigs in each pen.

Blood glucose data showed that the starch with the slowest rate of digestion/ glycemic index resulted in the lowest blood glucose level (4.0 mM), while the fastest starch resulted in the highest blood glucose (4.7 mM), as expected. Given that blood glucose affects satiety and, through this, feed intake, it follows that pigs ate 9% more of the feed based on the slowest starch than that based on the fastest starch. Since high glucose (through insulin) stimulates fat accretion, the effect on feed efficiency was even stronger; feed efficiency was 14% better for the diet based on the slowly digesting starch (Fig. 28.8). Correlation analysis between performance (treatment means) and starch characteristics showed that feed intake x glycaemic index of the feed was the strongest predictor of final weight (r=-0.86, p=0.03). Numerically, feed intake x glycemic index was the best predictor of blood glucose (r=0.67, p=0.15) and nearly the best predictor of feed efficiency (r=-0.58, P=0.22; feed intake x resistant starch had a slightly better P value of 0.18).

Total tract digestibility tests showed that all these sources of starch were completely digested. However, the slow degrading starch did result in wetter feces, suggesting that it stimulated fermentation in the large intestines. In contrast, this diet resulted in a slightly better fat digestion probably because rate of passage rate was slowed down.

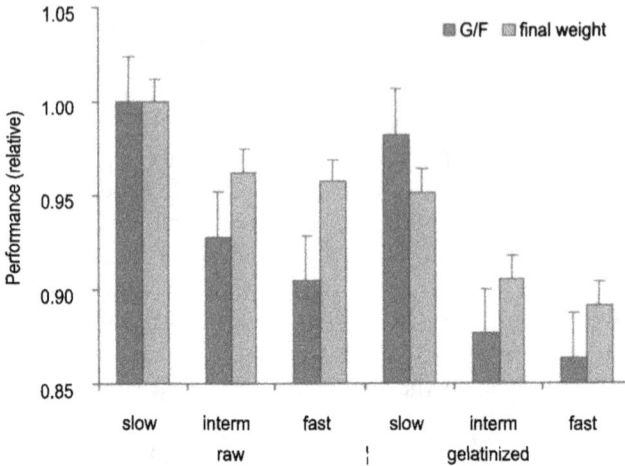

Figure 28.8 Relative animal performance as observed in piglets fed feeds based on 6 different starch sources.

Based on these data, it can be concluded that the rate of starch digestion has a strong effect on animal performance. For nursery pigs, a slowly digesting starch is preferred probably because it results in a slow and steady release of glucose which does not strongly stimulate insulin. This improves feed intake and possibly reduces the utilization of glucose for lipogenesis (fat accretion).

The finding that a slowly digestible starch source is preferred for livestock feeding is in line with findings from Weurding *et al.* (2003). It should be pointed out that the nutritional preference for slowly digestible (raw) starch does not mean that heat treatment of cereals should be avoided for diets for young animals, as heat treatment affects many other parameters besides starch digestion kinetics. Numerous in-house studies have shown that heat-treated cereals result in better animal performance for young piglets.

Conclusions

Starch is a dominant energy source in animal feeds which, between ingredient classes, is quite heterogeneous in the rate and extent of digestion. The rate and extent of digestion affects animal performance, with slowly/steadily digestible starch sources being preferred for maximal growth rate and efficiency of growth.

References

Björck, I., H. Liljebert, and E. Õstman (2000) Low glycaemic-index foods. Brit. J. Nutr. 83 (suppl. 1):S149-S155.

Buléon, A., P. Colonna, V. Planchot, and S. Ball (1998) Starch granules: structure and biosynthesis. Int. J. Biol. Macromol. 23:85-112.

Englyst, K.N., G.J. Hudson, and H.N. Englyst (2000) Starch analysis in Food. In: Encyclopedia of Analytical Chemistry (R.A. Meyers, Ed.). John Wiley & Sons, Chichester.

Englyst, K.N., S. Vinoy, H.N. Englyst, and V. Lang (2003) Glycaemic index of cereal products explained by their content of rapidly and slowly available glucose. Br. J. Nutr. 89:329-40.

Fekedulegn, D., M.P. Mac Siurtain, and J.J. Colbert (1999) Parameter estimation of nonlinear growth models in forestry. Silva Fennica 33:327-336.

Galant, D.J., B. Bouchet, and P.M. Baldwin (1997) Microscopy of starch: evidence of a new level of granule organization. Carb Polymers 32:177-191.

Garsetti, M., S. Vinoy, V. Lang, S. Holt, S. Loyer, and J.C. Brandmiller (2005) The glycemic and insulinemic index of sweet biscuits: relationship to in vitro starch digestibility. J. Am. Coll. Nutr. 24:441-447.

Jenkins, D.J., T.M. Wolever, R.H. Taylor, H. Barker, H. Fielden, J.M. Baldwin, A.C. Bowling, H.C. Newman, A.L. Jenkins, and D.V. Goff (1981) Glycemic index of foods: a physiological basis for carbohydrate exchange. Am. J. Clin. Nutr. 34:362-366.

Svihus, B., A.K. Uhlen, and O.M. Harstad (2005) Effect of starch granule structure, associated components and processing on nutritive value of cereal starches: A review. Anim. Feed Sci. Techn. 122:303-320.

Stevnebø, A., S. Sahlström, and B. Svihus (2006) Starch structure and degree of starch hydrolysis of small and large starch granules from barley varieties with varying amylose content. Anim. Feed Sci. Techn. 130:23-28.

Vandeputte, G.E., and J.A. Delcour (2004) From sucrose to starch granule to starch physical behaviour: a focus on rice starch. Carb. Polymers 58:245-266.

Weurding, R.E., H. Enting, and M.W.A. Verstegen (2003) The relationship between starch digestion rate and amino acid level for broiler chickens. Poultry Sci. 82:279-284.

Zhang, G., M. Venkatachalam, and B.R. Hamaker (2006) Structural basis for the slow digestion property of native cereal starches. Biomacromolecules 7:3259-3266.

List of Delegates

Anderson, Mr K	Novartis Animal Health, Frimley Business Park, Frimley, Camberley GU16 7SR, UK
Bates, Mr M	British Pig Association, Trumpington Mews, 40b High St, Trumptington, Cambs CB2 2LS, UK
Beckett, Miss N	A-One Feed Supplements Ltd, North Hill, Dishfield Airfield, Thirsk YO7 3DH, UK
Benson, Ms T	BPEX, Winterhill House, Snowdon Drive, Milton Keynes MK6 1AX, UK
Berkeveld, Ir M	Universiteit Utrecht, Faculty of Veterinary Medicine, Yalelaan 7, Utrecht 3584 CL, Netherlands
Best, Mr P	Pig International, Lavant House, Lavant Street, Petersfield GU32 3EL, UK
Bikker, Dr P	Schothorst Feed Research, PO Box 533, Lelystad 8200, The Netherlands
Blaken, Ms C	PIC UK, 2 Kingston Business Park, Kingston Bagpuize, Oxon OX13 5AS, UK
Boyd, Dr J	BOCM Pauls Ltd, Tucks Mill, Burston, Diss IP22 5TJ, UK
Brameld, Dr J	University of Nottingham, Sutton Bonington Campus, Loughborough, Leics LE12 5RD, UK
Burton, Mr H	ABN, ABN House, Oundle Rd, Peterborough PE2 9PW, UK
Chang, Miss J	British Trade and Cultural Office, Unit D, 7F, 95, Ming Tzu, 2nd Rd, Kaohsiung , Taiwan
Chaosap, Miss C	University of Nottingham, Sutton Bonington Campus, Loughborough, Leics LE12 5RD, UK
Chen, Dr S	Ming Dao University, 369, Wen-Hua Rd, Peetow, Chang Hua , Taiwan

Cliff, Ms Angela	BPEX, Winterhill House, Snowdon Drive, Milton Keynes MK6 1AX, UK
Close, Dr W	Close Consultancy, 129 Barkham Road, Wokingham RG41 2RS, Berks
Cole, Mr J	BFI Innovations Ltd, 1 Telford Court, Chester Gates, Dunkirk Lea, Chester CH1 6LT, UK
Coleman, Miss J	Division of Animal Physiology, University of Nottingham, Sutton Bonington Campus LE12 5RD, UK
Coma, Mr J	SCA Iberica S.A., Poliguno Industria Riols SN, Mequinenza 50170, Spain
Courtenay, Miss A	Veterinary Laboratories Agency, The Elms, College Road, Sutton Bonington, Loughborough LE12 5RB, UK
Crespo, Mr J	SCA Iberica S.A., Poliguno Industria Riols SN, Mequinenza 50170, Spain
Cunnah, Mr D	Janssen Animal Health, P O Box 79, Saunderton, High Wycombe HP14 4HJ, UK
David, Ms P	IATC, Royal Agricultural Society of England, Stoneleigh Park, Warwickshire CV8 2LZ, UK
Davies, Miss H	University of Nottingham, Sutton Bonington Campus, Loughborough, Leics LE12 5RD, UK
De Vos, Dr S	INVE Technologies N.V, Hoogevepol 93, Dendermonde 9200, Belgium
Donadeu, Ms M	PIC, 2 Kingston Business Park, Oxfordshire OX13 5FE, UK
Drewett, Miss M	MLC/BPEX, Winterhill House, Snowdon Drive, Milton Keynes MK6 1AX, UK
Durand, Mr H	Lallemand, 19 rue des Briquetiers, BP59, 31703 Blagnac Cedex , France
Enting, Dr I	Wageningen UR, Animal Science Group, P O Box 65, Lelystad 8200 AB, The Netherlands
Farquharson, Miss K	University of Nottingham, Sutton Bonington Campus, Loughborough, Leics LE12 5RD, UK
Ford, Mr S	PIC, UK, 2 Kingston Business Park, Kingston Bagpuize, Oxon OX13 5AS, UK
Foster, Mr C	Baby Bacon Inc, 944 Inley Road, Amboy, Illinois 61310, USA

Foxcroft, Prof G R	University of Alberta, 310-C Agric-Forestry Centre, Edmonton T6G 2P5, Canada
Gardner, Dr D	University of Nottingham, Sutton Bonington Campus, Loughborough, Leics LE12 5RD, UK
Gerique, Mr N	SCA Iberica S.A., Poliguno Industria Riols SN, Mequinenza 50170, Spain
Gerritsen, Miss R	Wageningen University, Marijkeweg 40, Wageningen 6709 P6, The Netherlands
Gill, Dr P	Meat & Livestock Commission, P O Box 44, Winterhill House, Winterhill, Milton Keynes HK6 1AX, UK
Gillies, Mr I	Rattlerow Farms, Hillhouse Farm, Stradbroke, Eye, Suffolk IT21 5ND, UK
Girdler, Dr C	Novartis Animal Health, Frimley Business Park, Frimley, Camberley GU16 7SR, UK
Glaves, Mrs F	PIC UK, 2 Kingston Business Park, Kingston Bagpuize, Oxon OX13 5AS, UK
Golds, Mrs S P	University of Nottingham, Sutton Bonington Campus, Loughborough, Leics LE12 5RD, UK
Gregson, Miss E	University of Nottingham, Sutton Bonington Campus, Loughborough, Leics LE12 5RD, UK
Hall, Mr J	ACMC Ltd, Upton House, Beeford, Driffield, East Yorkshire YO25 8AF, UK
Hamid, Mr N	University of Nottingham, Sutton Bonington Campus, Loughborough, Leics LE12 5RD, UK
Harrison, Miss L	Novartis Animal Health, Frimley Business Park, Frimley, Camberley GU16 7SR, UK
Hartley, Mr M	Whole Hog, 11 Church Street, Northborough PE6 9BN, UK
Hazeleger, Dr W	Wageningen University, Marijkeweg 40, Wageningen 6709 P6, The Netherlands
Hindson, Mr E	SCA Iberica, Pol Ind Riols, Mequinenza, SPAIN
Hong-Ji, Mr L	Evaglow International Co Ltd, 10F-1, 26, Sec 3, Chungshan North Rd, Taipei, Taiwan
Hooley, Mrs E	University of Nottingham, Sutton Bonington Campus, Loughborough, Leics LE12 5RD, UK

Hoste, Dr S	Quantech Solutions, Opari Cottage, Howe Farm, Malton YO17 6RG, UK
Hoving, Miss L	Wageningen University, Marijkeweg 40, Wageningen 6709 P6, The Netherlands
Howe, Ms T	Meat & Livestock Commission, P O Box 44, Winterhill House, Snowdon Drive, Milton Keynes MK6 1AX, UK
Hsia, Prof L	National Pingtung Univ of Sc & Tech, 1 Hsueh-Fu Rd, Lao-Pi Nei-Pu, Pintung, TAIWAN
Huang, Mr J	Orange Mart, 36, Lane 162, Science Park Rd, Hsin Chu, Taiwan
Huang, Mrs N	Danco Animal Health, 20, Sec 3, Chien Kuo North Rd, Taichung 403, Taiwan
Hughes, Miss A	Boehringer Ingelheim, Shanlis, Greenhill Road, Wicklow Co Wicklow, Ireland
Hughes, Prof P	SARDI, Univ of Adelaide, J S Davies Building, Roseworthy SA5371, Australia
Hung, Mr Y	Fwusow Industry Co, 45, Sha-Tyan Rd, Sha Lu, Taichung County, Taiwan
Hunter, Prof M	University of Nottingham, Sutton Bonington Campus, Loughborough, Leics LE12 5RD, UK
Hurdidge, Mr L	Biotal Ltd, Collivand House, Ocean Way, Cardiff CF24 5PD, UK
Ilsley, Dr S	Frank Wright Ltd, Blenheim House, Blenheim Road, Ashbourne DE6 1HA, UK
Jackson, Mr C	British Pig Association, Trumpington Mews, 40b High St, Trumpington, Cambs CB2 2LS, UK
Jager, Ms H C	University of Cambridge - CIDC, Madingley Road, Cambridge CB3 OES, UK
Jagger, Dr S	ABN, Oundle Road, Peterborough, PE29 2PW, UK
Johnson, Ms Kayt	BPEX, Winterhill House, Snowdon Drive, Milton Keynes MK6 1AX, UK
Jones, Dr M A	University of Nottingham, Sutton Bonington Campus, Loughborough, Leics LE12 5RD, UK
Jordan, Ms J	5M Publishing, 4 Hayward House, Hydra Business Pk, Nether Lane, Sheffield S35 9ZX, UK

Jorgensen, Dr H	Aarhus University, Dept of Animal Health, Welafare & Nut, Blichers Alle 20, P O Box 50 DK-8830, Tjele, Denmark
Kemp, Prof B	Wageningen University, Marijkeweg 40, Wageningen 6709 PG, The Netherlands
Koeppel, Dr P D	Chemforma Ltd, Rheinstrasse 28 - 32, Augst 4302, Switzerland
Kyriazakis, Prof I	SAC, Animal Nutrition and Health Dept, West Mains Rd, Edinburgh EH9 3JG, UK
Langendijk, Dr P	SARDI Livestock, Roseworthy Campus, Roseworthy 537 1SA, Australia
Le Treut, Mr Y	Lallemand, 19 rue des Briquetiers, BP59, 31703 Blagnac Cedex, France
Lear, Dr J	Biotal Ltd, Collivand House, Ocean Way, Cardiff CF24 5PD, UK
Lewis, Mr C R G	Roslin Institute, Roslin Biocentre, Roslin EH25 9PS, UK
Longthorp, Mr R W	NPA, Burland, Holme Rd, Howden, Goole DN14 7LI, UK
Lumb, Mr S	Stuart Lumb Assoc (Press), 12 Rawdale Close, S. Cave,, Brough, E Yorks HU15 2BT, UK
MacDonald, Mr P	Janssen Animal Health, P O Box 79, Saunderton, High Wycombe HP14 4HJ, UK
Mackinnon, Mr J	Pig Health & Production Consult., Cheneys Cottage, East Green, Kelsale, Saxmundham JP17 2PH, UK
Madeloso, Mr F	Robina Farms, Philippines
Marier, Ms E	VLA Weybridge, New Haw, Addlestone, Surrey KT15 3NB, UK
Masey O'Neill, Miss H	University of Nottingham, Sutton Bonington Campus, Loughborough, Leics LE12 5RD, UK
McArdle, Mr T	Alltech (UK) Ltd, 7 Turnberry Close, Kirkam, Preston, Lancs PR4 2TE, UK
McClean, Miss C	Vitrition, Grange Farm Buildings,Grange Farm,, Milby, Boroughbridge YO51 9HQ, UK
Mcorist, Dr S	University of Nottingham, Sutton Bonington Campus, Loughborough, Leics LE12 5RD, UK
Mellits, Dr K	University of Nottingham, Sutton Bonington Campus, Loughborough, Leics, LE12 5RD, UK

Miller, Prof H	University of Leeds, Biological Sciences, Woodhouse Lane, Leeds LS2 9JT, UK
Mohr, Dr M	Murphy Brown, 316 Charity Road, Rose Hill, NC 28458, USA
Morgan, Mr W	Writtle College, Writtle, Chelmsford CM1 3RR, UK
Morillo, Mr A	SCA Iberica S.A., Poliguno Industria Riols SN, Mequinenza 50170, Spain
Nieuwhof, Mr G	MLC, Winterhill House, Snowdon Drive, Milton Keynes MK6 1AX, UK
Odle, Prof J	North Carolina State University, Box 7621, Raleigh 27695, USA
Oksbjerg, Prof N	Aarhus University, Faculty of Agric Sciences, P O Box 50, Tjele 8830, Denmark
Oliveira, Dr S	University of Minnesota, 1333 Gormer Ave, Saint Paul MN S5108, USA
Pan, Mr D	Danco Animal Health, 20, Sec 3, Chien Kuo North Rd, Taichung 403, Taiwan
Parkes, Miss K	RSPCA, Farm Animals Dept, Science Group,Wilberforce Way, Southwater, Horsham, W Sussex RH13 9RS, UK
Parr, Dr T	University of Nottingham, Sutton Bonington Campus, Loughborough, Leics LE12 5RD, UK
Rachmawati Siswadi, Ms	Indonesia Monogastric Assoc, Unsued, Purwokerto, Indonesia
Rathje, Dr T	Danbred North America, 3220 25th St, Columbus NE 68601, USA
Ravn, Miss L	BPEX, 46 Claremont Terrace, Blyth, Northumberland NE24 2LE, UK
Richardson, Mr J	Intervet UK Ltd, Walton Manor, Walton, Milton Keynes MK7 7AJ, UK
Rodriguez-Martinez, Prof H	Swedish Univ of Agric Sciences, Ullsvaag 14C, Box 7054 Uppsala SE 750 07, Sweden
Roger, Mr L	Centralys, 9-11 Av F Arago BP 108, Trappes 78190, France
Roppa, Mr L	Provimi, Rua Maria Nassif Mokarzel, 375, Campinas 13.084-757, Brazil
Sanderson, Mr R	PIC, UK, 2 Kingston Business Park, Kingston Bagpuize, Oxon OX13 5AS, OX13 5AS

Sands, Dr J S	Danisco, P O Box 777, Marlborough SN8 1XN, UK
Smith, Prof A	Biotal Ltd, Collivand House, Ocean Way, Cardiff CF24 5PD, UK
Smith, Mr R P	Veterinary Laboratories Agency, Woodham Lane, New Haw,, Addlestone, Surrey KT15 3NB, UK
Smits, Mr R	QAF Meat Industries, P O Box 78, Corowa 2646 NSW, Australia
Stewart, Mr A	Harper Adams University College, Newport, Shropshire TF10 8NB, UK
Stickney, Dr K	Harbro, Markethill, Turriff, Aberdeenshire AB53 4PA, UK
Taylor-Pickard, Dr J	Alltech Ltd, Sarney, Summerhill Road, Dunboyne Co, Meath, Ireland
Theunissen, Mr H C H	Vion Food Group, Boseind 10, 5281RM Boxtel , The Netherlands
Thoday, Ms H	BPEX, Winterhill House, Snowdon Drive, Milton Keynes MK6 1AX, UK
Thomsett, Miss A	George Veterinary Group, High Street, Malmesbury, Wiltshire SN16 9AU, UK
Tibble, Mr S	SCA Iberica S.A., Poliguno Industria Riols SN, Mequinenza 50170, Spain
Toplis, Mr P	Primary Diets, Melmerby Industrial Estate, Melmerby, Ripon, N Yorks HG4 5HP, UK
Totemeyer, Dr S	University of Nottingham, Sutton Bonington Campus, Loughborough, Leics LE12 5RD, UK
Town, Dr S	SARDI, University of Adelaide, J S Davies Building, Roseworthy, S Australia 5371, Australia
Tseng, Mr T	Great Wall Enterprised Co Ltd, 3 Niao Sung 2nd St, Yung Kang City, Tainan County , Taiwan
Turnley, K Mr	Alltech Ireland, Sarney, Summerhill Rd, Dunboyne, Co. Meath, Ireland
Van den Brink, Mr P	Provimi BV, Veerlaan 17-23, Rotterdan 3072 AN, UK
van Kempen, Dr T	Provimi/NCSU, Lenneke Makelaan 2, Brussel 1932, Belgium
Vanasse, Mrs A	Universite Laval, 1316 Rue des Seigneurs, Quebec , Canada
Varley, Dr M	SCA NUTec, SCA Mill, Dalton, Thirsk, North Yorks YO7 3HC, UK

Vignola, Mr M	Shur-Gain/Landmark Agresearch, 1900, 2ieme Rue, Centre Industriel, St-Ropmuald, Quebec G6W 5M6, Canada
Voets, Mr Harm	Boehringer Ingelheim, Animal Health GMBH, Bingerstrasse 173, Ingelheim 55216, Germany
Voigt, Dr J P	University of Nottingham, Sutton Bonington Campus, Loughborough, Leics LE12 5RD, UK
Wade-West, Mr G	Novartis Animal Health, Frimley Business Park, Frimley, Camberley GU16 7SR, UK
Waldo, Mr M	Waldo Farms, 14144 West Dogwood Road, Dewitt, NE 68341, USA
Walling, Dr G	JSR Genetics, Southburn, Driffield, East Yorkshire YO25 9ED, UK
Webb, Prof R	University of Nottingham, Sutton Bonington Campus, Loughborough, Leics LE12 5RD, UK
White, Mr G	University of Nottingham, Sutton Bonington Campus, Loughborough, Leics LE12 5RD,
Whitehead, Mr J	Chemforma Ltd, Rheinstrsse 28-32, Augst 4302, Switzerland
Williams, Mr J	Boehringer Ingelheim, Shanlis, Greenhill Road, Wicklow Co Wicklow, Ireland
Wilson, Mr M	BPEX, Winterhill House, Snowdon Drive, Milton Keynes MK6 1AX, UK
Wilson, Mr S	BOCM Pauls Ltd, Tucks Mill, Burston, Diss, Norfolk IP22 5TJ, UK
Wiseman, Prof J	University of Nottingham, Sutton Bonington Campus, Loughborough, Leics LE12 5RD,
Wonnacott, Miss K	University of Nottingham, Sutton Bonington Campus, Loughborough, Leics LE12 5RD, UK
Woolfenden, Mr N	Bishopton Veterinary Group, Mill Farm, Studely Rd, Ripon HG4 2QR, UK
Wynn, Dr P	University of Sydney, Faculty of Veterinary Science, PMB P.O., Camden NSW 2570, Australia
Yang, Mr H	K C Nutrition, A4 17th F, 6, Sihwei 3rd Rd, Kaohsiung , Taiwan
Zarkos-Smith, Mr A	Zarkos-Smith Associates, The Spinney, 5 Park Lane, Histon, Cambridge CB4 9JJ, UK

INDEX